Science of Crystal Structures

István Hargittai • Balazs Hargittai
Editors

Science of Crystal Structures

Highlights in Crystallography

Springer

Editors
István Hargittai
Department of Inorganic and Analytical Chemistry
Budapest University of Technology and Economics
Budapest, Hungary

Balazs Hargittai
Department of Chemistry
Saint Francis University
Loretto, PA, USA

ISBN 978-3-319-19826-2 ISBN 978-3-319-19827-9 (eBook)
DOI 10.1007/978-3-319-19827-9

Library of Congress Control Number: 2015950211

Springer Cham Heidelberg New York Dordrecht London

On the cover and title page, the illustration is a detail of Figure 5 (Integrated microclusters; method of connecting superclusters into a layer) in Vladimir Ya. Shevchenko, V. A. Blatov, and G. D. Ilyushin, "Intermetallic compounds of the $NaCd_2$ family perceived as assemblies of nanoclusters." *Structural Chemistry* 20(6), 2009, 975–982.

Printed on acid-free paper

Springer International Publishing AG Switzerland is part of Springer Science+Business Media (www.springer.com)

To the pioneers of crystallography,
to the authors of Structural Chemistry,
and to the users of structural information.

Also by the Editors

Balazs Hargittai and István Hargittai, Eds., *Culture of Chemistry: The Best Articles on the Human Side of 20th-Century Chemistry from the Archives of the* Chemical Intelligencer (Springer, 2015)

István Hargittai and Magdolna Hargittai, *Budapest Scientific: A Guidebook* (Oxford University Press, 2015)

Balazs Hargittai, Magdolna Hargittai, and István Hargittai, *Great Minds: Reflections of 111 Top Scientists* (Oxford University Press, 2014)

István Hargittai, *Buried Glory: Portraits of Soviet Scientists* (Oxford University Press, 2013)

István Hargittai, *Drive and Curiosity: What Fuels the Passion for Science* (Prometheus, 2011)

István Hargittai, *Judging Edward Teller: A Closer Look at One of the Most Influential Scientists of the Twentieth Century* (Prometheus, 2010)

Magdolna Hargittai and István Hargittai, *Symmetry through the Eyes of a Chemist,* 3rd Edition (Springer, 2009; 2010)

Magdolna Hargittai and István Hargittai, *Visual Symmetry* (World Scientific, 2009)

István Hargittai, *The DNA Doctor: Candid Conversations with James D. Watson* (World Scientific, 2007)

István Hargittai, *The Martians of Science: Five Physicists Who Changed the Twentieth Century* (Oxford University Press, 2006; 2008)

István Hargittai, *Our Lives: Encounters of a Scientist* (Akadémiai Kiadó, 2004)

István Hargittai, *The Road to Stockholm: Nobel Prizes, Science, and Scientists* (Oxford University Press, 2002; 2003)

Balazs Hargittai, István Hargittai, and Magdolna Hargittai, *Candid Science I–VI: Conversations with Famous Scientists* (Imperial College Press, 2000–2006)

István Hargittai and Magdolna Hargittai, *In Our Own Image: Personal Symmetry in Discovery* (Plenum/Kluwer, 2000; Springer, 2012)

Ronald J. Gillespie and István Hargittai, *The VSEPR Model of Molecular Geometry* (Allyn & Bacon, 1991; Dover Publications, 2012)

Aldo Domenicano and István Hargittai, Eds., *Strength from Weakness: Structural Consequences of Weak Interactions in Molecules, Supermolecules, and Crystals* (Dordrecht: Kluwer, 2002)

Aldo Domenicano and István Hargittai, Eds., *Accurate Molecular Structures. IUCr Monographs on Crystallography*, Vol. 1. (Oxford University Press, 1992)

István Hargittai, Ed., *Quasicrystals, Networks, and Molecules of Fivefold Symmetry* (New York: VCH Publishers, 1990)

István Hargittai and B.K. Vainshtein, Eds., *Crystal Symmetries: Shubnikov Centennial Papers* (Oxford: Pergamon Press, 1988)

István Hargittai and W.J. Orville-Thomas, Eds., *Diffraction Studies on Non-Crystalline Substances* (Amsterdam: Elsevier, 1981)

Preface

During the first 25-year existence (1990–2015) of the international journal *Structural Chemistry*, it has published hundreds of articles in crystallography. From the start, there has been a characteristic direction in this periodical to bring out reports related to generalized crystallography, quasicrystals, icosahedral packing, other packing considerations, extended structures, data treatment and data mining, and what we would call "living history."

Several of the contributions came from the schools of such trend-setting crystallographers as J. Desmond Bernal and Aleksandr I. Kitaigorodskii. Internationally renowned scientists have figured as authors, among them, Tom L. Blundell, Johann Jacob Burckhardt, John L. Finney, Jenny P. Glusker, Nobel laureate Herbert A. Hauptman, the 2014 Ewald Prize winner A. Janner, Aminoff Prize winner Isabella Karle, Nobel laureate Jerome Karle, the late doyen of Chinese crystallography K. H. Kuo, Buckley Prize winner Alan L. Mackay, Ewald Prize winner David Sayre, Vladimir Shevchenko, and King Faisal Prize winner J. Fraser Stoddart.

A few frontier topics dominate the selected material in this volume. Pioneers of the direct methods describe the phase problem and how it was solved, including the mathematical approach and the utilization of experience with gas-phase electron diffraction. The reviews by Herbert Hauptman, Jerome and Isabella Karle, and David Sayre reach to the present day in assessing the possibilities of X-ray crystallography.

Another focus topic is the investigation of systems that are outside the so-called classical system of crystals. They include quasicrystals, imperfect and very small crystals, supramolecular species, crystal structures without lattice, clusters, and nanomaterials, among others. The application of synchrotron and cryoprotection techniques, the free-electron laser flash technique, and others is among the techniques of investigation in addition to X-ray crystallography.

The relationship between structural and materials properties is examined and uncovered. The broader topics include polymers, clusters, polydisperse chain assemblies, and giant icosahedral fullerenes. There are some key contributions related to the structural investigation of biological macromolecules. There are packing considerations and work toward one of the ultimate goals of crystallography as a science of structures: to predict crystal structures from the chemical composition of new substances.

A modern tool in structure analysis is data mining and data management. *Structural Chemistry* has paid attention to and attracted authors utilizing the possibilities of data banks, in particular the Cambridge Data Files and the inorganic crystal structure database (ICSD in Karlsruhe). Selected articles in this compilation demonstrate the potentials of the utilization of data banks.

We have arranged the 33 articles selected for inclusion in four parts. However, the classification is blurred as authors do not write articles to correspond to clear-cut classifications. Thus, the distribution of the contributions in this compilation is approximate, and there are overlaps.

The selection process which led us to the 33 articles reproduced here happened in three cycles during which we consulted with others as we were narrowing our choice in a gradual manner.

Only as an aside we note that the authors of the articles come from at least 14 countries (USA, UK, Russia, Hungary, India, Italy, Switzerland, Canada, China, France, Germany, Israel, Mexico, and the Netherlands). That the 33 articles originated from at least 14 countries reflects the international character of the collection although we paid no attention to this aspect in the process of compilation.

Budapest, Hungary
Loretto, PA, USA
Summer 2015

István Hargittai
Balazs Hargittai

Contents

Preface ix

Part I Direct Methods

X-Ray Crystallography: The Past and Present of the Phase Problem 3
David Sayre

History of X-Ray Crystallography 19
Herbert A. Hauptman

Gas Electron Diffraction and Its Influence on the Solution of the Phase Problem in Crystal Structure Determination 23
Isabella Karle and Jerome Karle

Part II Beyond the Classical System

Generalized Crystallography 37
Alan L. Mackay

Mackay, Anti-Mackay, Double-Mackay, Pseudo-Mackay, and Related Icosahedral Shell Clusters 43
K.H. Kuo

Crystallography Without a Lattice 53
John L. Finney

Polygrammal Symmetries in Biomacromolecules: Heptagonal Poly $d(As^4T) \cdot$ poly $d(As^4T)$ and Heptameric α-Hemolysin 69
A. Janner

Helical Structures: The Geometry of Protein Helices and Nanotubes 81
Eric A. Lord

From Waxes to Polymers—Crystallography of Polydisperse Chain Assemblies 91
Douglas L. Dorset

Structure, Chirality, and Formation of Giant Icosahedral Fullerenes and Spherical Graphitic Onions 101
Mauricio Terrones, Guillermo Terrones, and Humberto Terrones

A Generalized Model for the Shell Structure of Icosahedral Viruses 113
V. Ya. Shevchenko, A.E. Madison, and A.L. Mackay

Intermetallic Compounds of the $NaCd_2$ Family Perceived as Assemblies of Nanoclusters 117
V. Ya. Shevchenko, V.A. Blatov, and G.D. Ilyushin

Linus Pauling's Quest for the Structure of Proteins 125
István Hargittai

"There Is No Such Animal (אין חיה כזו)"—Lessons of a Discovery 133
István Hargittai

Quasicrystal Discovery—From NBS/NIST to Stockholm 137
Balazs Hargittai and István Hargittai

Atomic Structure of Quasicrystals 143
Marc de Boissieu

Some Properties of Three-Periodic Sphere Packings 155
Michael O'Keeffe

Cluster Packing from a Higher Dimensional Perspective 165
Walter Steurer and Sofia Deloudi

Part III Structural and Materials Properties

Magnetic Properties of Quasicrystals: Effects of Disorder on Fe Magnetic Moment Formation 173
R.A. Dunlap, R.C. O'Handley, M.E. McHenry, and V. Srinivas

Overlooked Opportunities in Stereochemistry. Part III: Alfred Werner's Awareness of Spontaneous Resolutions and of the Meaning of Hemihedral Faces in Optically Active Crystals 179
Ivan Bernal and George B. Kauffman

The Geometry of Intermolecular Interactions in Some Crystalline Fluorine-Containing Organic Compounds 187
Liat Shimoni and Jenny P. Glusker

Molecular and Crystal Structures of a Class of Push-Pull Quinonoid Compounds with Potential Nonlinear Optical Applications 205
M. Ravi, Shmuel Cohen, Israel Agranat, and T.P. Radhakrishnan

X-ray Crystallographic Studies on the Noncovalent Syntheses of Supermolecules 213
Matthew C.T. Fyfe, J. Fraser Stoddart, and David J. Williams

Asymmetry in the Multiprotein Systems of Molecular Biology 231
Tom L. Blundell, Victor Bolanos-Garcia, Dimitri Y. Chirgadze, Nicholas J. Harmer, Thomas Lo, Luca Pellegrini, and B. Lynn Sibanda

Molecular van der Waals Symmetry Affecting Bulk Properties of Condensed Phases: Melting and Boiling Points 239
Yuri L. Slovokhotov, Andrei S. Batsanov, and Judith A.K. Howard

The Role of Symmetry in Building Up Zeolite Frameworks from Layered Zeolite Precursors Having Ferrierite and CAS Layers 255
W.J. Roth and D.L. Dorset

Part IV Data Mining and Management

The Correspondence Fedorov-Schoenflies: The Groups Pm, Pc, Cm, Cc 263
Johann Jakob Burckhardt

Prediction of H-Bonding Motifs for Pyrazoles and Oximes Using the Cambridge Structural Database 269
Lourdes Infantes and Sam Motherwell

Hierarchies of Intermolecular Potentials and Forces: Progress Towards a Quantitative Evaluation ... 281
Angelo Gavezzotti

Where Are Genes in Paulingite? Mathematical Principles of Formation of Inorganic Materials on the Atomic Level ... 291
V. Ya. Shevchenko and S.V. Krivovichev

Supramolecular Architectures of Metal–Organic Host–Guest Compounds ... 299
Alessia Bacchi and Mauro Carcelli

What Is a Chemical Substance and How Is It Formed? ... 309
Vladimir Ya. Shevchenko

Investigation of Structural Relations Among the Compounds in the ICSD Using Geometry Based Comparison Techniques ... 323
M. Sultania, J.C. Schön, D. Fischer, and M. Jansen

Index ... 333

Part I

Direct Methods

X-Ray Crystallography: The Past and Present of the Phase Problem[a]

David Sayre[b]

Abstract

The story of the progress thus far made on the phase problem of X-ray crystallography is briefly reviewed.

Keywords

X-ray crystallography • Phase problem

Introduction

In 1912–1913 there occurred (Laue in Munich, W.L. Bragg in Cambridge) one of the key ideas of twentieth-century experimental science: that of placing a small crystal as specimen in an appropriately short-wavelength (~1 Å) X-ray beam, recording the pattern and intensities of the diffraction spots produced, and using those data to accept or reject proposed atomic arrangements inside the specimen. By 1929 (again due largely to Bragg), it was recognized that if the diffraction spot phases could also be supplied, the phased magnitudes of the spots could be regarded as Fourier coefficients, whose Fourier sum would directly image the internal structure of the specimen; this would immediately give the atomic arrangement as well as other important information. From that time to the present, the problem of how to supply the phases has been the principal theoretical problem of X-ray crystallography. Progress has been such that today reasonably accurate phases and, therefore, images can usually be arrived at, even for extremely complex structures, provided that crystals are available and provided that desire is strong enough to produce continued trying of the methods which now exist for handling the problem.

In this paper, I give a brief and I hope illuminating account of the post-1929 history. For this purpose, I have divided the history into two parts: the original and still vigorous approach based upon crystals and Bragg spots, and a newer approach, based—in its major form—on noncrystalline specimens. These call for substantially higher exposure than crystals but give a continuous diffraction pattern, from which intensities can be measured on a finer sampling lattice[1] than that given by Bragg spots, permitting more data to be collected. (The specimen in the noncrystalline case may be as small as an individual protein molecule or as large as an individual biological cell, making possible an enormous increase in the range of specimens available to the crystallographic methodology. Also, in theory, it is possible to obtain a continuous pattern from a sufficiently small crystal.) This increase in data is proving in current testing to be often—perhaps always—sufficient to permit a rapid and easy technique of phasing and imaging. Thus the problems of crystallization and phasing promise to disappear in the newer technique, while the problem of damage, due to the increased exposure, will become more important.

It will be pointed out, as the paper progresses, that two general types of phasing methods exist: methods that are physical in nature and succeed in making actual phase

[a] *Structural Chemistry* 2002, 13(1):81–96.

[b] Department of Physics and Astronomy, SUNY (State University of New York), Stony Brook, NY 11794, USA (deceased)

[1] Specifically, sampling at intervals somewhat finer (for more detail see Part II) than the Nyquist sampling interval. The Bragg sampling interval in diffraction space applies to crystals and is the inverse of the unit cell diameter. The Nyquist sampling interval applies to any specimen and is the inverse of the specimen diameter.

I. Hargittai, B. Hargittai (eds.), *Science of Crystal Structures: Highlights in Crystallography*,
DOI 10.1007/978-3-319-19827-9_1

measurements; and methods that fundamentally operate by specifying constraints that may justifiably be imposed (usually in image space), and through mathematics applying the corresponding constraints which must exist on the phases in diffraction space. Both methods fundamentally aim at finding ways to repair an initial information deficit. The latter methods, in this day of high-speed computation, tend to be fast and convenient, but can also be subject to problems of nonuniqueness and possible unwanted biasing of solutions. The former are often more costly in time and trouble, but when an answer is found it is, within the limits of precision of the measurements, correct.

Part I: Crystals and Bragg Spots

Fourier Refinement (Early 1930s)

The major accomplishment of these first years was to bring out a valuable characteristic of the Fourier-coefficient approach to structure solving: its ability to find the correct phases and electron-density map when provided with merely an initial approximation to those quantities. The technique, called Fourier refinement, is an iterative one, in which current phases are used with experimental magnitudes to give a current image or map; the map suggests revisions in the atomic arrangement; the new atomic arrangement, through inverse Fourier transformation, generates new phases; and the process repeats. This back-and-forth ability immediately lent added strength to the 1913–1929 technique, in which possible atomic arrangements were tested for consistency with the observed magnitudes without, however, a method of corrective feedback. Now, with the feedback described above, if one got reasonably close to the structure, refinement would complete the job. In addition, well-refined electron-density maps, available now for the first time, provided genuine images of structure, giving important new information about bond shapes, presence and location of hydrogen atoms, etc. The technique, of course, is in use to this day.

It will be seen that this technique does not, in itself, contribute information. How then does it help to overcome the underlying information deficit present in the phase problem? The answer is that in each cycle the technique presents the known and guessed information in a form (the guessed map) which allows the investigator to add information to the situation. In practice, what the investigator does is to try to collect the electron density into about the right number of distinct atoms in a reasonable spatial configuration; the method is thus of the constraint-based, rather than phase-measuring, type mentioned above. One may legitimately argue that this may have the fault of biasing the technique toward an atomic theory of matter, or toward an incorrect atomic arrangement, but I will disregard that here. What can be stated is that degree of information insertion usually suffices to reach a satisfactory phasing and structure. An important conclusion thus appears: that there is at least a rough equivalency between the information carried in atomicity and the deficit of information represented by the missing phases.

The Patterson Map (1934), Heavy Atoms (Mid- and Later 1930s)

In 1934 the next step was taken by Arthur Lindo Patterson, a young crystallographer then at M.I.T., who saw a way to obtain benefit from the Fourier-coefficient approach *ab initio*, i.e., even before an initial approximation to phases and image is available. If the symbol F (F complex) designates the correctly phased Fourier coefficients, then even without the phases the quantities $|F|^2$ (i.e., the intensities) are known. It was Patterson's idea to begin by using the $|F|^2$ as Fourier coefficients. This, it is easy to show through the convolution theorem, yields a map which, if the true F-map shows a structure containing N atomic peaks, will show a related but more complex map (the Patterson map) consisting of N shifted and superposed copies of the true N-peak structure; equivalently, the Patterson map displays the N^2 interatomic vectors of the true map.

Disentangling the N copies is a formidable task if N is large, although some crystallographers (e.g., Dorothy Hodgkin at Oxford), gifted with great chemical insight and remarkable spatial perception, became adept at doing just that; some complex structures in the next two decades were, in fact, solved by a heroic process of guessing fragments of structure by partial Patterson disentanglement, and submitting those to Fourier refinement, until at last the true details of the structure began to emerge.

A much more convenient Patterson-based method, however, was discovered quite early. At about this time, the special case of structures in which a small number (say M) of the atoms were markedly heavier (higher atomic number Z) than the rest was beginning to attract attention (J. M. Robertson at the Royal Institution in London); if just the M-atom structure could be found or guessed, submitting it to Fourier refinement would often reveal the remainder of the structure. Now the Patterson method and this "heavy-atom" method could combine beautifully to make, for the first time, a systematic and fairly reliable *ab initio* structure-solving method. Looked at from the interatomic vector point of view of the Patterson map, the weight of a Patterson peak is proportional to the product of the weights of the two atoms giving rise to it; hence, it will usually be possible to recognize in the complicated Patterson map of the full structure the simpler Patterson map of the M heavier atoms, allowing the

M-atom structure to be solved first. Submitting the M-atom structure to Fourier refinement will then often allow the remainder of the structure to be solved. This use of the so-called heavy-atom Patterson method (Huse & Powell) became, for organic structures, the workhorse method of the 1940s and early 1950s. Given an organic molecule which it was desired to study, one employed chemical methods to attach, for example, a bromine atom to it, crystallized it, and then solved it by the heavy-atom Patterson method.

At this time, the era of solving protein structures was still years in the future. Yet in 1935, and directly related to the above developments, there was already recognition of what the future might bring. In 1934–1935, J. D. Bernal and Dorothy Hodgkin (then Dorothy Crowfoot), working in Cambridge, obtained the first good diffraction photographs of a crystalline protein (pepsin); in February, 1935, Crowfoot, now in Oxford, repeated the success using zinc insulin. When she informed Bernal of this, he replied with a brief note:

> Dear Dorothy:
>
> | | |
> |---|---|
> | Zn | 0.52 wt. per cent. |
> | Cd | 0.77 wt. per cent. |
> | Co | 0.44 wt. per cent. |
>
> This gives slightly less than 3 in all cases.

What Bernal was saying was that he had looked up the work of the Canadian chemist D. A. Scott, who had prepared zinc insulin crystals and measured their zinc content, and that for zinc, as well as for cadmium and cobalt, the data indicated the presence of three heavy atoms per insulin molecule. Bernal, in other words, was making the suggestion that despite their low concentrations, the heavy atoms might offer, through the methods outlined above, at some future date, a solution to the phase problem, even for proteins. That insight was ultimately to turn out to be 100 % correct.

It will be seen that, in a manner reminiscent of Fourier refinement, the Patterson method, while not in itself adding information, does assist the investigator in doing so. The adding of the heavy atoms, however, is at least a step toward a true phase-measuring method. For once the arrangement of the heavy atoms is known, they become (in the language of holography) the source of a known reference signal, which mixes coherently with the signal from the remainder of the structure and assists in phasing it. As stated, the early technique relied simply on Fourier refinement to carry out the phasing and imaging. Since then, under the driving force of protein crystallography, that technique has given way to better ones (see sections below on macromolecules), but the idea that this might someday be done was already present at the time being discussed. It may also be noted how very neat an arrangement it is to have the reference signal source (i.e., the M special atoms) repeated in every unit cell of the crystal. Because of the strength of the Bragg spots, a strong reference signal is needed; the repeated arrangement produces that strong coherent signal exactly at the points where it is needed, i.e., at the diffraction spots.

Three Brief Digressions

The State of Affairs at the Start of the Postwar Period (1946)

It may be useful to summarize the overall situation in crystallography as the world prepared to return to more normal conditions after World War II. The field was small but healthy, with a considerable record already of elucidating the atomic structure of matter in inorganic compounds, minerals, metals and alloys, and organic compounds, in the latter area reaching to sterols and (within the next several years) to penicillin. For its time, the field had an adequate X-ray source (the metal target X-ray tube) and an adequate X-ray detector (X-ray film); in the Patterson map with and without heavy atoms, and Fourier refinement, it had a workable system of solving structures; and it had a seemingly inexhaustible supply of interesting crystalline materials to study. It even had (although only a handful were as yet thinking about it) the key to the future imaging of the highly complex structures which make up the living cell.

Crystallography did have a problem, however, and it lay in the area of computation. For the carrying out of Fourier sums (from $|F|^2$'s to Patterson maps and from F's to Fourier maps), the slowness of the prevailing method— the use of hand-operated office-type mechanical adding machines plus Beevers–Lipson or Patterson–Tunell strips (sine–cosine tables formatted for the Fourier operation)—was such that only two-dimensional (2D) maps were normally done, and even those would take a day or two each. As long as crystallography could not freely enter three-dimensional (3D) work, its potential for the future would be severely limited. Fortunately, the wartime advances in electronics would quite soon change this situation.

A Bit of Personal History

In 1943 I emerged from Yale University with an undergraduate degree in physics, and in 1943–1946 I did wartime electronics research in radar at the Radiation Laboratory of M.I.T. In 1946, guessing that biology might become the exciting field of the next half-century, I became a graduate student in biology at Pennsylvania and then Harvard, but did not find myself responding with interest to what I was learning. However, in June 1947, I happened upon J. M. Robertson's review article on X-ray crystallography [1], written shortly before the war began, which changed my life.

I joined the crystallography laboratory of Raymond Pepinsky at Alabama Polytechnic Institute, where I participated in the design of X-RAC (X-Ray Automatic Computer), an electronic analog computer which, once it had been given the $|F|^2$'s or F's, could compute 2D Patterson and Fourier maps in 1 s. (That machine remained in productive operation for some years, until it was supplanted by digital computation). When in 1949 the machine was successfully operating, my wife Anne and I went to Oxford, so that I could be a Ph. D. student with Dorothy Hodgkin.

The Beginnings of Digital Computation (Late 1940s, Early 1950s)

Digital computation, of course, was now just around the corner. In 1949 England was a little in the lead in that field, and so, not long after arriving in Oxford, I went to Manchester to visit Prof. Fred Williams (with whom I had become acquainted through radar work during the war) and to see his computer, MADAM, which had just recently started to work, with the idea that I might perhaps do crystallographic computing on that machine as my thesis work at Oxford. That did not happen and it was in fact Durward Cruickshank in Leeds who beautifully carried out such a program on the Manchester computer a little later (as did Bennett and Kendrew on EDSAC in Cambridge). In one of the major moments of my life, however, Williams handed me over to Alan Turing, who was then part of the Manchester effort, to discuss the idea. Instead, as it turned out, we talked for the better part of 2 days about the phase problem, which clearly intrigued him. What he gave me, in return, was an early look at the Shannon sampling theorem, which led me to some early thoughts on oversampling in crystallography [2] and undoubtedly helped prepare my mind for the Nyquist-basis concept many years later. As for crystallographic programming of a digital machine, I did do some later, but not until 5 years had gone by, when I did it on the IBM 701 machine, by which time the ability of the digital computer to handle crystallography's computing needs was quite well established.

Direct Methods (Late 1940s–Present)

In 1948, 14 years after Patterson, a major new advance in the phase problem occurred when David Harker and John Kasper of the General Electric Research Laboratory applied Schwarz and Cauchy inequalities to the expressions for the F's and derived inequality relationships which must exist between the magnitudes of some F's and the phases of others; 2 years later they used these inequalities to determine the values of enough phases to allow a quite difficult structure, dekaborane, to be solved. These papers [3] caused a major stir in crystallography; for 20 years it had been considered that the phases are lost, but now it appeared that they are not entirely lost, but are in some degree capable of being deduced from the magnitudes.

For several years, it was not really known how best to understand the Harker–Kasper work. Then in January 1950, in an important paper [4], Jerome Karle and Herbert Hauptman at the Naval Research Laboratory in Washington gave a general form of inequality (the Karle–Hauptman inequality) which provides all the inequalities of the Harker–Kasper type that can exist; in addition they showed that what those inequalities actually express is the nonnegativity of the electron density map. To me, now a student at Oxford, this was important news. It said that whereas the positivity condition had been captured in the Harker–Kasper and Karle–Hauptman inequalities, the atomicity condition (remember that the information in atomicity roughly balances the information in the missing phases) had not as yet been caught. To find a way to express mathematically, in terms of the F's, the concept of atomicity, became my goal.

We now know that this mathematization of atomicity has been done essentially twice, in two different ways, I did it first, by adopting an idealization of the problem in which the atoms are equal and resolved; this permits a complete mathematization, in the form of a set of equations to be satisfied by the F's. Solving the phase problem then reduces to inserting the measured |F|-values in the equations (called the squaring-method equations) and solving for the phases. The latter is not a trivial problem, but based on the structure of the equations, I developed a solution process consisting of three main steps.[2] I gave these results orally at the Third International Congress of Crystallography in Stockholm in June 1951 and in a paper [5], which appeared in *Acta Crystallographica* in January 1952. Then, in 1972, for the field of protein crystallography, I developed a second solution process, in the form of finding (by least squares) an extremum of a functional based on the equation system, and applied it successfully to the structure of rubredoxin in 1974 [6]. Finally, in 2000, Richard Rothbauer, still within the framework of exact mathematization, showed [7] that the extremum method can be extended in such a way as to allow the initial idealization to be dropped.

Returning now to the early 1950s, Karle and Hauptman sought to dispense with the idealization which I had used, so as to bring the study closer to real life. Their work extended over numerous papers, beginning [8] with Monograph #3 of the American Crystallographic Association, which appeared in September 1953, and over many years. The mathematics

[2]The steps are: (1) identification and tentative phasings of strong terms of the equations (these terms, which involve three F's whose Fourier indexes sum to zero, are known as "triplets"); (2) extension of these phasings to many or most of the weaker F's through further use of the equations; (3) use of an equation-based figure of merit and choice of a final phasing.

which they chose for this purpose is the mathematics of probability, giving their work a very different character from mine. Nevertheless, when it came to the application of the results in the form of a computational procedure, the Karle–Hauptman approach gradually came to a closely similar three-step process, their version of the second step (which is really the critical one of the three) appearing under the name "tangent formula" in a paper which they published in 1956. Much later, in the 1980s, the introduction of a newer probabilistic machinery (maximum likelihood) to replace the original machinery used by Karle and Hauptman has been carried out by Bricogne and others.[3] Also Hauptman and his co-workers, starting in 1990, have developed an extremum methodology for protein crystallography ("Shake'n Bake"), which resembles the 1972–1974 work which I did, but with the very useful insertion of automated Fourier refinement into the process as a means of securing good final atomicity [10].

Comparing the two lines of work, there is no doubt that the equal-atom idealization disqualifies the Sayre approach in some cases (especially in the field of inorganic structures) where the Karle–Hauptman approach may succeed. With organic structures, however, idealization does not severely penalize the Sayre approach, while its equational form adds significant general strength, so that, in this area, the difference, although there is a shortage of serious testing, is probably not great. In any event, the true choice today, I think, lies between the maximum likelihood approach on the probabilistic side and the methodology of Rothbauer on the exact side.

In protein studies, the subject of structural genomics is now producing an urgent desire to see a large productivity increase come into being, similar to the one produced by the atomicity-based direct methods in small-structure crystallography. Atomicity-based methods are thus starting to play a small role in protein crystallography, but their role may be fundamentally limited by the fact that protein data sets in which the |F|'s extend to atomic resolution are as yet rather rare.[4] In anticipation of the Nyquist-based technique, one of its most important potential advantages is that it can use, but does not require, atomic resolution data.

Finally, as between phase-measuring and constraint-based methods, the direct methods are, of course, in the latter group.

Macromolecules: DNA (1953), Isomorphous Replacement (1953–Present)

In 1953, only a little more than a year after the introduction of the atomicity-based direct method into small-structure crystallography, came the first two decisive events for macromolecules. Both took place in Cambridge, England.

The first, the working-out of the double-helical structure of DNA, did not really mark an advance in the treatment of the phase problem; however it is so important that it cannot be overlooked. It was in fact an outstanding example of the 1913–1929 method of structure solving, in which Crick and Watson utilized several lines of work—the classical chemistry on DNA, the 1950 work on DNA base ratios (Chargaff), and the 1952 theoretical work on the diffraction patterns of helical arrangements of atoms (Cochran, Crick, and Vand)—to propose a structure (the double-helical based-paired DNA structure) which beautifully accounted for the critical 1951 diffraction data obtained at Kings College London on the B-form of DNA (Franklin). Phase-finding, in other words, was not the solution method here (although Wilkins and his colleagues at Kings did subsequently move on to Fourier refinement of the DNA structure).

The second event was the 1953 discovery by Perutz and his co-workers of the phenomenon of isomorphous replacement in protein crystals, i.e., the ability of proteins to attach heavy atoms or heavy-atom complexes to specific protein sites, and to do so with little or no alteration of the crystal structure as a whole. The ability largely rests on the fact that protein crystals contain considerable regions of solvent, meaning that minor rearrangement of solvent can often succeed in maintaining crystal structure, even after the heavy atom addition. An important consequence of this mechanism is that *several* attachment patterns, often involving different types of heavy atoms, can often be prepared; this is the phenomenon of *multiple* isomorphous replacement (MIR).

The important effect of all this is that in the protein setting, the Patterson heavy-atom method is strengthened to the point where it can operate successfully even with the very low heavy-atom concentrations first noticed in insulin by Bernal in 1935.

This strengthening arises in two ways. There is, first, the fact that, by studying the *difference* map between the Pattersons of a native protein structure and of an isomorphous heavy-atom structure, most of the background material disappears, allowing the heavy-atom peaks belonging to the isomorphous structure to show up sufficiently well to permit the heavy-atom portion of that structure to be solved. Then, second, the existence of multiple isomorphous derivatives means

[3] The original machinery suffered from probability distributions (e.g., of phases of strong triplets) being computed too locally (e.g., from the three |F|'s only), and not being recomputed as phasing proceeded. Hauptman introduced strengthenings of the machinery in the late 1970s through his "Neighborhood Principle," but subsequently seems to have ceased from that work. (See Bricogne [9c] for details of the newer techniques.)

[4] In the extremum methodology work which I did in 1974 on rubredoxin, the data set had 1.5 Å resolution, whereas both maximum likelihood and Shake'n Bake today are reported to require somewhat higher resolution data. A Shake'n Bake, but with the direct method phasing being done by a Rothbauer process, might thus prove to be of interest. Another possible direction (Bricogne [9b]), but probably a very difficult one, would be to develop direct methods in which larger entities such as the amino acid residue, rather than the single atom, play the central role.

that, in the different diffraction patterns, different reference signals have been mixed holographically with the signal from the native protein; in 1956 it was shown (Harker) that with as few as two isomorphous derivatives, this is sufficient, in principle, to determine uniquely the phases of the native protein. With further derivatives present, the accuracy of the phases can usually be further improved, a subject on which much work has been done (Blow and Crick; Hendrickson and Lattman). There has also been much work on obtaining the coordinates of the heavy atoms in the various derivatives as exactly as possible (Blow and Matthews; Bricogne). Progress in these topics has by no means stopped today.

Perutz and his co-workers were working on hemoglobin, but in fact it was Kendrew (again in Cambridge) and his co-workers, working with MIR on myoglobin, who, in 1957 first succeeded in fully carrying out the technique; this success was then soon repeated with other proteins, including hemoglobin. Thus it was through MIR phasing that a collection of solved protein structures first began to be accumulated.

Macromolecules: Anomalous Dispersion (1949–Present)

Valuable as MIR phasing is, it has the disadvantage that, even when starting with good crystals of the native protein, months or years can be spent in finding good isomorphous heavy-atom derivative crystals. Fortunately, at about the same time as MIR was being developed, an alternative was also coming into view. Isomorphous replacement operates by changing the chemical identity of a few of the scattering atoms; anomalous dispersion accomplishes essentially the same thing (and with perfect isomorphism), using just a single crystal form, by wavelength rather than chemical change. The day would come when, through synchrotron X-ray sources, accurately choosable wavelengths for this purpose would be quite readily available.

The observation (Nishikawa; Knol and Prins) that the diffraction patterns of crystals present additional data through development of Friedel nonsymmetry at wavelengths near X-ray absorption edges goes back to the same period in the late 1920s as the emerging of the phase problem. As nearly as I can tell, the first utilization of this effect in connection with the phase problem occurred 20 years later (Bijvoet) with the settling of the choice of the two possible enantiomorphic phasings for a rubidium salt of tartaric acid. (This established, for the first time, the absolute handedness of an optically active molecule). Then, in the same year that Harker gave the underlying analysis establishing the MIR technique, a sort of parallel analysis (Okaya and Pepinsky) gave some of the underlying ideas of what would later become the multiple-wavelength anomalous dispersion technique (MAD).

Despite that paper, the lack of an appropriate X-ray source delayed the application of the ideas for a quarter-century; interest, however, revived again with the promise of the synchrotron as a suitable source (Karle, papers from the Dutch school following Bijvoet, etc.). Finally the MAD method appeared essentially in its present form (Hendrickson), and is now the leading phasing method in macromolecular crystallography. The method calls for a specimen crystal containing a substructure of a moderate number of atoms of a species having an X-ray absorption edge in the general 1 A wavelength region; such a specimen can often be made by the usual heavy-atom addition methods, but a favorite method for protein studies is to substitute selenium methionine for normal methionine in an organism's diet, causing the organism to produce protein which differs from the normal protein only in the fact that the methionine sulfur atoms have been replaced by selenium. The MAD method is rapid and produces maps of excellent quality, and (as stated above) is now the dominant method in protein crystallography.

Both anomalous dispersion and isomorphous replacement, by providing for the coherent mixing of the diffraction signal from the major structure with known signals from one or several reference structures, fall in the favorable category of actual phase-measuring techniques.

Macromolecules: Additional Methods

In isomorphous replacement and anomalous dispersion we have now treated the key methods in macromolecular crystallography, but large numbers of structures have been worked out by other methods as well. In this section, I will briefly touch on some of these methods.

1. Once a protein structure has been solved, the structure of the same protein after addition e.g. of an inhibitor or substrate molecule becomes of interest. That structure can usually be found quite easily by adding to the original map a further (so-called difference-Fourier) map, which can be calculated with fair accuracy from the diffraction intensity differences between the native and modified proteins. The technique, first used by Stryer, Kendrew, and Watson at Cambridge, is in a sense an inverted version of MIR, and its theory resembles the Harker theory for MIR.
2. Similarly, given that the structure of a particular protein is known, there are usually numerous other proteins with closely related amino acid sequences that are of interest as well. Operating on the assumption that similarity of sequence may imply similarity of structure, it is reasonable to determine whether the known structure can serve as a starting point in solving the new structure. In the computer, the known structure is tried in varying orientations and positions in the unit cell of the new crystal; if, at some

orientation and position, the computed diffraction pattern or its Patterson resembles the new observed pattern or its Patterson, a starting point that will successfully refine to the new structure may have been found. This method, sometimes known as molecular replacement, also originated in Cambridge (Rossmann and Blow).

3. Protein crystallography presents a few special situations in which, even within the constraint of observing at Bragg spots only, a limited degree of extra sampling of a diffraction pattern can occur. Crystals may contain considerable solvent, separating the molecules by more than their natural diameters, and causing the Bragg spots to sample at higher fineness than would otherwise be the case; in addition, there may be swelling and shrinking of a crystal form with changing amounts of solvent, causing the Bragg spots to shift their positions along the pattern, and in that way providing further sampling; or a protein may crystallize in several different crystal forms, thus providing several different Bragg-spot samplings of its pattern; or a protein, and especially a virus, may be large, with correspondingly fine-spaced Bragg spots, and yet really consist of N identical subunits, meaning that the *subunit* pattern is being sampled N times more frequently than would otherwise occur. Methods by which the increased information obtained in these situations can be used to assist in handling the phase problem are the subject of a considerable literature, once again due largely to Cambridge, or formerly Cambridge, personnel, including such names as Main, Rossmann, Blow, Bricogne, and Crowther. (Terms applied to these methods include solvent methods, change of space group methods, and noncrystallographic symmetry methods; a unified maximum-entropy treatment of many of these methods has been given by Bricogne [9a].) The methods bear a relationship to that which will be discussed in Part II, but are limited and made more complicated by the Bragg-spot constraint. Somewhat similarly, Szoke [11a] has developed a number of additional structure-solving techniques based on the viewpoint that a diffraction pattern can always be regarded as a hologram, with the signal from one part of the structure serving as reference signal for the remainder of the structure.
4. In addition to the above, there are still other worthwhile concepts which are attracting effort (see summary of a conference [12] on new phasing approaches held May 17–19 2001 at the Lawrence Berkeley Laboratory).

One More Method: Three-Beam Diffraction (1977–Present)

A very attractive conceptual approach to the phase problem in crystals is to excite two different reflection wavefields which will propagate in the same direction, allowing them to interfere coherently; then, from the resultant observable intensity, it may be possible to infer phase information about the crystal. As was pointed out by Lipscomb in 1949, the excitation can be realized by having two reflections h and g simultaneously on the Ewald sphere; these will generate reflected wavefields in directions K(h) and K(g); examination will show that through geometrical identities the reflections g–h and h–g will be on the Ewald spheres for the K(h) and K(g) wavefields; these will produce two additional, doubly reflected, wavefields in the original directions K(h) and K(g), bringing about the desired situation in both of those directions. Experimental evidence for this was then provided by Hart and Lang in 1961. It remained for Post and Colella, working independently in 1977, to suggest the details of extracting the phase information. They suggested fixing h while slowly causing g to move through the Ewald sphere; as this occurs the doubly reflected (g, h–g) wavefield passes through a resonance, which, in turn, produces an intensity trace for the combined K(h) signal; from the shape of this trace the sum of the phases of the –h, g and h–g reflections can be read off. But $-h+g+h-g=0$, so the three reflections constitute a "triplet" in the sense described in the section on Direct Methods (this paper). Thus three-beam diffraction allows, in principle, the experimental measurement of the same quantities (triplet phases) as can be estimated from direct-method arguments in cases where atomicity is present. From that point on, the details merge with those of direct methods. The method falls, however, in the true phase-measurement category.

Practical experience with the method is fairly encouraging in that triplet phases have been successfully measured for a number of small molecules and proteins. Proteins, however, pose several problems: their large unit cell means that care must be taken to avoid having several g's contributing overlapping interference traces; mosaicity is desired to be low for the three-beam method, but tends to be high in proteins; and cryoprotection, which is needed because of the long exposures involved, tends to be ruled out because of its effect in increasing mosaicity still further. Nevertheless, groups in Karlsruhe (Weckert et al.) and Cornell (Shen et al.) have, in several cases, succeeded in overcoming these difficulties.

The Sparse or "Dream" Crystal

Should it ever become possible to stabilize a crystal by forces other than contact forces, it might be possible to create the ideal for X-ray crystallography; this would be a crystal in which the structure is regularly repeated at distances somewhat (perhaps 30–40 %) greater than contact distances.

This would provide oversampling for phasing (see Part II), while retaining the large amplification enjoyed by Bragg spots (permitting high resolution at moderate exposure levels).

An apparatus for virtually instantaneous formation and stabilization of 2D crystals in a standing-wave field has been described by Golovchenko and his colleagues [13]; in addition, orientational alignment of molecules by the use of elliptically polarized laser fields has recently been demonstrated by Larsen and his colleagues [14]. If such apparatus could be developed to generate accurate 3D crystals, then in situations like protein crystallography, where large numbers of exact copies of a structure are available, a high-throughput (i.e., free of slow crystallization, free of phasing problems) and high-resolution crystallography would immediately become possible.

Part II: Finer Sampling (1980–Present)

However, although the apparatus just noted would thus be extremely desirable, it may never come about. More serious still, many of the most important structures do not exist in multiple exact copies (e.g., major cellular constituents and whole cells in biology, amorphous structures and structures with imperfections in materials science). Fortunately, however, it is beginning to appear that there exists an adaptation of present-day crystallography which can operate successfully even in those situations. This is the subject of the remainder of this article.

Overview and Brief History

In Part I, the attitude taken was that the diffraction pattern consists solely of spatially discrete Bragg spots. In Part II we drop this attitude and consider instead that diffraction intensity may exist at every point of diffraction space. This view is mathematically correct whenever the diffracting specimen is of finite size (and when "point" in the sentence above is replaced by "elementary volume"). It is nearest to being incorrect (i.e., Bragg spots are most dominant) when the finite specimen is a sizable and well-formed crystal such as is assumed in Part I; it becomes more correct as we move to imperfect or very small crystals; and it is most correct (to the point where Bragg spots may give way entirely to continuous pattern) when the specimen is noncrystalline. Thus, in Part II, we are specifically considering specimens which are *not* excellently crystalline. Three major things result:

1. Our base of usable specimens expands enormously and we are relieved of the necessity of crystallization;
2. The diffraction features we now seek to record include many which are much weaker than those in Part I, making necessary a much higher level of X-ray exposure, with consequent specimen damage and danger of lowered image quality; and
3. The patterns, being continuous and capable of being finely sampled, yield more information than the Bragg-spot-only patterns of Part I; current evidence is that the added information allows the phase problem to largely, if not fully, disappear.

Points (1) and (3) are major gains for X-ray crystallography, point (2) is a major, but probably not disabling, problem.

These perceptions, although they seem obvious now, came slowly and with difficulty. I have mentioned (see section on Three Brief Digressions) a paper [2] which grew out of my conversation with Turing; it pointed out that if sampling could be carried out at twice the Bragg fineness, the phase problem (at least for centrosymmetric crystals) would effectively go away. The paper did not, however, suggest an effective way of obtaining such sampling; the situation was still conceived of as referring to crystalline matter and sampling at Bragg positions only. The suggestion that crystallographers might put a specimen like a single biological cell in an X-ray beam and nevertheless treat it essentially as if it were a crystal—collecting, phasing and Fourier inverting its *continuous* diffraction pattern—was not made[5] until a paper [15] written in 1980.

Starting, therefore, in 1980, the question became whether that process could actually be carried out. Here the principal issues were whether the diffraction patterns, which would be very weak, could be successfully observed, and whether they could be phased. At that time the answer to both questions was probably no; X-ray sources and detectors, and, similarly, the computers of the time, were not adequate to the tasks. Fortunately, however, the Brookhaven National Laboratory, only 20 miles distant from Stony Brook, decided shortly before 1980 to build a major synchrotron X-ray facility, with the result that it has been possible for us to enjoy the full advantage of the subsequent advances in

[5] From 1955 to 1970, I did not do much work in crystallography, working instead on early computer projects at IBM. In 1970, however, still at IBM, I returned to crystallography, and did the work on the extremum formulation of direct methods mentioned earlier [see section on Direct Methods]. Following that, and feeling that an extension to noncrystalline objects would be an important future need in crystallography, I began to work in that direction, soon forming a close collaboration with Janos Kirz at SUNY Stony Brook. Our initial approach, which has been excellently realized by Janos and his students, was for an X-ray microscope, based on a Fresnel zone plate as an image-forming device. It was not until about 1979 that it occurred to me to turn to crystallography itself for a methodology.

needed technologies. Stages along the way have included: the demonstration, using early bending-magnet radiation, that the diffraction pattern from a large and contrasty single cell (diatom) could be recorded [16]; demonstration, using undulator radiation, that the pattern from a small typical single cell could be recorded [17] and proposal[6] of oversampling of the continuous pattern as a possible phasing method [17]; mathematical treatment of the exposure levels required for pattern observation and first tests of oversampling as a phasing method [18]; general article expressing the probability that this extended version of crystallography could exist [19]; and improved theory of oversampling [20]. A decisive point was achieved with the successful demonstration of the complete process in a 2D nonbiological situation [21]. Following that, there have been three pieces of work emphasizing potential applications: demonstration, with simulated data, of oversampling phasing for amorphous specimens, very small crystals, and imperfect crystals [22]; preparation of a detailed grant proposal for a full 3D project on a cryoprotected yeast cell; and [23] successful acquisition (simulated) of a 3D high-resolution large-protein data set without need for crystallization by the method of femtosecond-scale FEL laser pulses [24], followed by phasing and imaging by oversampling.

With that as background, there follows a brief discussion of some of the issues involved in collecting the diffraction pattern (next Section), and in phasing it (see section on Phasing the Pattern).

The Two Probable Experimental Regimes

The radiation dosage needed to obtain a statistically accurate diffraction pattern rises steeply (roughly 8th-power law) with the desired imaging resolution, rough numbers for organic materials being 10^8, 10^{16}, and 10^{24} rads for 20, 2, and 0.2 nm resolution, respectively [18]. Especially at the higher resolutions, these dosages can vastly exceed the specimen's ability to tolerate dosage. Our major defenses today against this problem are crystallinity, replaceability (ability to replace the specimen with an identical fresh one), cryo, and flash. Crystallinity, in conjunction with Part I phasing techniques, works excellently; it reduces the dosage for Bragg spot collection by some 12 orders of magnitude, i.e., to about 10^{12} rads for 0.2-nm resolution work, which, with the assistance of replaceability and cryo, is acceptable; this is the basis for present-day protein crystallography. Where crystallinity does not exist or is intentionally bypassed, one can turn to Part II phasing, but it has only replaceability, cryo, and flash to work with. The result for Part II is the probable emergence of two distinct experimental regimes. With replaceability, flash can be used and it promises to allow the very large dosage needed for atomic or near-atomic resolution, accomplishing the exposure before damage becomes evident [24]. Without replaceability there is only cryo, which should, however, allow dosages to roughly 10^9 rads, or roughly 15 nm resolution, on micrometer size-range specimens (such as the 3000-nm frozen-hydrated yeast cell of our planned project), hopefully making it possible for crystallography to contribute to cell as well as molecular biology. There follows a little more detail on these two regimes.

The Free-Electron Laser-Flash Technique for Molecules

As mentioned above, a recent paper [24] reports computer simulation studies indicating that if the needed dosage for obtaining the diffraction pattern from a protein molecule is delivered to the molecule in a time period of 5×10^{-15} s or less, significant structural alterations in the molecule will not have occurred by the time the pattern has been generated. Furthermore, it appears [25] that it should, in time, be possible to build a free electron laser (FEL) of that pulse width and of the needed photon flux at 1.5 A wavelength. As a result, a consortium of research teams (including our own) under the leadership of Prof. Hajdu has submitted a research proposal for a flash-technique project on high-resolution imaging of large biomolecules at the Stanford Linac Coherent Light Source, based on the possibility that such a laser may in the future be constructed. The research proposal envisages the delivery of single molecules (or very small molecular clusters), in random orientation and with no other matter nearby to confuse the pattern, into the laser beam. In further support of this, in a recent paper [23] it was assumed that the 2D diffraction patterns from a succession of 10^6 such randomly oriented molecules of the protein rubisco delivered into the beam of an assumed 1.5 A wavelength FEL laser had been assembled into a 3D oversampled data set at 2.5 A resolution; that dataset was then successfully phased by the method to be described in the section on Phasing the Pattern. At the planned repetition rate of 120 Hz for the flashes, the 10^6 2D patterns would be acquired in about 2.3 h.

A Synchrotron/Cryoprotection Technique for Supramolecular Specimens

This technique, as currently employed by us on the X1B undulator beamline at the National Synchrotron Light Source, is diagrammed in Fig. 1. The undulator, entrance slit, grating, and exit slit deliver tunable monochromatic photons to the diffraction experiment, the tuning range

[6] This proposal was developed in a conversation in the late 1980s with Gerard Bricogne.

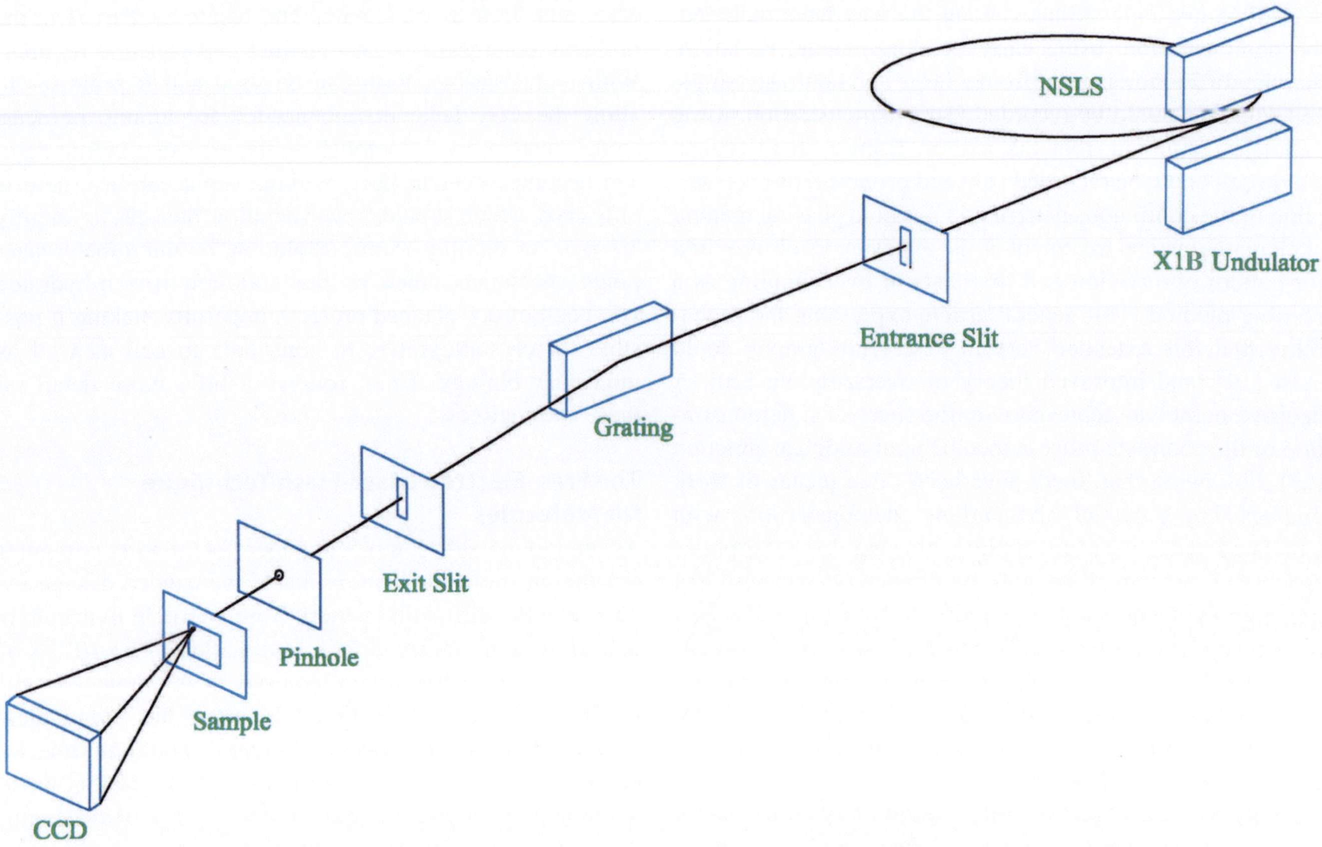

Fig. 1 Simplified schematic of the synchrotron-based experiment at NSLS for supramolecular specimens.

being approximately 1.5–5 nm wavelength.[7] The experiment begins with a 10-μm pinhole approximately 10 cm upstream from the specimen, narrowing the beam so that a single specimen can be illuminated on the specimen holder, and also ensuring that all photons reaching the specimen are monodirectional to 0.1 mrad. (The monochromaticity and monodirectionality produce a sharp Ewald sphere, which will allow future software to place each pixel of the observed 2D patterns accurately in the final 3D pattern). In Fig. 1, the specimen is shown as being close to a corner of the specimen holder; this effectively screens three quadrants of the detector from any edge-scatter or Airy pattern arising from the pinhole; i.e., in those three quadrants, photons emanating from the specimen are essentially falling into a totally dark area of the detector, allowing the detection of very weak pattern. A few inches further downstream is the detector, a back-thinned liquid-nitrogen cooled CCD detector. To catch a 2.34-nm pattern extending to, e.g., 15-nm resolution, the detector must subtend a full angle of about 18° at the specimen. Not shown in the diagram are a beam catch just forward of the detector (to protect the central area of the CCD from the intense central beam), and a silicon nitride window just upstream of the pinhole (to protect the UHV of the beamline from the moderate vacuum of the experiment). Also not shown is the apparatus for positioning and setting the specimen to its successive orientations and, for cryo work, for cooling it. A variant arrangement (but as yet untried) would use long-throw focusing instead of the pinhole to accomplish the beam narrowing.

In addition to subtending a sufficient angle at the specimen, the detector must have enough recording pixels to allow for sufficiently fine sampling of the pattern falling on it. If, in

[7]This region, falling in the soft X-ray region, is not mandatory (there is no reason why X-rays down to 1 Å wavelength cannot be used), but it gives adequately short wavelengths for the resolution expected in this technique, and offers valuable contrast mechanisms for biological work at the absorption edges of low-Z atoms. Thus, for the yeast cell project, we plan to use 2.34-nm photons—just on the low-energy side of the oxygen absorption edge—for best transparency through the ice portion of the frozen-hydrated cell.

the example above, the specimen has a diameter of 3000 nm, sampling at twice the Nyquist fineness—this is usually somewhat finer than necessary; see next section—means sampling at intervals of 1/6000 nm, while the extent of the pattern is 2/15 nm. Thus, 800 × 800 pixels on the CCD will suffice. Back-thinned CCDs meeting these requirements are available today. The full 3D data set ultimately assembled, in this case, could approach 800 × 800 × 800 numbers.

A problem of systematically missing data arises with the technique as described. The need for a beam catch to protect the CCD produces missing data near the central beam; the loss could be reduced if one or a few pixels at the center of the CCD were removed or made transparent. In addition, in 3D work the planar specimen mount produces in the data the undesirable effect of a missing double cone normal to the plane of the mount. At present (see next section), we rely on the phasing algorithm to overcome these data losses.

Phasing the Pattern

It remains to explain how the finer sampling can be used in a phasing process. Turning back to the start of Part II, we pass by the first case of the finite specimen (the sizable well-formed crystal) as providing no opportunity for finer sampling[8] and as having been the subject-matter of Part I. We next defer, for a time, the second case (the imperfect or very small crystal), and turn directly to the major case (the non-crystal). Fourier theory then tells us that if the diffraction pattern sampling is at the Nyquist fineness, and the sampled values (correctly phased) are Fourier summed, the result will be correct images of the specimen indefinitely repeated and just in contact with each other. If the sampling is finer than Nyquist, and the values (again correctly phased) are Fourier summed, the images are still correct and repeated, but are now not in contact, thus giving rise to zero regions between them. It is the presence of these regions, which should be zero, which drives the phasing. The sampling is made sufficiently fine that the number of voxels in image space between the specimen envelopes somewhat exceeds the number of voxels inside the specimen envelopes. Then, starting with e.g. random phases, an iterative algorithm, repeatedly moving via Fourier transformation between the two spaces, alternately (1) in image space pushes the voxels between specimen envelopes toward zero and (2) resets pattern magnitudes in diffraction space to their experimentally observed values. Our experience is that usually, after several hundred to several thousand iterations, the interenvelope voxels are essentially zero and the intraenvelope voxels and the phases are essentially correct (see Figs. 2 and 3). Arriving at the correct phasing can be assisted by adding appropriate positivity constraints on the real and imaginary parts of intraenvelope voxels [20]. The algorithm is basically a Fienup-type algorithm [26] and is attractively simple and fast, being an alternation of simple processing in one space followed by high-speed Fourier transformation to the other space.

This process clearly falls in the constraint-based category of phasing methods, and is thus subject to the concern that it may produce incorrect, or near-correct but inaccurate, solutions; this concern is perhaps here further heightened by the fact that in all real cases there can be data inadequacies, such as data error, too much missing data, possibly insufficiently fine sampling, and inaccuracies in the assumed specimen envelope. I think a fair summary of what has been learned thus far—mainly through computer simulation studies[9]—is that with mild data inadequacies (levels that can probably be reached with careful experimentation) the probability of correct and reasonably high-quality phasing and imaging is high, but that there may not be a large margin of safety[10] in this respect. At this point, however, these matters require more study and experience.

Finally, in a recent paper [22], it is shown that the phasing process also works for the second case, the imperfect or very small crystal. (Insofar as this case has a difficulty, it lies in the experimental portion of the work, and arises from the Bragg spots still being strong and taking intensity away from the non-Bragg Nyquist points, making the latter more difficult to measure.) Here the process could be of considerable

[8] However, see Szoke [11b] who suggests that the use of partially coherent radiation in the crystal case will cause the Bragg spots to become broader and allow additional information to be observed.

[9] The simulation studies of which I am aware at present are slightly flawed in that they assume pointwise sampling at the sampling points, whereas experimental values will normally involve integration over a detector pixel size.

[10] The smallness of the margin of safety may be important in our planned yeast cell work, where we have significant missing data in both the central region and double cone (see section on Two Probable Experimental Regimes: A Synchrotron/Cryoprotection Technique for Supramolecular Specimens). We have shown through simulation studies that the algorithm can start with arbitrary magnitudes as well as phases in those regions and refine both successfully, but that trouble starts to develop as the amount of missing data approaches that which we may have to face in practice. With this in mind, we are considering an alternative algorithm, resembling a crystallographic least-squares model refinement but with values of Nyquist-fineness pixels (rather than atom positions) being the quantities refined in image space, and with observed finer-than-Nyquist diffraction magnitudes (unobserved magnitudes omitted) being the data to be fitted.

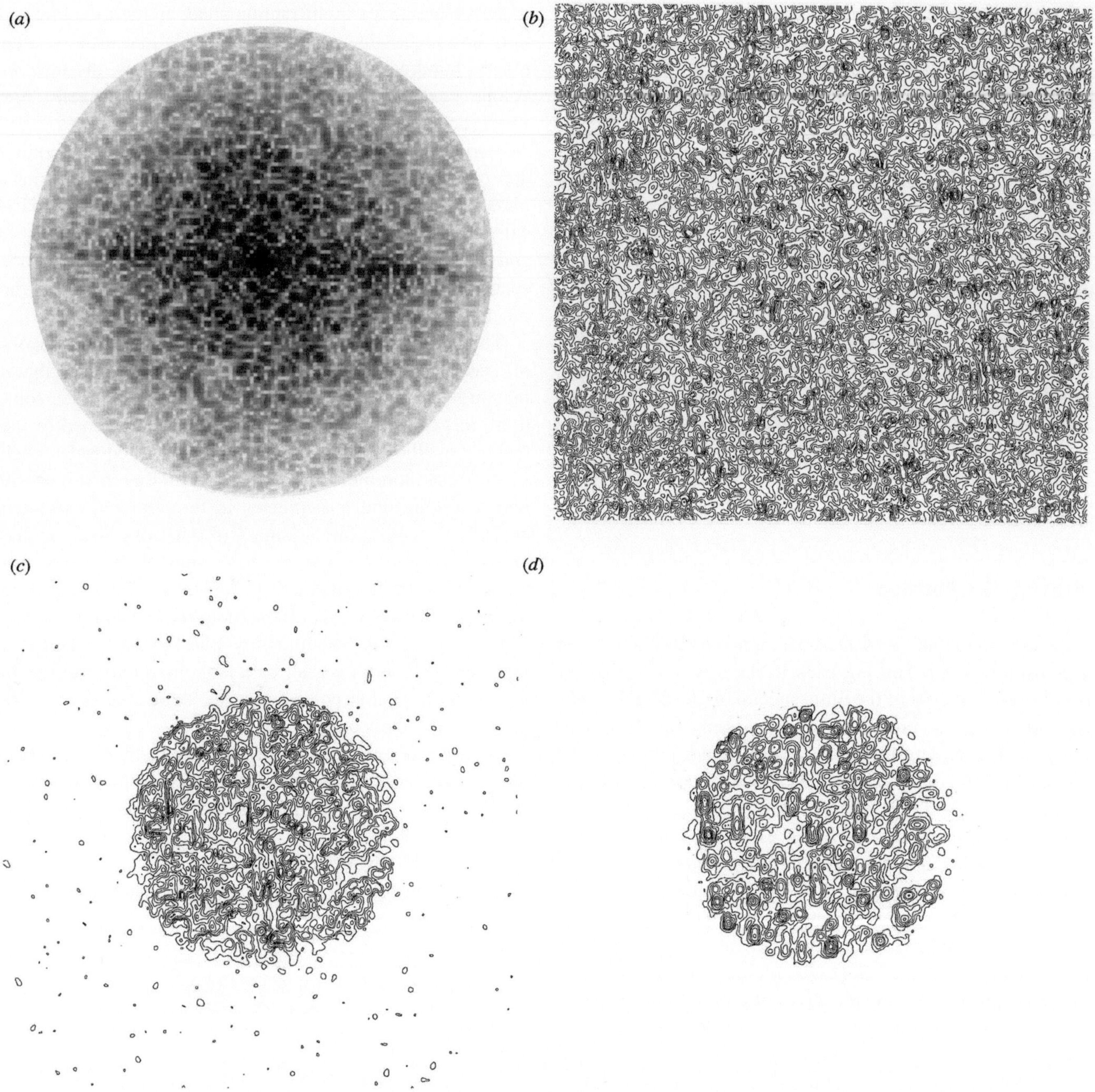

Fig. 2 An example of phasing by oversampling. (**a**) A set of 13,965 discrete magnitudes of a 2D computer-generated diffraction pattern. The magnitudes were oversampled by 2 times in each dimension. (**b**) The structure image corresponding to the initial random phasing. (**c–e**) The images after 50, 100, and 200 phasing cycles, the cycles primarily devoted to pushing the electron density outside the roughly known specimen envelope to zero. Note that after 50 cycles, the zeroing has advanced but is not complete. After 100 cycles, the contents of the envelope are taking form, but are not complete. After 200 cycles, the content has reached a final form. (**f**) The actual content on which (**a**) was based. Note that (**e**) has correctly found the structure. (**g, h**) Like (**e, f**) except that 17 % noise (peak-to-peak) was added to the data in (**a**) and 425 cycles were needed to complete the phasing (Taken from Ref. [19]).

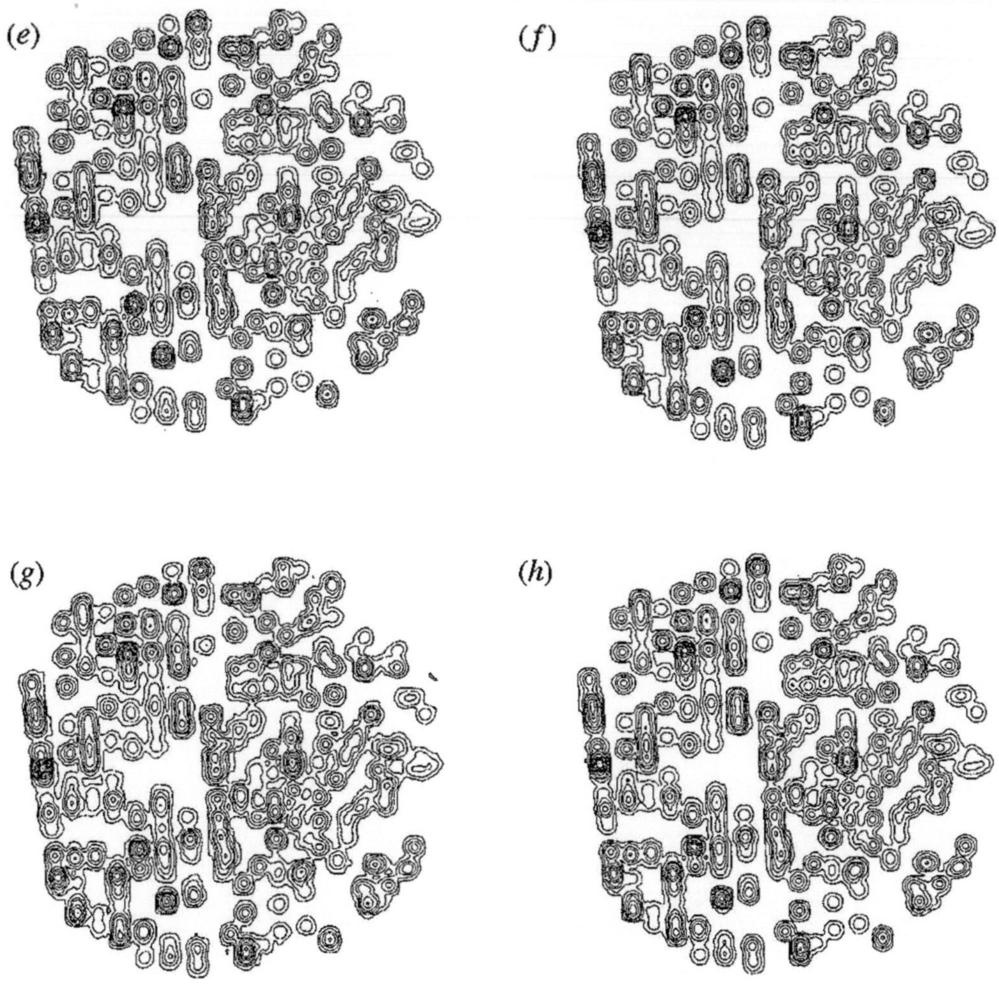

Fig. 2 (continued)

value in the study of crystal imperfections (Fig. 4) and possibly also in protein crystallography, where it is not infrequent for crystallization attempts to result in the production of numerous very small crystals.

Summary

The phase problem, which might have stunted the growth of crystallography, and held it to not much more than its 1929 dimensions, did not do so, and the subject has (through the work of many) been allowed to grow freely to wherever the ability to grow crystals may take it. One path to yet further growth is thus through more powerful crystal-growing techniques, and one version of this, which would be a highly attractive one, is mentioned as a future possibility in the last section of Part I. However, a second path also may exist and that is the use of new technology to extract copious diffraction data from arbitrary (including noncrystalline) specimens. This path, only now approaching its first major tests, is briefly set forth in Part II.

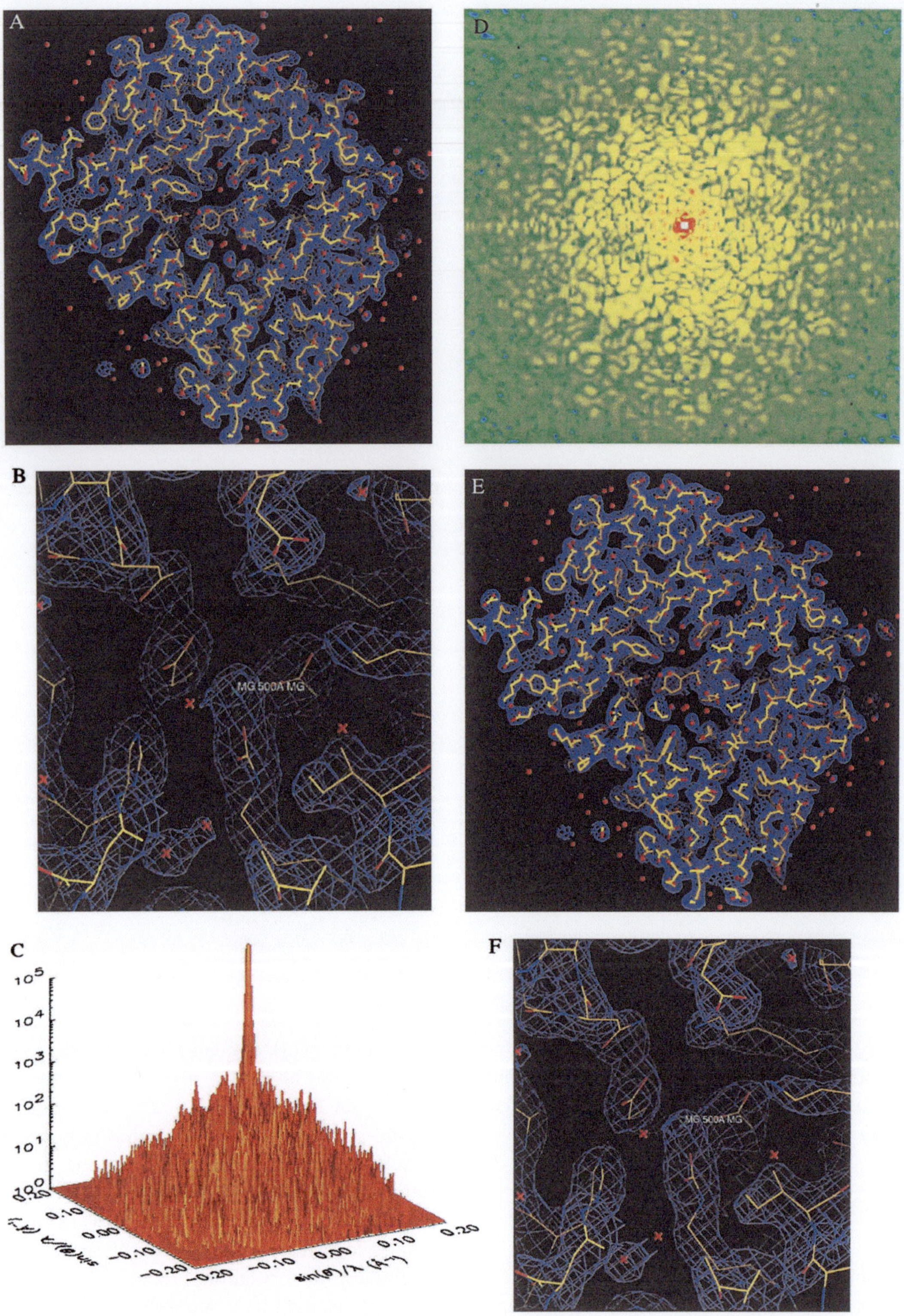

Fig. 3 Simulation of a possible future method of protein structure determination. (**c**) and (**d**) Simulation of one section of the oversampled 3D dataset of the protein rubisco, which could be obtained using 10^6 individual randomly oriented rubisco molecules successively illuminated (and destroyed) by 10^6 fs-scale 1.5 Å wavelength FEL X-ray pulses. The cumulative intensity of the pulses is sufficient to yield a 3D pattern to 2.5 Å resolution with reasonable Poisson noise statistics. The 10^6 exposures could reasonably be carried out in approximately 2.3 h. (**e**) and (**f**) show the molecule and its active site as found after oversampling phasing of the dataset, while (**a**) and (**b**) show the corresponding views of those structures as found in the original normal protein crystallographic investigation [27] (Taken from Ref. [23]).

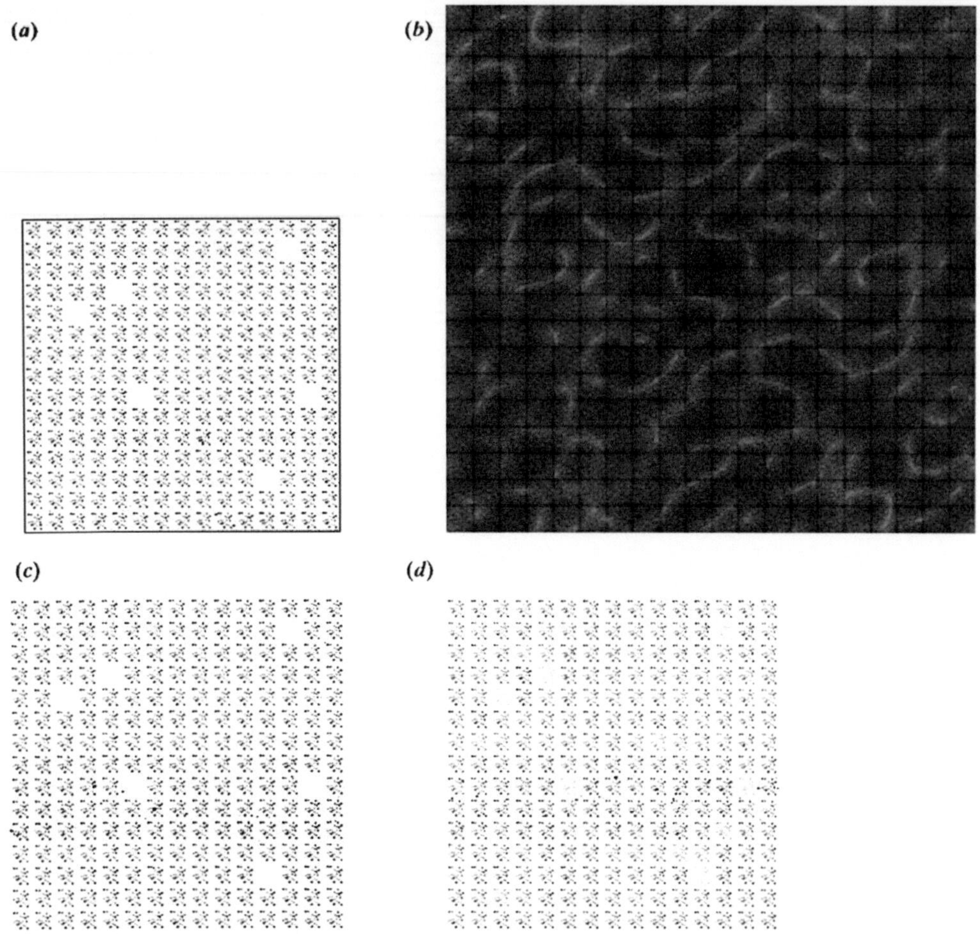

Fig. 4 Example of the possible value of finer sampling in the study of crystal imperfections. (**b**) A simulated strongly exposed diffraction pattern of a small 2D crystal. The pattern is continuous and permits oversampling. (**c**) The result of phasing and imaging. The crystal is found to be imperfect, with six missing unit cells. (**a**) The actual structure from which (**b**) was generated. (**d**) The result with random noise added to (**b**). The imaging is less clear (Taken from Ref. [22]).

Acknowledgments It is evident that the progress briefly recorded in this article is the work of very many people. I hope that I have given a brief but fair indication of that. At the same time I do not wish to close this article without acknowledging the special help given to me by certain individuals in the course of my lifetime. Gerard Bricogne of Cambridge has been, for many years, a companion and often guide in thinking about the phase problem. John Miao, graduate student at Stony Brook and now friend at Stanford, played a vital part in advancing the work reported in Part II of this article. Eaton Lattman of Johns Hopkins has often been my advisor in crystallographic matters, and was so in the writing of this article. Last, but also first, is Janos Kirz of Stony Brook, who more than any other made possible the work of Part II, with untiring support and accurate advice. Finally, my wife Anne Sayre was everything to me in the work, as well as the living, of life.

References

1. Robertson, J. M. *Rep. Progr. Phys.* **1937**, *4*, 332.
2. Sayre, D. *Acta Crystallogr.* **1952**, *5*, 843.
3. Harker, D.; Kasper, J. S. *Acta Crystallogr.* **1948**, *1*, 70; Kasper, J. S.; Lucht, C. M.; Harker, D. *Acta Crystallogr.* **1950**, *3*, 436.
4. Karle, J.; Hauptman, H. *Acta Crystallogr.* **1950**, *3*, 181.
5. Sayre, D. *Acta Crystallogr.* **1952**, *5*, 60.
6. Sayre, D. *Acta Crystallogr.* **1972**, *A28*, 210; *Acta Crystallogr.* **1974,** *A30*, 180.
7. Rothbauer, R. *Z. Kristallogr.* **2000**, *215*, 157.
8. Hauptman, H.; Karle, J. *ACA Monograph #3*; Edwards: Ann Arbor, **1953**.
9. (a) Bricogne, G. *Acta Crystallogr.* **1988**, *A44*, 517; (b) *Trans. ACA* **1994**, *30,* 163; (c) *Methods Enzymol.* **1997**, *276*, 361.
10. Weeks, C. M.; DeTitta, G. T.; Hauptman, H. A.; Thuman, P.; Miller, R. *Acta Crystallogr.* **1994**, *A50*, 210.
11. (a) Szoke, A. *Acta Crystallogr.* **1993**, *A49*, 853.; Somoza, J. R.; Szoke, H.; Goodman, D. M.; Beran, P.; Truckses, D.; Kim, S.-H.; Szoke, A. *Acta Crystallogr.* **1995**, *A51*, 691. (b) Szoke, A. *Acta Crystallogr.* **2001**, *A57*, 587.
12. Spence, J. C. H. *Ultramicroscopy*, in press.
13. Burns, M. M.; Fournier, J.-M.; Golovchenko, J. A. *Science* **1990**, *249*, 749.
14. Larsen, J. J.; Hald, K.; Bjerre, N.; Stapelfeldt, H.; Seideman, T. *Phys. Rev. Lett.* **2000**, *85*, 2470.

15. Sayre, D. In *Imaging Processes and Coherence in Physics*; Schlenker, M.; Fink, M.; Goedgebuer, J. P.; Malgrange, C.; Vienot, J. Ch.; Wade, R. H., Eds.; Springer-Verlag: Heidelberg, **1980**; p. 229.
16. Yun, W. B.; Kirz, J.; Sayre, D. *Acta Crystallogr.* **1987**, *A43*, 131.
17. Sayre, D. In *Direct Methods of Solving Crystal Structures*; Schenk, H., Ed.; Plenum Press: New York, **1991**; p. 353.
18. Sayre, D.; Chapman, H. N. *Acta Crystallogr.* **1995**, *A51*, 237.
19. Sayre, D.; Chapman, H. N.; Miao, J. *Acta Crystallogr.* **1998**, *A54*, 232.
20. Miao, J.; Sayre, D.; Chapman, H. N. J. *Opt. Soc. Amer.* **1998**, *A15*, 1662.
21. Miao, J.; Charalambous, P.; Kirz, J.; Sayre, D. *Nature* (*London*) **1999**, *400*, 342.
22. Miao, J.; Sayre, D. *Acta Crystallogr.* **2000**, *A56*, 596.
23. Miao, J.; Hodgson, K. O.; Sayre, D. *Proc. Natl. Acad. Sci.* **2001**, *98*, 6641.
24. Neutze, R.; Wouts, R.; v. d. Spoel, D.; Weckert, E.; Hajdu, J. *Nature* (*London*) **2000**, *406*, 752.
25. Arthur, J.; Bane, K.; Bharadwaj, V.; Bowden, G.; et al. *Linac Coherent Light Source (LCLS) Design Study Report*; Stanford Linear Accelerator Center, **1998.**
26. Fienup, J. R. *Appl. Opt.* **1982**, *21*, 2758.
27. Lundqvist, T.; Schneider, G. *Biochemistry* **1991**, *30*, 904.

History of X-Ray Crystallography[a]

Herbert A. Hauptman[b]

Abstract
A brief account is given of the history of X-ray crystallography from its beginning in 1912 to the present time. Particular emphasis is placed on the phase problem, the major obstacle in the path leading from the observed X-ray diffraction pattern to the desired crystal structure. The essential role which mathematics plays in resolving this problem is also stressed.

In this brief sketch of the history of X-ray crystallography, I would like to emphasize the important role played by the development of the direct methods which were devised to solve the central problem of X-ray crystallography, the so-called phase problem. I shall also stress the importance of cross disciplinary research, in particular the essential role which mathematics played in this development.

In 1895 Wilhelm Röntgen discovered X-rays. With this discovery, the stage was set for the creation of the modern science of X-ray crystallography.

In 1912 Paul Ewald was completing his doctoral dissertation about the optical properties of a medium consisting of a regular arrangement of isotropic resonators. A crystalline solid which, on the sub-microscopic level, consists of a triply periodic, regular arrangement of atoms, or molecules, is therefore precisely the kind of medium with which Ewald was concerned. Since the smallest interatomic distances in a crystal are of the same order of magnitude as the wavelengths of X-rays, it occurred to Max von Laue, upon learning of Ewald's results, that a crystal might serve as a three-dimensional diffraction grating for X-rays. In order to test this hypothesis, he prevailed upon the younger physicists Walter Friedrich and Paul Knipping to perform the necessary scattering experiment.

The scattering experiment indeed showed that when a beam of X-rays strikes a crystal, the crystal scatters the incident beam in many different directions and with different intensities. If these scattered X-rays strike a photographic plate they will blacken the plate at those points where the scattered rays strike the plate. In this way one obtains the so-called diffraction pattern. This experiment marked the birth of the science of X-ray crystallography and, because of its fundamental importance in determining crystal and molecular structures, must be regarded as a landmark event in twentieth century science. The major obstacle in the path leading from the observed diffraction pattern to the desired crystal structure is known as the phase problem, for reasons to be given shortly. I propose to give here a brief historical account of the methods devised to overcome this obstacle, the so-called direct methods of X-ray crystallography.

It has already been remarked that a crystal may be regarded as a regular triply periodic arrangement of an array of atoms. One imagines three families of planes, the planes in each family being parallel to and equidistant from one another. In this way, one obtains a tiling of the crystal space by means of congruent parallelepipeds, each one of which is said to be a fundamental parallelepiped, or unit cell, of the crystal.

If each unit cell contains a molecule—a collection of atoms—in its interior, and if the atoms are arranged in precisely the same way in all the unit cells, then each unit cell and its contents are indistinguishable from every other unit cell and its contents.

There corresponds to each atom an electron density function; hence, by superposition of the individual atomic

[a] *Structural Chemistry* 1990, 1(6):617–620.

[b] Medical Foundation of Buffalo, Inc., 3 High Street, Buffalo, NY 14203, USA (deceased)

I. Hargittai, B. Hargittai (eds.), *Science of Crystal Structures: Highlights in Crystallography*,
DOI 10.1007/978-3-319-19827-9_2

electron density functions, one obtains an overall electron density function $\rho(\mathbf{r})$, a nonnegative function of the position vector $\mathbf{r}$ which gives the number of electrons per unit volume at the position $\mathbf{r}$. It is clear from the geometric construction that the electron density function in any unit cell is identical to that in every other unit cell. Hence $\rho(\mathbf{r})$ is a triply periodic function of position, and this property may be taken as the mathematical definition of a crystal.

If on the other hand we choose to regard a crystal as a triply periodic arrangement of an array of atoms, or molecules, then by a crystal structure we mean simply the arrangement and identities of the atoms in the unit cell and by a molecular structure the arrangement and identities of the atoms in the molecule.

It was recognized almost from the beginning that the diffraction pattern, the directions and intensities of the X-rays scattered by a crystal, is uniquely determined by the crystal structure; which is to say that if one knew the crystal structure—the arrangement of the atoms in the crystal—then one could calculate the diffraction pattern completely. It turns out that, conversely, diffraction patterns determine, in general, unique crystal and molecular structures, although this fact was not known until many years later. In short, the information content of a typical molecular structure coincides precisely with the information content of its diffraction pattern. It is a measure of the great advances made by the new science of X-ray crystallography that, nowadays, one can routinely transform the information content of a diffraction pattern into a molecular structure, at least for the so-called "small" molecules, that is those consisting of some 150 or fewer non-hydrogen atoms.

Since X-rays, like ordinary visible light, are electromagnetic waves, they have a phase as well as an intensity, just as any other wave disturbance. In order to work backwards, from diffraction patterns to crystal and molecular structures, it turns out to be necessary to measure not only the intensities of the X-rays scattered by the crystal but their phases as well. However, the phases cannot be measured in the ordinary kind of diffraction experiment; they appear to be irretrievably lost. Only the intensities can be directly measured. This then gives rise to the central problem of X-ray crystallography, the so-called phase problem, how to deduce the values of the phases of the X-rays scattered by a crystal when only their intensities are known. For some 40 years after the landmark experiment of Friedrich and Knipping, all attempts to find a general method for going directly from the diffraction pattern, that is measured intensities alone, to the crystal structure, with or without the intervention of the phases (a method that would be useful for the complex structures of interest to chemists, biologists, and mineralogists) were defeated.

Because the needed phase information was lost in the diffraction experiment, it was thought that one could use arbitrary values for the phases associated with the measured intensities of the scattered X-rays. In this way one obtains a myriad of different crystal structures, all consistent with the known intensities. It therefore came to be generally believed that a procedure for going directly from the measured intensities to crystal structures could not, even in principle, be devised. By the same thinking, the problem of deducing the values of the individual phases from the diffraction intensities, the so-called phase problem, was also though to be unsolvable, even in principle. It wasn't until the early 1950s, through the exploitation of special properties of molecular structures and through a simple mathematical argument, that these erroneous conclusions were finally refuted.

The special property that all crystal and molecular structures possess may be summed up in one word: atomicity. Thus the electron density function $\rho(\mathbf{r})$ in a crystal takes on large positive values at the atomic position vectors and drops to small values between the atoms. If our goal is merely to determine the positions of the atoms—the positions of the maxima of $\rho(\mathbf{r})$—rather than the much more complicated electron density function associated with the distribution of atoms in the crystal, then our problem is greatly simplified; it turns out to be not only determinate but actually greatly overdetermined by the available X-ray diffraction intensities.

This is most easily seen by eliminating the lost phase information from the relationships between the diffraction pattern and the crystal structure. Doing this results in a system of equations relating the diffraction intensities alone with the atomic position vectors. Because the number of these relationships far exceeds (by a factor of ten or so) the number of unknown position vectors needed to define the crystal structure, our problem is greatly overdetermined. Thus it is clear that there exist relationships between the measured diffraction intensities and the lost phases that may be exploited. It follows that the phases of the scattered X-rays are also determined by their intensities. In short, the lost phase information is to be found among the available intensities, and the phase problem is therefore a solvable one, at least in principle. There remains the task of devising numerical algorithms leading from the abundance of experimentally measured diffraction intensities to the values of the individual phases. The techniques of X-ray crystallography that deduce the individual phases by exploiting relationships between measured diffraction intensities and phases are known as direct methods.

The argument just presented was in fact anticipated in 1927 by Heinrich Ott [1], who showed by algebraic analysis and applications that the method is capable of solving simple centrosymmetric structures, in which all phases must be either 0 or π. The method was further elaborated by Kedareswar Banerjee in 1933 [2] and Melvin Avrami in 1938 [3] but was clearly of only limited value in applications. While this early work of Ott, Banerjee and Avrami shed important light on the more general phase problem, it attracted little attention at the time and was not further developed; it appears now to be all but forgotten.

My work on this problem started in 1948 about a year after I joined the Naval Research Laboratory in Washington, D.C. and initiated my collaboration with Jerome Karle. It had been some 35 years since Friedrich and Knipping had carried out their famous experiment, and by 1947 the phase problem, the central problem of X-ray crystallography, was still unsolved and generally regarded as unsolvable. The central importance of this problem and its strong mathematical component combined to provide a challenge that could not be denied.

Then too, there was a certain air of mystery surrounding the problem. On the one hand the simplicity and logic of the argument "proving" its unsolvability, even in principle, appeared to be overwhelming. On the other hand, crystal and molecular structures were being solved, although the structures studied were almost always very simple ones involving a small number of atoms or larger structures containing one or a small number of heavy atoms, for which special techniques had been devised. It had not yet been generally understood that the implicit assumption of atomicity and the concomitant trial-and-error approach to most structure solutions had imposed a powerful restriction on the permitted values of the phases.

The first important contribution that Karle and I made was the recognition that it would be necessary to exploit prior structural knowledge to transform the phase problem from an unsolvable one to one that was solvable, at least in principle. Our first step in this direction was to exploit the nonnegativity of the electron density function $\rho(\mathbf{r})$. Before our analysis was complete, however, David Harker and John Kasper published their famous paper [4] in which they derived inequalities in which the measured intensities restrict the possible values of the phases. This was a very mysterious paper, because nowhere in it does there appear any explicit mention of the basis for the inequality relations, and indeed the most important fact is conspicuous by its absence. It is simply that the electron density function is nonnegative everywhere. This fact is, however, implicit in Harker and Kasper's work. In very short order Karle and I completed our own analysis and derived the complete set of inequality relationships based on the nonnegativity of the electron density function [5]. It includes the Harker-Kasper inequalities as a special case, and many others besides. Although the complete set of inequalities greatly restricts the values of the phases, the relations appear to be too intractable to be useful in applications, except for the simplest structures, and their potential has never been fully exploited.

The recognition in 1950 and 1951 that molecules consist of atoms that, to a good approximation, may be regarded as points transformed completely the nature of the phase problem. While it meant accepting as fact that the observed diffraction intensities by themselves were indeed not sufficient to determine a unique electron density function, it also meant that they were more than sufficient, by far, to determine the atomic position vectors [6]. It meant as well that the phases corresponding to the point atom structure were greatly overdetermined by the available intensities. Finally, it meant that a formidable psychological barrier had been removed, because it now made sense to look for a solution to the phase problem, for numerical algorithms leading from measured intensities to individual phases. In hindsight it is perfectly clear that owing to the great overabundance of diffraction data, a probabilistic approach is called for; some 40 years ago, however, this was not so apparent.

Before we could even get started, an unexpected complication arose. It turned out that because the values of the individual phases clearly depend not only on the crystal structure but also on the choice of origin, they are not uniquely determined by the crystal structure alone. It followed that the diffraction intensities alone do not determine unique values for the phases. The process leading from diffraction intensities to phases would have to include a recipe for specifying the origin. This required that we separate out two contributions to a phase, one due to the crystal structure alone and one due to the choice of origin. We clearly needed to study how a phase is transformed when the origin is shifted, a problem that was complicated by the fact that the permissible origins depend on the crystallographic elements of symmetry, which were usually known in advance.

The solution was made easier by the discovery that there are always certain linear combinations of the phases, the so-called structure invariants, that are uniquely determined by the crystal structure alone and are independent of the choice of origin. It is therefore only the values of the structure invariants that we can hope to estimate from the measured intensities. Once we have estimated a sufficient number of these we can then hope to evaluate the individual phases by a process that incorporates a recipe for specifying the origin.

What was clearly called for was devising a method for identifying the structure invariants, and then using these to come up with recipes for fixing the origin appropriate to the different elements of crystallographic symmetry that may be present. Once this was done there would remain the task of estimating the values of the structure invariants by means of their conditional probability distributions, assuming that an appropriately chosen set of diffraction intensities is known.

Beyond any doubt our most important contribution during the early 1950s was the introduction of probabilistic techniques—in particular, use of the joint probability distribution of several diffraction intensities and the corresponding phases—as the central tool in the solution of the phase problem [7]. We assumed to begin with that all positions of the atoms in the unit cell of the crystal were equally likely, or, in the language of mathematical probability, that the atomic position vectors were random variables, uniformly and independently distributed. With this assumption the intensities and phases of the scattered X-rays, as functions of the atomic position vectors, are also random variables, and one can use the methods of modern mathematical probability theory to calculate the joint probability distribution of any collection of intensities and phases. A suitably chosen joint probability distribution leads directly to the conditional probability distribution of a specified structure invariant, assuming again an appropriately chosen set of diffraction intensities. The conditional distribution in turn leads to the structure invariant, an estimate of which is given, for example, by its most probable value. Once one has a sufficiently large number of sufficiently reliable estimates of structure invariants, one can use standard techniques to calculate the values of the individual phases, provided that the process incorporates a recipe for specifying the origin.

Although probabilistic methods played an essential role in the development of the direct method and provided it with its logical foundation, it must also be pointed out that non-probabilistic methods also played an important part. In this connection the early work of Sayre [8], Zachariasen [9], Cochran [10] and Woolfson [11] should be mentioned. In particular the well known Sayre equation, a relationship of fundamental importance among measured magnitudes and unknown phases, continues to be useful to the present day and lies at the heart of some of the more successful computer programs for solving crystal structures.

I cannot conclude this brief account of the early history of the direct methods of X-ray crystallography without also describing the reception this work received at the hands of the crystallographic community. This was, simply, extreme skepticism, if not outright hostility. In hindsight I think this reaction was due, first, to the strong mathematical flavor of this early work, not well understood by most crystallographers, as well as the ingrained and almost universal belief that the phase problem was unsolvable in principle and that any claim to the contrary must therefore be flawed. This nearly universal skepticism and inability to understand the proposed solution no doubt explains why so few early attempts to apply the new methods were made. It wasn't until the 1960s when easy to use computer programs became available that widespread applications were made.

Today some 100,000 molecular structures are known, most determined by the direct methods, and about 5,000 new structures are added to the list every year. It is no exaggeration to say that modern structural chemistry owes its existence to this development.

Although no equations are shown in this article, it should be clear that the developments described here would not have been possible without strong dependence on mathematical techniques, in particular the modern theory of mathematical probability, and it is this interaction between mathematics and the phase problem of X-ray crystallography which I have tried to emphasize in this article. Work on the phase problem continues to this day and applications to structures of ever increasing complexity continue to be made. It still appears that progress is made only in proportion to our ability to bring more powerful mathematical techniques to bear on this fascinating problem.

Acknowledgements This work was supported by the National Science Foundation grant number CHE-8822296.

References

1. Ott, H. *Z. Kristallogr.* **1927**, *66*, 136.
2. Banerjee, K. *Proc. R. Soc. London, Ser. A* **1933**, *141*, 188.
3. Avrami, M. *Phys. Rev.* **1938**, *54*, 300.
4. Harker, D.; Kasper, J. S. *Acta Crystallogr.* **1948**, *1*, 70.
5. Karle, J.; Hauptman, H. *Acta Crystallogr.* **1950**, *3*, 181.
6. Hauptman, H.; Karle, J. *Phys. Rev.* **1950**, *80*, 244.
7. Hauptman, H.; Karle, J. *Solution of the Phase Problem I. The Centrosymmetric Crystal*; American Crystallographic Association Monograph No. 3, Polycrystal Book Service: Dayton, Ohio, 1953.
8. Sayre, D. *Acta Crystallogr.* **1952**, *5*, 60.
9. Zachariasen, W. H. A. *Acta Crystallogr.* **1952**, *5*, 68.
10. Cochran, W. *Acta Crystallogr.* **1952**, *5*, 65.
11. Woolfson, M. *Acta Crystallogr.* **1954**, *7*, 61.

Gas Electron Diffraction and Its Influence on the Solution of the Phase Problem in Crystal Structure Determination[a]

Isabella Karle[b] and Jerome Karle[b,c]

Abstract

We studied, and performed research for our Ph.D. degrees in the area of gas electron diffraction. Our mentor was Lawrence Brockway, a pioneer in this subject. At that time, research in gas electron diffraction was in its early stages of development. In 1941, the distinguished Peter Debye wrote a theoretical paper concerning gas electron diffraction which challenged ones capability to develop the necessary experimental equipment and to further advance the theoretical developments so as to greatly extend the science of gas electron diffraction. We carried these thoughts in mind when we joined the Naval Research Laboratory, where the opportunity to design and produce excellent equipment was readily available. In the course of pursuing this research area, one of the findings was the existence of non-negativity as a condition for the results of a diffraction experiment for gaseous substances. When we became interested in the field of crystal structure determination, the familiarity with non-negativity which was needed in the study of gases, led to a search for the necessary and sufficient condition for a Fourier series to be non-negative. The search was successful and has played an important part in crystal structure determination. Some early applications to complicated structures are described.

Keywords

Gas electron diffraction • Crystal structure determination • Non-negativity

Introduction to Gas Electron Diffraction

We were introduced to electron diffraction by gaseous molecules as graduate students at the University of Michigan during the early 1940s. Except for being attracted to the field of physical chemistry, we did not enter graduate school with any strong convictions concerning a particular area of specialization for our thesis research. However, as a consequence of being students in the excellent physical chemistry course of L. O. Brockway (Photo 1), we sought to carry out our research program under his direction.

During our graduate student days, gas electron diffraction research was performed in circumstances that, in time, changed considerably. High-frequency electronics had not been commercially developed and high potential was achieved by use of greatly overpowered equipment designed for use with X-ray tubes. Electronic control of the high potential, beam current and focusing systems were quite excellent, but the circuits were homedesigned and homemade. Pump design and vacuum measurements, while adequate, and in principle, similar to present systems, had not begun to attain current sophistication. Perhaps, one of the greatest differences in the technical aspects of conducting gas diffraction research was in computing. Molecular intensity functions for comparison with visually estimated data were calculated by making use of tabulations of the function sine *sr*/*sr* for various values of *r*. The tabulations were in the form of paper strips with each strip associated with a

[a] *Structural Chemistry* 2005, 16(1):5–16.

[b] Laboratory for the Structure of Matter, Naval Research Laboratory, CODE 6030 Washington, DC, USA

[c] Deceased

I. Hargittai, B. Hargittai (eds.), *Science of Crystal Structures: Highlights in Crystallography*,
DOI 10.1007/978-3-319-19827-9_3

Photo 1 Lawrence Brockway, Isabella Karle and Jerome Karle at the University of Michigan, Ann Arbor in the early 1940s.

particular value of *r* and a range of values for *s*. Calculations were performed in a fashion similar to those for electron densities in X-ray crystal structure analysis that employed Beevers–Lipson strips. With the selection of appropriate strips to correspond to the distances in a model structure and a simple estimate of their coefficients, the computation could proceed. The calculations were first performed with a hand-cranked calculator that was subsequently replaced with the type driven by an electric motor.

We both investigated the structure of numerous molecules by these means in the course of pursuing our graduate research. During this time, Jerome became quite interested in the theoretical paper of Debye [1] that displayed a deep insight into the effects of molecular motion on the molecular intensity function and their expression in terms of the Fourier transform of this function, the radial distribution function. Although it did not appear that experimental data at the time would be readily applicable to this theory, it was thought that it might afford some insight into the internal rotation occurring in the molecule trifluoroacetic acid, that Jerome was investigating at that time. The paper of Debye made a profound impression and, as it turned out, had a strong influence on our research program after we joined the Naval Research Laboratory.

Getting Started at NRL

We completed our graduate studies in 1943 within some months of each other and went off to Chicago to work on a war-related research project. Some time later we returned to the University of Michigan where Jerome became associated with a research project sponsored by the Naval Research Laboratory. After the war had ended we needed to seek employment. Jerome's project had ended and Isabella's teaching position at the University had no potential for future permanence. The unique opportunity for us both to join a laboratory and pursue a research program in subjects of interest was offered to us by the Naval Research Laboratory and we joined this institution in 1946.

A valuable feature of this Laboratory was its extensive and very fine machine shop facilities. It was therefore not only readily possible to construct the necessary equipment for gas electron diffraction research but also to develop and introduce innovative features. As we entered into gas electron diffraction again after the hiatus that was filled by war-related research, we recalled the potential for advances in this field implied by an ability to implement the various theoretical aspects discussed in the previously mentioned paper by Debye [1]. The decision was

Photo 2 The instrument built at the Naval Research Laboratory for gas-phase electron diffraction studies (late 1940s).

soon made not only to construct a state-of-the-art diffraction instrument but also to develop the data reduction process and the theoretical tools in an attempt to fulfill this promise.

A main feature of the diffraction instrument that was built [2], was a double electromagnetic lens system permitting a beam crossover after the first lens, enhancing the quality of the beam passing through the second one before reaching the sample. The apparatus also had the first rotating sector used in the United States and a relatively large sample chamber. The construction proceeded quite readily. In fact, we built two instruments at the same time in order to obtain one suitably leak-proof instrument more quickly. The completed instrument is shown in Photo 2.

We made plans to use microphotometry for measuring the scattered electron intensities rather than depend upon visual estimates. The great capabilities of a well-trained eye in eliminating changing backgrounds, pinpointing positions of maxima and minima and approximating their relative levels made this a not-so-obvious decision at the time. We were aware of the use of rotating sectors and microphotometry already initiated in Norway. However, among our concerns was the relatively weak signals that occur with increasing scattering angle and the interference that arises in microphotometry from the granular character of the photographic emulsions. It occurred to us that spinning the photographic plate while tracing it should be quite helpful and we fitted out a Leeds and Northrup microphotometer with a rotating attachment that worked rather well. Before we submitted a little note on this subject for publication in 1947 [3], we learned that the idea had been pursued more than a decade earlier by Degard and colleagues [4] in Liege, in a laboratory that was under the direction of P. Debye.

Problems to Be Overcome

If we were to fulfill the hope of not only obtaining interatomic distances and bond angles of enhanced accuracy but also information concerning the average internal molecular motion, problems had to be overcome. If successful, it would be possible to obtain a radial distribution function that would represent accurately the probability distributions for the interatomic distances. The problems to be solved were

1. The molecular intensity function, whose Fourier transform would produce the radial distribution function of interest, had to be obtained from the total intensity of scattering in such a way that the interference functions (sine sr_{ij}/sr_{ij}, for example, where s is a function of the scattering angle and r_{ij} is the distance between the ith and jth atoms) would have coefficients that arise from essentially point atoms, i.e., the coefficients would be, ideally, independent of the variable s.
2. The experimental molecular intensity function is known only over a limited angle range, whereas the calculation of the Fourier transform of this function requires it to be known over an infinite range or, at least, to a large enough value of s, so that the omitted region is negligible. This is ordinarily not attainable, hence the calculation of the Fourier transform had to be modified to avoid "termination errors."
3. In the course of extracting the molecular intensity from the total intensity function, it is necessary to generate a suitable background scattering function. There were difficulties with the theory for the background scattering as well as added difficulties arising from the recording of the total intensities on photographic plates. As a consequence, adjustments were required for the theoretical background function in order to avoid serious errors in the derived molecular intensity function.
4. The microphotometry of the photographic images of the diffraction patterns had to be optimized. As mentioned previously, the rotating sector and plate spinning helped considerably. In addition, the densitometry and transformation of photographic density to intensity had to be made as accurate as possible.

These various problems were overcome. For example, by dividing the total intensity of scattering by an appropriate background function, a molecular intensity function could be obtained whose coefficients varied little over the experimental range of s [5, 6]. A suitable damping function introduced by Degard in 1937 [7] and Schomaker in 1939 [8] took care of termination errors and we contributed a further analysis based on the known character of the probability distributions for interatomic distances that took account of the spreading effect of the damping function on the probability distribution [9].

The solution of the problem of developing a suitable background scattering function was facilitated by the use of the rotating sector. Adjustments to optimize the shape of this function were based on a principle that has found extended application to other areas of structure research. The principle concerns the introduction of constraints into the analysis of diffraction data to satisfy the non-negativity of certain structural characteristics. The non-negativity of the Fourier transform of the molecular intensity function representing modified probability distributions was used to produce an accurate background function.

It was therefore possible by various means to solve the four problems listed above and develop a molecular intensity curve whose Fourier transform could be interpreted in terms of probability distributions for the distances between pairs of atoms in a molecule. Along with this came enhanced accuracy that permitted not only the determination of interatomic distances, but also the possibility of investigating the internal motion of the molecules. The procedures for performing such analyses were described in 1949 and 1950 [6, 9, 10].

Early Studies

In the course of these developments, we began to make use of IBM punch card machines for our computing. The operation of addition proceeded with tolerable speed, but multiplication in the early machines proceeded at the rate of about 2 multiplications per minute, too slow to be useful. Then about 1950, IBM came out with a multiplier that could perform 100 multiplications per minute. For those of us who had initially performed calculations with a crank-handle computer, computing paradise had indeed arrived.

Confidence in the increased accuracy with which interatomic distances could be determined, as well as the possibility of obtaining reliable values for root mean squared amplitudes of vibration, was enhanced by satisfactory comparisons with results from infrared and microwave spectroscopy. An incident that occurred soon after our first work using the new analytical techniques was published provided a poignant example of the value of such comparisons. One of the first molecules investigated by the new analytical techniques was CO_2 [6]. We reported a value of 0.040 ± 0.007 Å for the rms amplitude for the O···O distance. However, we had calculated a value of 0.029 Å from spectroscopic data which indicated a discrepancy somewhat larger than our reported limit of error for the electron diffraction experiment. It was soon shown by Yonezo Morino [11] that the formula that we were using was in error and that after correction the computed value for spectroscopic data was 0.041 Å, in very good agreement with the result from the electron diffraction experiment.

The study of CO_2 [6] also provided a first experimental insight into the concept of 'shrinkage'. The value of the O···O distance was observed to be smaller than twice the value of the C—O distance. Owing to the early stage of the development of the new techniques, no particular notice was taken of the result in the publication. However, it did catch the attention of Lawrence Brockway and Jerome recalls discussing the matter with him during a visit to Ann Arbor. The observed shrinkage can be explained in terms of a bending mode such that in the course of the vibration, the O···O is equal to the sum of the C—O distances only when the molecule passes through the equilibrium position. It is otherwise always smaller. It was noted that the motions of this type could cause large shrinkage effects for trans distances. An experimental application of the theory was soon made to 1,2-dichloroethane [12].

Internal motion has remained a problem of interest theoretically and experimentally over the years and has been extended to investigations of double rotors [13]. Theoretical studies coupled with electron diffraction experiments afford rough insights into potential barriers that hinder internal rotational motion.

Implications

Our work in electron diffraction of gases provided the foundation concepts that led us into other fields of structural research of which the investigation of crystalline materials has occupied the largest part of our time. The broad concept concerns the observed importance in structure determination of imposed mathematical and physical constraints on the analytic procedure. The possibility of making advances in gas diffraction was realized because such constraints were appropriately implemented. Perhaps the most fruitful one in its implications to other fields of structure analysis was the non-negativity criterion. It arose in the attempt to obtain an appropriate experimental function for the background intensity on the basis that the resulting radial distribution function had to be non-negative since it was interpretable in terms of probability functions.

Non-negativity

The non-negativity criterion was impressively useful in analyzing gas diffraction data to such an extent that it motivated the search for other areas of application. One such area that attracted our interest was atomic structure in which there was the possibility of obtaining more accurate electron distributions around atoms on the basis that the electron density is a non-negative function. Since the electron density is expressible as a Fourier transform of the diffraction data, the possibility exists that the non-negativity criterion could be used to improve the experimental data. This was done by developing an extrapolation procedure for the data, which are terminated at some finite scattering angle, in such a way that the Fourier transform of the measured and extrapolated data combined is non-negative. This was followed by a very fine electron diffraction investigation of the electron distribution in argon by Lawrence Bartell in 1953 [14] that made use of this non-negativity criterion.

The area that has found the most extensive application of the non-negativity criterion is crystal structure analysis. With the phase problem in mind, it seemed quite appropriate to consider the possibility that there would be constraints imposed on the magnitude and phases of the structure factors by the fact that the electron density in a crystal had to be a non-negative function. Such constraints do indeed occur and are given in terms of an infinite sequence of determinantal inequalities that contain the structure factors as elements, Karle and Hauptman, 1950 [15]. The implications of these inequalities for crystallography were pursued. The third order inequality and its probabilistic characteristics have provided the main mathematical tools for phase determination procedures [16]. The higher order inequalities associated with their general joint probability distributions [17] have been of interest as a potential means for facilitating further advances in structure determination and refinement. It is most fascinating that the procedures used for a very large number of routinely performed crystal structure determinations had their origins in the fact that there was a problem concerning an appropriate background intensity function in the analysis of gas electron diffraction patterns, which was solved by means of the non-negativity criterion.

The Period Between the 1940s and 1963

An article, "The Determination of Phase Angles," concerning the efforts made to determine phase angles from crystallographic data in the period from the late 1940s to 1963 has been written by Jerome Karle [18]. The first and second topics and the last topic of the above article is now reproduced here. If the interested reader can not find a copy of the above publication, we will send a copy of the article.

Introduction

There is the temptation to delay the preparation of a manuscript describing the present status of the phase problem, owing to the current rapid rate of development in this stimulating field of research. However, the same comment could have been made at any time during the last decade. The efforts expended during this period are in marked contrast to the relative inactivity prior to this time when it was generally believed that a solution to the phase problem did not exist, even in principle. Certainly the great increase in research activity on this problem was motivated by the realization that the values of the phases were in fact derivable from the measured X-ray intensities.

It is well known that the ideal diffraction pattern can be calculated for a given structure consisting of N atoms. For this purpose the structure factor equations,

$$F_h = |F_h| e_h^{i\varphi} = \sum_{j=1}^{N} f_{jh} \exp\left(2\pi\ ih \cdot r_j\right) \tag{1}$$

are employed, where F_h is the crystal structure factor associated with the Miller index triple h=(h, k, l), φ_h is the phase angle associated with F_h, f_{jh} is the atomic scattering factor for the jth atom and $r_j=(x_j, y_j, z_j)$ is a vector whose components are the coordinates of the jth atom. The intensities of the diffraction maxima are proportional to $|F_h|^2$ and their positions may be calculated from geometric considerations based on a knowledge of the unit cell parameters. The question of importance to structure determination concerns, of course, the converse, namely if the diffraction pattern is given, can the structure be computed? The answer to this question reveals the origin and the nature of the phase problem.

Equation (1) may be more generally written as an integral

$$F_h = \int \rho(r) \exp(2\pi\ ih \cdot r)\ dr. \tag{2}$$

This equation expresses the structure factor F_h in terms of the electron density distribution function $\rho(r)$. By means of the Fourier inversion theorem $\rho(r)$ may be expressed in terms of F_h, giving,

$$\rho(r) = \frac{1}{V} \sum_{h_{-\infty}}^{\infty} F_h \exp(-2\pi\ ih \cdot r), \tag{3}$$

the maxima of which represent the positions of the atoms. The symbol V represents the volume of the unit cell. Thus if the Fourier series could be calculated directly from the experimental data, the display of the crystal structure would be immediate. However, only the magnitudes of $|F_h|$ are directly derivable from experiment and the phases φ_h are not, giving rise to the so-called phase problem. Somehow we

Photo 3 Isabella and Jerome Karle in Washington, D.C. 2000 (Photograph by Magdolna Hargittai).

must find the φ_h if (3) is to be calculated. The misconception that the phase problem was unsolvable in principle arose from the formulation of the structure problem in terms of (2) and (3). It would appear from (3) that, corresponding to any given set of magnitudes $|F_h|$ there may be associated an arbitrary set of phases φ_h, each giving rise to an acceptable electron density function $\rho(r)$. What has been overlooked in this argument, however, is the fact that $\rho(r)$ is severely restricted. Not only is $\rho(r)$ non-negative everywhere, but it is representable as a superposition of a known number of discrete atomic electron distributions, each of which is known with good accuracy. It turns out that this additional information, together with the known magnitudes $|F_h|$ of the structure factors, is sufficient to determine uniquely the values of the phases φ_h (Photo 3).

Inspection of the structure factor Eq. (1) shows that we have a system of simultaneous equations in which the $|F_h|$ are known from experiment, the f_{jh} are known from tables and the phases φ_h and atomic positions r_j are the unknown quantities. As has been pointed out [19], this system of equations is ordinarily greatly overdetermined by the number of experimental data. It may therefore be concluded that the phases φ_h are expressible in terms of the measured $|F_h|$ and thus a solution to the phase problem, at least in principle, exists. This conclusion is clearly dependent upon the recognition that the atomic electron distributions in a crystal are known fairly accurately and that therefore the atomic scattering factors f_h may be treated as known quantities.

Of the two types of unknown quantities in (1), the atomic positions r_j are the ones which describe the structure and are clearly the ones which are ultimately desired. The phases φ_h act as intermediaries, leading to the values of the r_j most conveniently through the calculation of (3). Attempts have been made to obtain the r_j more directly from the $|F_h|$. This has met with partial success through the use of the Patterson function (4) [20, 21], although, generally, the latter function has been most successfully used to obtain initially a portion of the r_j from which approximate φ_h are computed, leading finally to values for all the r_j (Photo 4).

In this article we shall view the main approaches to the phase problem and the status of the current developments.

Historical Background

Ott in 1928 [22], derived relationships among the structure factors and atomic positions by means of algebraic manipulations on the structure factor Eq. (1). In some simple cases Ott showed that these equations could be used to obtain the atomic coordinates directly. Banerjee in 1933 [23] applied Ott's results to obtain a trial and error self-consistency routine for finding the signs of structure factors. The final form of the equations still contained both the known magnitudes and the unknown phases in a way that seemed to limit their applicability. Not only did the number of trials increase rapidly with complexity, but errors in the observed data also entered to such an extent as to render the proce-

Photo 4 Isabella and Jerome Karle in 2004.

dure ineffective except for the simpler structures. Avrami in 1938 [24] worked with equations relating the magnitudes of the structure factors to the interatomic vectors. Such calculations can be formed by multiplying (1) with its complex conjugate. In carrying out the solution of these equations, it was necessary to obtain the roots of a polynomial equation whose degree increased with extreme rapidity as the crystal became more complex. This fact and accuracy problems involving the experimental data appear again to have limited the general applicability of this procedure. Except for these few isolated instances, the algebraic approach to the structure problem initiated by Ott was not further pursued for 20 years.

An alternative approach to the structure problem was developed by Patterson in 1934 [20]. He introduced the Fourier series, generally known as the Patterson function,

$$P(r)=\sum_{h_{-\infty}}^{\infty}|F_h|^2\ \exp(-2\pi\ ih\cdot r), \tag{4}$$

the maxima of which represent the interatomic vectors. Since the coefficients of this series are proportional to the measured X-ray intensities, this function can be readily calculated for any crystal. A great many structures have been solved with the help of the Patterson function, some of great complexity, for which the additional aid of heavy atom substituents in the crystal was employed.

Titles of Additional Topics

3. Invariants and Seminvariants
4. Positivity, Inequalities, and Equalities
5. Heavy Atom and Isomorphous Substitution Methods
6. Probability Methods
7. Algebraic and Additional Probability Formulas
8. Phases and the Patterson Function
9. A General Addition Formula
10. Concluding Remarks

Several general statements can be made which are justified by the foregoing discussion. A great deal of progress has been made in the problem of determining phase angles and many of the theoretical studies have been brought to the stage of practical application. At the present time many approaches are receiving further study, both with regard to their theoretical and experimental aspects. Several structures have been solved using various direct procedures for phase determination. In the case of some complex crystals such as the protein, the isomorphous substitution method has afforded a very powerful procedure for determining the phases. The outstanding problem however is that of developing a generally practical method for determining the phases in non-centrosymmetric crystals. The high rate of progress to date together with the continuing development of high speed computers led this author to believe that the practical problem will be solved.

Note added in proof (August 1963): Within the past year and one-half an especially simple and reliable phase-determining procedure has been developed for centrosymmetric crystals. It involves the application of the Σ_2 formula $sE_h \sim s\sum_k E_k E_{h-k}$ which is valid for all the space groups, and is based on the observation that a successful phase-determination procedure employing this formula with a three-dimensional set of E_h magnitudes[1] requires a basic set of phases which are relatively few in number. It is thus possible to form the basic set from those phases which are assigned to specify the origin, and a few additional ones which are labeled with unknown symbols. If p symbols are used, 2^p Fourier maps must in principle be computed. In the course of numerous applications of this procedure it was found that relations are often found among the assigned symbols which reduce their number. In all cases the phase determination was completed for a sufficient number of the largest E_h magnitudes leaving no more than four unknown symbols.

[1] The E_h values are normalized F_h values as described in Karle, J. [25]. One of their properties is that $\langle|E_h|\rangle_h = 1.0$ further property is that for centrosymmetric crystals $\langle E_h^2 - 1\rangle_h = 0.968$ and for noncentrosymmetric crystals $\langle|E_h^2 - 1|\rangle_h = 0.736$ values that are very useful for distinguishing the type of crystal involved.

Many other considerations such as probability theory and direct phase determining relations employing the structure factor magnitudes may be used to further reduce the number of unknown symbols.

Since the procedure employs unknown symbols and an addition formula, $sE_h \sim s\sum_k E_k E_{h-k}$, it has been termed the symbolic addition method. The method has been described in several forthcoming structure investigations, e.g., cyclo-hexaglycyl [26], 3-indole acetic acid [27], and an alkaloid $C_{21}H_{35}N_3$ [28]. A generalization of the symbolic addition procedure is now under development for non-centrosymmetric crystals. The method has been applied to determine the structure of L-arginine dihydrate [29] that crystallizes in space group $P2_12_12_1$. From the point of view of the symbolic addition method, the role of many of the phase determining relations discussed in this article has been reversed. Instead of using these formulas at the start of the procedure in order to obtain a basic set of phases with which to implement formulas such as $sE_h \sim s\sum_k E_k E_{h-k}$, they may be used at the end of the procedure to reduce the number of unknown symbols.

The Remaining Problem

The search for a method for solving the structures of non-centrosymmetric crystals containing only light atoms, such as C, H, N, and O, directly from the measured intensities of X-ray diffraction data, in the early 1960s, was stimulated by the success of solving the structure of crystals containing a center of symmetry. The consequence of having a center of symmetry means that each of the phases assumes one of only two values, specifically 0 or π. The structure of cyclic hexaglycyl [26] with four molecules in the asymmetric unit, each with a different conformation, received particular attention. The symbolic addition procedure was used for the first time to solve the hexaglycyl structure. Not only was the number of atoms whose *x*, *y*, and *z* positions were determined impressive (there were 98 independent atoms to be placed in the hexaglycyl crystal), but also that a concrete example established that peptides could exist in more than one conformation. Furthermore, the hairpin beta-bends, accompanying hydrogen bonds, and their molecular parameters established in hexaglycyl served as a prototype for beta-bends, now called Type I [30, 31].

The solution of the structure of crystals that did not possess a center of symmetry presented an additional challenge since the phase associated with a particular X-ray reflection could have values anywhere from 0 to 2π, rather than just one of the two values, 0 or π, that is the case for centrosymmetric crystals. (The molecules themselves in a centrosymmetric crystal need not, and generally do not, possess a center of symmetry). Molecules in a non-centrosymmetric crystal usually are chiral and they include most molecules of biochemical significance, such as amino acids, nucleic acids, antibiotics, etc. Hence it was important to find a straightforward method for solving non-centrosymmetric crystal structures for both practical and theoretical reasons.

The question arose as to whether the modus operandi, developed for centrosymmetric crystals could be extended or adapted for the much more complex problem existing in non-centrosymmetric space groups. The well-known answer is affirmative, although some additional features had to be introduced.

The infinite set of inequalities that relate unknown phases to the experimentally measured intensities in an X-ray diffraction pattern, established by Karle and Hauptman in 1950 [15], apply to centro- and non-centrosymmetric crystals. The unknown phases cannot be determined algebraically directly or by using simultaneous equations since the unknowns are buried in transcendental functions (that is, in the arguments of a sine or cosine function). Certain simplifications and approximations can be made in the inequalities, under special conditions, provided that they are accompanied by an appropriate probability measure. The third inequality of the infinite set has proved to be particularly useful. The higher inequalities become more and more complex in the number of terms involved and may be difficult to handle or impossible to handle because of computer limitations or because the number of experimental data as compared to the number of atoms is restricted. The simplest approximation from the third order inequality is the relationship

$$\phi_h \approx \phi_k + \phi_{h-k} \tag{5}$$

when each of the terms is associated with a strong X-ray reflection, which is related to a high probability that the single numerical phase indication for a reflection with indices *h* is correct within certain limits, perhaps within ±30° range. Every crystal produces a unique set of strong reflections giving many combinations of indices *k* and $h-k$ whose sum is a particular *h* value associated with a strong reflection. Hence, it follows that

$$\phi_h \approx \langle \phi_k + \phi_{h-k} \rangle_{kr} \tag{6}$$

where k_r implies that an average is taken over "known" phases for strong reflections. The average of phase values results in a phase value for ϕ_h that is closer to the correct value than a single term.

The immediate question was what to use for the initial "known" values. There are three main sources for "known" values. The origin of a unit cell can be fixed by assigning phase values of 0, π, and/or $\pm\pi/2$ to three (sometimes two, or one) strong reflections of specific type of indices, depending upon the crystallographic space group which is determined from experiment, see Karle, 1974 [32]. In addition, in most non-centrosymmetric space groups, the enantiomorph can be specified by assigning the value $+\pi/2$ to a specific type of reflection. The second source of initial known phases are various other relations among indices, also derived from the third inequality, that depend only on the magnitude of reflections, not the value of their phases. These relationships are applied to special reflections that must have restricted values, such as 0 or π, in non-centrosymmetric space groups such as $P2_12_12_1$. Detailed examples are shown in the phase determination procedure for the structure of arginine [29, 33] and for enkephalin [34]. The third source is the assignment of symbols as phases to several of the strong reflections that have many interactions with other strong reflections (see the above references).

The phase determining procedure is one of a cascading development, during which relationships occur among the symbols. For example, for a particular ϕ_h different combinations of $\phi_{h-k}+\phi_k$ may indicate the phase to be a or $\pi/2+b$. Phase indications for another ϕ_h may indicate its phase to be $\pi+b$ or $2a-\pi/2$. If all these phase indications arose from relationships with a high probability of being approximately correct, then it could be expected that $a\approx 0$ and $b\approx +\pi/2$. Interestingly, there was a degree of "fuzzy logic" in assigning numerical values to the symbols that had some similarity to, but independent of, the early papers on fuzzy logic that appeared in the later 1960s. For any remaining phase symbols whose approximate values were not evaluated, several trials were made with assignment of 0, $+\pi/2$, π, and $-\pi/2$ for phases attached to hkl reflections, or two values, 0 and π, or $+\pi/2$ and $-\pi/2$ for special reflections depending upon the crystallographic space group. The approximate phase values were efficiently refined, and subsequently new phases were evaluated for additional reflections by reiteration in the tangent formula

$$\tan\phi_h \approx \frac{\sum_{kr}\left|E_k E_{h-k}\right|\sin\left(\phi_k+\phi_{h-k}\right)}{\sum_{kr}\left|E_k E_{h-k}\right|\cos\left(\phi_h+\phi_{h-k}\right)} \quad (7)$$

The tangent formula was derived by Karle and Hauptman in 1956 [35] from their third-order inequality. It lay dormant until 1963 when it was applied to the structure of L-arginine dihydrate [29]. The effectiveness of the application was considerably enhanced by an appropriate protocol. The three-dimensional E-map computed using the phase values as obtained above, yielded the complete non-centrosymmetric structure without any spurious peaks.

The lessons that we learned from this initial experience were several-fold, namely

1. It was possible to solve a non-centrosymmetric structure containing only C, N, O, and H atoms by obtaining the phase values directly from the measured X-ray intensities in a direct, organized procedure.
2. We had established appropriate approximations and restraints and devised a proper process which soon proved to be successful in space groups other than $P2_12_12_1$ with some minor modifications.

In the 1960s, computers were just becoming available to a number of institutions. Most of the phase determinations were performed manually. A full set of three dimensional X-ray data was still being collected photographically on a Weissenberg camera. During the period of 5 years (1964–1968), we published seven non-centrosymmetric structures (see Table 1). These were the only light atom non-centrosymmetric structures published that were obtained by direct phase determination until Oh and Maslen in Australia published the structure of iso-eremolactone in 1968 [42], using the methodology described above with some variations.

One more important feature was introduced into the direct structure solution of noncentrosymmetric crystals by Karle in 1968 [43]. A partial structure rather than a complete structure, is often obtained from an initial phase determination. It was therefore desirable to have a procedure for generating the complete structure once a fragment of the structure had been located in a unit cell. The tangent formula, Eq. (7) can readily serve this purpose. Phases generated by the partial structure are used as input for the tangent formula (7) to extend phase evaluation to the stronger structure factors as yet with unassigned phases. Acceptance criteria were developed for accepting new phases that are an essential part of the protocol. For example, $|F_h|_{calc} > p|F_h|_{obs}$ for a new ϕ_h, where p is related to the fraction of the structure already

Table 1 Noncentrosymmetric structures solved with tangent formula 1964–1968

Substance	Space group	Number C, N, O atoms	Reference
1. Arginine $2H_2O$	$P2_12_12_1$	14	[29]
2. Panamine $2ClO_4$	$P2_1$	31+2Cl	[36]
3. 6-Hydroxycrinamine	$P2_12_12_1$	46[a]	[37]
4. Photolysis product	$P2_12_12_1$	17	[38]
5. Substituted triazine	$P2_12_12_1$	23	[39]
6. Reserpine	$P2_1$	44	[40]
7. Tetracyclic indole	$P2_12_12_1$	16	[41]
8. Iso-eremolactone	$P2_12_12_1$	22	[42]

[a]Two molecules per asymmetric unit

known. The expanding of the phase determination by adding the new atoms that appeared in calculated electron density maps and recycling with the tangent formula usually results in a complete structure in a few cycles. The tangent formula and the recycling procedure are the mainstay of modern computerized programs for structure determination.

Some Later Structure Analyses

A few examples of crystal structures of interesting molecules or assemblies of molecules, solved in later years by direct phase determination in our laboratory follow. These exemplary structures do not contain any atoms heavier than oxygen. Experience has shown that the presence of even moderately heavier atoms, such as P, S, or Cl provides a tremendous boost to the phase determining process due to the stronger probabilities so that individual terms of phase relationships yield a more correct evaluation of a phase value (Photo 5).

Valinomycin, a dodecadepsi peptide that encapsulates a K+ ion and transports it through cell membranes, was an early subject for structure analysis by various physical and computational means. The crystal structure analysis of uncomplexed valinomycin [44], crystallized from a nonpolar solvent, (156 light weight atoms in two independent molecules in the asymmetric unit) showed that the bracelet conformation established in the K^+ ion complex [45] collapsed into a very flattened tube, replacing the circular volume occupied by the K^+ ion by interior cross-tube hydrogen bonds. A subsequent analysis of uncomplexed valinomycin crystallized from DMSO (dimethylsulfoxide) corroborated NMR, IR and ORD measurements that indicated that the conformation of uncomplexed valinomycin in solution was solvent dependent. Indeed, the conformation of uncomplexed valinomycin crystallized from DMSO had nearly threefold symmetry and was poised in a form ready to attract and encapsulate a K^+ ion [46].

Endogenous enkephalin, a linear pentapeptide functioning as a natural analgesic with opiate-like activity, occurs in the brain. The extreme flexibility of peptides as compared to the more rigid molecules of morphine and other opiates, necessitates the delineation of probable conformations for structure activity studies. Leu^{5}enkephalin, crystallized from dimethylformamide (DMFA) and water, occurs with four different conformers side-by-side in the asymmetric unit (208 C, N, and O atoms) [34]. The four conformers with extended backbones form an infinite antiparallel beta-sheet by extensive hydrogen bonding. The beta sheets are separated from each other by a distance of 12 Å without any direct peptide-to-peptide interaction. The large volume between the peptide sheets is filled with many water and DMFA molecules, somewhat disordered. The assembly can be compared to a club sandwich where the peptide beta-sheets are the bread and the solvent layers are the jelly. The integrity of the unit cell is maintained under tenuous conditions. The X-ray data were collected with the crystal sealed in a capillary with mother liquor.

Ions are transported across cell membranes by several different mechanisms, such as the carrier mechanism in which a

Photo 5 The Karles and the Hargittais at the conference on diffraction studies on non-crystalline substances, Pécs, Hungary, 1978.

charged ion is encapsulated in a carrier molecule, usually with a cyclic backbone, that is polar on the inside and has a hydrophobic exterior to facilitate passage through a lipid membrane. Valinomycin falls into that category. Another mechanism is the formation of channels that span a bilayer membrane through which ions can pass from one side of the membrane to the other. The channels are made by the insertion of linear peptides that are amphiphilic. They need to fold so that a nonpolar surface makes contact with the membrane, and a polar surface, compatible with a charged ion, lines the channel wall. The 16-residue Leu-zervamicin, extracted from a soil fungus, serves such a function. The crystal structure showed that the peptide folded into a helix with a severely bent helix axis [47]. The helix type changed from an α-helix in the upper portion to a β-ribbon type helix in the lower portion. The exciting feature of the structure is the formation of amphiphilic, hour-glass shaped channels by the assembly of three bent helices. Further, at the waist line of the channel, the crystal structure contained two conformations. One was an open form just large enough for the passage of a potassium ion, the other a closed form in which a glutamine side chain rotates to close the passageway with a hydrogen bond. A successful passage of an ion is suggested by one opening and one closing for each ion, a condition entirely compatible with a doublegating mechanism. The spatial arrangements in the irregular channel are consistent with double gating for a single K^+ ion passage at a time, as indicated by measurements of conductance through artificial membranes [48]. Subsequently, we were fortunate in co-crystallizing antiamoebin with *n*-octanol. Antiamoebin, an ionophore, is found in the same source as zervamicin. Despite a difference in sequence for residues 1–8, antiamoebin folds and assembles to form hour-glass shaped ion channels in the same conformation as zervamicin, although only the closed form was observed in these crystals. The exciting feature, however, is the flanking of the channel assembly, on the hydrophobic exterior, with *n*-octanol molecules as if they were mimicking a lipid membrane [49].

Stereochemical control over peptide folding has been achieved by incorporation of conformationally constrained residues such as α-aminoisobutyric acid (Aib) that promotes helix formation and D-Pro that nucleates β-hairpin structures. The crystal structure of synthetic Boc-Val-Ala-Leu-Aib-Val-Ala-Leu-Gly-Gly-Leu-Phe-Val-D-Pro-Gly-Leu-Phe-Val-OMe shows independent domains that consist sequentially of an N-terminal extension, a helical region, a helix termination segment, and a β-hairpin turn. It is interesting that in the crystal a more extensive β-sheet is formed by the two strands of the β-hairpin in one molecule and the N-terminal extension of an adjacent molecule [50].

Isabella has extended her interests to conformational properties of peptides containing beta-, gamma-, and delta-amino acid residues, as well as to the physical properties of naturally occurring peptides such as the mechanism for the elasticity in elastin analogs that occur in connective tissue and in walls of blood vessels.

As time goes on and opportunities to investigate new scientific areas develop from ones of earlier interest, the decision to proceed to the new areas often occurs. This has been true for us. Jerome has extended his interests in structural research combined with some aspects of quantum mechanics. The main objective of his current research is to calculate properties such as accurate energies for very large molecules with great speed and high accuracy. With the collaboration of two colleagues, good progress has been made.

References

1. Debye, P. *J. Chem. Phys.* **1941,** *9*, 55.
2. Karle, J. In *Determination of Organic Structures by Physical Methods*; Nachod, F. C.; Zuckerman, J. J., Eds.; Academic Press: New York, 1973.
3. Karle, I. L.; Hoober, D.; Karle, J. *J. Chem. Phys.* **1947,** *15*, 765.
4. Degard, C.; Piérard, J.; van der Grinten, W. *Nature* **1935,** *136*, 133.
5. Karle, J.; Karle, I. L. *J. Chem. Phys.* **1947,** 15, 764.
6. Karle, I. L.; Karle, J. *J. Chem. Phys.* **1949,** *17*, 1052.
7. Degard, C. *Bull. Soc. Roy. Sci. Liége* **1937,** *12*, 383.
8. Schomaker, V. Paper presented at April meeting of the American Chemical Society, Baltimore, 1939.
9. Karle, J.; Karle, I. L. *J. Chem. Phys.* **1950,** *18*, 957.
10. Karle, I. L.; Karle, J. *J. Chem. Phys.* **1950,** *18*, 963.
11. Morino, Y. *J. Chem. Phys.* **1950,** *18*, 395.
12. Ainsworth, J.; Karle, J. *J. Chem. Phys.* **1952,** *20*, 425.
13. Karle, J. In *Diffraction Studies on Non Crystalline Substances*, Hargittai, I.; Orville-Thomas, W. J.; Eds.; Akademiai Kiado: Kiado, Budapest, 1981.
14. Bartell, L. S. *Phys. Rev.* **1953,** *90*, 833.
15. Karle, J.; Hauptman, H. *Acta Cryst.* **1950,** *3*, 181.
16. Karle, J. *Proc. Natl. Acad. Sci. USA* **1978,** *75*, 3540.
17. Karle, J. *Proc. Natl. Acad. Sci. USA* **1978,** *75*, 2545.
18. Karle, J. In *Fortschritte der Strukturforschung mit Beugungsmethoden*; Brill, R., Ed.; F. Vieweg und Sohn: Braunschweig, 1963; Vol. 1.
19. Hauptman, H.; Karle, J. *Phys. Rev.* **1950,** *80*, 244.
20. Patterson, A. L. *Phys. Rev.* **1934,** *46*, 372.
21. Patterson, A. L. *Z. Krist.* **1935,** *90*, 517.
22. Ott, H. *Z. Krist.* **1928,** *66*, 136.
23. Banerjee, K. *Proc. Roy. Soc. A*, **1933,** *141*, 188.
24. Avrami, M. *Phys. Rev.* **1938,** *54*, 300.
25. Karle, J. In *Advances in Chemical Physics*; Prigogine, I; Rice, S. A. Eds.; Wiley, 1969; Vol. XVI, pp. 131–222.
26. Karle, I. L.; Karle, J. *Acta Cryst.* **1963,** *16*, 969.
27. Karle, I. L.; Britts, K.; Gum, P. *Acta Cryst.* **1964,** *17*, 496.
28. Karle, I. L.; Karle, J. *Acta Cryst.* **1964,** *7*, 1356.
29. Karle, I. L.; Karle, J. *Acta Cryst.* **1964,** *17*, 835.
30. Ramachandran, G. N.; Sasiekharan, V. In *Advances in Protein Chemistry*; Anfinsen, C. B., Jr.; Anson, M. L.; Edsall, J. T.; Richards, F. M., Eds.; Academic Press: New York, 1968; 283–438 (specifically pp. 339–342).
31. Venkatachalam, C. M. *Biopolymers* **1968,** *6*, 1425.
32. Karle, J. (1974) "Direct Method Structure Determination: Origin Specification, …" In *International Tables for X-RAY Crystallography*; Ibers, J. A.; Hamilton, W. C., Eds.; Knoch Press: Birmingham, England, 1974; Vol. IV, pp. 339–358.
33. Karle, J.; Karle, I. L. *Acta Cryst.* **1966,** *21*, 849.

34. Karle, I. L.; Karle, J.; Mastropaolo, D.; Camerman, A.; Camerman, N. *Acta Cryst*. **1983,** *B39*, 625.
35. Karle, J.; Hauptman, H. *Acta Cryst*. **1956,** *9*, 635.
36. Karle, I. L.; Karle, J. *Acta Cryst*. **1966,** *21*, 860.
37. Karle, J.; Estlin, J.; Karle, I. L. *J. Am. Chem. Soc*. **1967,** *89*, 6510.
38. Karle, I. L.; Karle, J.; Estlin, J. *Acta Cryst*. **1967,** *23*, 494.
39. Karle, J.; Flippen, J.; Karle, I. L. *Z. f. Krist*. **1967,** *125*, 1.
40. Karle, I. L.; Karle, J. *Acta Cryst*. **1968,** *B24*, 81.
41. Yonemitsu, O.; Okuno, Y.; Kanaoka, Y.; Karle, I. L.; Witkop, B. *J. Am. Chem. Soc*. **1968,** *90*, 6522.
42. Oh, Y. L.; Maslen, E. N. *Acta Cryst*. **1968,** *B24*, 883.
43. Karle, J. *Acta Cryst*. **1968,** *9*, 635.
44. Karle, I. L. *J. Am. Chem. Soc*. **1975,** *97*, 4379.
45. Pinkerton, M.; Steinrauf, L. K.; Dawkins, P. *Biochem. Biophys. Res. Commun*. **1969,** *35*, 512.
46. Karle, I. L.; Flippen-Anderson, J. L. *J. Am. Chem. Soc*. **1988,** *110*, 3253.
47. Karle, I. L.; Flippen-Anderson, J. L.; Agarwalla, S.; Balaram, P. *Biopolymers* **1994,** *34*, 721.
48. Balaram, P.; Krishna, K.; Sukumar, M.; Mellor, I. R.; Sansom, M. S. P. *Eur. Biophys. J*. **1992,** *21*, 117.
49. Karle, I. L.; Perozzo, M. A.; Mishra, V.; Balaram, P. *Proc. Natl. Acad. Sci. USA* **1998,** *95*, 5501.
50. Karle, I. L.; Das, C.; Balaram, P. *Proc. Natl. Acad. Sci. USA* **2000,** *97*, 3034.

Part II

Beyond the Classical System

Generalized Crystallography[a]

Alan L. Mackay[b]

Nulla cogente natura, sed concursu quodam fortuito atomorum
Cicero [2] (106–43 BC) De Natura Deorum 1:24
(Not by teleology, but by the chance association of atoms).

Abstract
X-ray crystal structure analysis can now be seen as a special kind of microscopy which is being extended to the recognition and examination of many kinds of ordered structure more general than crystals and which leads to their synthesis or construction by various methods. Electron microscopy and many other techniques now combine to give a coherent science of structure at the scale range of Ångstroms to microns, atoms to assemblies visible to the eye, which should continue to be called "crystallography" although it overlaps with nanotechnology, molecular biology, and solid state physics. Most generally, "a crystal is a structure the description of which is much smaller than the structure itself" and this view leads to the consideration of structures as carriers of information and on to wider concerns with growth, form, morphogenesis, and life itself.

Keywords
Crystallography • Hierarchy • Morphogenesis • Microscopy • Information

Introduction

Democritos of Abdera (ca. 400 BC) first formulated the idea that the properties of all matter, including life and mind, might be understood in terms of an underlying fine structure. This approach was, however, actively suppressed by the propagators of the view that things are determined from above [3] (rather than from below), in a hierarchy of scales. The text of Lucretius' Latin poem, expanding on Democritos, laying out a complete vision of how the macro world might be explained as a consequence of the micro world, was first printed in Brescia in 1473, although it must have been written about 60 BC. The confusion between the real world and the noumenal world, propagated by Plato [4] and numerous religious and philosophical leaders, is still with us, made more acute by information, dysinformation, ideology and postmodernism [5].

The aim of **generalized crystallography** is to understand the properties of matter, inert and living, at our human scale, in terms of the arrangement and operation of atoms at a level which, until X-ray crystal analysis, remained unobservable. The techniques for the observation of atoms are now so numerous that the term "crystallography" is perhaps no longer suitable and might now be replaced by "structural chemistry" [6]. The manipulation in space of structures at the level of atoms and molecules is now the science of "nanotechnology," but it was only in 1874 that van 't Hoff [7] began to

[a] *Structural Chemistry* 2002, 13(3/4):215–220.

[b] School of Crystallography, Birkbeck College, University of London, Malet Street, London WC1E 7HX, UK
e-mail: alanlmackay@gmail.com

I. Hargittai, B. Hargittai (eds.), *Science of Crystal Structures: Highlights in Crystallography*,
DOI 10.1007/978-3-319-19827-9_4

draw the consequences of the spatial arrangement of atoms in with the postulation of the tetrahedral carbon atom, although Archibald Scott Cooper (1831–1892) had discussed tetrahedral carbon atoms in 1858 and Pasteur had made the crucial discovery of molecular chirality in 1848 [8]. René-Just Haüy had noted the left- and right-handed forms of quartz crystals and was responsible for the hypothesis that macroscopic crystals were made up of identical unit cells, which were below the level of direct observation, thus giving a special case of the connection between the microscopic and macroscopic worlds. In 1934, J. D. Bernal and Dorothy Crowfoot (Hodgkin) showed that a protein actually had a structure, with each atom in its proper place, and today an immense project of "proteomics," to find the structure of all proteins involved in human life, is under way. The relation of DNA to protein, the dialectical relation between description and referent, was the key discovery of the last century and has meant that crystallography is also now concerned with information, informational structures, growth and form, and morphogenesis generally, involving dynamic as well as static structures in space. Classical crystallography has given access to what atoms do. Crystallographers, for example, A. D. Booth at Birkbeck College, were foremost in the development of the electronic computer and had a sharp appreciation of the problems of information handling.

The Reality of Spatial Structure

Isaac Newton (1642–1727) [9], being occupied with alchemy, suspected that the chemical properties of matter depended on atoms interacting in space, like the planets, but he had no access to this level.

> I suspect that they may all depend upon certain forces by which the particles of the bodies, by some causes hitherto unknown, are either mutually impelled towards one another, and cohere in regular figures, or are repelled and recede from one another. These forces being unknown, philosophers have hitherto attempted the search of Nature in vain (Preface to the Principia)

These forces are now well understood and their use in computing the properties of real crystals and molecules verifies their accuracy, and gives every confidence in predictions for hypothetical systems, although the prediction of the topology of complex crystal structures from composition alone is still elusive.

The empirical model-making tradition of William Barlow [10] and W. J. Pope, W. H. and W. L. Bragg, Kelvin (William Thomson (1824–1907)) [11], Bernal, Crick, and Watson and many others, can be traced to the Enlightenment and the escape, started by Galileo, through experimental testing of theories, from the religious world outlook, through the Reformation and Dissent to modern humanism [12]. Model-making has recently been enormously enhanced by the use of the computer.

Computer graphics can do almost everything, but our tactile senses are coupled to our vision and our mental manipulation of space, so that physical models still help the understanding. Stereolithography is a rapidly developing technique, which fabricates three-dimensional models from plastic, providing effectively a general 3-D computer printer [13]. It can present the results of 3-D image processing from microscopy as solid models for the committee table. Model-making remains essential in thinking about the microcosmos [14], strange as quantum mechanics has now revealed it to be.

The particular British tradition in crystallography has been characteristically empirical, model-making and engineering, flowing from the Industrial Revolution. In contrast, only recently, a Franco-German physicist and academic, confirmed the characteristics of the rather different continental tradition: "... in the case of conflict between theory and practice, the Gallo-Germans tend to choose theory" [15]. However, the quantum world is not like the world at the human scale and only very sophisticated experiments can elucidate it. The electronic devices of nanotechnology are now so small that quantum effects cannot be neglected [16]. Theory and experiment now intimately interact and, for example, it has been confirmed that carbon-60 molecules are diffracted in the same way as electrons and that the traditional billiard ball models of molecules are now not enough. The problems of reductionism are answered immediately from the consideration of the wave equation for two electrons where there is a cross-term so that two electrons together are more than the "sum" of the parts.

Immense progress has been made in understanding our human perception of spatial structures developed in the course of our evolution, although, "It is because nothing required us to apprehend atomic structure during our evolutionary development that we are incapable of understanding what it is that quantum physics describes" [17] (David Mermin). Now we have to use physical models in more subtle ways to exploit what we can do well as animals in order to understand the counterintuitive.

The Industrial Revolution in Britain required the mastery of reality [18] and people, such as Boris Mikhailovich Hessen (1883–1938) [19] showed, in 1931, that even Newton was greatly affected by the technological problems and needs of the day. Today the connection between science and technology is much closer, being recognized in the 1960s as the scientific and technological revolution, which now provides economic, military, and political power.

Generalised Microscopy

It is somewhat of an accident that the X-ray diffraction analysis was the first technique to give direct access to structure at the scale of atoms [20]. W. H. Bragg first added sine waves together linearly by photography to give a picture of the positions of the atoms in diopside. The phases in which the waves were added came from the comparison of the observed and calculated structure factors. However, a real crystal is the nonlinear superposition of waves, the connections between the phases contain the chemistry, and the nonlinearity is expressed as the Karle–Hauptman determinant. Later, W. L. Bragg, and separately Martin Buerger at MIT, came close to the invention of the hologram, which was achieved by Denis Gabor at Imperial College as a result of considering the electron microscope. In attempts to implement the techniques of X-ray structure analysis, B. K. Vainshtein greatly deepened the understanding of electron microscopy, which now complements X-ray techniques [21].

The emphasis on infinite periodicity, which is what a Bragg diffraction spot provides, has somewhat distorted the development of crystallography. It led to the use of the 14 Bravais lattices and the 230 crystallographic space groups, which presume infinite crystals. Development could have taken another emphasis, namely, the ways in which space-filling polyhedra pack together. This approach, pioneered by B. N. Delone [22], started with 24 space-filling polyhedra and asked about local order, developing a theory which asked "out to what range are identical points required to have identical surroundings to generate crystallinity?" rather than, given an infinite repeating structure, asking "under what operations is this whole object invariant?" Delone's approach is more suitable for a unified theory of crystals, quasicrystals, hierarchic and amorphous, or statistically ordered materials.

Of course, the development of other instruments, especially the atomic force microscope and its many variants, the synchrotron and the neutron generator, has enormously extended the experimental examination of crystals and other materials and has made many kinds of observation and experiment possible in combination with microscopy. Solid (and, indeed, liquid and gas) state physics has been greatly developed and merged with crystallography.

Generalized Geometric Algebra (Clifford Algebra)

It is clear, particularly with the entry of considerations of six-dimensional geometry into analysis of the structure of quasicrystals, that Clifford [23] algebra is the appropriate language for generalising crystallography [24]. However, although present practices are adequate, the notation needs popularizing and it would be desirable to recast the formalism of classical crystallography in Clifford form, although nothing new would emerge and the results would be valuable chiefly for extension to N dimensions.

Barycentric coordinates, developed by Möbius, where the coordinates of a point are given as the weighted sum of the coordinates of four other points is also a useful way of describing structures free of arbitrary axial systems, but the mathematical description of structure, independent of arbitrary axes, is still not perfected. It would be desirable, as Lazare Carnot perceived, to be able to describe spatial structures entirely in terms of chemically significant structure invariants (bond lengths = distance geometry; bond angles, and torsion angles), but there is still some way to go.

Altogether, the development of computer systems, most obviously "Mathematica," has greatly extended our mathematical capabilities at all levels.

The Description of Structure

With the discovery of the genetic code and the problems of the transmission and encryption of data, the whole topic of information has become connected with that of structure. DNA describes a protein, if we know the language, but both descriptor and referent are made of atoms, which have to follow the laws of chemistry. We can now ask how much information a structure contains, what is its complexity and, very significantly, following G. Chaitin, what is the length of the smallest computer programme necessary to generate or describe it. For example, the number of operations necessary to sort a sequence of N numbers into a specified order is about $N \log_2 N$. The analogy with natural language thus enters crystal chemistry. We might note that Charles Babbage invented a "mechanical notation," an abstract general descriptive language, which he used a great deal, and that this would bear investigation, since the complex machines, which he designed with its aid, have proved successful. The descriptions of inorganic crystal structures are still rather rudimentary and without great predictive value. Enormous databases are being compiled and await exploitation in unexpected directions. Questions of the mathematical description of shape and form [25] lead to questions as to whether corresponding natural laws operate. There is still much to gain from a dialog between mathematics and crystallography.

The computer has made possible the exploration of spatial structures of various kinds, especially enabling the construction and use of huge databases. For example, certain categories of networks related to those involved in zeolites can be exhaustively enumerated [26] and periodic minimal surfaces have entered the vocabulary of the description of the surfaces that separate networks. In the crystallographic domain, Elke Koch and Werner Fischer, in particular, have extended traditional crystallography by the discovery and enumeration of surfaces, which are useful for the description and understanding of structures and which supplement the traditional polyhedra as referents. Liquid crystals, mesoporous zeolites, photonic band gap structures, and Fermi surfaces can be comprehended within such descriptions, but they may occur on all scales from chemistry to mega-engineering.

Crystal Chemistry and Informatics

The development of gigantic computational power, and the appearance of the first generations of mathematicians able to use it, has begun to change crystal chemistry. It has been remarked that Linus Pauling, who formulated the principles of structural crystal chemistry, used only a perhaps a thousandth of the information now available, but nevertheless produced rules that have not required significant correction. Data bases have been compiled.

The IUC have adopted a somewhat retrograde definition of crystal depending on diffraction spots which gives an undue emphasis to Fourier analysis and periodicity. The most general definition, echoing work by G. Chaitin, might be that "a crystal is a structure the description of which is much smaller than the structure itself." Information has now entered explicitly into the structure of matter. Some structures composed of atoms contain information about other structures made of atoms. In the rush to unravel genetics, the role of shape is overshadowed. The DNA and protein systems evolve together. Genomics is only half of the secret of life; the other half of the secret is shape [27].

Recognition of Higher Level Structure

N. V. Belov [28] made the perceptive comment that:

> The exclusion of a fivefold axis for crystals, as well recognised, results from the impossibility of reconciling it (and axes of order greater than 6) with the "lattice state" of crystalline matter. It would appear, then, that for small organisms the fivefold axis represents a distinctive instrument in their struggle for existence, acting as insurance against petrifaction, against crystallisation, in which the first step would be their "capture" by a lattice.

The golden number entered crystallography rather dramatically, because it is the most irrational number, that is the furthest from a rational ratio. Thus, strongly icosahedral local order is able to prevail over the pressure for periodicity and produce the surprise of quasicrystals. This represented a break with traditional crystallography, although foreshadowed by the discovery of incommensurable magnetic structures and modulated structures.

The present challenge is to deal with materials which have hitherto been characterized as poorly crystallized, but can now be seen to have order of a different kind. For example, crystals with incommensurate modulations gave X-ray powder diagrams, which could not be satisfactorily indexed as crystals and were thus thought to be mixtures. Indeed, a material now recognized to be quasicrystalline was reported many years before the actual discovery [29].

In general, for crystal structure analysis of organic compounds and biomolecules, the structure of the molecule is primary and the crystallization is simply a means to this end. In inorganic and biological materials (such as fibers, membranes, and networks), the properties of interest are often at the level of the crystal aggregate. For example, in zeolites, the pore structure is the major concern. Now, however, the assembly of complex components into crystals and other types of order is of great technical concern.

Hierarchic Structures

Fractal structures, where there are self-similarity operations, were proposed by Shubnikov [30]. These are clearly incompatible with atomicity, because of the infinity regress, but many real structures are hierarchic with perhaps six discernable levels, the rule of composition at each level being different. Bernal drew attention to the hierarchic universe proposed by Charlier, where the local surroundings could be everywhere the same but where the density remained small. Attempts to pack icosahedra hierarchically, giving clear rules for filling in the gaps recursively, led to a model for the quasicrystals, which were discovered about 1982. This model, depending on local interactions, has eventually prevailed over the global ordering model.

Synthesis on Growth and Form

Could X-ray diffraction be played backward to produce periodic nodal surfaces like Lippmann films, where standing waves of light are used to produce periodic layers of silver atoms in a photographic emulsion as in an early form of color photography? Methods of building structures, such as atomic beam deposition, manipulation by atomic force microscopy, electron beam, and X-ray etching have almost reached atomic resolution. Chladni plates, investigated by Faraday, produce structures at nodes or antinodes with the wavelengths of sound waves. Nodal surfaces in three dimensions are, at present, being actively investigated [31] mathematically with potential applications in devices for photonics.

In crystals, nodal surfaces are by way of being the materializations of high-order structure invariants, sums of symmetrically related density waves where the sum of the reciprocal lattice vectors is zero. Triple invariants are central to the direct methods of crystal structure determination. Tortuously, Cyril Smith (1968) expressed the wave/particle duality as: "Matter is a holograph of itself in its own internal radiation."

There is great interest at present in photonic band gap materials, a bibliography obtainable on the Internet having some 2000 entries. The aim is the construction, for use with digital opto-electronics devices, of triply periodic dielectric networks having an optical band gap at about 1.5 μm. It is not clear yet whether chemical, physical or engineering methods will be the most successful, but the question of synthesizing, for example, opal crystals, has moved to the forefront, now for engineering purposes as well as for jewellery. The very early work of H. Zocher [32] and later S. Hachisu and the examination of *tipula iridescent virus* [33] might be recalled. Crystallography may occur at all scales.

Production of such large period structures might be done as liquid crystals, as microdeposition on suitable substrates, by ion beam, electron beam or laser beam manipulation or etching or by mechanical drilling.

The whole field of growth and form is now very active, with the synthesis of shapes more general than those of crystals [34] and has a new coherence as nanotechnology.

Many of the themes mentioned here were first articulated in a paper [35] presented to the Yugoslav Centre for Crystallography in June 1975, where the situation was summarized as: "Crystallography is only incidentally concerned with crystals … crystallography is rapidly becoming the science of structure at a particular level of organisation, being concerned with structures bigger than those represented by simple atoms but smaller than those of, for example, the bacteriophage. It deals with form and function at those levels, particularly with the way in which large-scale form is the expression of local force."

Acknowledgments On this special occasion it is appropriate to acknowledge, as well as my immediate colleagues, especially those who have given me a congenial home at Birkbeck, the long-range influences of three of the great polymaths and encyclopedists who continue to affect us with their perceptions of science and civilisation. These are, Titus Lucretius Carus [36] (ca. 99–55 BC), Denis Diderot (1713–1784), and John Desmond Bernal (1901–1971). Modern crystallography has taken place within the Republic of Science, the community of those who seek to acquire "reliable knowledge" about the natural world by verifiable experimentation rather than by revelation and it has been a great privilege to have taken a small part in it.

References

1. A. L. Mackay, 2001.
2. Cicero, better known for his literary writings, was the editor of the text of Lucretius' "De rerum natura," ["On the Nature of Things," a title also used by William Bragg for a lecture series on the structure of matter at the Royal Institution]. He also found and restored the tomb of Archimedes in Sicily and perhaps also saw the anti-Kythera machine, a complex analog astronomical computer, or something like it. Our school teaching has underestimated the scientific activities of antiquity.
3. Plato, in particular, in spite of the reputed legend "let no one ignorant of geometry enter here," is supposed to have suppressed the works of Democritos and propagated the technology of state myths to which would be attributed the social order.
4. George Sarton, doyen of historians of science, labels the influence of Plato's "Timaeus" as an evil one.
5. Sokal, Alan; Bricmont, Jean, *Intellectual Impostures*; Profile: London 1998.
6. However, the social structure of the those dealing with crystallography is so coherent and is formalized by many national crystallographic societies and by the International Union of Crystallography, that a change of name would be quite undesirable, although the subject matter and techniques may change. See *Fifty Years of X-ray Diffraction*; Ewald, P. P., ed; IUC, in 1992.
7. van 't Hoff, J. H. *Stereo-isomerism*, *Encyclop. Brit.*, 11th edn., **1910**, 25, 890.
8. Bernal, J. D. *Science and Industry in the Nineteenth Century*; London, 1953. Routledge, Kegan Paul, London.
9. Newton was born, a posthumous child, on Christmas day in the year (1642) that Galileo died.
10. Tandy, P. *Crystallogr. News* (BCA), 15–23 (June 1999) and earlier in *Geolo. Curator*, **1998**, *6*, 333. Groth, P. *Physikalische Krystallographie*; Engelmann, Leipzig, 1895 (contains a large section of advertisements for crystallographic models).
11. See the important essay by Ludwig Boltzmann *Models*, Encyclop. *Brit.* 11th edn., 1910.
12. P. Redondi, in "Galileo, Heretic" suggests that Galileo was induced to plead guilty to the heresy of heliocentricism so that the more serious crime of atomism (which would question the nature of transsubstantiation) should not be ventilated.
13. There are about four different types of machine. It is used for making prototypes and models in the size range of 1 to 30 cm, but is as yet not economic for serial production.
14. Microcosmic salt, an ammonium sodium hydrogen orthophosphate, was so named because it is found in decomposing human urine, "the microcosmos."
15. Blanchard, P. *Alliage* (*Nice*), **1999**, p. 47.
16. At the other end of the scale of sizes, the devices for using Global Positioning System (GPS), now extremely cheap through nanotechnology, apply the nonintuitive equations of general relativity.
17. *Nature* (*London*), **1996**, *383*, 772, (1996).
18. Rudyard Kipling, in "The Secret of the Machines," wrote "But remember please the Law by which we live, / We are not built to comprehend a lie, / We can neither love nor pity nor forgive, / If you make a slip in handling us you die." (Not really well-designed poetry, but like the first machines themselves!)
19. See the essay on Hessen by Paul Josephson. *Introducer of Quantum Mechanics and Relativity into Russia*, Hessen's fate showed the tight connections between science, technology, ideology, and power.
20. James Clerk Maxwell's essay on atoms (from the 9th edn.) is to be found quoted in the 11th edn. of the Encycl. Brit. under "Molecule." The scale of molecular structure was then (about 1900) estimated at 10^{-8} cm.
21. Zou, X. *Electron Crystallography of Inorganic Structures*; Stockholm University: Stockholm, 1995.
22. Delone, B. N.; Padurov, N.; Aleksandrov, A. *Matematicheskie Osnovy Strukturnogo Analiza Kristallov*; ONTI: Leningrad-Moscow, 1934.
23. William Kingdon Clifford (1845–1879) (whose tomb is in Highgate Cemetery) was also concerned with the application of mathematics to chemical structure as "chemico-algebraic theory."
24. A Mathematica package "Clifford Algebra Calculations" has been prepared by Jose Luis Aragon, who has also demonstrated crystallographic calculations in several papers; email: joseluis@iec.csic.es.

"Modern geometric calculations in crystallography" Aragón, G.; Aragon, J. L.; Davila, F.; Gómez, A.; Roderíguez, M. A., Geometric Algebra: A Geometric Approach to Computer Vision, Neural and Quantum Computing, Robotics and Engineering. (eds.) E. BayroCorrochano and G. Sobczyk, Birkhäuser, Boston, 2001 [ISBN-081-76-4199-8].
25. Lord, E. A.; Wilson, C. B. *The Mathematical Description of Shape and Form*; Wiley: New York, 1984; Thompson, D'Arcy Wentworth *On Growth and Form*, Cambridge, 1917.
26. Friedrichs, Olaf Delgado; Dress, Andreas W. M.; Huson, Daniel H.; Klinowski, Jacek; Mackay, Alan L. *Nature* **1999**, *400*, 644.
27. Needham, J. *Order and Life*; Yale University Press: New Haven, CT, 1936.
28. "Essays on structural crystallography," *XIII Mineral. Sb. L'VOV Geol. Obschch.*, **1962**, *16*, 41.
29. Hardy, H. K.; Silcock, J. M. *J. Inst. Met.*, **1955–1956**, *24*, 423.
30. Shubnikov, A. V. *Kristallografiya*, **1960**, *5*, 489.
31. Klinowski, J.; Mackay, A. L. *Phil. Trans. Roy. Soc. London*, **1996**, *A 354*, 1069.
32. Zocher, H.; Török, C. *Acta Crystallogr.*, **1967**, *22*, 751.
33. Klug, A.; Franklin, R. E.; Humphreys-Owen, S. P. F. *Biochim. Biophys. Acta* **1959**, *32*, 203.
34. Ball, P. "*The Self-Made Tapestry. Pattern Formation in Nature*"; Oxford Univ. Press: Oxford, 1999; Ozin, G. A. *Acc. Chem. Res.* **1997**, *30*, 17; Hyde, S.; Andersson, S.; Larsson, K.; Blum, Z.; Landh, T.; Lidin, S.; Ninham, B. W. *The Language of Shape*; Elsevier: Amsterdam, 1997.
35. Mackay, A. L. *Izv. Jugosl. Centr. Kristallogr.* (Zagreb), **1975**, *10*, 15; *Comp. Math. Appl.* **1986**, *12*B, 21; *Proc. R. Soc. London A* **1993**, *442*, 47.
36. R. E. Latham's translation in Penguin Classics: "Lucretius: On the Nature of the Universe" (rev. edn, 1994) provides understanding, but I happen to have the 2nd edition of Thomas Creech (1659–1700), London, 1717, which is a marvel of scholarship, and of the technology of printing, and which provides direct access to the actual words of the author formulated more than two thousand years ago. Lucretius final words on chemical bonding: "When the textures of two substances are mutually contrary, so that the hollows in the ones correspond to full sections in the other and vice versa, then connection between them is most perfect. It is even possible for some things to be coupled together, as though linked by rings and hooks" is immediately followed by a horrific description of the medical and social effects of the plague in Athens. Both topics are still worth our attention.

Mackay, Anti-Mackay, Double-Mackay, Pseudo-Mackay, and Related Icosahedral Shell Clusters[a,b]

K.H. Kuo[c]

Abstract

Mackay introduced two important crystallographic concepts in a short paper published 40 years ago. One is the icosahedral shell structure (iss) consisting of concentric icosahedra displaying fivefold rotational symmetry. The number of atoms contained within these icosahedral shells and subshells agrees well with the magic numbers in rare gas clusters, $(C_{60})N$ molecules, and some metal clusters determined by mass spectroscopy or simulated on energy considerations. The cluster of 55 atoms within the second icosahedral shell occurs frequently and has been called Mackay icosahedron, or simply *MI*, which occurs not only in various clusters, but also in intermetallic compounds and quasicrystals. The second concept is the hierarchic icosahedral structures caused by the presence of a stacking fault in the fcc packing of the successive triangular faces in the iss. For instance, a fault occurs after the *ABC* layers resulting an *ABCB* packing. This is, in fact, a hierarchic icosahedral structure of a core icosahedron connected to 12 outer icosahedra by vertex sharing, or an icosahedron of icosahedra (double *MI*). Contrary to Mackay's iss, a faulted hierarchic icosahedral shell is, in fact, a twinlike face capping of the underlying triangles; it is, therefore, called an anti-Mackay cluster. The hierarchic icosahedral structure in an Al-Mn-Pd icosahedral quasicrystal has a core of body-centered cube rather than an icosahedron and, therefore, is called a pseudo-Mackay cluster. The hierarchic icosahedral structures have been studied separately in the past in the fields of clusters, nanoparticles, intermetallic compounds, and quasicrystals, but the underlying geometry should be the same. In the following a unified geometrical analysis is presented.

Keywords

Mackay polyhedra • Icosahedral shell clusters • Hierarchic structures

Introduction

From the point of view of generalized crystallography, Alan L. Mackay in 1962 published a short paper entitled "A Dense Noncrystallographic Packing of Equal Spheres" [1], which had made tremendous impact on particle, cluster, intermetallics, and quasicrystal researches in the past 40 years. With the advent of nanoscience and nanotechnology, the importance of this paper will become even more profound in the coming years.

[a] *Structural Chemistry* 2002, 13(3/4):221–230.

[b] This contribution is part of a collection titled Generalized Crystallography and dedicated to the 75th anniversary of Professor Alan L. Mackay, FRS.

[c] Institute of Physics, Beijing Laboratory of Electron Microscopy, Chinese Academy of Sciences, P.O. Box 2724, 100080 Beijing, China (deceased)

I. Hargittai, B. Hargittai (eds.), *Science of Crystal Structures: Highlights in Crystallography*, DOI 10.1007/978-3-319-19827-9_5

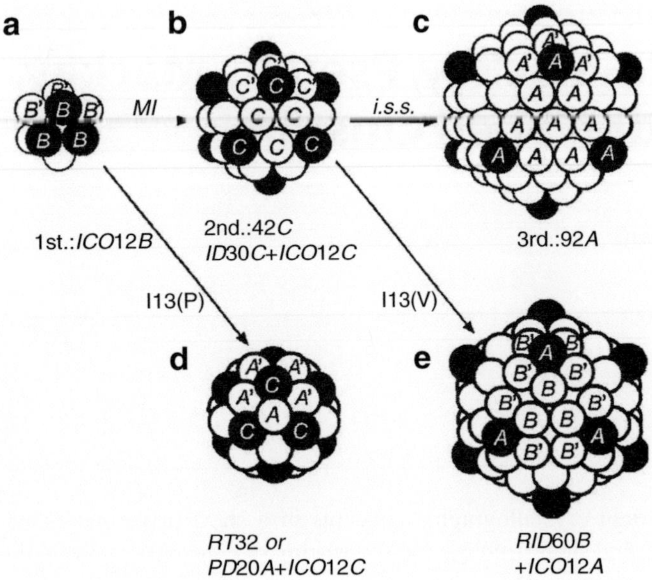

Fig. 1 Mackay icosahedral shell (**a**–**c**) and faulted (**d**, **e**) structures (the origin is an *A* sphere). *Solid circles* are the spheres at the icosahedral vertices, except for the core icosahedron in (**a**) in which only three spheres are drawn in *solid circles*. The *A′* spheres are equivalent ones of the *A* sphere and both of them are written *A* in the present paper; so are the *B′* and *C′* spheres. (**a**) First shell, 12 *B* spheres, the core icosahedron; (**b**) second shell, 42*C* spheres, the Mackay icosahedron (*MI*); (**c**) third shell, 92 *A* spheres; (**d**) a fault occurs in the interior of the *triangular faces* in the second shell, an *A* sphere atop three *B* spheres in (**a**); (**e**) *B* sphere in the third shell atop three *C* spheres in (**b**).

The main part of Mackay's paper is the icosahedral shell structure (iss) consisting of a set of concentric icosahedra of increasing sizes. Icosahedron is a regular polyhedron with the $\overline{235}$ rotational symmetry, whose fivefold rotation is incompatible with the periodic translation of a crystal lattice and, therefore, noncrystallographic. Figure 1a–c shows, respectively, the sphere model of the first three icosahedral shells. From the packing of spheres in a triangle of these shells (one A sphere at the center), it is clear that they are the (111) planes of the face centered cubic (fcc) closely packed sequence *ABCA*…. The spheres along the fivefold axes are shown in solid circles, except for Fig. 1a in which only three spheres are drawn in solid circles. The *A′* spheres in Fig. 1 are the equivalent ones of the *A* sphere and both of them are written *A* in the present paper; so are the *B′* and *C′* spheres. In addition to the 12 spheres (solid circles) at the vertices in the second shell (Fig. 1b), there is enough room at the mid-point of the 30 edges to accommodate further 30 spheres (open circles), adding to a total of 42 spheres. The third shell again has 12 spheres at the vertices, 60 spheres on the edges, and 20 more spheres at the centers of the 20 triangular faces, adding to 92 spheres (Fig. 1c). Using the cage notation @ introduced in the C_{60} literature, the first three icosahedral shells can be written as 1*A*@12*B*@42*C*@92*A*. Mackay showed that the number of spheres on the *n*th icosahedral shell is $10n^2+2(n\geq 1)$ and the total number of spheres within the *n*th shell $N=\{10(n+1)^3-15(n+1)^2+11(n+1)-3\}/3$ for $(n\geq 1)$ is, in fact, the magic numbers of 13, 55, 147, 309, 561, 923, 1415, 2057, 2869, etc. In this context, it should be pointed out that earlier Pauling [2] has considered the icosahedral coordination with a coordination number CN12 in metal crystals and Frank [3] in liquid metals. Both of them emphasized the higher packing density and also the higher stability of the icosahedral coordination in comparison with the fcc and hcp (hexagonal close-packed) structures, also with CN12.

The second and minor part of this paper, and nevertheless of equal importance, is the concept of the hierarchic icosahedral structure reported in the last paragraph. According to Mackay [1], Bernal had communicated to him in 1960 the hierarchic icosahedral structure: "The unit of packing is the arrangement of 13 spheres as an icosahedron making a quasi-sphere unit, 13 of which packed together to make a quasi-sphere of next order." In other words, this is an icosahedron of icosahedra and this can extend hierarchically in space. Mackay made it clear that "However, there is a clear relationship between the early stages of such a hierarchic and the packing described above" and explained further that the hierarchic structure could be obtained from the iss by the introduction of a stacking fault in the fcc packing sequence of the latter, as shown in Fig. 1d, e. In other words, hierarchy could be an alternative of growth process to periodic translation. This brief hierarchic concept has inspired many research workers to probe it in greater depth as well as in broader scope.

The icosahedral multishell and hierarchic structures have been studied separately in the fields of particle science, cluster science, intermetallic compounds, and quasicrystals with different emphases and different terminology, although the underlying geometry should be the same. The present paper is trying to fill this gap by giving a unified geometrical presentation, especially the hierarchic icosahedral structures, and will begin with a brief description of some regular and semiregular polyhedra of icosahedral symmetry, followed by Mackay's icosahedral shell structures, and Bernal–Mackay's hierarchic icosahedral structures. For a recent review on the structure of atomic clusters, see the contribution of Dr. Echt in this volume.

Icosahedron and Related Polyhedra

Both of icosahedron (*ICO*) (Fig. 2a), and pentagonal dodecahedron (*PD*) (Fig. 2b), is platonic regular polyhedra consisting of regular faces and regular vertices. Their characteristics can best be represented by the Schläfli symbols of their vertices, 3^5 and 5^3, respectively. The superscript shows the

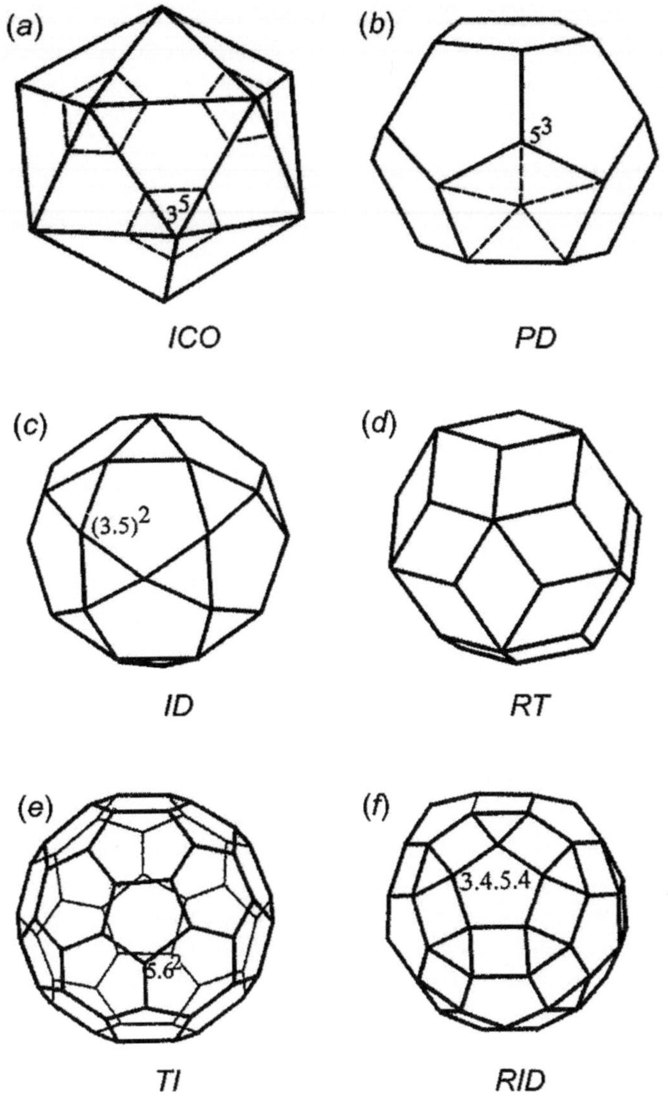

Fig. 2 Polyhedra of icosahedral symmetry: (**a**) icosahedron (*ICO*), 3^5; (**b**) pentagonal dodecahedron (*PD*), 5^3; (**c**) icosi-dodecahedron (*ID*), 3.5.3.5. or $(3.5)^2$; (**d**) rhombic triacontahedron (*RT*); (**e**) truncated icosahedron (*TI*), 5.6^2; (**f**) rhombic icosidodecahedron (*RID*), 3.4.5.4.

number of faces, triangle (3) or pentagon (5), surrounding a vertex. The symbols 3^5 and 5^3 imply an exchange of face and vertex between an *ICO* and a *PD*. In other words, they are dual to each other. Such a dual relationship is important in constructing the iss. If atoms are added to cap the pentagons of a *PD* to form pentagonal pyramids (one of which is outlined in dotted lines in Fig. 2b), their apices may form an *ICO* and vice versa.

Figure 2c is a polyhedron consisting of 20 triangles and 12 pentagons, i.e., a composite of *ICO* and *PD*. Hence, it is called icosi-dodecahedron (*ID*). Their faces are regular, but the 30 3.5.3.5 or $(3.5)^2$ vertices are not. This is one of the Archimedian semiregular polyhedra with regular faces described by Kepler through the truncation of a regular polyhedron. Truncating the vertices of either an *ICO* or a *PD* at the midpoint of edges, an *ID* will result. Through dual operation, the 20 triangles and 12 pentagons of an *ID* are transformed into 20 regular threefold and 12 regular fivefold vertices, and a rhombic triacontahedron (*RT*) consisting of 30 rhombi will result (Fig. 2d). This can also be seen in Fig. 1d, where a pair of *A* spheres and the two *C* spheres on their two sides form a rhombus. In contrast to the semiregular polyhedra with regular polygons such as *ID*, *RT* belongs to the family of semiregular polyhedra with regular vertices formed by either three or five rhombi. This can also be obtained by capping the pentagons of a *PD* (Fig. 2b), so that the neighboring triangles sharing an edge are coplanar forming a rhombus. Needless to say, truncating and capping are dual operations.

If an *ICO* is truncated at one-third of the edges (see the dotted lines in Fig. 2a), 12 regular pentagons will appear normal to the fivefold axes, while the 20 triangles become smaller hexagons (Fig. 2e). This truncated icosahedron (*TI*) has 60 5.6^2 vertices and can accommodate 60 atoms, such as the Buckminsterfullerene C_{60} molecule. If an *ID* is further truncated at the midpoint of edges, the 30 $(3.5)^2$ vertices become 30 rectangles or, after some readjustment, 30 squares. In the mean time, the 20 triangles and 12 pentagons are reduced to smaller ones. Such a 62-face polyhedron is called rhombic icosi-dodecahedron (*RID*) (see Fig. 2f and the 60 *B* spheres in Fig. 1e). Obviously, both *TI* and *RID* are semiregular polyhedra, consisting of 12 isolated regular pentagons with 60 equivalent vertices, but the polygons between the pentagons and thus the vertex types are different, one being 5.6^2 and the other 3.4.5.4.

All the polyhedra in Fig. 2 have icosahedral symmetry, because neither truncating the vertices of an icosahedron nor capping the faces of a pentagonal dodecahedron will change the inner symmetry of them, although their outer forms are different. Moreover, the number of faces (*F*), edges (*E*), and vertices (*V*) obeys the Euler rule of convex polyhedron, namely, $F+V=E+2$.

Icosahedral Shell Structures

An icosahedron consists of 20 equilateral triangles and, therefore, can be visualized as 20 slightly distorted tetrahedra sharing a common vertex at its center. The edge of an icosahedron is about 1.05 times longer than its radius (the distance between its center and a vertex) and, hence, the vertex-center-vertex angle is 63.43°, rather than 60° of an equilateral triangle in a regular tetrahedron. In Fig. 1a–c, the spheres in the successive three shells are in *B*, *C*, and *A* positions, respectively. Mackay iss beautifully explained the icosahedral growth morphology [4–6] of the nanometer-sized icosa-twins (or, more generally, the multiple twin particles) of fcc metals as well as Si, diamond, TiC, etc. (for a recent review, see [7]). Recently, Volkov et al. [6] have observed by high-resolution electron microscopy a Pd particle consisting of five icosahedral shells with $N=561$ atoms, exactly as Mackay predicted. In cubic crystals, successive twinning on {111} planes around a common [110] axis will yield fivefold twins. The dihedral angle between two {111} planes is about 71.5°, so that there is a gap of about 2.5° after five successive twinning operations around one <110>axis. This will produce a noticeable strain in the icosahedral particle during growth and eventually it will change into a cuboctahedral particle. In other words, these cubic icosatwins are metastable and can only exist in nanometer scale. On the other hand, icosa-twins of rhombohedral crystals with=63.43° would fill the three-dimensional space without leaving any gap and, therefore, could grow infinitely, or at least to a size of micron scale. Indeed, this was confirmed recently in the 30 m icosa-twin aggregates of the rhombohedral B_6O with=63.1° [8, 9], very close to the theoretical value of 63.43°. However, this came almost 40 years after Mackay's prediction [1].

Cluster scientists became interested in Mackay's icosahedral shells because they found icosahedral clusters of rare gases, especially the $n=2$ and $N=55$ cluster, showed minima of Lennard-Jones and Morse potentials [10, 11]. Moreover, this theoretical simulation was also supported by gas electron diffraction experiments made on argon clusters [12]. Hoare and Pal [10] called the $N=55$ cluster "Mackay icosahedron" or simply *MI*. Hoare [11] concluded that the Mackay iss "far from being merely a crystallographic curiosity, have now been shown beyond doubt to be the dominant motif in the growth process of rare-gas microclusters in the important size range of $N=50$ to $N=1000$." Several icosahedral shells with the number of atoms agreeing with Mackay's 13, 55, and 147 have been identified in the mass spectroscopy of xenon and these numbers are called magic numbers in the cluster literature [13]. Such a study was later also extended to metal clusters, such as Na, Mg, Ca, and Sr clusters (for a recent review, see [14]). In Mg clusters, mass peaks occurred between the third shell of 147 atoms and the ninth shell of 2869 atoms have been observed [14, 15]. However, this does not guarantee an icosahedral structure because the cuboctahedral shells also have these magic numbers. Therefore, Martin et al. [14, 16], following Northby's subshell growth concept [17], looked at the moderate mass peaks of the subshells between the fifth shell of 561 atoms and the sixth shell of 923 atoms in Ca clusters. They found that the number of atoms in the subshells agrees with the successive covering of the triangular faces of an icosahedron.

Smalley was interested in the studying of heavy metal (Mo) and semiconductor (Si, GaAs) clusters and had constructed a powerful laser-vaporization time-of-flight mass spectrometer. Kroto, Curl, Smalley, and their coworkers [18] used the strong laser beam to bombard a graphite target and surprisingly obtained a strong mass peak corresponding to 60 carbon atoms. They ingeniously suggested a spherical cage model of 60 carbon atoms located on the 60 vertices of a *TI* (see Fig. 2e) consisting of 12 pentagons and 20 hexagons, the molecular cluster of Buckminsterfullerene C_{60}. Later, Martin et al. [14, 19] succeeded in producing even larger molecular clusters of (C60)*N* with $N=13$, 55, and 147. Once more, they encountered the problem of differentiating between the icosahedral and the cuboctahedral shell packing and again used the mass peaks of subshells to prove the face capping of the icosahedral subshells.

Continuing his study of generalized crystallography, Mackay [20] independently developed a two-dimensional hierarchic pattern of pentagons similar to the celebrated Penrose pattern at about the same time as Penrose did [21] (see [22]). Mackay [23] called it "quasilattice" and produced its optical transform displaying clearly fivefold rotational symmetry. Following Ammann's suggestion, Mackay constructed the three-dimensional (3-D) quasilattice built of rhombohedra of=63.43° and 117.57°, respectively. Hargittai [22] considered this "predicting quasicrystal," which was discovered independently by Shechtman et al. [24] without knowing Mackay's work. Elser and Henley [25] as well as Guyot and Audier [26] suggested the *MI* in the cubic -(Al-Mn-Si) [27] with a composition of $(Al,Si)_{42}$ Mn_{12} (Mn atoms occupy the vertices of the outer icosahedron, solid circles in Fig. 1b), while the center of the core icosahedron is vacant, as the structural motif of the Al-Mn-Si icosahedral quasicrystal. Since then *MI* was universally accepted as the fundamental structural motif for the family of Al-TM (transitional metal) icosahedral quasicrystals, which were called later the Mackay-type of quasicrystals [28].

Figure 3 is taken from Sung et al. [29] and modified to include the hierarchic icosahedral shells. Figure 3a shows one of the triangular faces of the first two icosahedral shells (*MI*) formed by adding spheres (solid circles) along the extended fivefold axes so that the successive shells are icosahedra of increasing sizes.

The origin is an *A* sphere at the center and the first shell is the core *ICO* composed of *B* spheres. In the second icosahedral shell, however, the *C* spheres are not of the same distance from the origin, the spheres (solid circles) on the fivefold axes being slightly further away from the center than those at the midpoints of the edges (open circles). Thus, the second icosahedral shell can be divided into two subshells. One is the second *ICO* composed of 12 *C* spheres at the vertices (solid circles) and the other is an *ID* consisting of 20 triangles (open circles) and 12 pentagons [five spheres surrounding a fivefold axis (see Fig. 1b), although only two spheres are shown in Fig. 3a]. The pentagons in the first and second shells have the same orientation (along the $\bar{5}_a$ axis, the triangle of the *B* spheres in the first shell and that of the *C* spheres in the second shell are parallel). This can be seen more clearly in Fig. 1a, b in which the *B* and *C* pentagons are

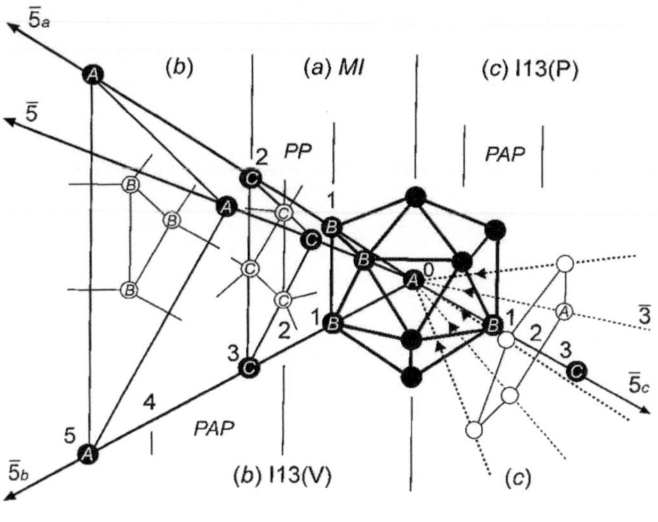

Fig. 3 Schematic diagram of successive triangles of the icosahedral shells. (**a**) The first two shells of the *MI* cluster with an *ABC* sequence (see Fig. 1a, b); (**b**) stacking fault occurs after the *ABC* layers resulting in an *ABCB* sequence in the interior of the *triangular faces* (see Fig. 1e). An outer *ICO* centered at a *C* sphere along the fivefold axis $\bar{5}_b$ shares a vertex *B* with the core *ICO*, thus producing an I13(V) cluster; (**c**) stacking fault occurs in the second shell resulting in an *ABA* sequence in the interior of triangles (see Fig. 1d). An outer *ICO* centered at a *B* sphere forms and penetrates into the core *ICO*, thus producing an I13(P) cluster. Spheres at the icosahedral vertices are drawn in *solid circles* (Taken from Sung et al. [29] and modified).

parallel oriented. Two neighboring pentagons of the same orientation form a pentagonal prism (*PP*) with a relative large volume. Surrounding this *MI*/*ID* there is another *ID* about times larger, consisting of 156 atoms in the cubic -(Al-Mn-Si) [30, 31]. Yang [30] called it "double-Mackay icosahedron" or simply *DMI*. In the pseudo-body-centered cubic structure of -(Al-Mn-Si), a *DMI* is located at the origin and a *MI* at the body center so that they share a triangular face along the <111> directions [30, 31]. One set of the three orthogonal twofold axes of both the *MI* and *DMI* clusters is parallel to the cubic axes of -(Al-Mn-Si).

In fact, this *DMI* is nothing but the hierarchic icosahedral structure, suggested by Bernal to Mackay in a private communication reported in [1], an icosahedron of icosahedra. Mackay [1] mentioned the possibility of the occurrence of stacking faults in the fcc packing *ABCA* … of the triangular planes, for instance, the packing sequence of *ABCB* consisting of a fault plane of *B* spheres or a twin/hexagonal packing sequence of *BCB*. This is shown in Fig. 1a → Fig. 1b → Fig. 1e.

Hierarchic Icosahedral Structures

In Fig. 3b, the *ID* and *ICO* subshells of the second icosahedral shell in Fig. 3a become the second and third shells, respectively. In the Mackay iss, the next layer will be an *A* layer (see Fig. 1c). However, if a stacking fault is introduced in the *ABC* sequence so that the interior of a triangle in the fourth shell is made of *B* spheres (open circles), then a hexagonal *BCB* sequence instead of the fcc *BCA* sequence will be created. This is shown clearly in Fig. 1e in contrast to Fig. 1c. Twenty such *B* triangles (open circles in Fig. 3b) form 12 vertex-sharing pentagons around the fivefold axes, as well as 30 squares across the edges (Fig. 1e). This is the *RID* shown in Fig. 2f. The pentagon in the fourth shell is quite open and a sphere can be put above its center, namely, in the *A* position along an extended fivefold axis (solid circle), as in the Mackay iss. Thus, these 12 *A* spheres form the third *ICO* of the fifth shell. It is of interest to note that the five *C* spheres (open circles) surrounding the $\bar{5}_b$ axis in the second shell and the five *B* spheres in the fourth shell have opposite orientations. This can also be seen in the orientation of the five *C* spheres (solid circles) in Fig. 1b and that of the five *B* spheres surrounding an *A* sphere in Fig. 1e. These two pentagons form a pentagonal antiprism (*PAP*) between them and this *PAP*, together with spheres on the $\bar{5}_b$ axis in the first, third, and fifth shells, namely, *B*, *C*, and *A* spheres, form an *ICO*. This *ICO* is centered at the *C* sphere (solid circle) of the third shell and shares a *B* sphere in the first shell with the core *ICO*. In other words, they are a pair of *ICO*s sharing a common vertex along the $\bar{5}_b$ axis. Thus, there are 12 outer icosahedra connected to the core *ICO* by sharing vertices, or a larger icosahedron consisting of 13 smaller *ICO*s. The hierarchic icosahedral shell sequence of 13 vertex-sharing *ICO*s in Fig. 3b, called I13(V), is:

$$I13(V): \underbrace{1A@ICO12B@ID30C@ICO12C}_{MI55}@RID60B@ICO12A \quad \text{Total 127 spheres}$$

Obviously, a stacking fault occurs after the *MI*55. If the stacking fault of the twin/hexagonal sequence occurs in the second layer, as suggested by Northby [17], then an *ABA* sequence will occur, i.e., Fig. 1a → Fig. 1d. In Fig. 1d, an *A* sphere (open circle) is atop the center of three *B* spheres of the first shell in Fig. 1a, forming a tetrahedron. This is also shown in Fig. 3c in which there is an *A* sphere along the threefold axis capping the triangular face of three *B* spheres. This sphere *A*, together with other four equivalent ones, form a pentagon around the $\bar{5}_c$ axis and 12 such pentagons form a *PD* of the second shell (see Fig. 1d). Moreover, this pentagon has an opposite orientation with respect to that in the core *ICO*, thus forming a *PAP*. Adding a *C* sphere along the $\bar{5}_c$ axis atop of this pentagon forms a small *ICO* centered at the *B* sphere of the core *ICO* and, at the same time, a large *ICO* of the third shell. The *ICO*s centered at the origin and at the vertex *B ICO* are fused together forming an inter*penetrated* pair. The icosahedral shell sequence in Fig. 3c, called I13(P), is:

$$I13(\mathrm{P}):\ 1A@ICO12B@PD20A@ICO12C$$

Total 45 spheres

It is clear from Fig. 3 that a stacking fault in Mackay iss is necessary, either in the second or the third shell, to form hierarchic icosahedral structures. Such a fault is also called "twinning" [32, 33], "face capping" [17], or anti-Mackay packing [34] in cluster science. Hoare [32] obtained the multiple-twinning structure by a simulation based on Lennard-Jones potential and considered this "an indication of the nature of the free-energy barrier to crystallography." Northby [17] gave this, besides a theoretical calculation, also a naive explanation. A sphere on the edge of an *ICO*, as in Fig. 1b, makes contact with only two spheres underneath it, whereas in the interior of a triangular face, as in Fig. 1d, it falls in the cavity of three spheres forming a tetrahedron, which is more stable. After completing an icosahedral shell, face capping of triangular faces will be preferred when atoms are added to form new subshells in the next shell. Based on these two kinds of stacking faults, two kinds of hierarchic icosahedral structures, vertex sharing and interpenetrating, respectively, of the core *ICO* and the 12 outer *ICO*s result. These two, as well as the 13 isolated *ICO*s, will be discussed below.

I13(I) Cluster

This term is coined to represent the icosahedrally connected 13 "isolated" *ICO*s. Earlier, Kreiner and Franzen [35] have used I13 to denote icosahedrally connected 13 *ICO*s. Here one letter in parentheses is added to indicate the type of linkage between the core *ICO* and the 12 outer *ICO*s. "Icosahedrally" means the centeres of these *ICO*s themselves form a hierarchic large *ICO*. "Isolated" or (I) means the core *ICO* forms bonds with the outer *ICO*s, but without any sharing of spheres between them. Figure 4a shows the connection of the core *ICO* and an outer *ICO*, all centeres being vacant. The order of the icosahedral shells is marked and the number *N* of spheres within the *n*th shell is given in parentheses. The first shell is the core *ICO*, and the second shell is a large *ICO*, consisting of one sphere from each of the 12 outer *ICO*s. The third shell consists of 12 pentagons forming a *TI* with 60 spheres (see Fig. 2e). This icosahedral cluster has 84 spheres, such as the B_{84} cluster in the structure of -B [36] and YB_{55} [37]. The pentagon in the third shell and the apex sphere in the second shell form an inverted pentagonal pyramid. The centeres and the outer pentagons of the outer

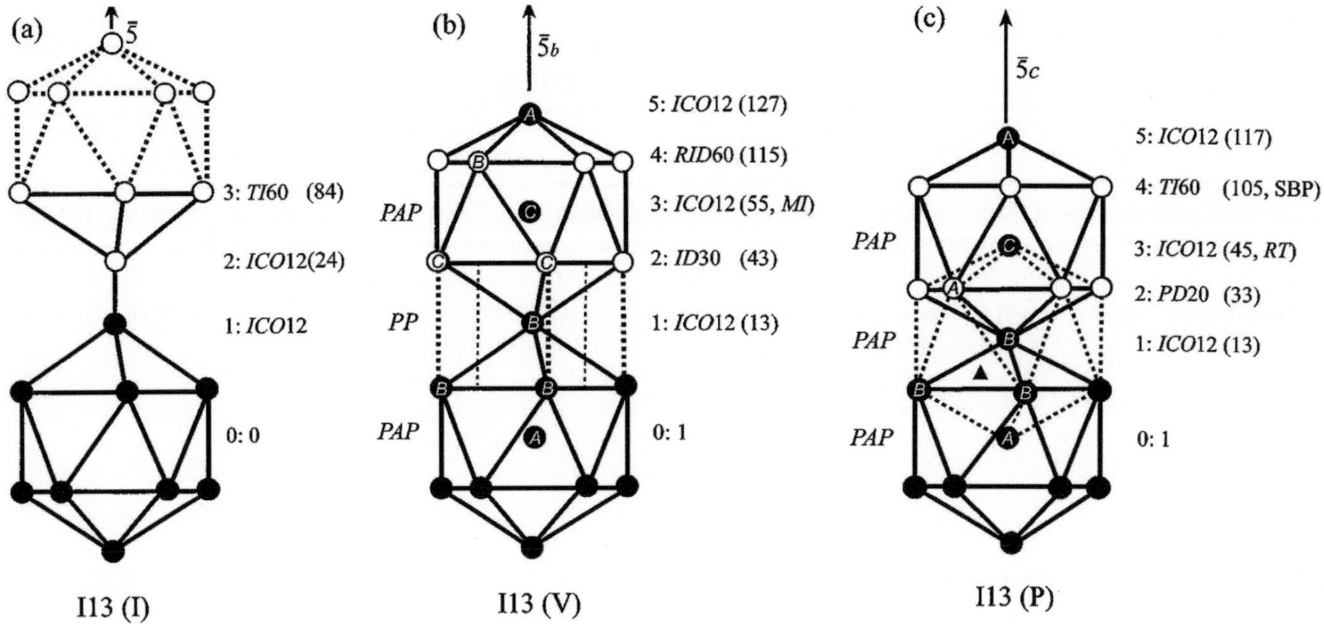

Fig. 4 The core and outer *ICO*s of three types of hierarchic icosahedral clusters. (**a**) I13(I), 13 isolated *ICO*s; (**b**) I13(V), 13 vertex-sharing *ICO*s; (**c**) I13(P), the similarly oriented core and outer *ICO*s form an intermediate *ICO* (*dotted lines*) between them, thus producing a chain of three mutually interpenetrated *ICO*s along a fivefold axis. The number in *parentheses* is the total number of spheres *N* within the *n*th shell.

*ICO*s in the next two shells are far away from each other and do not form any meaningful icosahedrally connected shell. The first three icosahedral shells for the B84 cluster with a vacant center (□) are:

$$B_{84}: \; \square @ ICO12 @ ICFO12 @ TI60 \quad \text{Total 84 } atoms$$

The hierarchical I13(I) cluster of -B can be written as $B_{12}(B_{12})_{12}$ and a schematic drawing of it has been given in Fig. 4b in [9].

I13(V) Cluster

In this hierarchic cluster, the 12 *ICO*s are linked to the core *ICO* through vertex sharing, as shown schematically in Fig. 4b. It has the same shell sequence as in Fig. 3b, but it shows more clearly the vertex sharing of two *ICO*s along a fivefold axis. In this case the core and outer *ICO*s are differently oriented, otherwise they will form a chain of three interpenetrated *ICO*s, as in Fig. 4c. Consequently, a *PP* exists between them. As shown in Fig. 3c, the spheres of the upper pentagon are *C* spheres. The first icosahedral shell is the core *ICO* and the second shell is an *ID* of 30 *C* spheres formed by sharing a vertex between two neighboring, lower pentagons of the outer *ICO*s (see the open circles in Fig. 1b). Obviously, all these 13 *ICO*s are linked by vertex sharing between the neighboring *ICO*s. The third shell is the second, large *ICO* composed of the central *C* spheres of the 12 outer *ICO*s. This is, in fact, the *MI* cluster of 55 atoms. The fourth shell consists of the 12 upper pentagons of the outer *ICO*s. Since these pentagons are further away from the center, compared with the lower pentagons, they are separated from each other and interlaced with squares and triangles forming a *RID* of 60 spheres (see Fig. 2f). A stacking fault occurs in this shell so that a *RID* composed of *B* spheres is formed and the packing sequence of the four shells becomes *ABCB*. Finally, the fifth shell is the third, even larger *ICO* of 12 *A* spheres.

Cenzual et al. [38] have considered the vertex sharing of neighboring *ICO*s in the I13(V) cluster in the structure of the cubic Sc_{57} Rh_{13} $\left(Pm\overline{3},\ a = 1.4405\ \text{nm}\right)$ and the orthorhombic Hf_{54} Os_{17} (*Immm*, $a=1.3856$, $b=1.4104$, $c=1.450$ nm). They used the general formula I_{114} C_{13} to represent this family of phases (I denotes the atoms in the icosahedral shell and C those at the center). It gives also 127 atoms in this I_{13}(V) cluster. A schematic drawing of the 13 vertex-sharing *ICO*s in these compounds has been given in Fig. 1 in [38].

Yang [30] showed that in -(Al-Mn-Si) 30 more atoms are added above the squares of the *RID*, forming a second, larger *ID* in the sixth shell, so that the total number of atoms within the sixth shell is 156 (center vacant). This is his *DMI*. Tamura [39] has extended the icosahedral shell in -(Al-Mn-Si) from *DMI* (he called it large *MI* or *LMI*) to the seventh and eighth shells and calculated the deviation of the experimental atom positions in various shells from the theoretical positions in a hierarchic icosahedral model. The deviations in the center-vertex positions in the first three shells, i.e., the *MI*, are 0.0015 – 0.0087 nm, but increase to 0.021–0.0334 nm in the fourth and fifth shells of the *DMI*. Kreiner and Franzen [35] have also considered the giant $(MI55)_{13}$ cluster, i.e., an icosahedron of *MI*. The hierarchic icosahedral shells in Sc_{57} Rh_{13} and -(AlMnSi) are:

$$\overbrace{}^{MI55}$$

$$Sc_{57}Rh_{13}: \quad 1@ICO12@ID30@ICO12@RID60@ICO12 \qquad \text{Total 127 atoms}$$

$$(AlMnSi): \quad O@ICO12@ID30@ICO12@RID60@ICO12@ID30 \qquad \text{Total 156 atoms}$$

$$\underbrace{}_{MI54}\ \underbrace{}_{DM156}$$

I13(P) Cluster

In the hierarchic icosahedral cluster I13(P) shown in Fig. 4c, the core and outer *ICO*s not only share a vertex but also have the same orientation so that the two neighboring pentagons of the first and second shells are differently oriented. Thus, a *PAP* or a new *ICO* (drawn in dotted lines) centered at the common vertex *B* is formed between these two *ICO*s. The core, intermediate, and outer *ICO*s interpenetrate into the neighboring one, forming a chain of three *ICO*s or of three centered *PAP*s capped at both ends. The first two shells are the same as shown in Fig. 3c. The *PAP* layer block here is more compact than the *PP* layer block in Fig. 4b; thus, the second shell in I13(P) must be smaller than the *ID* in I13(V), but also consisting of 12 pentagons. The only possible pentagonal shell is the *PD* of 12 edge-sharing pentagons. The third shell is the second *ICO*, whereas the fourth shell is again composed of pentagons. Comparing with the pentagon shell of *RID* of 60 *B* spheres in I13(V), the pentagon shell in I13(P) should be somewhat smaller. The only possible case is the *TI* with 12 pentagons and 20 hexagons. Finally, there are 12 *A* spheres capping the pentagons forming the third *ICO* of the fifth shell. The hierarchic icosahedral shell structures of I13(P) together with those of the cubic $Al_5 Mg_2Cu_6$ [40] (also the isostructural Mg_2Z_{11} [41]) and $(Al,Zn)_{49} Mg_{32}$ [42–46] (also the structural $Al_5 Li_3 Cu$ [47, 48]) are:

Al_5Mg2Cu_6 :	1@*ICO*12@*PD*20@*ICO*12	Total 45 atoms
I13(P):	1@*ICO*12@*PD*20@*ICO*12@*TI*60@*ICO*12	Total 117 spheres
$(Al,Zn)_{49} Mg_{32}$:	O@*UCI*12@*PD*20@*ICO*12@*TI*60@*ICO*12@*PD*20	Total 136 atoms
	*RT*44 104 *LRT*136	

In 1949, Samson first determined the crystal structure of the cubic $Al_5 Mg_2 Cu_6$ (*Pm3*, a=0.8311 nm, atoms) [40] and its isostructural Mg_2Z_{11} [41]. Figure 5 shows the [001] projected view of the structure of $Al_5 Mg_2Cu_6$: 6 Cu at (e), 12 Cu at (k), 1 Al at (b), 6 Al at (g), 8 Al at (i), and 6 Mg at (f). After the discovery of the Al-ZnMg quasicrystal, Pauling recalled [44] how he found the icosahedral shell structure of the cubic $(Al,Zn)_{49} Mg_{32}$. "In 1950, I recognized that the 39-atom cubic structure of $Al_5Mg_2Cu_6$ and Mg_2Z_{11} (Samson[19,20]) could be obtained from a central icosahedron surrounded by a shell of 32 atoms." From Fig. 5, it can be seen that around the central Al(b) atom there are 12 Cu(k) atoms of the core *ICO* (the three orthogonal golden rectangles are shaded). Adding 12 Mg(f) and 8 Al(i) atoms to cap the triangular faces of the core *ICO* will result in a *PD* [for the sake of clarity, only one of the 8 Al(i) atoms is shown at the bottom-left corner and 8 Mg(f) atoms are shown on the {100} planes in Fig. 5]. Thus, only tetrahedral interstices result by adding the second shell of a *PD*. A third shell of 12 Al(g) atoms (only 8 are shown in Fig. 5) are added above the pentagons of the *PD*. They lie on the extended lines (fivefold axes) connecting the central Al(b) atom to the 12 Cu(k) atoms of the core *ICO*, thus forming the outer *ICO* [see Fig. 4c]. This will produce a pentagonal bipyramid with an apex at Al(b) and another one at Al(g). Each pentagonal bipyramid can be visualized as 5 deformed tetrahedra sharing a common edge, namely, the fivefold rotational axis. Thus the 45-atom shell structure is a tetrahedrally close-packed (tcp) structure. According to Frank [3] and Boerdijk [49], tcp clusters are stable and occur frequently in metallic compounds, such as the Frank–Kasper phases [50], the tcp phases of the Shoemakers [51, 52], and the giant cubic phases of Samson [53, 54].

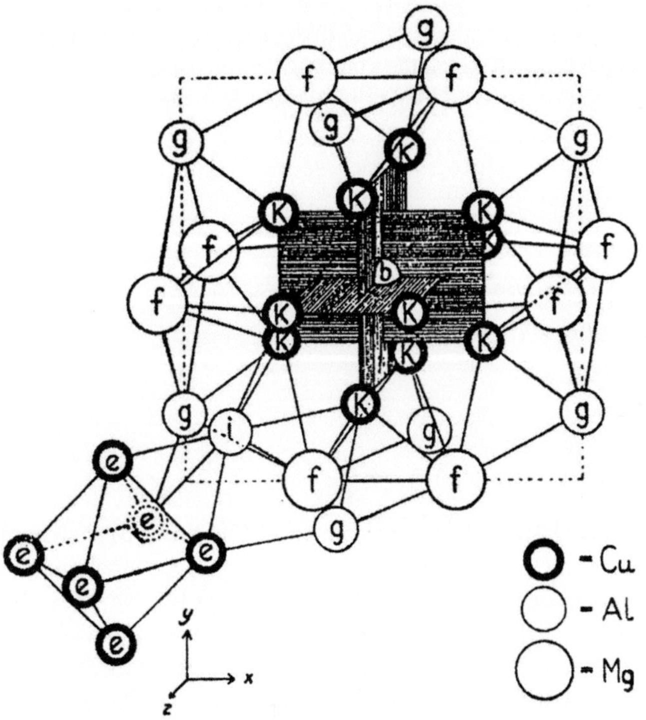

Fig. 5 Samson's schematic diagram of the cubic structure of $Al_5Mg_2Cu_6$ [40] and its 45-atom, three-shell icosahedral cluster analyzed by Pauling [44, 45]; 1Al(b)@*ICO*12[12Cu(k)]@*PD*20[12 Mg(f),8Al(i)]@*ICO*12[12Al(g)], with the 12Al(g) and 12 Mg(f) atoms on {100} planes (Taken from Samson [40]).

The outer shell of the 45-atom cluster is, in fact, a *RT* of 30 rhombi (Fig. 2d) or a capped *PD* of 60 triangles (Fig. 2b). Continuing the tetrahedral packing to obtain the structural model of $(Al,Zn)_{49} Mg_{32}$, Pauling [45] added 60 more atoms atop the 60 triangles and finally obtained the fourth shell of a *TI*. Adding 12 more atoms above the pentagons to form the fifth shell and completes the I13(P) cluster. This 105-atom soccer ball-shaped icosahedral cluster as well as the 127-atom structural model of $(Al,Zn)_{49} Mg_{32}$ obtained first by the stochastic method [45] are included in Pauling's famous book "*The Nature of the Chemical Bond and Structure of Molecules and Crystals: An Introduction to Modern Structural Chemistry*" [55]. Following Pauling's suggestion, Bergman et al. solved the complicate crystal structure of $(Al,Zn)_{49} Mg_{32}$ first by X-ray powder method in 1952 [42] and later by single crystal diffraction in 1957 [43]. It turned out that the center of the core is 80 % occupied. Recently, Sun et al. [46] found the center is empty in an accurate redetermination of this structure. Therefore, this four-shell icosahedral cluster in fact has only 104 atoms. Cenzual et al. [38] denoted it by $I_{104}C_{13}$, where C13 means the 13 atoms at the centers of 13 *ICO*s and also gave a schematic drawing of the I13(P) cluster in their Fig. 2, in which the 12 outer *ICO*s are themselves edge-sharing between neighboring ones. Immediately after the first discovery of the Al-Mn icosahedral quasicrystal [24], Mackay [56] suggested that an alloy of the composition of $(Al,Zn)_{49}Mg_{32}$ might be a suitable candidate for finding new quasicrystals. Indeed, such a quasicrystal was found independently at about the same time by Ramanchandrarao and Sastry [57], based on the icosahedral shell structure of the cubic $(Al,Zn)_{49}Mg_{32}$. Since then the 104-atom hierarchic shell is often called the Bergman cluster and the icosahedral quasicrystals of this family the Bergman type quasicrystals [58, 59]. Considering the pioneer work of Samson on $Al_5Mg_2Cu_6$ [40] and Pauling's [45] stochastic model of $(Al,Zn)_{49}Mg_{32}$, perhaps this family of quasicrystals might better be called Samson–Pauling–Bergman or SPB type of quasicrystals.

Instead of icosahedron, Samson [53, 54] likes to use truncated tetrahedron (*TT*) with four triangular faces opposite to four hexagonal faces to describe complicated metallic structures. As the icosahedron, this polyhedron can also be used to describe the structures of $MgCu_2$, $MgZn_2$, and $MgNi_2$. Capping the four hexagonal faces of a *TT* with four more atoms yields a CN16 Kasper polyhedron with only tetrahedral interstices. Twenty such *TT*s can condense into a *TI* if each *TT* is sharing a hexagon with each of its three *TT* neighbors. The remaining hexagons of these *TT*s will form an outer shell of 20 hexagons with 12 pentagonal cavities or 12 inverted pentagonal pyramids. The remaining 20 triangles, one from each *TT*, form the core *ICO* with an empty center. Samson [53, 54] used the 104-atom *TI* unit to discuss not only the structure of $(Al,Zn)_{49}Mg_{32}$ but also the giant cubic $NaCd_2$ (a=3.056 nm) and -Mg_2A_{13} (a=2.8239 nm). Tillard–Charbonnel, Belin, and their co-workers [60–62] later used the 104-atom shell unit, called Samson polyhedron by them, to describe the structures of a number of ternary Ga compounds found by them. For instance, in the body-centered cubic $Li_{13}Cu_6Ga_{21}$ [60], structurally similar to Al_5Li_3 Cu and $(Al,Zn)_{49}Mg_{32}$, the Samson polyhedron at the center is connected to the other eight Samson polyhedra by sharing a hexagon with each of them along the <111> directions. Such a chain of hexagonsharing Samson polyhedra can produce a hexagonal structure, such as the Li_{68} Zn_{16} Ga_{133}, with a=1.3657 nm and c=23435 nm [61]. In the cubic $Na_{35}Cd_{24}Ga_{56}$ of a=2.1286 nm, however, two Samson polyhedra are connected by sharing a pentagonal face [62]. Thus, it is suggested to call this 104-atom cluster the Samson–Pauling– Bergman cluster or SPB cluster.

Since the first discovery of the icosahedral quasicrystal in rapidly solidified Al-Mn alloys, both the *MI*54 of the cubic -(Al-Mn-Si) structure [25, 26] and the 104-atom Samson–Pauling–Bergman *TI* of the cubic $(Al,Zn)_{49}Mg_{32}$/Al_5Li_3 Cu structure [58, 59] have been considered to be the main building blocks of icosahedral quasicrystals [28]. Moreover, Elser and Henley [25, 58] have further shown that substituting rational ratios of two successive Fibonacci numbers for the irrational golden number of a quasicrystal will yield an approximate crystalline structure, called approximant. A similar idea has also been put forward by Mackay [56] using 5/3, 8/5, and 21/13 of the ratios of Fibonacci numbers to substitute for the irrational in the aperiodic structure of quasicrystals to obtain the Hume–Rothery compounds. This is the fundamental crystallographic connection between a quasicrystal and its crystalline approximant.

Using X-ray and neutron diffraction made on a single grain of Al-Mn-Pd icosahedral quasicrystal, Janot and his co-workers [63, 64] obtained an incomplete *MI* with only 9 atoms in the core icosahedron in the form of a body-centered cube. They called this 51-atom cluster "pseudo-*MI*" or simply *PMI* because it is almost, but not exactly, icosahedral. Recently, this has found support in the structures of many crystalline approximants of icosahedral quasicrystals, in which a 9-atom body-centered cubic or a 15-atom centered rhombic dodecahedron core has been found [65–67]. Recently, Tsai et al. [68, 69] found binary $Cd_{5.7}Yb$ and $Cd_{5.7}Ca$ icosahedral quasicrystals close to the composition of the body-centered cubic Cd_6Yb [70] and Cd_6Ca [71], respectively. The 66-atom cluster in the structure of these cubic phases is a tetrahedral core surrounded successively by *PD*, *ICO*, and *ID*. This is another cluster of predominately, but not exactly, icosahedral symmetry. There will be more such new findings in the coming years which will enrich the structures of icosahedral quasicrystal and its approximants, but the icosahedral shell and hierarchic icosahedral structures first suggested by Mackay 40 years ago will remain the guiding concepts in the future.

References

1. Mackay, A. L. *Acta Crystallogr.* **1962**, *15*, 916.
2. Pauling, L. *J. Amer. Chem. Soc.* **1947**, *69*, 542.
3. Frank, F. C. *Proc. Roy. Soc. London* **1952**, *A215*, 43.
4. Ino, S. *J. Phys. Soc. Jpn.* **1966**, *21*, 346.
5. Allpress, J. G.; Sanders, J. V. *Surf. Sci.* **1967**, *7*, 1.
6. Volkov, V. V.; Van Tendeloo, G.; Tsirkov, G. A.; Cherkashina, N. V.; Vargaftik, M. N.; Moiseev, I. I.; Novotortsev, V. M.; Kvit, A. V.; Chuvilin, A. L. *J. Cryst. Growth* **1996**, *163*, 377.
7. Hofmeister, H. *Cryst. Res. Technol.* **1998**, *33*, 3.
8. Mackay, A. *Nature (London)* **1998**, *391*, 324.
9. Hubert, H.; Devouard, B.; Garvie, L. A. J.; O'Keeffe, M.; Buseck, P. R.; Petuskey, W. T.; McMillan, P. F. *Nature (London)* **1998**, *391*, 376.
10. Hoare, M. R.; Pal, P. *J. Cryst. Growth* **1972**, *17*, 77.
11. Hoare, M. R. *Advan. Chem. Phys.* **1979**, *40*, 49.
12. Farges, J.; Raoult, B.; Torcet, G. *J. Chem. Phys.* **1973**, *59*, 3454.
13. Echt, O.; Sattler, K.; Recknagel, E. *Phys. Rev. Lett.* **1981**, *47*, 1121.

14. Martin, T. P. *Phys. Rep.* **1996**, *273*, 199.
15. Martin, T. P.; Bergmann, T.; Göhlich, H.; Lange, T. *Chem. Phys. Lett.* **1991**, *176*, 343.
16. Martin, T. P.; Näher, U.; Bergmann, T.; Göhlich, H.; Lange, T. *Chem. Phys. Lett.* **1991**, *183*, 119.
17. Northby, J. A. *J. Chem. Phys.* **1987**, *87*, 6166.
18. Kroto, H. W.; Heath, J. R.; O'Brien, S. C.; Curl, R. F.; Smalley, R. E. *Nature* (*London*) **1985**, *318*, 162.
19. Martin, T. P.; Näher, U.; Schaber, H.; Zimmermann, U. *Phys. Rev. Lett.* **1993**, *71*, 3079.
20. Mackay, A. L. *Izv. Jugosl. Centr. Krist.* (*Zagreb*) **1975**, *10*, 15.
21. Penrose, R. *Bull. Inst. Math. Appl.* **1974**, *10*, 266.
22. Hargittai, I. *Chem. Intell.* **1997**, *3*, 25; Hargittai, I.; Hargittai, M. *In Our Own Image, Personal Symmetry in Discovery*; Kluwer/Academic: New York, 2000; p. 152.
23. Mackay, A. L. *Physica* **1982**, *114A*, 609.
24. Shechtman, D.; Blech, I.; Gratias, D.; Cahn, J. W. *Phys. Rev. Lett.* **1984**, *53*, 1951.
25. Elser, V.; Henley, C. L. *Phys. Rev. Lett.* **1985**, *55*, 2883.
26. Guyot, P.; Audier, M. *Phil. Mag. B* **1985**, *52*, L15.
27. Cooper, M.; Robinson, K. *Acta Crystallogr.* **1966**, *20*, 614.
28. Henley, C. L. *Comments Condens. Matter Phys.* **1987**, *13*, 59.
29. Sung, M.-W.; Kawai, R.; Weare, J. H. *Phys. Rev. Lett.* **1994**, *73*, 3552.
30. Yang, Q. B. *Phil. Mag. B* **1988**, *58*, 47.
31. Sugiyama, K.; Kaji, N.; Hiraga, K. *Acta Crystallogr.* **1998**, *C54*, 445.
32. Hoare, M. *Ann. N.Y. Acad. Sci.* **1976**, *279*, 186.
33. Farges, J.; de Feraudy, M. F.; Raoult, B.; Torchet, G. *J. Chem. Phys.* **1986**, *84*, 3491.
34. Doye, J. P. K.; Wales, D. J.; Berry, R. S. *J. Chem. Phys.* **1995**, *103*, 4234.
35. Kreiner, G.; Franzen, H. F. *J. Alloys Comp.* **1995**, *221*, 15.
36. Hoard, I. L.; Sullenger, D. B.; Kennard, C. H. L.; Hughes, R. E. *J. Solid State Chem.* **1970**, *1*, 268.
37. Higashi, I.; Kobayashi, K.; Tanaka, T.; Ishizawa, W. *J. Solid State Chem.* **1997**, *133*, 16.
38. Cenzual, K.; Chabot, B.; Parthé, E. *Acta Crystallogr.* **1985**, *41*, 313.
39. Tamura, N. *Phil. Mag. A* **1997**, *76*, 337.
40. Samson, S. *Acta Chem. Scand.* **1949**, *3*, 809.
41. Samson, S. *Acta Chem. Scand.* **1949**, *3*, 835.
42. Bergman, G.; Waugh, J. L. T.; Pauling, L. *Nature* (*London*) **1952**, *169*, 1057.
43. Bergman, G.; Waugh, J. L. T.; Pauling, L. *Acta Crystallogr.* **1957**, *10*, 254.
44. Pauling, L. *Phys. Rev. Lett.* **1987**, *58*, 365.
45. Pauling, L. *Amer. Scientist* **1955**, *43*, 285.
46. Sun, W.; Lincoln, F. J.; Sugiyama, K.; Hiraga, K. *Mater. Sci. Eng.* **2000**, *294–296*, 327.
47. Cherkashin, E. E.; Kripyakevich, P. I.; Oleksiv, G. I. *Sov. Phys. Crystallogr.* **1964**, *8*, 681.
48. Audier, M.; Pannetier, J.; LeBlanc, M.; Janot, C.; Lang, J.-M.; Dubost, B. *Physica B* **1988**, *153*, 136.
49. Boerdijk, A. H. *Philips Res. Rept.* **1952**, *7*, 303.
50. Frank, F. C.; Kasper, J. S. *Acta Crystallogr.* **1958**, *11*, 184; *Acta Crystallogr.* **1959**, *12*, 483.
51. Shoemaker, P. D.; Shoemaker, C. B. *Acta Crystallogr. B*, **1986**, *42*, 3.
52. Shoemaker, P. D.; Shoemaker, C. B. *Mater. Sci. Forum* **1987**, *22–24*, 67.
53. Samson, S. In *Structural Chemistry and Molecular Biology*; Rich, A.; Davidson, N., eds.; Freeman: San Francisco, CA, 1968; p. 687.
54. Samson, S. *Mater. Sci. Forum* **1987**, *22–24*, 83.
55. Pauling, L. *The Nature of the Chemical Bond and Structure of Molecules and Crystals: An Introduction to Modern Structural Chemistry*, 3rd edn.; Cornell University Press: Ithaca, New York, 1963.
56. Mackay, A. L. *Nature* (*London*) **1985**, *315*, 636.
57. Ramanchandrarao, P.; Sastry, G. V. S. *Pramana* **1985**, *25*, L225.
58. Henley, C. L.; Elser, V. *Phil. Mag. B* **1986**, *53*, L59.
59. Audier, M.; Sainfort, P.; Dubost, B. *Phil. Mag. B* **1986**, *54*, L105.
60. Tillard-Chabonnel; Belin, C. *J. Solid State Chem.* **1991**, *90*, 270.
61. Tillard-Chabonnel; Chahine, A. Belin, C. *Mater. Res. Bull.* **1993**, *28*, 1285.
62. Tillard-Chabonnel; Belin, C. *Mater. Res. Bull.* **1992**, *27*, 1277.
63. Boudard, M.; Boissieu, M. D., Janot, C.; Heger, G.; Beeli, C.; Nissen, H.-U.; Vincent, H.; Ibberson, R.; Audier, M.; Dubois, J. M. *J. Phys. Condens. Matter* **1992**, *4*, 10149.
64. Janot, C.; de Boissieu, M. *Phys. Rev. Lett.* **1994**, *72*, 1674.
65. Mahne, S.; Steurer, W. *Z. Kristallogr.* **1996**, *211*, 17.
66. Edler, F. J.; Gramlich, V.; Steurer, W. *J. Alloys Comp.* **1998**, *269*, 7.
67. Hiraga, K.; Suiyama, K.; Ohsuna, T. *Phil. Mag. A* **1998**, *78*, 1051.
68. Tsai, A. P.; Guo, J. Q.; Abe, E., Takakura, H.; Sato, T. J. *Nature* (*London*) **2000**, *408*, 537.
69. Takakura, H.; Guo, J.; Tsai, A. P. *Phil. Mag. Lett.* **2001**, *81*, 411.
70. Palenzona, A. *J. Less-Common Met.* **1971**, *25*, 367.
71. Bruzzone, G. *Gazz. Chim. Italy* **1972**, *102*, 234.

Crystallography Without a Lattice[a,b]

John L. Finney[c]

Abstract

Molecular structure determination in crystals depends on the presence of the crystal lattice. In liquids, there is no underlying lattice, so phase relationships between scattered X-rays or neutrons are not preserved. Hence, only pair-distance distributions can be obtained from simple diffraction experiments. Advances in the past 30 years in radiation sources, instrumentation, and computing have enabled us to go beyond this apparent limitation. We can now obtain detailed structural information on relatively complex liquids and thus see clearly for the first time how molecules actually interact in solution, and how the solvent is perturbed by the presence of the solute. Using as an example recent work on aqueous solutions of tertiary butanol, the ways in which we can obtain high quality structural information in the absence of a lattice are summarized. Not only can we see how intermolecular interactions are influenced by changes in temperature and concentration, we can also begin to see where the structural source of the entropic driving force for the hydrophobic interaction may be found.

Keywords

Liquid structure • Neutron scattering • Aqueous solutions • Hydrophobic interactions

Prologue

In the late summer of 1972, half way across the Pacific Ocean, I got into conversation with David Harker. We were both returning on a flight chartered by the American Crystallographic Association, to facilitate attendance at the 1972 Congress of the International Union of Crystallography in Kyoto, Japan. Many of my memories of this first foray into involve Alan Mackay—including staying in a ryokan with him, and watching him gleefully embarrass several of our hosts with "difficult" Japanese characters.

Reconsidered in retrospect, the discussion with Dave Harker highlights the progress over 30 years in tackling one of the most fascinating aspects—to me—of Generalized Crystallography, to which both Alan and myself were exposed when working with J. D. Bernal in the Birkbeck Crystallography Department in the 1960s. I was just beginning my scientific career in the study of liquids. And liquids, of course, cannot be treated by standard structural techniques as their molecules do not helpfully line up on a lattice as they do in crystals. Hence, the whole standard crystallographic technique armory is not available for the structural study of liquids. Apparently, the best we could do with diffraction techniques was to obtain the radial distribution function—the distribution of pair distances in the liquid. Interestingly, this function has a strong formal relationship with the Patterson function then used extensively by crystallographers in the solution of the phase problem. Harker had worked with Patterson and was, of course, responsible for the Harker

[a]*Structural Chemistry* 2002, 13(3/4):231–246.

[b]This contribution is part of a collection titled Generalized Crystallography and dedicated to the 75th anniversary of Professor Alan L. Mackay, FRS.

[c]Department of Physics and Astronomy, University College London, London WC1E 6BT, UK
e-mail: j.finney@ucl.ac.uk

I. Hargittai, B. Hargittai (eds.), *Science of Crystal Structures: Highlights in Crystallography*,
DOI 10.1007/978-3-319-19827-9_6

section of the Patterson function that was particularly useful. Talking with Harker, I took the opportunity to discuss the problems of liquid structure determination in the hope that his insight into the development and use of the Patterson function would suggest how we might move forward to try to obtain data on liquid structures that could be comparable in detail to that obtainable on crystals.

In the absence of a lattice, this seemed to me a tall order. Although having no clear ideas at the time as to how the problem of the absence of a lattice could be overcome, I argued with Harker that there must be some way of extracting more information than the rdf and, ultimately, obtain really detailed structural information on those relatively complex liquids that are of biological and chemical importance, for example, in obtaining information on the mechanisms of protein association and enzyme-ligand binding in solution. I felt that making significant progress in this direction would be a scientific career well spent. He very patiently listened, although I think he felt I was on a wild goose chase.

Thirty years on, things have changed dramatically. The major developments we have seen in radiation sources, instrumentation, and computational techniques have transformed the situation. I believe we have now achieved what I then saw as a dream. We really can do liquid state crystallography and obtain remarkably detailed structural information in relatively complex liquid structures. We can see how amphiphilic molecules interact with each other as temperature, pressure, and concentration are varied. We can determine the structure of the hydration shell of a nonpolar molecule in aqueous solution and see how it is perturbed by adding salt. And we are, perhaps, on the threshold of understanding the structural basis of crucially important phenomena, such as, the hydrophobic interaction.

In this paper, I try to summarize how we obtain this detailed structural information in the absence of a lattice. In addition to the technical advances mentioned above, leaves have also been taken out of the crystallographer's technique book to enable the detailed information to be obtained in a way that perhaps parallels standard crystallographic refinement techniques. I take as an example aqueous solutions of an amphiphile—tertiary butanol—which has a large nonpolar head group and a polar tail and is a system whose thermodynamics is classically hydrophobic [1]. Not only can we see details of the structures of the associations of these molecules in solution, we can see also how these differ from those predicted by computer simulations and, hence, begin to see how the standard potential functions used in such simulations might be significantly improved. Finally, we can perhaps glimpse the structural origin of the entropic driving force for the hydrophobic interaction.

Introduction

In the absence of a lattice, our description of the structure of a liquid has to be a local one. For a single component liquid, this structure is described by the *radial distribution function* $g(r)$ (the rdf), which quantifies the ratio between the local density of atoms at a distance r from an atom at the origin to the average number density of atoms in the system. A two-dimensional analogy of the function is shown in Fig. 1 and indicates how useful structural information can, in principle, be obtained from this function.

A diffraction experiment uses the known wavelength of the (usually X-ray or neutron) scattering probe to obtain the information on pair-distance distributions contained in the radial distribution function. The scattering experiment measures the intensity scattered as a function of the scattering vector **Q**, defined simply as the difference between the scattered and incident wave vectors, with a magnitude $|\mathbf{Q}| = 4\pi \sin\theta/\lambda$, where 2θ is the scattering angle and λ the wavelength of the radiation used. From the scattered intensity, the structurally significant *structure factor* $S(Q)$ can be extracted,

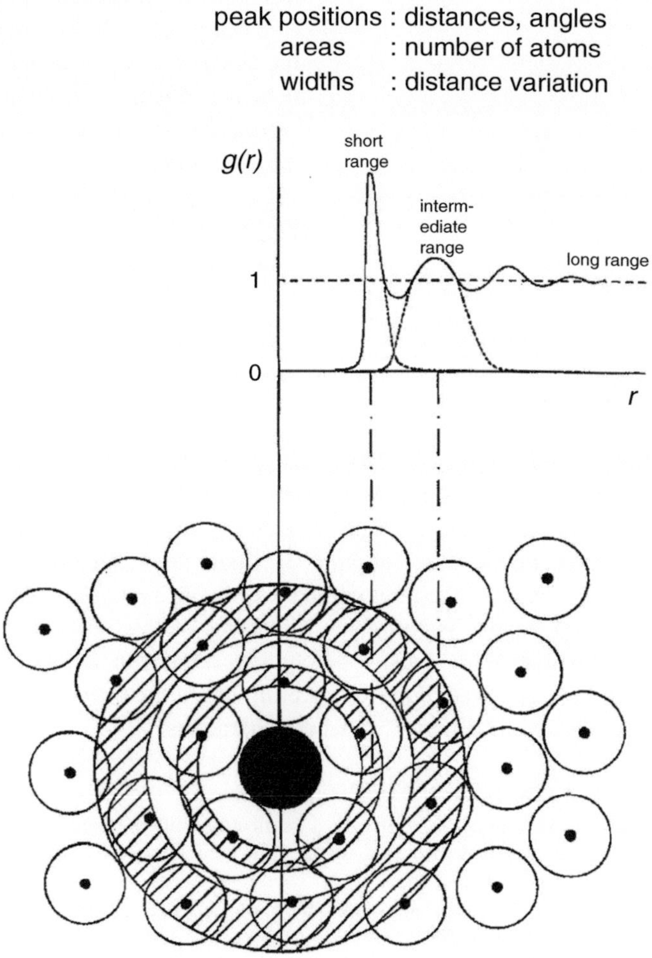

Fig. 1 The radial distribution for a liquid in two dimensions.

which can then be related to the radial distribution function through a Fourier transform relationship:

$$\left[S(Q)-1\right]=4\pi\rho\int_0^\infty r^2dr\left[g(r)-1\right]\frac{\sin Qr}{Qr} \qquad (1)$$

Single component liquids are, however, of very limited interest when trying to understand interactions in solution. For any solution, there must be more than one different atomic component, so the above functions must be generalized. Consider a two-component system, with components α and β, with N_α, N_β (where $N_\alpha+N_\beta=N$) atoms of each, and the atomic fractions of each component defined as $c_\alpha=N_\alpha/N$, $c_\beta=N_\beta/N$. The structure factor can be split into three terms—partial structure factors $S_{\alpha\beta}$—each relating to different pairs of interacting atoms $\alpha\alpha$, $\beta\beta$, and $\alpha\beta$. Thus, *partial radial distribution functions* $g_{\alpha\beta}(r)$ are defined similarly to the radial distribution function, but with the added identification of the type of atom both at the local origin and at the distance r. We can then rewrite Eq. (1) as

$$\left[S_{\alpha\beta}(Q)-1\right]=4\pi\rho\int_0^\infty r^2dr\left[g_{\alpha\beta}(r)-1\right]\frac{\sin Qr}{Qr} \qquad (2)$$

For a two-component system, we require three partial radial distribution functions to describe the structure of the system. For an n component system, $n(n+1)/2$ partials are needed. Each of these partials will be related to a partial structure factor that is embedded in the measured diffraction data and, taken together, comprise the *total* structure factor

$$F(Q)=\sum_{\alpha,\beta}\left(2-\delta_{\alpha\beta}\right)c_\alpha c_\beta b_\alpha b_\beta\left[S_{\alpha\beta}(Q)-1\right] \qquad (3)$$

where b_α b_β are the scattering lengths of the components α and β.

A single diffraction experiment gives us access to only the total radial distribution function $g(r)$, which is a weighted sum of the partial rdfs. This is not very useful information for a solution: we need to try to get closer to determining the partial structure factors. The key to achieving this is given in Eq. (3): the fact that the total structure factor *is* a weighted sum gives us a way forward. One of the weighting factors is the scattering length of each component. If we could perform an experiment on the same system, but somehow change the scattering length, we could perform more than one experiment on chemically similar systems and begin to extract the kind of more detailed information we really want.

That we can do this is made possible by the fact that the neutron is scattered by the nucleus and that many elements have different isotopic forms. If these isotopic forms had different neutron scattering lengths, then we could perform an additional experiment on a system in which we change only the isotope of one of the components. Referring to Eq. (3), we could then obtain from a second scattering experiment a different structure factor $F'(Q)$ in which, say b_α, is replaced by $b_{\alpha'}$. Now performing a third experiment replacing the isotope of the second component, b_β is replaced by $b_{\beta'}$. Alternatively, isotope α could be replaced by either a third isotope α'' (or what is, in effect, the same, a mixture of isotopes α and α'). We now would have three equations with the three partial structure factors as the three unknowns. In principle, these could be solved for and the corresponding partial radial distribution functions obtained.

For relatively simple two-component liquids and glasses, such "full" partial structure factors and, hence, "full" partial radial distribution functions, have been extracted (see, for example, Ref. [2]). For the relatively complex systems in which we are here interested, performing the much larger number of isotopically distinct scattering experiments that would be necessary to extract all the partial rdfs looks a tall order. In our example case of *t*-butanol in water, there are seven chemically distinct atoms in the system (methyl carbon, methyl hydrogen, central carbon, alcohol oxygen, alcohol hydroxyl hydrogen, water oxygen, water hydrogen). We would require $7\times8/2=28$ isotopically distinct experiments to yield the 28 partial radial distribution functions (or "partials" for short). Setting aside whether an appropriate set of isotopically distinct solutions could be prepared (not all elements have isotopes with significantly different scattering lengths), the requirements for neutron beam time, which is not cheap, would be extremely high. As we shall see below, there are other ways round the problem that enable us to obtain not only all the partials for a system such as this, but even to go *beyond* the partials to obtain even more detailed structural information on relatively complex solutions.

Experimental

For the *t*-butanol–water system, performing 28 isotopically distinct diffraction experiments is not a realistic possibility. The "best" kind of isotopic substitution we can do on this system is to deuterate the hydrogens on (a) the methyl groups of the alcohol head group and (b) the water. In fact, for each substitution, we work with (a) fully deuterated, (b) fully hydrogenated, and (c) a 50:50 hydrogen/deuterium mixture. This gives three values of the neutron-scattering length of the substituted sites and, hence, for each triplet of substitutions, enables us to obtain three sets of (partial) radial distribution functions. We will call these $g_{HH}(r)$, $g_{HX}(r)$, and $g_{XX}(r)$. Here the subscript H refers to a site whose hydrogens are substituted, while the X subscript refers to all the nonsubstituted sites. Thus, for a set of substitutions on the methyl hydrogen sites of *t*-butanol, H refers to all the substituted methyl hydrogen sites, while X refers to all other sites.

The corresponding set of experiments can now be specified as follows.

1. Solvent–solvent distribution functions are probed through substitution on the water hydrogen sites to yield the (partial) radial distribution functions $g_{HH}(r)$, $g_{HX}(r)$, and $g_{XX}(r)$ for the solvent–solvent correlations. The three experiments are made with (a) $(CD_3)_3COD$ in D_2O, (b) $(CD_3)_3COH$ in H_2O, and (c) 50:50 $(CD_3)_3COH$/$(CD_3)_3COD$ in 50:50 H_2O/D_2O.
2. Solute–solvent distribution functions are probed through substitution on the alcohol methyl hydrogen and the water hydrogen sites. In combination with the solute–solute and solvent–solvent distribution functions, this yields the (partial) radial distribution functions $g_{HH}(r)$, $g_{HX}(r)$, and $g_{XX}(r)$ for the solute–solvent correlations. The three experiments are made with (a) $(CH_3)_3COH$ in H_2O, (b) $(CD_3)_3COD$ in D_2O, and (c) 50:50 $(CD_3)_3COD$/$(CH_3)_3COD$ in D_2O.
3. Solute–solute distribution functions are probed through substitution on the alcohol methyl hydrogen and the water hydrogen sites. In combination with the solute – solute and solvent – solvent distribution functions, this yields the (partial) radial distribution functions $g_{HH}(r)$, $g_{HX}(r)$, and $g_{XX}(r)$ for the solute – solvent corrections. The three experiments are made with (a) $(CH_3)_3COH$ in H_2O, (b) $(CD_3)_3COD$ in D_2O, and (C) 50:50 $(CD_3)_3COD$/$(CH_3)_3COH$ in 50:50 H_2O/D_2O.

Thus we appear to have specified nine isotopically distinct experiments. However, we note that samples 1(a), 2(a), and 3(b) are identical. Therefore, we perform only seven experiments to yield the nine (partial) radial distribution functions specified above, namely $g_{HH}(r)$, $g_{HX}(r)$, and $g_{XX}(r)$ for each of the solvent–solvent, solute–solute, and solute–solvent cases considered.

Theory

Despite the fact that we have only performed seven different experiments, we can use these data, together with other known chemical information, to extract the partial radial distribution functions we require. This is not a case of getting something for nothing—it is doing the same sort of thing as a normal crystallographer would do in refining a crystal structure in which the number of independent reflections is—as is usually the case for large molecules—less than the number of structural parameters that need to be refined. That refinement is constrained by what we know of the chemistry of the system (e.g., standard bond lengths and angles, non-overlap of atoms). In effect, we do something similar in the liquid case.

The procedure used—the Empirical Potential Structure Refinement (EPSR) technique developed by Soper [3]—produces model ensembles of molecules that are consistent with the observed scattering. These ensembles can be interrogated to see the detailed geometries of the various intermolecular interactions in a given solution, just as sets of coordinate data from a crystalline refinement can be similarly mined. In summary, the procedure starts by setting up a Monte Carlo simulation of the system using a set of standard potential functions $U^0_{\alpha\beta}(r)$. The Monte Carlo simulation is then run to equilibrium, from which the various partial radial distribution function $g_{\alpha\beta}(r)$ are estimated. This, in general (in all cases we are aware of so far), fails to give adequate agreement with experiment, indicating in passing that the starting potentials are inadequate to reproduce the experimental data. Thus, the initial potential energy function is then modified by adding sets of potentials of mean force between the various different sites. These are derived from comparing the experimental $g^D_{\alpha\beta}(r)$ and the simulated $g_{\alpha\beta}$ (r) radial distribution functions. Thus we obtain a modified potential set:

$$U^N_{\alpha\beta}(r) = U^0_{\alpha\beta}(r) + kT \ln g_{\alpha\beta}(r) / g^D_{\alpha\beta}(r) \qquad (4)$$

This new potential is fed into the system and a further Monte Carlo simulation performed. The above procedure is then repeated until $U^0_{\alpha\beta}(r) \approx U^N_{\alpha\beta}(r)$ and $g_{\alpha\beta}(r) \approx g^D_{\alpha\beta}(r)$ for all r and for all pairs of atoms α, β. The resulting experimentally consistent ensembles can then be interrogated to obtain site–site partial radial distributions to examine the detailed structure of the system. In passing, we can note that the way in which the potential function is modified to bring the simulation into agreement with the experiment may be able to guide us in improving the potential functions originally used in the system—or at least give us an idea of what aspects of the starting potentials need modification. Note that what is described above is an earlier implementation of EPSR, which operated in real space, i.e., the comparisons were made between partial radial distribution functions. The normal implementation is now in reciprocal space, i.e., using the partial structure factors $S_{\alpha\beta}$ (Q) to modify the potential.

Results

Partial Radial Distribution Functions

Now that we have these ensembles of molecules, we can extract whatever structural information we want. The 28 partial radial distribution functions for *t*-butanol water at three concentrations (0.06, 0.11, 0.16 mole fractions) [4] are shown in Figs. 2 and 3. The various site labels are: CC, central carbon; C, methyl carbon; M, methyl hydrogen; O, alcohol oxygen; H, alcohol hydrogen; O_w, water oxygen; H_w,

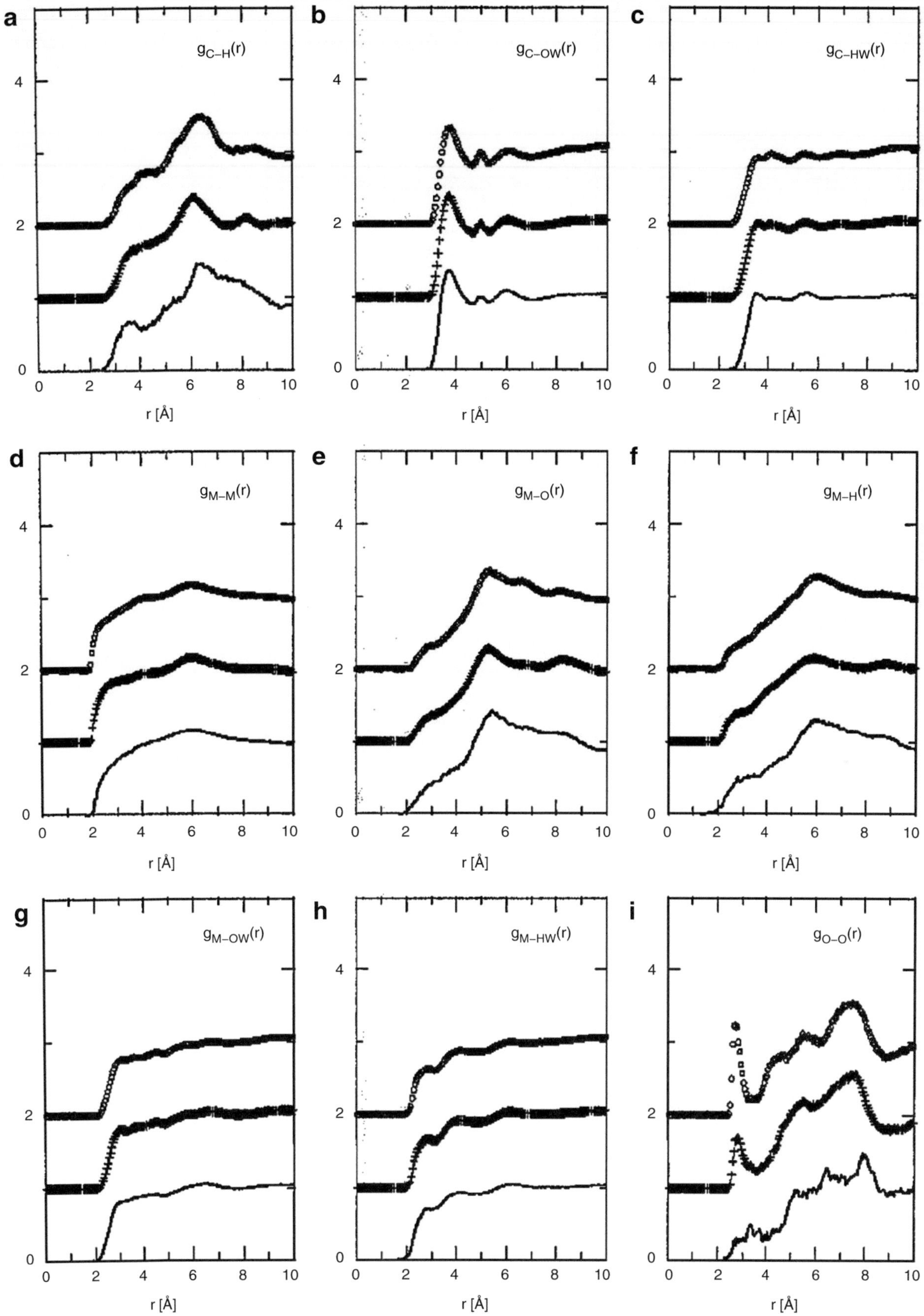

Fig. 2 Intermolecular partial rdfs for 0.16 (O, *top*), 0.11 (+, *middle*), and 0.06 (–, *bottom*) mole fraction *t*-butanol in water.

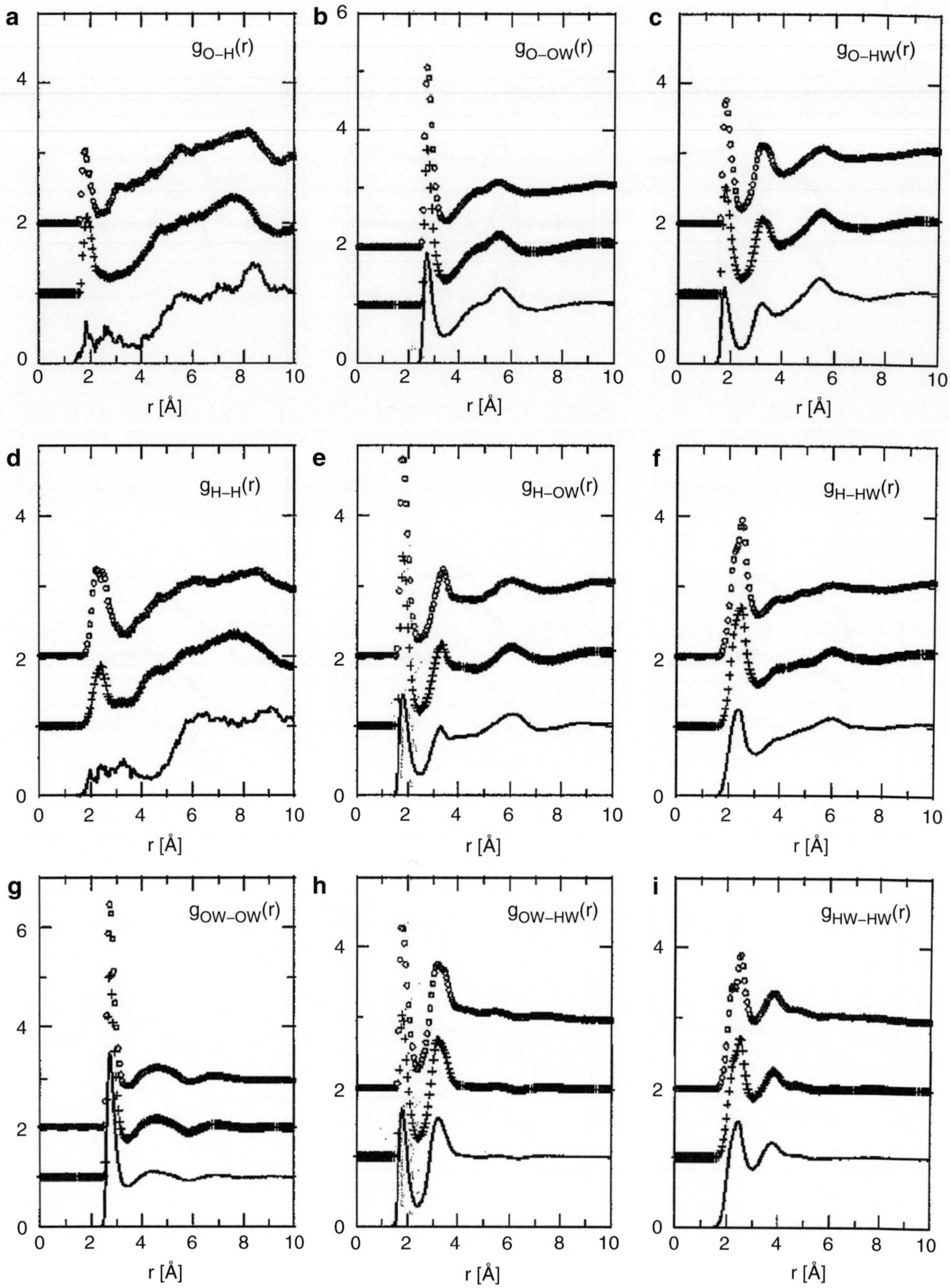

Fig. 2 (continued)

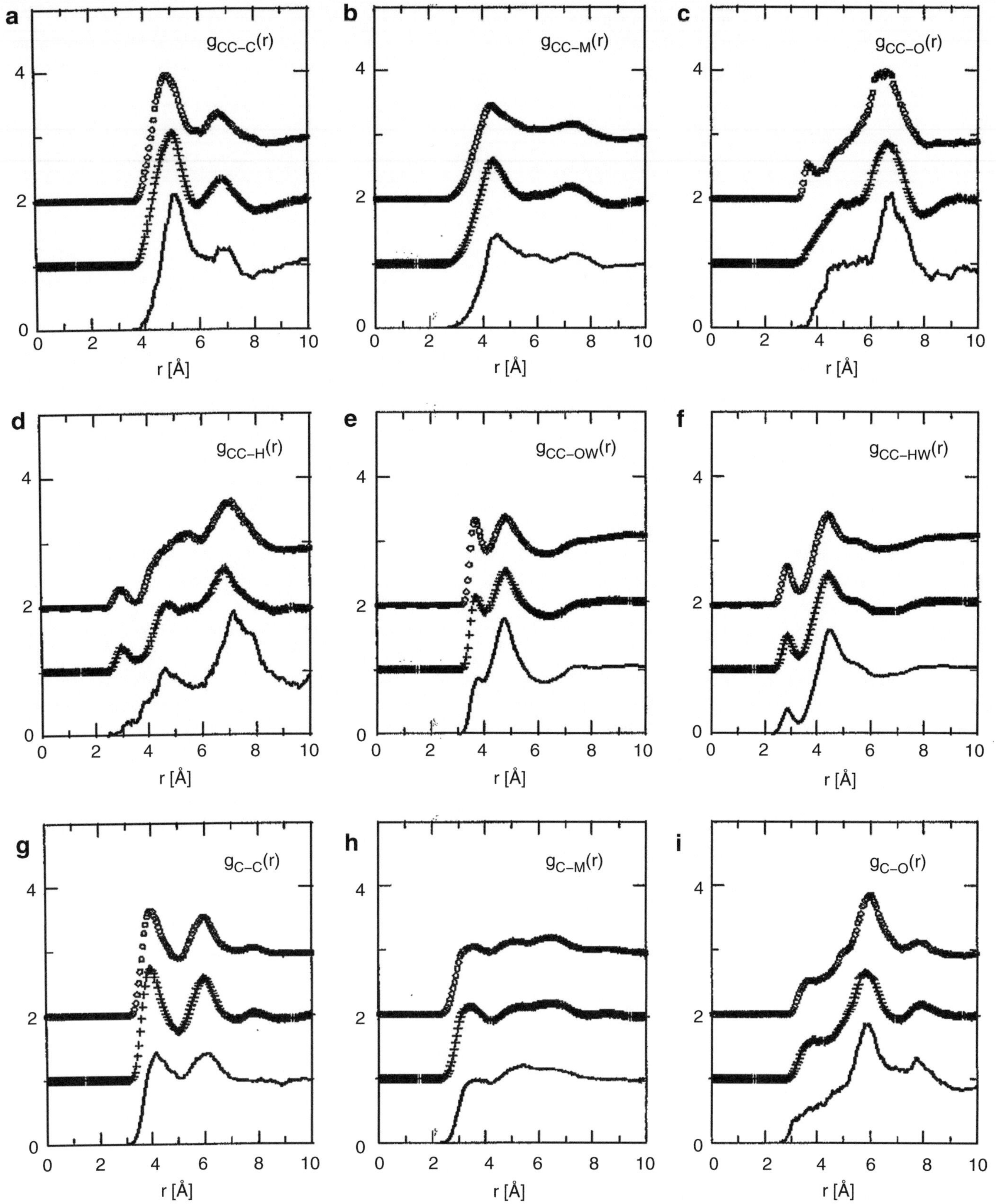

Fig. 2 (continued)

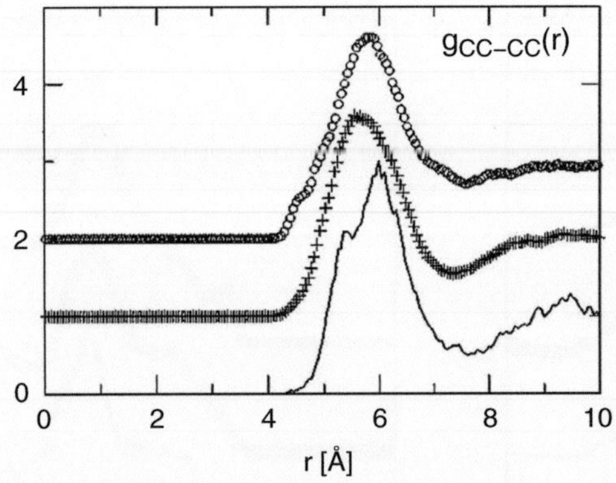

Fig. 3 The molecular centers function for 0.16 (O), 0.11 (+), and 0.06 (–) mole fraction *t*-butanol in water.

Table 1 Selected coordination numbers for aqueous *t*-butanol solutions

Correlation	R_{min} (Å)	R_{max} (Å)	0.06 Mole fraction	0.11 Mole fraction	0.16 Mole fraction
CC-CC	4.0	7.0	2.8±0.6	4.4±0.6	5.8±0.6
CC-O_w	3.0	4.0	2.1±0.4	2.2±0.3	2.0±0.2
	4.0	6.2	21.0±1.2	16.4±0.9	12.8±0.6
O-O_w	2.3	3.3	2.2±0.4	2.3±0.3	2.1±0.3
O-H_w	1.4	2.5	1.2±0.3	1.3±0.2	1.2±0.2
H-O_w	1.4	2.5	0.8±0.2	0.9±0.2	0.9±0.2
O-O	2.4	3.25	–	0.1±0.1	0.2±0.1
O-H	2.5	3.5	–	0.1±0.1	0.1±0.1
H-H	1.6	3.0	–	0.2±0.1	0.3±0.1

water hydrogen. There is clearly a tremendous amount of information in these functions. We now proceed to look at some of this information.

First, looking at the central carbon–central carbon radial distribution function (Fig. 3), we see for all three concentrations a broad peak centered between 5.5 and 6.0 Å. The peak position tells us immediately that even at the lowest concentration there is direct contact between neighboring *t*-butanol molecules: the molecules are not—as suggested in earlier models of the hydrophobic interaction [5]—separated by an intervening solvent layer. The area under this peak tells us how many molecules, on average, surround a central solute molecule. This and other coordination numbers are given in Table 1: it rises from 2.8±0.6 at 0.06 mole fraction, through 4.4±0.6 at the intermediate concentration, to 5.8±0.6 at 0.16 mole fraction. The fact that significant solute association is found at the lowest concentration was perhaps unexpected, suggesting some tendency to microscopic phase separation.

The CC-O_w partial in Fig. 3 gives us information on how the water molecules are arranged around the solute molecule from the point of view of the *t*-butanol's central carbon. We note two peaks at around 3.7 and 4.8 Å, respectively. The shorter distance is consistent with a water oxygen hydrogen bonding to the hydroxyl group of the alcohol, while the broader peak is at a distance consistent with waters surrounding the nonpolar head group, i.e., those in the hydration shell of the nonpolar head. Integration under this latter peak tells us (see Table 1) that the number of water molecules in this hydration shell falls from about 21 at the lowest concentration to about 13 at the highest—water molecules are, as expected, displaced from the hydration shell as the solute–solute coordination increases with concentration in the manner described above.

Integration under the first, hydrogen-bonding peak gives an essentially concentration invariant figure of 2. This tells us that each alcohol molecule hydrogen bonds to two water molecules through the polar tail. We can compare this with the O-O_w, O-H_w, and H-O_w coordination numbers in Table 1: not only are the values fully consistent with the alcohol hydroxyl group making two hydrogen bonds to water, but they also demonstrate that one of these waters donates a hydrogen to the alcohol oxygen, while the second water accepts a hydrogen from the alcohol.

The bottom three lines of Table 1 give information on the solute–solute hydrogen bonding interactions through the polar tails. For the lowest concentration, as is shown in the appropriate O-O, O-H, and H-H partials in Fig. 2, there is not even a peak visible at the appropriate hydrogen-bonding distance, showing effectively zero intermolecular hydrogen bonding. Even at the two higher concentrations, the peak areas are very small, suggesting that solute–solute hydrogen bonding is extremely limited. These results seem to be telling us that the *t*-butanol alcohol group preferentially bonds to water molecules rather than to other *t*-butanols. Thus, we can conclude quite clearly that the experimental data is not consistent with a significant amount of alcohol–alcohol hydrogen bonding at these concentrations. As the initial Monte Carlo simulations performed as part of the EPSR refinement process suggested a larger degree of intermolecular hydrogen bonding, there are indications here that the model potentials used for this system overemphasize the strength of the intermolecular hydrogen bonding interaction at the expense of the water–alcohol interaction. As the stability of a protein is a delicate balance between protein–protein and protein–water hydrogen bonds, this result indicates that work needs to be done on these hydrogen bonding potentials if they are to be seriously used in modeling biomolecular interactions.

Spatial Distribution Functions

Although all the information we should need to understand structure is, in principle, included in the full set of partial radial distribution functions, these functions are still averages

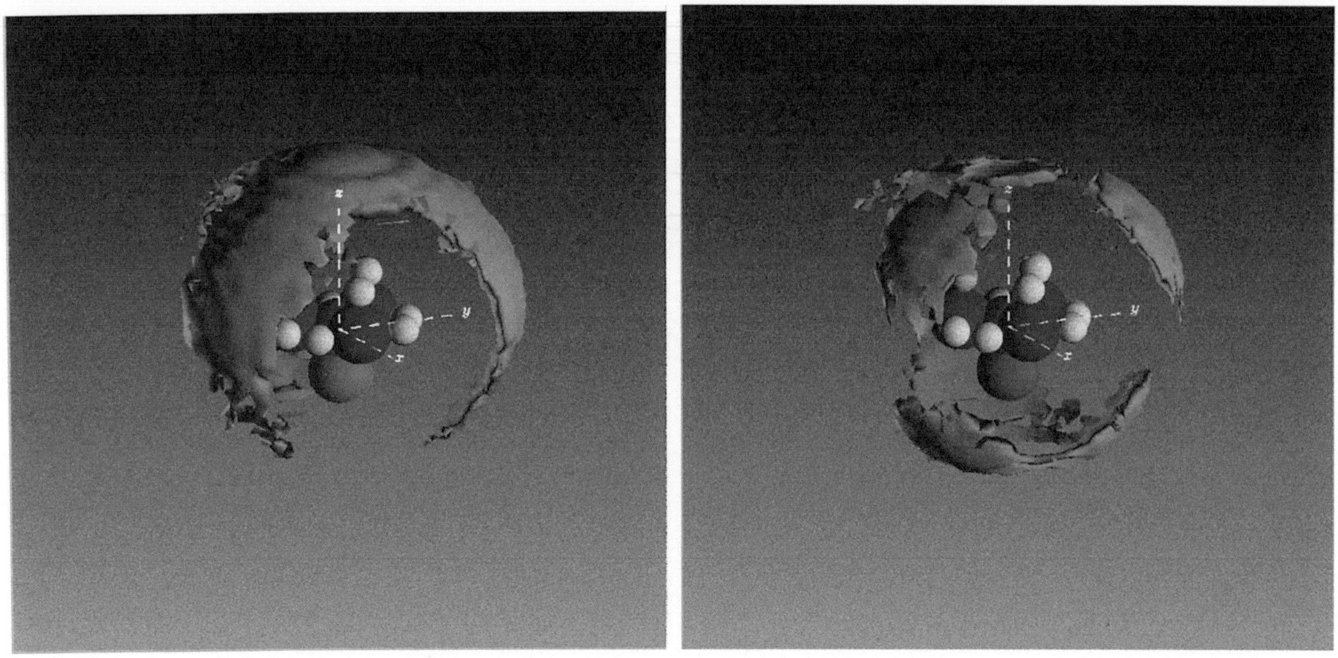

Fig. 4 The spatial density function of 0.06 mole fraction aqueous *t*-butanol at 25 °C (*left*) and 65 °C (*right*).

over orientations of both the central molecule at the origin of the function, and the neighboring molecule at a distance *r*. As we have complete data on the coordinates of all the sites on all the molecules in the model ensembles we are interrogating, we can construct further functions that enable us to look at the relative orientations of molecules.

First, we imagine ourselves sitting on a molecular site and looking at the various neighbors at different distances. We begin to note that there is a preference for a certain atom not only at a given radial distance corresponding to a peak in the partial radial distribution function, but also that peak tends to be in a particular direction. We thus can construct a further pair-distribution function that tells us the orientational preference of a particular site as a function of distance from the central site. This function, the *spatial density function*, gives us orientational information on the nature of the interactions between the various molecules in our solution. For example, referring again to Fig. 3, we have a peak that tells us that there are neighboring *t*-butanols (specified by the positions of their central carbon atoms) at a distance centered between 5.5 and 6 Å. We do not know from this orientational average the directions relative to the geometry of the central *t*-butanol in which these molecules are to be found. Do they cluster around the head groups or are there some contacts through the tails? What we really want to do is to assign a location on a surrounding sphere—in essence a latitude and a longitude—for each of the neighboring molecules included under the peak shown in Fig. 3. This is what the spatial density function $g\alpha\beta\,(r, \Omega)$ does [6]. It represents a three-dimensional (3-D) map of the density of β sites as a function of radial distance, *r*, and orientation *Q* about a central site α.

The spatial density function of *t*-butanol molecules around a central *t*-butanol molecule, defined with reference to the central carbon sites of each, is shown in Fig. 4 (left panel) for the 0.06 mole fraction concentration at 25 °C. The continuous density above the central *t*-butanol molecule and the three sections curving down show the angular distributions around the central molecule of the molecules that are under the broad peak in Fig. 3. We thus have more information than in that orientationally averaged partial radial distribution function: in addition to the preference for the "capping" position, there is greater density of neighboring molecules in three orientations separated by about 120°. Interestingly, these directions correspond to the positions between the three methyl groups on the head group of the central molecule, which enables a neighboring molecule to approach slightly closer than if it were approaching along a direction in which a methyl group was located. In fact, if we consider the concentration dependence of the peak position in Fig. 3, we see a slight outward move as the concentration is reduced. This outward move—which separates out a low-*r* shoulder at the lowest concentration—appears to be related to an apparent compressive driving force at the higher concentrations, which "pushes in" the neighboring *t*-butanol molecule. To anticipate the following section, the methyl groups on neighboring molecules seem to be engaging with each other like two cogwheels.

Orientational Distribution Functions

In the discussion of the previous paragraph, we are able to comment only on the positional distributions of the β sites around the central α site. We do not know how the neighboring molecule is oriented with respect to the central one. This information would obviously be useful in understanding the chemical nature of the interactions between the molecules in the various orientations around the central molecule: for example, are the nonpolar groups of the neighboring *t-butanol* in direct contact with the nonpolar groups of the central *t*-butanol, or are other orientations involved? Moreover, if we increase the temperature to 65 °C, how would an enhancement of the hydrophobic interactions driving the *t*-butanol association show up? Does the nature of the solute–solute interactions change? The spatial density function for 65 °C in the right-hand panel of Fig. 4 already shows that there is a shift in solute–solute interaction away from the head group to the polar tail: what is the chemistry of these interactions in the case?

Just as when moving from the partial radial distribution function to the spatial density function, we added orientational information to the molecules under the one-dimensional peak of the partial, so here we now need to add orientational information to the molecules in the lobes of the spatial density functions such as those in Fig. 4. As we have the detailed coordinate data in the ensembles generated by the EPSR refinement, this information can be accessed. The orientational distribution function reduced disorder in the head-group to head-group interaction, an interesting consequence of the increased temperature. Second, there is a significant feature around 12 o'clock, which is indicative of a polar to nonpolar interaction, i.e., the head group of one molecule is interacting with the polar tail of a neighbor. A further point to note is that this lobe is shifted out to a distance of about 8 Å, so this appears not to be a direct contact between the hydroxyl group and the methyl head groups. Rather there is space for an intervening water molecule. We can, therefore, perhaps consider this as a neighboring *t*-butanol hydrogen bonding to one of the water molecules in the nonpolar hydration shell of the central molecule.

Finally, the bottom panel shows the orientational plot in the direction of the hydroxyl group of the central *t*-butanol molecule. The most obvious change to note is the growth of intensity around the 6 o'clock position at 65 °C. This lobe corresponds to *t*-butanol to *t*-butanol contacts through direct hydrogen bonding, an occurrence that, we have already noted, is much rarer at 25 °C. A second point to note is a splitting of the lobe around 12 o'clock that denotes polar to nonpolar contacts; at 65 °C, the inner lobe comes in closer, the rise in temperature appearing to make easier a close, apparently chemically unfavorable, contact.

Raising the temperature in this system thus seems to do several interesting things. First, the straight nonpolar contacts seem to become more ordered. Second, there is an increase in the degree of direct hydrogen bonding between *t*-butanol molecules. Third, the temperature increase seems to make classically unfavorable interactions between polar and nonpolar groups easier. The "strengthening" of the nonpolar contacts might be expected from classical ideas of the temperature dependence of the hydrophobic effect. The other two effects seem less easy to relate directly to consequences of a hydrophobic driving force. $g(\mathbf{r}, \omega_1, \omega_2)$ relates the relative position vector **r** of two molecules 1 and 2, with their orientations ω_1, ω_2 in the laboratory reference frame. As this is strictly a function of nine variables (three positional, six angular), there are major computer memory limitations to storing this function, even when the variables are reduced by symmetry. An efficient way of dealing with this problem is to expand the orientational correlation function as a spherical harmonic expansion. The correlation function can then be stored as a series of spherical harmonic coefficients [7] $h(l_1 l_2 l;\ n_1\ n_2;\ r)$, where l_1, l_2, n_1, n_2 relate to the orientational coordinates ω_1 and ω_2, while l relates to the spatial distribution of molecule 2 about molecule 1.

Figure 5 shows the definition of the coordinate system used for both the orientational position of the neighbor with respect to the geometry of the central molecule and for the orientation of the neighboring molecule with respect to the first. The two angles θ_1 and φ_1 describe the direction of the neighbor (the central molecule being rotationally averaged about the Z-axis shown) and the three Euler angles φ_m, θ_m, and χ_m define the orientation of the neighboring molecule. As a first example of its use, we show in Fig. 6 correlation maps of θ_m for the 0.06 mole fraction concentration at 25 °C for three values of θ_1. The $\theta_1 = 0°$ plot shows high density for θ_m around 180°, i.e., the dominant interaction in this direction is with the two *t*-butanol molecules making contact through their nonpolar head groups. This is classically what we would expect for a system subject to a hydrophobic driving

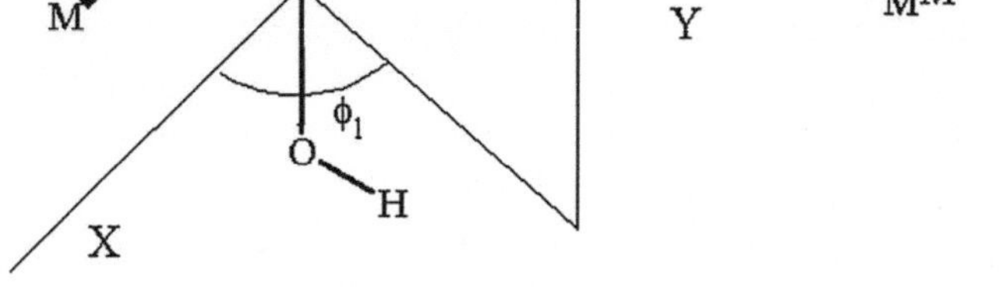

Fig. 5 Schematic diagram of the coordinate system used to define the orientational correlations between two *t*-butanol molecules.

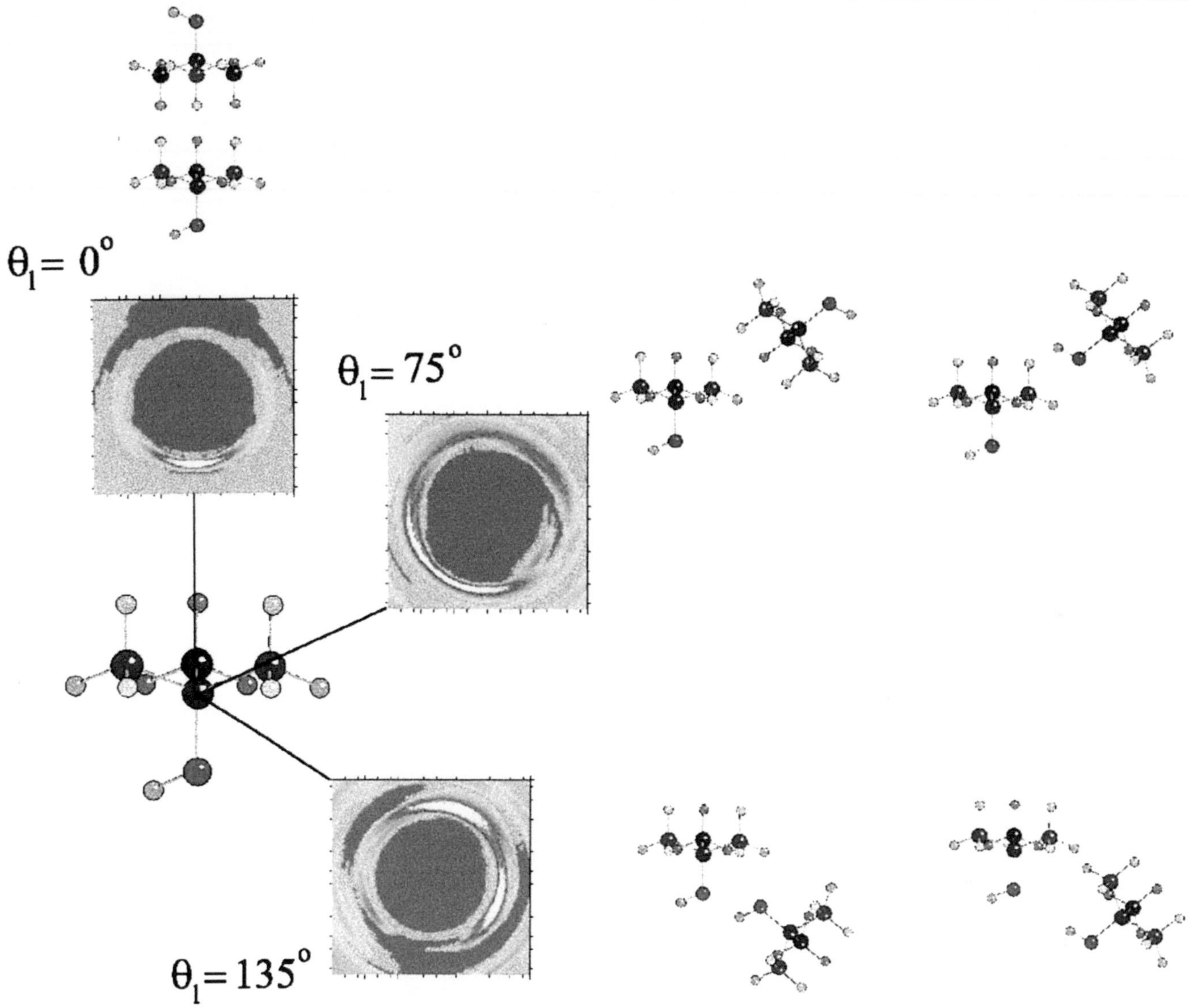

Fig. 6 Sections through the orientational correlation function (correlation maps) of two *t*-butanol molecules in 0.06 mole fraction aqueous solution at room temperature. The molecular arrangements shown are the dominant ones corresponding to the highest contour levels in the plots. For each plot shown, the angular variable plotted is θ_m.

force. Moving to the $\theta_1 = 75°$ plot, we see the highest intensity between the 6 o'clock and 9 o'clock positions, again consistent with a dominant nonpolar to nonpolar contact as shown in the figure. Here, however, there is some—though small—intensity around the 1–2 o'clock position, indicating some contribution from the polar to nonpolar contact, again indicated by the molecules in the figure. Finally, moving to the $\theta_1 = 135°$ plot, we note bright lobes in the two molecular orientations shown—clearly these correspond to intermolecular hydrogen bonding. We have already noted, however, that these contacts occur only rarely in this solution. What the orientational correlation functions tell us is that, even if the intermolecular contacts in this region of the central molecule are rare, when they do occur, they are of a typically hydrogen bonding nature.

A second example of the information in the orientational correlation function refers back to the spatial density functions shown for the 0.06 mole fraction concentration at the two temperatures of 25 and 65 °C (Fig. 4). What can we learn about the change in the nature of these intermolecular interactions as the temperature is raised in such a way as to (we expect) enhance the hydrophobic driving force [8]? We have already seen that at 25 °C the lobes in the spatial density function relate to nonpolar to nonpolar contacts through the head groups of neighboring molecules; at the higher temperature (Fig. 4, right-hand panel), these lobes shrink, and there is significant growth in contacts at the polar end of the molecule. What is the chemical nature of these new contacts?

We can answer all these questions by referring to the plots in Fig. 7. The left-hand panels relate to the 25 °C solution,

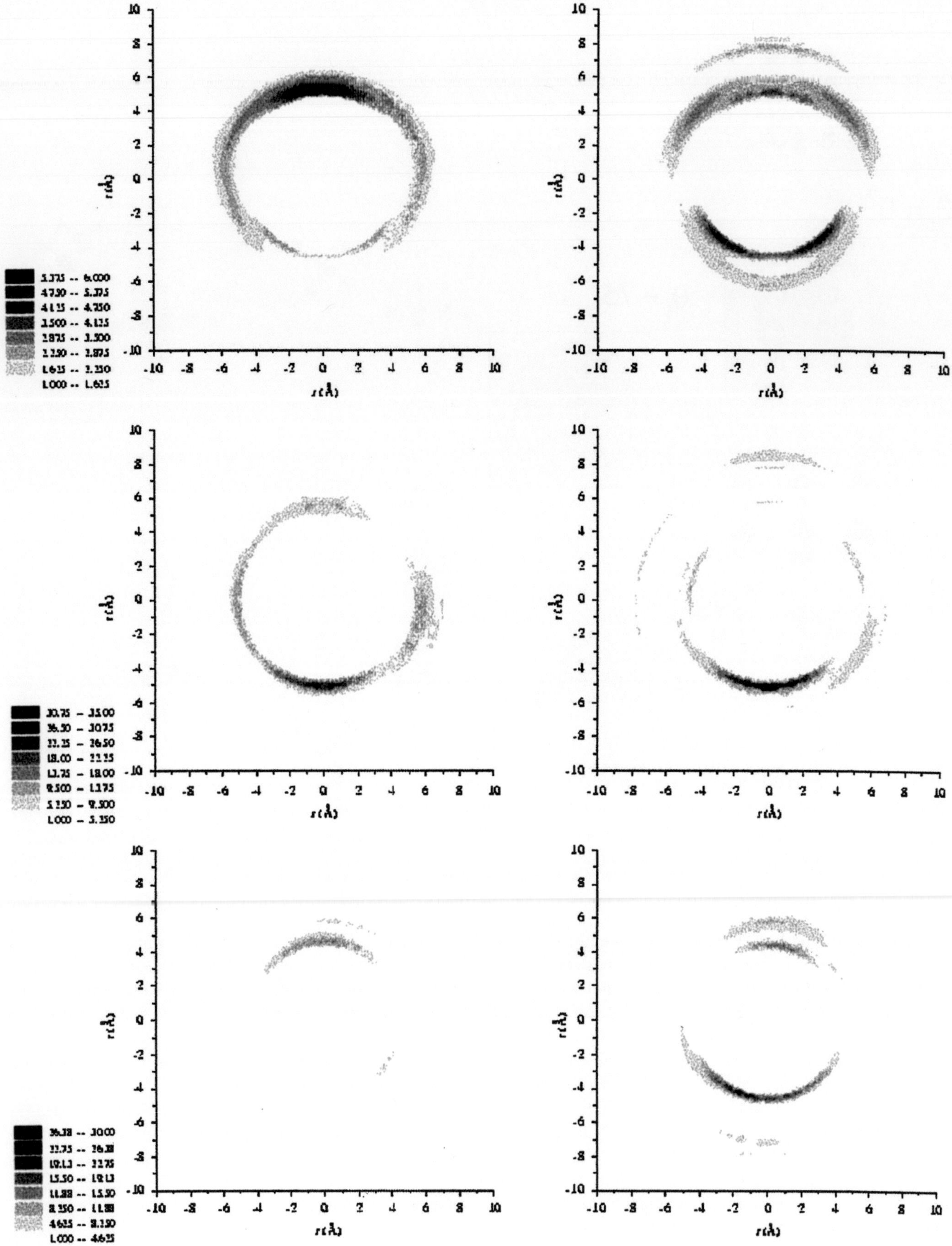
5.375 -- 6.000
4.750 -- 5.375
4.125 -- 4.750
3.500 -- 4.125
2.875 -- 3.500
2.250 -- 2.875
1.625 -- 2.250
1.000 -- 1.625
30.75 – 35.00
26.50 – 30.75
22.25 – 26.50
18.00 – 22.25
13.75 – 18.00
9.500 – 13.75
5.250 – 9.500
1.000 – 5.250
26.38 -- 30.00
22.75 -- 26.38
19.13 -- 22.75
15.50 -- 19.13
11.88 -- 15.50
8.250 -- 11.88
4.625 -- 8.250
1.000 -- 4.625
r (Å)

while the right-hand column relates to 65 °C. The top two panels are plots through the two spatial density functions of Fig. 4 and show that the dominant head-to-head contact indicated by the bright lobe around 12 o'clock for 25 °C is reduced in favor of increasing intensity around 5 and 7 o'clock in the 65 °C solution. The second panel shows orientational correlation function plots of θ_m. The dominant intensity around 180° again indicates a dominance of nonpolar contacts through the head groups of contacting molecules. Moving right to the 65 °C panel, we find two interesting points. First, the 6 o'clock feature is a little "tighter" indicating.

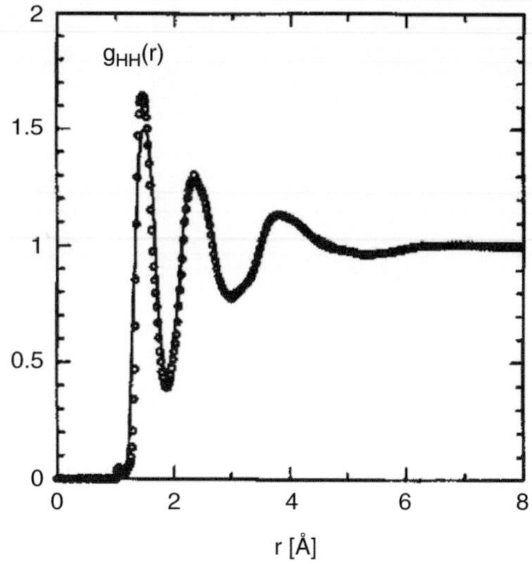

Fig. 8 The hydrogen–hydrogen partial rdf for a 1:9 mole fraction solution of methanol in water (*line*) compared to the same function for pure water.

Structural Basis of the Hydrophobic Interaction

The above examples show that we can now obtain detailed geometrical information on intermolecular interactions in solution. We can see how solute molecules interact and how these interactions might change under changed external conditions. We conclude by addressing some structural issues relating to the perturbation of the water structure in our *t*-butanol solution. These impinge directly on the nature of the solvent ordering that is thought to be the driving force for the hydrophobic interaction.

First, we note the classical view of the nature of the hydration shell of a nonpolar group and how this is thought to relate to the entropic driving force of the hydrophobic interaction. The conventional wisdom on this comes from the classic paper of Frank and Evans in 1945 [9]. In brief, this paper argues that the water in the hydration shell is somehow "more ordered" than that in the bulk. Bringing together two nonpolar surfaces in solution will then displace some of this "ordered" water into the bulk, with a consequent gain in entropy to the system. Hence, the source of the entropic driving force to hydrophobic association.

The detailed structural nature of this solvent ordering has been the subject of much discussion. The prevailing view seeming to prefer a picture something like the clathrate cage structure found in gas hydrate crystals [10]. Earlier work using the kind of techniques discussed here has partly verified this picture, although it also demonstrates that the simple cage model specifies too much order. Rather, we should think of the hydration shell as a *disordered* structure that has some relationship to the clathrate cage; we are, after all, dealing with a liquid in which such long-lived ordered structures are unlikely to exist. A more surprising conclusion of that work [10] concerned the structure of the water in the hydration shell. Figure 8 compares the hydrogen–hydrogen partial rdf for bulk water with that of the water in a 1:9 aqueous methanol solution [11]. The conclusion we can draw from the figure is both obvious and unexpected: there is no detectable difference between the two functions. As far as the water is concerned, when its structure is measured through the hydrogen–hydrogen radial distribution function, a sensitive structural measure that includes all-important orientational information, there is no observable perturbation of the structure in the direction of increased—or decreased—order. Similar conclusions have been drawn since in studies of a number of different systems containing nonpolar head groups [4, 8, 12]. Thus, these structural results do not support the traditional view that the water in the first hydration shell of a nonpolar group is structurally more ordered than in the bulk.

The clue to the actual source of the entropic driving force may be in Fig. 9, which shows the spatial density functions of the first (top panels) and second (bottom panels) neighbor shells *of water* in (from left to right) bulk water at 25 °C, 0.06 mole fraction *t*-butanol at 25 °C, and 0.06 mole fraction *t*-butanol at 65 °C. We emphasize that we are looking here at the neighbourhood *of the water*, *not* of the nonpolar group itself.

Fig. 7 The intermolecular orientation correlation maps corresponding to the solute–solute interactions as a function of molecular centers separation in 0.06 mole fraction *t*-butanol–water solution. The *left-hand column* represents the correlations prevalent at 25 °C, while the *right hand column* those at 65 °C. The *top panels* represent the probability of finding a neighboring alcohol molecule about an arbitrarily chosen central alcohol molecule as a function of the angle θ_1, i.e., the result is averaged over all orientations of the neighboring alcohol molecule. The *middle panels* illustrate the dominant orientation as a function of θ_m of a neighboring alcohol molecule in the direction above the methyl groups of the central molecule, i.e., where $\theta_1 = \phi_1 = \phi_m = \chi_m = 0°$. The *lower panels* show the relative orientations of alcohol molecules θ_m, in the direction of the hydroxyl group interactions where $\theta_1 = 135°$, $\varphi_1 = 60°$, and $\varphi_m = \chi_m = 0°$.

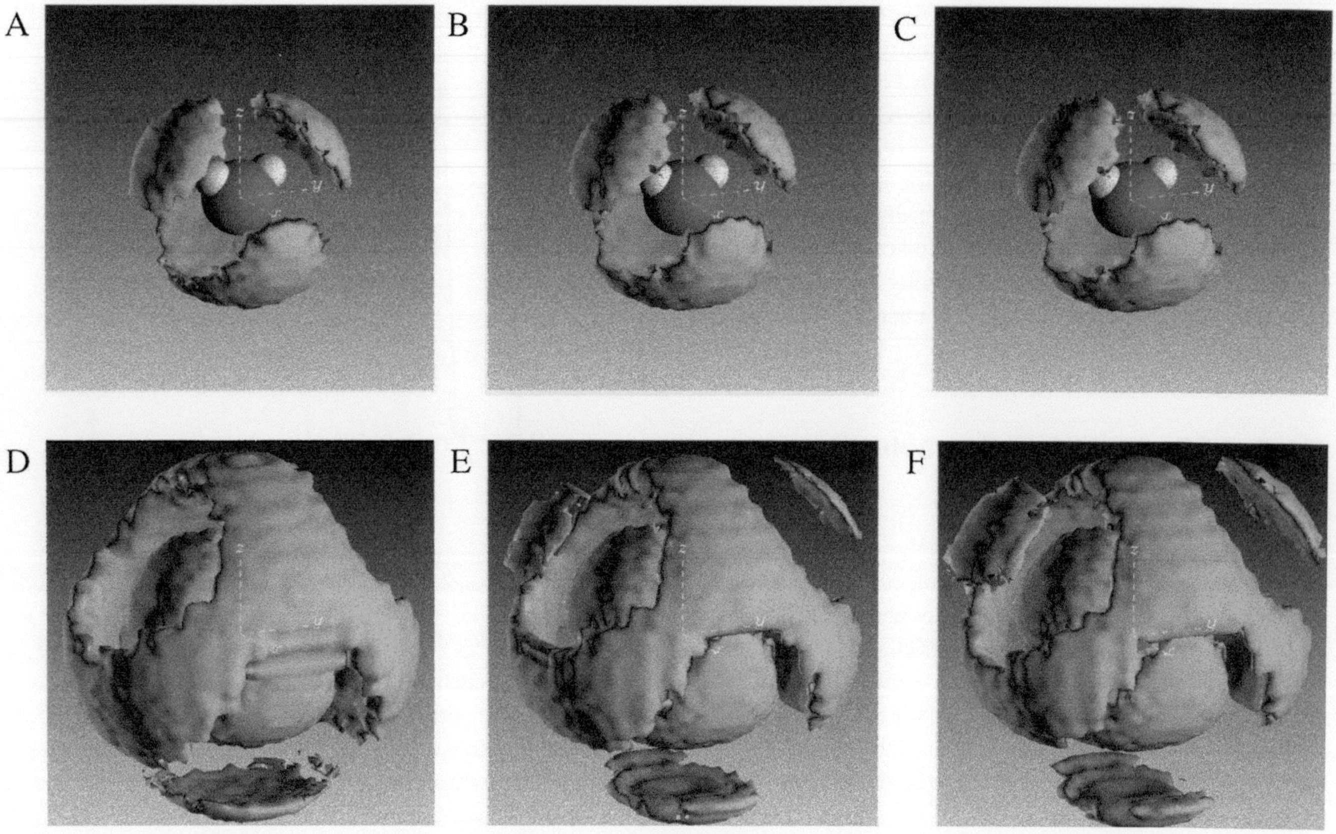

Fig. 9 The spatial density functions of water molecules around a central water molecule for (*top row*) first neighbor shell water molecules and (*bottom row*) first and second shell waters. In turn, the panels show the distributions for (*left*) pure water at 25 °C, (*center*) 0.06 mole fraction *t*-butanol–water solution at 25 °C and (*right*) 0.06 mole fraction *t*-butanol–water solution at 65 °C. The first shell of neighbors (*top row*) illustrates the invariance between the three systems of the short-range correlations that are of a predominantly tetrahedral local water coordination. The appearance of the third neighbor shell in the direction of the central molecule O-H bonds for the alcohol–water mixtures (*bottom center* and *bottom right*) visibly illustrates the solute-induced compression of the solvent molecular density, which appears to be enhanced by increasing temperature.

Focusing first on the first shells in the top panels, we see the standard spatial density function for the first neighbor shell in water. The two lobes above the central water relate to neighboring waters that accept hydrogens from the central water, while the broad lobe beneath indicates waters donating hydrogens to the lone pair region of the central water. This first neighbor shell appears unchanged in both *t*-butanol solutions.

When we look at the bottom panels, where the contour levels have been lowered to show up the second neighbor shell, we see some clear differences. Although the basic geometry of the second shell is similar in all four cases, there are very significant differences: the second shell is progressively "pulled in" toward the central molecule as we move to the 25 °C solution, and then to the 65 °C case. Consequently, lobes of spatial density corresponding to the third shell begin to appear in the two solution plots.

This "drawing in" of the second shell is accompanied by a significant ordering of the second shell. This is, indeed, visible in the progressive narrowing of the band in the center of the bottom panels of Fig. 9 as we move from left (bulk water) to right (65 °C 0.06*t*-butanol). It is perhaps more obvious, however, in the 2-D spatial density maps of Fig. 10, which are obtained by rotationally averaging $g_{oo}(r, \Omega)$ about the *z* axis (shown in Fig. 9). The second shell structural tightening induced by the presence of the alcohol molecule can be seen in the breaks of continuity in the 10 and 2 o'clock directions as we move from bulk water to the two solutions. Also visible is the small, but significant, inward shift of the lobes, which is particularly obvious in the 12 o'clock direction. Taking sections through the appropriate lobes allows us to quantify both these inward shifts and the narrowing of the shell widths. In terms of the water molecule's second neighbor shell, we have clear evidence that the presence of *t-butanol* increases structural order in the system, an ordering that is enhanced with the rise in temperature to 65 °C. Perhaps it is here—in the *water's* second shell rather than the *alcohol's* first hydration shell—that the entropic driving force for the hydrophobic interaction is to be found.

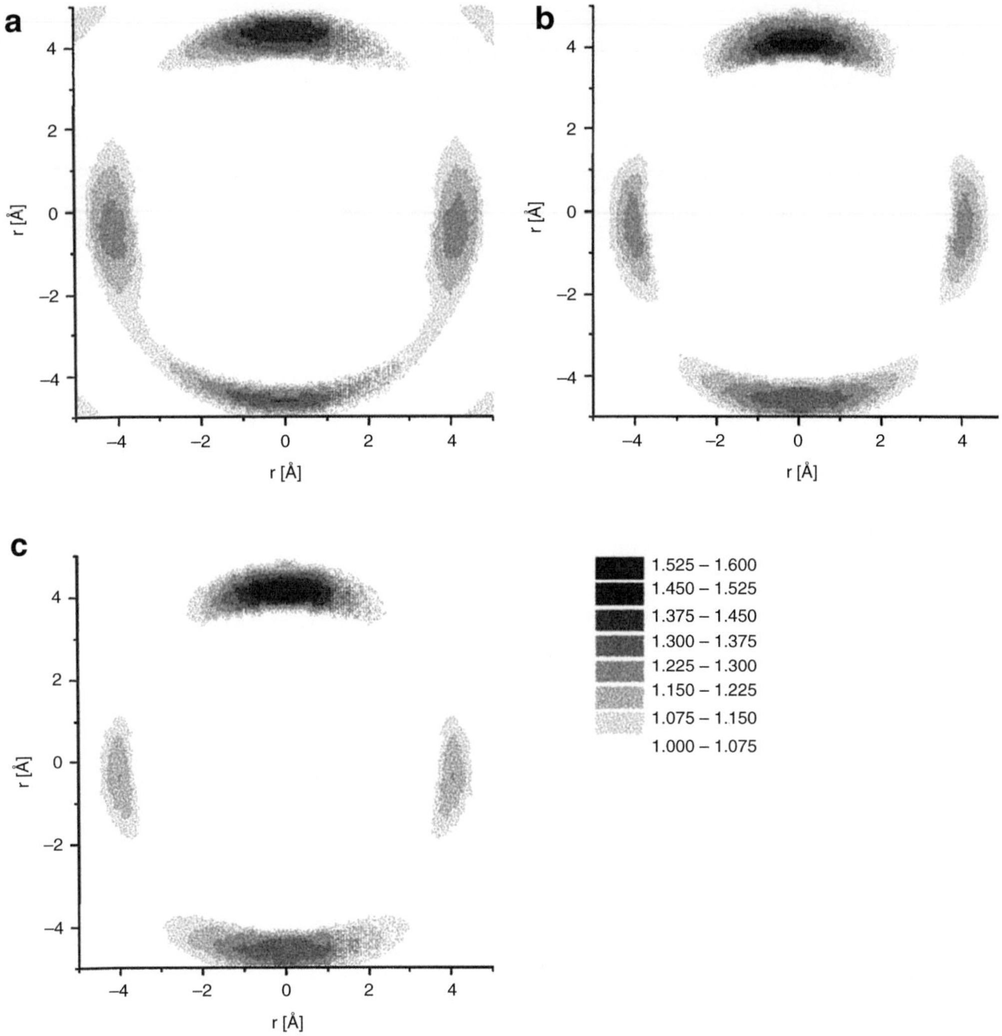

Fig. 10 The upper panels are 2-D maps of the second water shell spatial density functions of Fig. 9. These are obtained by rotationally averaging $g_{oo}(r, \Omega)$ about the z axis, i.e., about the ϕ coordinate. The map in (**a**) corresponds to pure water at 25 °C, in (**b**) to 0.06 mole fraction *t*-butanol–water solution at 25 °C, and in (**c**) to the alcohol solution at 65 °C. The disruption of the second shell water in the presence of the alcohol molecule can be seen in the 2 o'clock direction: this disruption is also marked by the inward migration of the shell, in particular, along the 12 o'clock direction.

Conclusions

"Normal" crystallography exploits the existence of the lattice to maintain phase relationships between scattered beams of X-ray, neutrons, or electrons, which enable us to determine detailed structures of even highly complex molecules within a crystal. Without the crystal lattice, these phase relationships are lost, and we are left with being able to determine only distributions of pair distances. When we try to generalize crystallography to disordered systems, in general, we have to develop ways of obtaining high-quality structural information in the absence of this lattice "crutch." If we do want to under-stand process in solution—which are central to much of chemistry and biology—we need to do structure determination without a lattice being present. We need to develop "no-lattice crystallography."

Taking advantage of the past 30 years' developments in neutron sources and instrumentation, allied to advances in both computer hardware and software, we have been able to make major progress in structure determination in the absence of the lattice. We can perhaps make a comparison with the advances in crystallography since the 1920s, when Kathleen Lonsdale's solution of the structure of hexamethylbenzene and hexachlorobenzene caused Christopher Ingold to comment that "the calculations must have been terrible … but one structure like this brings more certainty into organic chemistry than generations of activity by us professionals." From those early molecular structures, we have advanced—taking advantage of developments in radiation sources, instrumentation, and computing resources—to determining structures of macromolecular assemblies, such as the ribosome.

In the liquid case, we have similarly made major steps forward. Christopher Ingold's comment on molecular structure is paralleled in a way by the chemist David Feakins' 1993 (personal communication) comment on liquids: "(our understanding of solutions is) in the state that our knowledge of molecules would be if we only had, say, the thermodynamics and kinetics of reaction to guide us." Thirty years ago it was unimaginable to either me or to David Harker on that Air China charter flight that we could determine the structural details of the interaction between two amphiphiles in solution, let alone how that interaction might change with temperature or concentration. Or that we would be able to measure either the hydration shell of a nonpolar group, or how the solvent in an aqueous solution of molecules containing large nonpolar groups. We can now do these things, and more. At the turn of the century, liquid state crystallography—crystallography without a lattice—has perhaps come of age.

Acknowledgments The work described here, and building up to it, has been performed in collaboration with a number of colleagues, Daniel Bowron, Alan Soper, and Jacky Turner having been particularly closely involved. The early ideas developed owe much to many discussions with John Enderby, who pioneered isotope substitution methods in liquid structure studies. Alan Mackay has remained a stimulating colleague throughout.

References

1. Franks, F.; Desnoyers, J. E. *Water Sci. Rev.* **1985**, *1*, 171.
2. Edwards, F. G.; Enderby, J. E.; Howe, R. A.; Page, D. I. *J. Phys. C* **1975**, *8*, 3483.
3. Soper, A. K. *Chem. Phys.* **1996**, *202*, 295.
4. Bowron, D. T.; Finney, J. L.; Soper, A. K. *J. Phys. Chem. B* **1998**, *102*, 3551.
5. Pratt, L.; Chandler, D. *J. Chem. Phys.* **1977**, *67*, 3683.
6. Svishchev, I. M.; Kusalik, P. G. *J. Chem. Phys.* **1993**, *99*, 3049.
7. Gray, C. G.; Gubbins, K. E. *Theory of Molecular Liquids, Vol. I: Fundamentals*; Oxford University Press: New York, 1984.
8. Bowron, D. T.; Soper, A. K.; Finney, J. L. *J. Chem. Phys.* **2001**, *114*, 6203.
9. Frank, H. S.; Evans, M. W. *J. Chem. Phys.* **1945**, *13*, 507.
10. Davidson, D. W. In *Water A Comprehensive Treatise*; Franks, F., Ed.; Plenum Press, New York, 1973; Vol. 2, p. 115.
11. Soper, A. K.; Finney, J. L. *Phys. Rev. Lett.* **1993**, *71*, 4346.
12. Turner, J.; Soper, A. K.; Finney, J. L. *Molec. Phys.* **1990**, *70*, 679.

Polygrammal Symmetries in Biomacromolecules: Heptagonal Poly *d* (As^{4}T) · poly *d* (As^{4}T) and Heptameric α-Hemolysin[a,b]

A. Janner[c]

Abstract

Polygrammal symmetries, which are scale-rotation transformations of star polygons, are considered in the heptagonal case. It is shown how one can get a faithful integral six-dimensional (6-D) representation leading to a crystallographic approach for structural properties of single molecules with a sevenfold point symmetry. Two biomacromolecules have been selected in order to demonstrate how these general concepts can be applied: the left-handed Z-DNA form of the nucleic acid poly *d* (As^{4}T) · poly *d* (As^{4}T), which has the line group symmetry $7_6 22$, and the heptameric transmembrane pore protein *α*-hemolysin. Their molecular forms are consistent with the crystallo-graphic restrictions imposed on the combination of scaling transformations with sevenfold rotations. This all is presented in a two-dimensional (2-D) description, which is natural because of the axial symmetry.

Keywords

General crystallography • Molecular forms • Scale rotations • Poly *d* (As^{4}T) · poly *d* (As^{4}T) • *α*-hemolysin

Introduction

In a previous paper [1], a unifying general crystallography has been defined, characterized by the possibility of point groups of infinite order. Crystallography then becomes applicable even to single molecules, allowing a crystallographic interpretation of structural relations, which extend those commonly denoted as "*non-crystallographic*" in protein crystallography. In the same paper it is pointed out that molecular crystallography can be approached in a fairly similar way, as for quasicrystals. Alan Mackay, to whom this paper is dedicated on the occasion of his 75th anniversary, has given a pioneering and fundamental contribution to the problem of the relations between structure and symmetry in quasicrystals. For this occasion it seemed to me a good idea to focus attention on systems having a sevenfold point group symmetry.

Sevenfold self-similar quasicrystals, where crystallographic point groups of infinite order arise naturally, have been considered by various authors (see for some references [2–4]), but this field has not been developed because such quasicrystals have not been observed in nature. Observed in nature are, however, biomacromolecules with a rotation symmetry of order 7 and it is challenging to investigate the relevance of these point groups in the molecular case. A first attempt in this direction has been made for two hexagonal DNA's in terms of scaling properties of molecular forms reflecting an underlying point group of infinite order leaving the hexagonal lattice invariant [5]. In this paper, evidence is also given of similar behavior in B-DNA and in A-DNA,

[a] *Structural Chemistry* 2002, 13(3/4):277–287.

[b] This contribution is part of a collection entitled Generalized Crystallography and dedicated to the 75th anniversary of Professor Alan L. Mackay, FRS.

[c] Institute for Theoretical Physics, University of Nijmegen, Toernooiveld, Nijmegen 6525 ED, The Netherlands
e-mail: alo@sci.kun.nl

I. Hargittai, B. Hargittai (eds.), *Science of Crystal Structures: Highlights in Crystallography*,
DOI 10.1007/978-3-319-19827-9_7

which correspond to the tenfold and the elevenfold case, respectively.

It is essential to verify whether these scaling properties occur in nature as a rule rather than as an exception. From the point of view of molecular crystallography, this implies molecular forms with scaling factors critically related to the given point group symmetry. Typically, one should find τ, the golden mean, in the fivefold case, but this factor should be totally absent in the heptagonal case, because of the underlying number theoretical structure [4, 6].

The interest for the sevenfold case arose from the very nice axial view of α-hemolysin appearing in the cover picture of the book on Macromolecular Structures 1997 [7], which shows a nearly perfect sevenfold scalerotation symmetry of the heptamer. In order to distinguish between law and accident, nucleic acids have been considered as an alternative molecular case.

Nucleic acids occur as polynucleotides in different helical conformations, with a variety of axial symmetries. A full list of polynucleotides can be found in a historical survey of S. Arnott in the Oxford Handbook of Nucleic Acids edited by Neidle [8]. There, only one case is reported with sevenfold axial symmetry: the Z-DNA form of the left-handed poly $d(As^4T) \cdot$ poly $d(As^4T)$ which, therefore, has been selected together with α-hemolysin. A third case, the double chaperonin (a larger and a smaller one) complexed with ADP and denoted as GroEl-GroES-$(ADP)_7$ will be discussed elsewhere. The analysis of the molecular forms is limited here to the prismatic forms of the backbone formed by the C_α's in the protein and disregarding hydrogen in the nucleic acid. The prismatic forms can be viewed along the sevenfold axis, allowing a two-dimensional (2-D) description of their three-dimensional (3-D) scale-rotational properties.

Many of the concepts raised so far are certainly nonfamiliar to most readers. Before developing the subject, a visualization of what one is talking about is, therefore, important. Polygrams and star polygons are first introduced, because these geometrical objects allow a simple approach to the symmetry of the various molecular forms. A *star polygon* denoted by the Schläfli symbol $\{n/m\}$ is obtained from a regular polygon by connecting each of the n vertices with its mth-subsequent one [9]. The regular polygon itself represents the trivial case $\{n/1\}$. A *polygram* consists of star polygons. Known for more than 2,000 years are the *pentagram* $\{5/2\}$ and the *hexagram* $\{6/2\}$. In the sevenfold case, in addition to the regular heptagon $\{7/1\}$, there are the two star heptagons $\{7/2\}$ and $\{7/3\}$ (Fig. 1). The other possible ones coincide with one of these.

The morphological role of these star polygons consists in relating regular heptagons of various size characterizing a molecular form. In particular, the central hole can very often be obtained from the external polygonal boundary enclosing

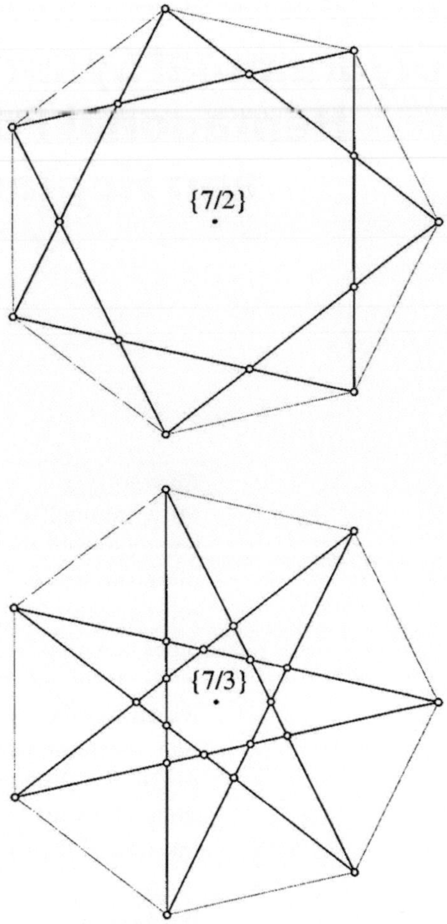

Fig. 1 The two regular heptagonal star polygons and their Schläfi symbol $\{7/2\}$ and $\{7/3\}$.

the biomacromolecule by star polygons, as it has already been shown in the hexagonal case [5].

In the case of α-hemolysin, the external boundary is a regular heptagon with vertices at the positions of the residue Glu71 in the cap domain of the seven chains, whereas the vertices of the central hole can be defined by the positions of the residue Lys147 of the transmembrane stem, which has the structure of a right-handed β-barrel. These boundaries are related by a $\{7/3\}$ star heptagon (Fig. 2). This implies that the two heptagons have the same orientation and that the axial distances of Glu71 and Lys147 are in the ratio $1{:}\mu$ determined by the symmetry of $\{7/3\}$, with $\mu = \cos(3\pi/7)/\cos(\pi/7) = 0.2469\ldots$.

The structure of poly $d(As^4T) \cdot$ poly $d(As^4T)$ projected along the helical axis is dominated by relations expressible in terms of $\{7/2\}$ star heptagons. The central molecular hole (due to s^4T, where s^4T indicates that the oxygen in the position 4 of the thymine is replaced by a sulfur atom) is so small with respect to the external heptagonal boundary, that at this stage no reliable scaling factor can be derived from polygrammal relations. Therefore, only the molecular form of the A subsystem is presented here (Fig. 3). One sees that,

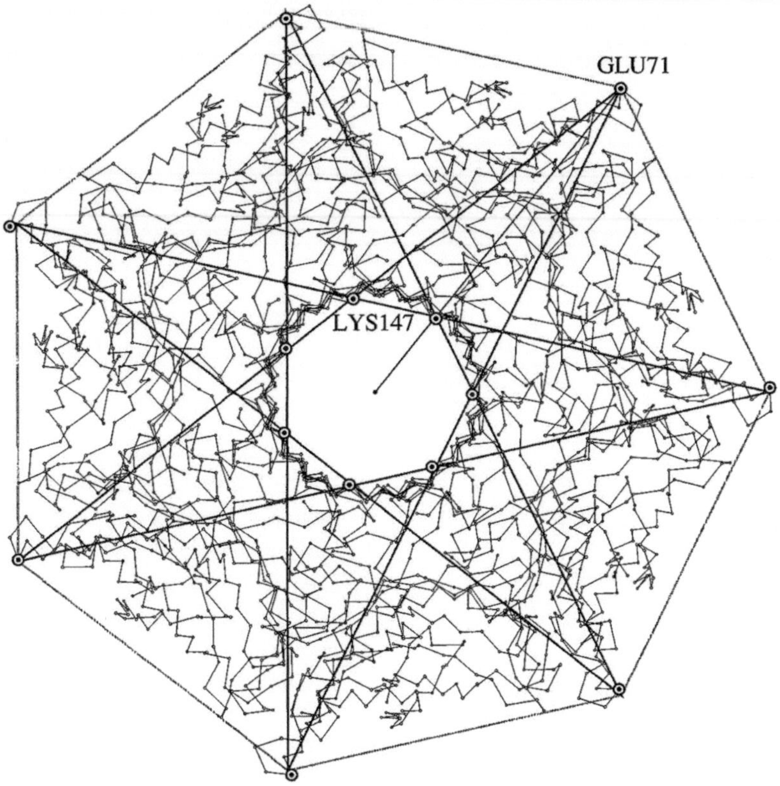

Fig. 2 Axial view of the heptameric protein α-hemolysin with a prismatic enclosing form. The two delimiting heptagons, with vertices at the projected positions of the residues Glu71 and Lys147, define the external boundary and the central hole, respectively. They are related by a scaling transformation with factor $\mu_3 = 0.2469\ldots$, which obeys the same crystallographic restriction as in the star heptagon {7/3}.

in this case, the central hole is scaled with respect to the external heptagon by a factor $-\mu_1^2$ with $\mu_1 = \cos(2\pi/7)/\cos(\pi/7) = 0.6920\ldots$. The minus sign implies that the two heptagons are in reverse orientation.

Following these preliminary observations, in the next section, an outline is given of a crystallographic characterization of star heptagons. In the other two sections, the molecular forms of α-hemolysin and of poly d(As^{4}T) · poly d(As^{4}T) are analyzed further. Additional morphological features (like the elliptic holes recognizable in Fig. 3) are discussed on the basis of crystallographic linear scalings, which are dilations (or contractions) in a given direction only.

Crystallography of Star Heptagons

The symmetry of a regular heptagon is given by the 2-D point group $K_0 = 7mm$ generated by a rotation R of order 7 and a reflection m:

$$K_0 = \langle R, m \mid R^7 = m^2 = (Rm)^2 = 1 \rangle \quad (1)$$

The frame of general crystallography adopted here [1] requires a faithful integral representation Γ of K_0, whose minimal dimension, in the present case, is 6.

$$K_0 \simeq \Gamma(K_0) \subset GL(6, \mathbb{Z}) \quad (2)$$

Where $\simeq$ denotes group isomorphism. This representation can be obtained by applying R and m to six of the seven radial vectors $a_1, a_2, \ldots, a_7$ of the heptagon. The orientation chosen is with a_7 along the positive direction of the x axis, and with the other ak as basis a for the representation Γ: $a = \{a_1, a_2, \ldots, a_6\}$. This basis is linearly dependent in the real R, but linearly independent in the rational integers Z. The components of these vectors with respect to an orthonormal basis $e = \{e_1, e_2\}$, with e_1 and e_2 in the direction of the x and y axis, respectively, are given by:

$$a_k = a_0\left[\cos(k\varphi), \sin(k\varphi)\right], \quad \varphi = \frac{2\pi}{7} \quad \text{and} \quad k = 1, 2, \ldots, 6 \quad (3)$$

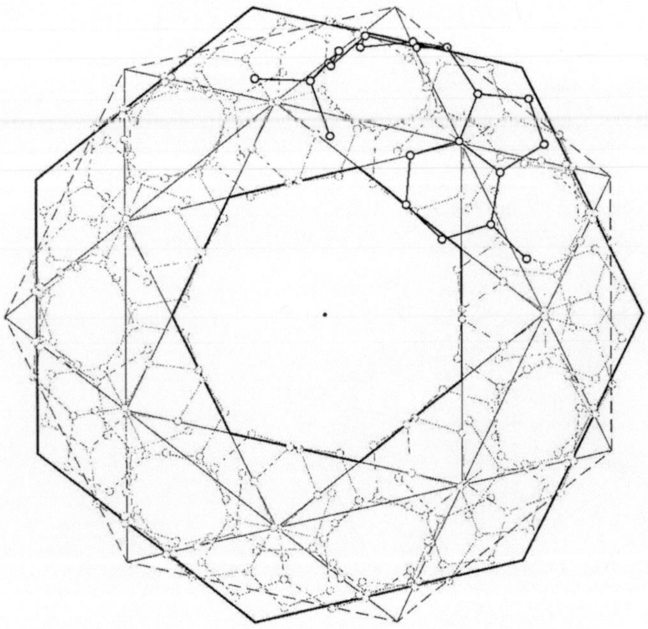

Fig. 3 Axial view of the A subsystem of the nucleic acid poly d(As[4]T)·poly d(As[4]T) in a Z-DNA conformation. As in the previous figure, the external heptagon (delimited by the sugar) and the central hole (due to the adenine) are in a scaling relation, which obeys a crystallographic restriction, as revealed by the star heptagons {7/2} inserted. The negative value of scaling factor $-\mu_1^2$, with $\mu_1=0.692\ldots$, implies that the central hole and the external boundary have a reverse orientation.

The representation matrices of the generators are then given by:

$$R(a)=\begin{pmatrix} 0 & 0 & 0 & 0 & 0 & -1 \\ 1 & 0 & 0 & 0 & 0 & -1 \\ 0 & 1 & 0 & 0 & 0 & -1 \\ 0 & 0 & 1 & 0 & 0 & -1 \\ 0 & 0 & 0 & 1 & 0 & -1 \\ 0 & 0 & 0 & 0 & 1 & -1 \end{pmatrix}$$
$$m(a)=\begin{pmatrix} 0 & 0 & 0 & 0 & 0 & 1 \\ 0 & 0 & 0 & 0 & 1 & 0 \\ 0 & 0 & 0 & 1 & 0 & 0 \\ 0 & 0 & 1 & 0 & 0 & 0 \\ 0 & 1 & 0 & 0 & 0 & 0 \\ 1 & 0 & 0 & 0 & 0 & 0 \end{pmatrix} \tag{4}$$

By considering $R(a)$ and $m(a)$ as orthogonal transformations R_s and m_s in a 6-D Euclidean space (the so-called *superspace*) one gets a formulation equivalent to a higherdimensional crystallography. Indeed, R_s and m_s leave invariant a

Table 1 Character table of the point group D_7 ($\varphi=2\pi/7$)

D_7	ε	α, α⁻¹	α², α⁻²	α³, α⁻³	βα^k
$\Gamma^{(1)}$	1	1	1	1	1
$\Gamma^{(2)}$	1	1	1	1	−1
$\Gamma^{(3)}$	2	$2\cos\varphi$	$2\cos 3\varphi$	$2\cos 3\varphi$	0
$\Gamma^{(4)}$	2	$2\cos 2\varphi$	$2\cos 3\varphi$	$2\cos\varphi$	0
$\Gamma^{(5)}$	2	$2\cos 3\varphi$	$2\cos\varphi$	$2\cos 2\varphi$	0

6-D lattice Σ, spanned in the superspace by a basis $a^s=\{a_1^s, a_2^s, \ldots, a_6^s\}$, with metric tensor

$$g_{ii}^s=6, \quad g_{ik}^s=-1 \quad \text{for} \quad i\neq k \tag{5}$$

left invariant by R_s and m_s. The projection $\pi: a_k^s \to a_k$ maps the lattice Σ into the Z-module $M=\langle a_1, a_2, \ldots, a_6 \simeq \mathbb{Z}^6\rangle$, of all integral linear combinations of these basis vectors. Accordingly, K_0 leaves M invariant: $K_0M=M$.

The relation between the two bases a and as can be derived from the decomposition of the representation Γ into irreducible components. Looking at the character table of $D_7 \simeq 7\,mm$ (Table 1) one finds:

$$\Gamma=\Gamma^{(3)}\oplus\Gamma^{(4)}\oplus\Gamma^{(5)} \tag{6}$$

where

$$\begin{gathered}\Gamma^{(3)}(\alpha)=R(e), \Gamma^{(4)}(\alpha)=R^2(e), \Gamma^{(5)}(\alpha)=R^3(e) \\ \Gamma^{(3)}(\beta)=\Gamma^{(4)}(\beta)=\Gamma^{(5)}(\beta)=m(e)\end{gathered} \tag{7}$$

with α and β the two generators of the abstract group D_7, as in Table 1, and e denoting the 2-D orthonormal basis introduced above. This implies the following embedding of the basis a in the 6-D basis as:

$$\begin{aligned} a_1^s&=(a_1, a_2, a_3) & a_4^s&=(a_4, a_1, a_5) \\ a_2^s&=(a_2, a_4, a_6) & a_5^s&=(a_5, a_3, a_1) \\ a_3^s&=(a_3, a_6, a_2) & a_6^s&=(a_6, a_5, a_4). \end{aligned} \tag{8}$$

Making use of the 2-D metrical relations between the elements of the basis a (with indices taken modulo 7):

$$a_i a_{i+k}=a_0^2\cos(k\varphi) \tag{9}$$

one verifies that the basis a^s form, indeed, the R_s-invariant metric tensor $(a_0^2/2)g^s$. The orthogonal projection π considered above from the superspace to the heptagonal plane and with corresponds to the homomorphism $\Gamma(K_0)\to\Gamma^{(3)}(K_0)$, is given by:

$$\pi\ a_k^s=a_k, \quad k=1,2,\ldots,6. \tag{10}$$

The idea on which a molecular crystallography approach is based is that there are possibly hidden molecular structural relations expressible as additional transformations leaving the Z-module M invariant. In the present case, this is indeed the case for the radial scaling symmetries of a self-similar star heptagon (obtained from a starting regular heptagon by successive direct and inverse star constructions).

In the case {7/2} the radial scaling transformation $S_{-\mu 1}$:

$$S_{-\mu 1}(a)=\begin{pmatrix} 2 & -2 & 1 & 0 & -1 & 1\\ 0 & 0 & -1 & 1 & -1 & 1\\ 2 & -2 & 1 & -1 & 0 & 1\\ 1 & 0 & -1 & 1 & -2 & 2\\ 1 & -1 & 1 & -1 & 0 & 0\\ 2 & -1 & 0 & 1 & -2 & 2 \end{pmatrix} \quad (11)$$

has a negative scaling factor $-\mu_1$ with

$$\mu_1=-1+2\cos\varphi-2\cos 2\varphi=0.692021\ldots,\quad \varphi=2\pi/7 \quad (12)$$

In the case {7/3}, there are two radial scaling transformations, with scaling factors indicated by $-\mu_2$ and μ_3, respectively

$$S_{-\mu_2}(a)=\begin{pmatrix} 1 & 1 & -2 & 0 & 2 & -1\\ 0 & 2 & -1 & -2 & 2 & 1\\ -1 & 1 & 0 & -1 & 0 & 1\\ 1 & 0 & -1 & 0 & 1 & -1\\ 1 & 2 & -2 & -1 & 2 & 0\\ -1 & 2 & 0 & -2 & 1 & 1 \end{pmatrix},$$
$$S_{\mu_3}(a)=\begin{pmatrix} -2 & 1 & 0 & 0 & 0 & -1\\ 0 & -1 & 1 & 0 & 0 & -1\\ -1 & 1 & -1 & 1 & 0 & -1\\ -1 & 0 & 1 & - & 1 & -1\\ -1 & 0 & 0 & 1 & -1 & 0\\ -1 & 0 & 0 & 0 & 1 & -2 \end{pmatrix} \quad (13)$$

where:

$$\mu_2=2\cos\varphi+4\cos 2\varphi=0.35689\ldots,$$

and

$$\mu_3=-1+2\cos\varphi=0.24697\ldots \quad (14)$$

These additional transformations allow the extension of the original Euclidean point group K_0 to infinite order, in a way still leaving the Z -module M invariant:

$$K_{\{7/2\}}=\langle R, m, S_{-\mu_1}\rangle,\quad K_{\{7/3\}}=\langle R, m, S_{-\mu_2}, S_{\mu_3}\rangle \quad (15)$$

Note the relations:

$$S_{-\mu}=-S_{\mu} \quad \text{and} \quad S_{-\mu_3}=S_{-\mu_1}S_{-\mu_2} \quad (16)$$

In a heptagram, however, the points at the intersection of the two different star polygons {7/2} and {7/3} have also to be considered for characterizing its self-similar symmetry. These points cannot be obtained by radial scaling from the heptagon considered but have, nevertheless, also integral coordinates (in the a basis). The reason is that they follow from integral linear scalings X_λ, Y_λ along the x and the y axis, respectively, whose product is a radial scaling radial scaling S_λ with the same scaling factor λ. This is a property that the radial scalings considered, so far, do not have. For fixing the ideas, consider a point P with coordinates $P(e)=(x, y)$. Then, for these linear scaling transformations one has:

$$X_\lambda(x,y)=(\lambda x, y),\quad Y_\lambda(x,y)=(x,\lambda y),\quad S_\lambda(x,y)=\lambda(x,y) \quad (17)$$

In particular, there are two sets of linear transformations ensuring the equivalence of the additional points with the previous ones:

$$X_{\mu_4}(a)=\begin{pmatrix} -4 & 3 & -1 & -1 & 3 & -5\\ -1 & 1 & 0 & 0 & 0 & -1\\ -3 & 2 & 0 & -1 & 2 & -3\\ -3 & 2 & -1 & 0 & 2 & -3\\ -1 & 0 & 0 & 0 & 1 & -1\\ -5 & 3 & -1 & -1 & 3 & -4 \end{pmatrix},$$
$$Y_{\mu_4}(a)=\begin{pmatrix} 0 & 1 & -1 & 1 & -1 & 1\\ 1 & -1 & 2 & -2 & 2 & -1\\ -1 & 2 & -2 & 3 & -2 & 1\\ 1 & -2 & 3 & -2 & 2 & -1\\ -1 & 2 & -2 & 2 & -1 & 1\\ 1 & -1 & 1 & -1 & 1 & 0 \end{pmatrix} \quad (18)$$

$$X_{-\mu_5}(a)=\begin{pmatrix} 0 & -3 & 2 & 2 & -3 & -1\\ -1 & -7 & 3 & 3 & -5 & -1\\ 0 & -1 & 1 & 0 & -1 & 0\\ 0 & -1 & 0 & 1 & -1 & 0\\ -1 & -5 & 3 & 3 & -4 & -1\\ -1 & -3 & 2 & 2 & -3 & 0 \end{pmatrix},$$
$$Y_{-\mu_5}(a)=\begin{pmatrix} -2 & 1 & 2 & -2 & -1 & 3\\ 1 & 0 & -1 & 1 & 1 & -1\\ 2 & -1 & -1 & 2 & 1 & -2\\ -2 & 1 & 2 & -1 & -1 & 2\\ -1 & 1 & 1 & -1 & 0 & 1\\ 3 & -1 & -2 & 2 & 1 & -2 \end{pmatrix} \quad (19)$$

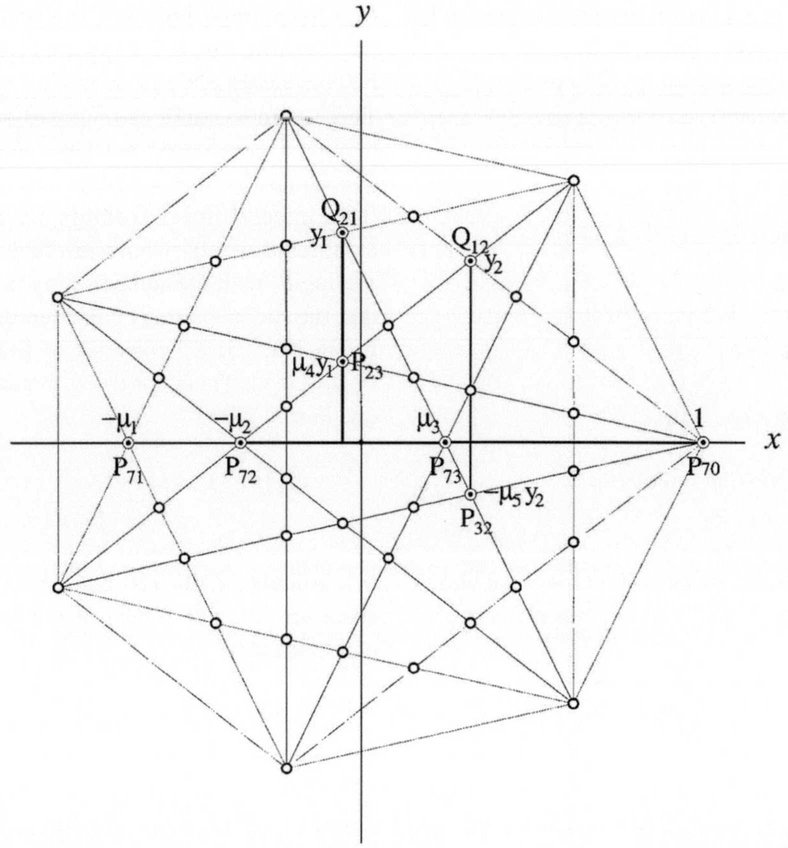

Fig. 4 Labeling convention of the intersections point in a heptagram consisting of the two star polygons {7/2} and {7/3} with examples of their mutual scaling relations. Thus, P_{71}, P_{72}, and P_{73} are obtained from P_{70} by radial scaling transformations S_{μ_i}, with scaling factors $-\mu_1$, $-\mu_2$, $-\mu_3$, respectively, whereas linear scalings Y_{μ_j} along the y axis and with factors μ_4 and $-\mu_5$, relate correspondingly Q_{21} with P_{23} and Q_{12} with P_{32}.

with scaling factors μ_4 and $-\mu_5$, where:

$$\begin{aligned}\mu_4 &= -3+4\cos\varphi-4\cos 2\varphi = 0.38404\ldots,\\ \mu_5 &= 1-4\cos\varphi-8\cos 2\varphi = 0.28620\ldots\end{aligned} \tag{20}$$

Note that $X_{-\lambda} \neq -X_\lambda$ and so also for $Y_{-\lambda}$.

These results allow the expression of all the intersection points of a heptagram as integral linear combination of the basis vectors and to assign to each point a set of *integral indices*, which are the components of the corresponding position vectors expressed in the basis a:

$$P(a) = (n_1, n_2, \ldots, n_6)$$

with

$$P = \sum_{k=1}^{6} n_k a_k \tag{21}$$

The following conventions for labeling the heptagrammal points have been adopted:

$$P_{ik} = R^i P_{0k}, \quad P_{ih} = S_{\mu_h'} P_{i0}, \quad k,h = 1,2,3,$$
$$i = 1,2,\ldots,7$$

$$\begin{aligned}Q_{21} &= Y_{\mu_4}^{-1} P_{23}, \quad Q_{12} = Y_{-\mu_5}^{-1} P_{32},\\ Q_{ij} &= R^i Q_{oj}, \quad j = 1,2\end{aligned} \tag{22}$$

where μ_h' denotes $(\pm)\mu_h$ taken with the appropriate sign. A relation between scaling factors and indices is visualized in Fig. 4. The missing indices are easily computed by applying $R^k(a)$ to those indicated. Table 2 gives the result.

The intersection points Q_{ij} of a heptagram represent one way to arrive at crystallographic linear scalings, but this way does not reveal their characteristic structural role. In order to find out the typical aspect, consider how the vertices of a regular heptagon (chosen in the x orientation) transform under the combination of linear scalings to first order X_λ, Y_λ, and the sevenfold rotations of the point group K_0 as in the double cosets $K_0\, X_\lambda\, K_0$ and $K_0\, Y_\lambda\, K_0$. The image points of the a_k one gets, all are on circles with center C_k at the mean distance between a_k and $S_\lambda\, a_k$, for $S_\lambda = X_\lambda\, Y_\lambda$. While X_λ gives rise to an off-center heptagon in the same x orientation, Y_λ yields an heptagon in the reverse orientation $-x$. Figure 5 illustrates these

Table 2 Indexed intersection points of the heptagram (see Fig. 4)

{7/1}	{7/2}	{7/3}	
P_{10} 100000	P_{11} 202112	P_{12} $10\bar{1}11\bar{1}$	P_{13} $\bar{2}0\bar{1}\bar{1}\bar{1}\bar{1}$
P_{20} 010000	P_{21} $\bar{2}0\bar{2}0\bar{1}\bar{1}$	P_{22} 121022	P_{23} $1\bar{1}1000$
P_{30} 001000	P_{31} $1\bar{1}1\bar{1}10$	P_{32} 210120	P_{33} $01\bar{1}100$
P_{40} 000100	P_{41} $01\bar{1}1\bar{1}1$	P_{42} $0\bar{2}\bar{1}0\bar{1}\bar{2}$	P_{43} $001\bar{1}10$
P_{50} 000010	P_{51} $\bar{1}\bar{1}0\bar{2}0\bar{2}$	P_{52} 220121	P_{53} $0001\bar{1}1$
P_{60} 000001	P_{61} 211202	P_{62} $\bar{1}11\bar{1}01$	P_{63} $\bar{1}\bar{1}\bar{1}\bar{1}0\bar{2}$
P_{70} $\bar{1}\bar{1}\bar{1}\bar{1}\bar{1}\bar{1}$	P_{71} $\bar{2}0\bar{1}\bar{1}0\bar{2}$	P_{72} $\bar{1}\bar{2}00\bar{2}\bar{1}$	P_{73} 211112
{7/2} ∩ {7/3}			
Q_{11} $010\bar{1}10$	Q_{21} $0010\bar{1}1$	Q_{31} $\bar{1}\bar{1}\bar{1}0\bar{1}\bar{2}$	Q_{41} 211121
Q_{51} $\bar{1}10001$	Q_{61} $\bar{1}\bar{2}0\bar{1}\bar{1}\bar{1}$	Q_{71} $10\bar{1}100$	
Q_{12} $\bar{1}\bar{1}\bar{1}0\bar{2}1$	Q_{22} $10001\bar{1}$	Q_{32} 121112	Q_{42} $\bar{2}\bar{1}0\bar{1}\bar{1}\bar{1}$
Q_{52} $1\bar{1}0100$	Q_{62} $01\bar{1}010$	Q_{72} $001\bar{1}01$	

properties for the case $\lambda = \mu_4^{-1} = 2.60388\ldots$ given above. The presence of off-center circles in the structure of a given molecular form strongly suggests the possible presence of linear scaling.

α-Hemolysin

α-Hemolysin is a heptameric transmembrane pore protein. The data for the atomic coordinates have been taken from PDB 7AHL by Song et al. [10]. The structure consists of seven chains labeled *A* to *G*. Before projecting the structure along the axis, the original data have been oriented and centered using the positions of the residue Glu71 in the chains *A*, *D*, *F*. The corresponding prismatic molecular form has already been presented in the introduction (Fig. 2).

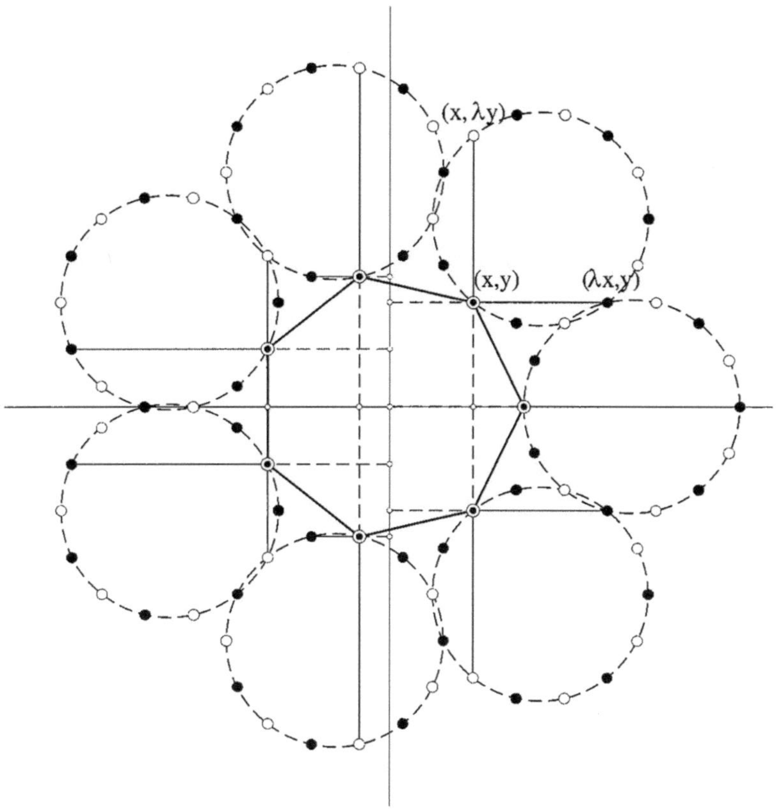

Fig. 5 Off-center circles obtained by applying linear scalings with given scaling factor to the vertices of a regular heptagon. Indicated is how the (x, y) coordinates of these points are transformed by X_λ and Y_λ to $(\lambda x, y)$ and $(x, \lambda y)$, respectively. The remaining points follow from the sevenfold rotational symmetry (leading to conjugated linear scalings $R^k X_1 R^{-k}$ and $R^k Y_1 R^{-k}$). Here, the scaling factor $1 = m_4^{-1} = 2.60388\ldots$ ensures that the matrices $X_\lambda(a)$ and $Y_\lambda(a)$ are integral, so that all these points can be indexed by integers.

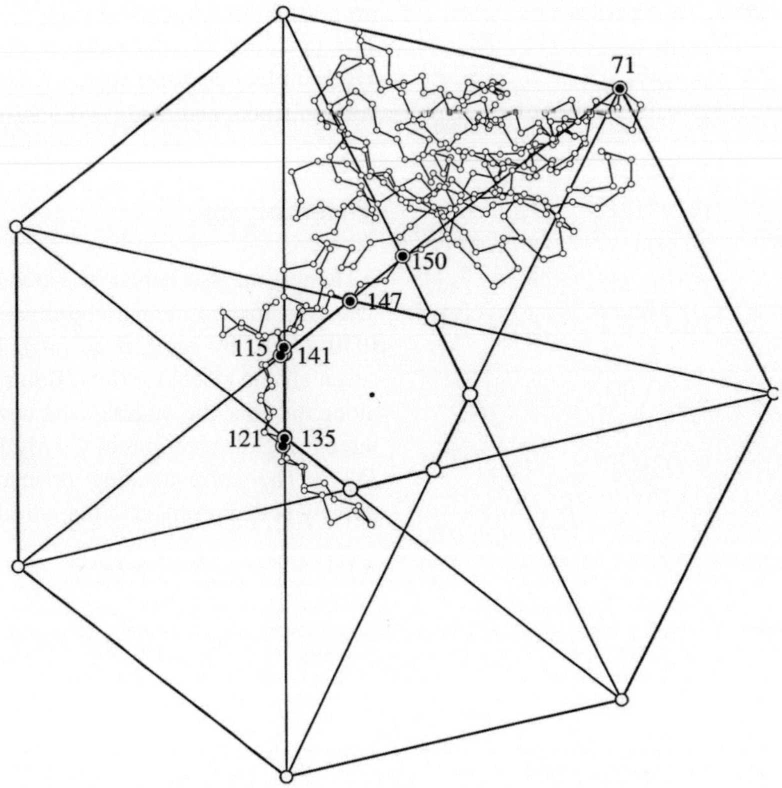

Fig. 6 A protomer of α-hemolysin is shown in a view along the central axis of the heptamer and in relation with the star heptagon {7/3} fitted at the position of the residue 71. The projected positions of a number of other residues (150, 147, 141, 135, 121, and 115) at points of the star heptagon are an indication that the protomer folds in a way predisposed not only to the sevenfold rotational symmetry (in the stem domain), but also to crystallographic scale rotations, which relate the stem domain with the cap domain, in a way still requiring further investigation.

The projection of a single chain shows that the scalerotation symmetry of the star polygon {7/3}, which characterizes the molecular form of the heptamer, is already predisposed in the folding of the monomer (Fig. 6). One sees that, in addition to the residues 71 and 147 indicated in Fig. 2, several more (150, 141, 135, 121, and 115) also fit fairly well with the intersection points of the star heptagon. The corresponding scale-rotation transformations, however, do not connect atomic positions, but only vertical lines (parallel to the central axis) through these positions. This does not exclude the possibility of a 3-D point group relating these same positions. In the spirit of molecular crystallography, one even expects the existence of such a group, which would explain the molecular form observed, but this investigation has not yet been carried out. From the molecular form one finds some evidence for additional linear scaling, but not strong enough to be presented here.

Poly *d*(As^4T) · poly *d*(As^4T)

The left-handed double helix formed by the two strands of poly d(As^4T) · poly d(As^4T) is characterized by a dinucleotide repeat with line group symmetry $7_6 22$ of which the details have been reported in *Nature* [11]. The corresponding data are listed as structure No. 16 in a survey by Chandrasekaran and Arnott, published in a volume of the Landolt-Börstein collection [12].

It is convenient to first consider the molecular form of each nucleotide subsystem separately. The form of the A-subsystem has already been presented in the introduction (Fig. 3), that of the s^4T subsystem is shown in Fig. 7. Both share the absolute value of scaling factor $\mu_1 = 0.6920\ldots$ of the star heptagon {7/2}, but not always the corresponding negative sign. This implies that the successively scaled heptagons may have the same or the reverse orientation. While in the case of adenine the central hole is already reached at the second contraction

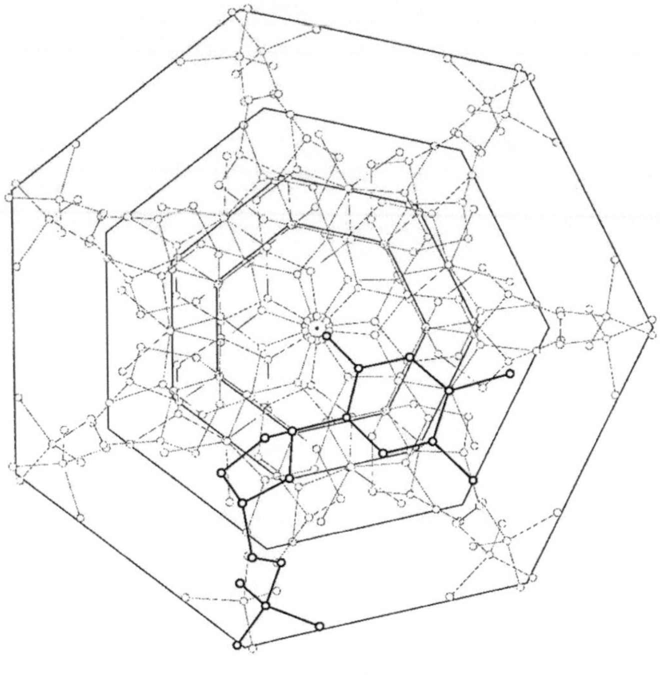

Fig. 7 Prismatic molecular form of the s^4T subsystem in poly d(As^4T) · poly d(As^4T) viewed along the helical axis. Shown are successive internal boundaries scaled from the external heptagon by μ_1, μ_1^2, μ_1^3, respectively, where $\mu_1 = 0.6920\ldots$ is the same factor on which the star heptagon {7/2} is based and which also plays a role in the A subsystem (Fig. 3). In the present case, the central hole and the external boundary are not related by a scaling factor obeying crystallographic restrictions. (See also Fig. 10).

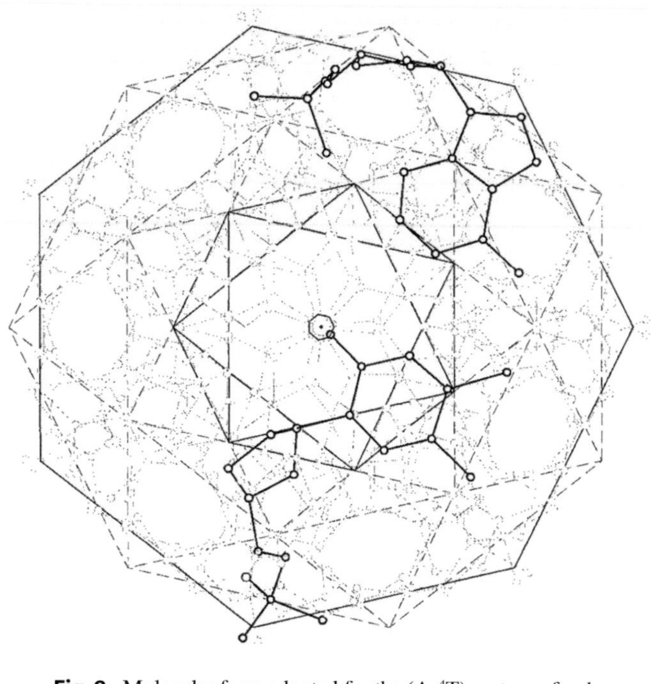

Fig. 8 Molecular form adopted for the (As^4T) system of poly d(As^4T) · poly d(As^4T), with the sugar of the A nucleotide delimiting the external boundary, leaving outside oxygens of the phosphate of the s^4T nucleotide. Despite this mismatch, the overall structure is still compatible with internal prismatic boundaries base on {7/2} star polygons.

level, that of the thymine is not simply scaled by a power of μ_1. The heptagons scaled by μ_1, μ_1^2, μ_1^3, show, however, the expected compatibility with morphological features.

This general behavior is confirmed when combining the two subsystems in the molecular form of the dinucleotide shown in Fig. 8. As the external boundaries of the two subsystems are slightly different in size, the combination requires an adaptation. The one of the A subsystem gives the best fitting for the whole, despite the fact that two of the phosphate oxygens of the s^4T nucleotide are no more inside the heptagon of the external boundary. The morphological importance of the star heptagon {7/2} is confirmed.

Looking at Fig. 8, one recognizes elliptically shaped holes tangent to the external heptagon at midedge positions. As already pointed out in the introduction, this indicates the possible presence of linear scalings which, as explained in the second, can give rise to off-center circular structures. In the present case, the natural candidates are the two linear scaling transformations X_{μ_1}, Y_{μ_1} whose product gives the radial scaling S_{μ_1}. These linear scalings have not been considered so far because they not integral in the heptagonal basis a. However, the appearance of midedge heptagonal

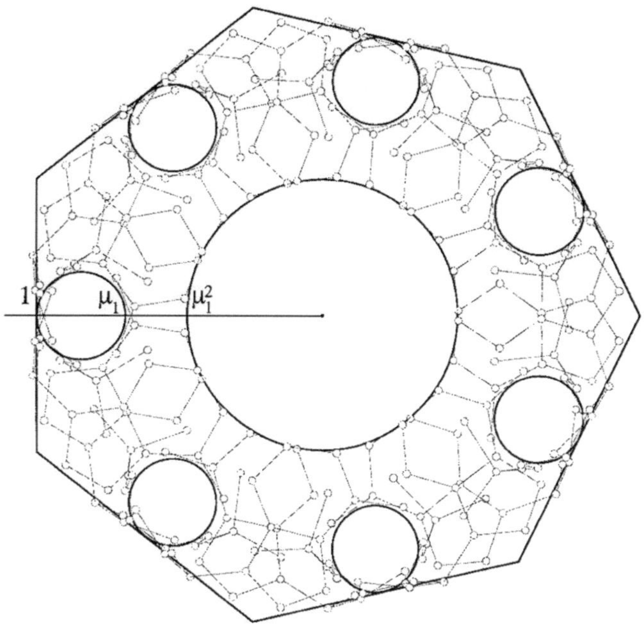

Fig. 9 The elliptical holes visible in the molecular form of the A subsystem (as shown in Fig. 3) are obtained as off-center circles from linear scalings X_{μ_1}, Y_{μ_1} (and their *R*-conjugated ones) applied to the midedges of the external heptagon, in the same way as illustrate in Fig. 5, but now for $\mu_1 = 0.692\ldots$ One sees that the central hole then follows by a further radial scaling transformation S_{μ_1}, and this is unexpected.

points with half-integer indices, suggests a corresponding centering of the 6-D lattice, leading to half-integers in the entries of the linear scaling when expressed in what one could denote as the conventional basis a. One finds indeed:

$$X_{\mu_1}(a) = \begin{pmatrix} -3/2 & 3/2 & -1/2 & -1/2 & 3/2 & -5/2 \\ -1/2 & 1 & 0 & 0 & 0 & -1/2 \\ -3/2 & 1 & 1/2 & -1/2 & 1 & -3/2 \\ -3/2 & 1 & -1/2 & 1/2 & 1 & -3/2 \\ -1/2 & 0 & 0 & 0 & 1 & -1/2 \\ -5/2 & 3/2 & -1/2 & -1/2 & 3/2 & -3/2 \end{pmatrix} \quad (23)$$

$$Y_{\mu_1}(a) = \begin{pmatrix} 1/2 & 1/2 & -1/2 & 1/2 & -1/2 & 1/2 \\ 1/2 & 0 & 1 & -1 & 1 & -1/2 \\ -1/2 & 1 & -1/2 & 3/2 & -1 & 1/2 \\ 1/2 & -1 & 3/2 & -1/2 & 1 & -1/2 \\ -1/2 & 1 & -1 & 1 & 0 & 1/2 \\ 1/2 & -1/2 & 1/2 & -1/2 & 1/2 & 1/2 \end{pmatrix} \quad (24)$$

with positive scaling factor $\mu_1 = 0.6920\ldots$. The off-center holes of the A subsystem are fairly well approximates by applying these linear scaling transformations to the mid-edges of the external heptagon, as shown in Fig. 9. Moreover, the central hole can also be described by a circle, which is a factor μ_1^2 smaller than the one inscribed in the heptagon. One could also start from the central hole and arrive at the off-center hole observed by applying the inverse radial scaling $S_{\mu_1}^{-1}$ first, and then the corresponding linear dilation $X_{\mu_1}^{-1}$, $Y_{\mu_1}^{-1}$ in combination with the appropriate sevenfold rotations.

This last interpretation opens the door to a corresponding characterization of the molecular form of the s⁴T subsystem, as represented in Fig. 10. Starting from the heptagon of the central hole with vertices at a mean O2 position of diadically related thymine molecules, one gets by a radial scaling $S_{\mu_1}^{-4}$ the heptagon formed by the corresponding mean C1 positions, and by a further $S_{\mu_1}^{-3}$ transformation those corresponding to the C6 ones. From this last heptagon, one obtains the circular off-center holes by the linear transformations $X_{\mu_1}^{-2}$, $Y_{\mu_1}^{-2}$ squares of the dilations given above. These off-center circles are thus related to the central hole by elements X_{μ_1}, Y_{μ_1}, R, and m of the point group, but no simple relation could be found with the heptagon delimiting the molecular form. The reason is perhaps that this heptagon does not represent the relevant molecular envelop for the s⁴T subsystem. This would explain why in this case there is no crystallographic scaling transformation relating the central hole and the external heptagon, as it is the case for the A subsystem.

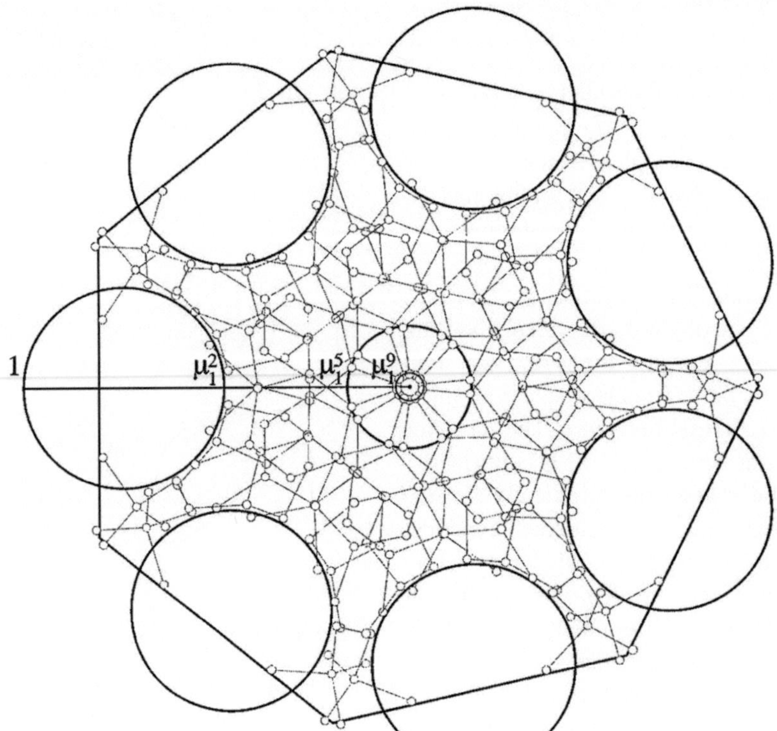

Fig. 10 The same linear scalings as in Fig. 9 allow to find a better envelop for the s⁴T subsystem than the regular heptagon. Now successive radial scalings $S_{\mu_1}^3$ and $S_{\mu_1}^4$ relate this curved boundary to the internal circular ones through the C1 and the O2 positions of the thymine, respectively. The latter positions define the central hole.

Acknowledgments Thanks are expressed to R. Chandrasekaran for kindly having made available to the author the structural data of poly d(As^4T)·poly d(As^4T) reported in Landolt-Börstein; to R. de Gelder and to B. Souvignier for pertinent remarks and valuable suggestions. Stimulating discussions with C. W. Hilbers are gratefully acknowledged. The graphical program XPS used has been originally designed by J. M. Thijssen.

References

1. Janner, A. *Acta Crystallogr.* **2001,** *A57*, 378.
2. Nischke, K.-P.; Danzer, L. *Discrete Comp. Geom.* **1996,** *15*, 221. (See also Materialien/Preprints LXXIX, University of Bielefeld, 1994.)
3. Barache, D. Ph. D. Thesis, University of Paris 7, 1995.
4. Gazeau, J. P. In *Symmetry and Structural Properties of Condensed Matter*; Lulek, T.; Florek, W.; Walcerz, S., Eds.; World Scientific: Singapore, 1995; p. 369.
5. Janner, A. *Crystal Eng.* **2001,** *4*, 119.
6. Janner, A. In *GROUP21 Physical Applications and Mathematical Aspects of Geometry, Groups and Algebras.*; Doebner, H.-D.; Scherer, W.; Schulte, C., Eds.; World Scientific: Singapore, 1997; p. 949.
7. Hendrickson, Wayne A.; Wüthrich, Kurt. *Macromolecular Structures 1997. Atomic Structures of Biological Macromolecules Reported during 1996*; Current Biology Ltd.: London, 1997.
8. Arnott, S. In *Oxford Handbook of Nucleic Acid Structure*; Neidle S., Ed.; Oxford University Press: Oxford, 1998; p. 1.
9. Coxeter, H. S. M. *Introduction to Geometry*; Wiley: New York, 1961.
10. Song, L.; Hobaugh, M. R.; Shustak, C.; Cheley, S.; Bayley, H.; Gouaux, J. E. *Science* **1996,** *274*, 1859.
11. Arnott, S.; Chandrasekaran, R.; Birdsall, D. L.; Leslie, A. G. W.; Ratliff, R. L. *Nature* (*London*) **1980,** *283*, 743.
12. Chandrasekaran, R.; Arnott, S. In *Landolt-Börstein Numerical Data and Functional Relationships in Science and Technology* (*Group VII, Biophysics*); Springer: Berlin, 1989; Vol. VII, 1b, p. 31.

Helical Structures: The Geometry of Protein Helices and Nanotubes[a,b]

Eric A. Lord[c]

Abstract

In nature, helical structures arise when identical structural subunits combine sequentially, the orientational and translational relation between each unit and its predecessor remaining constant. A helical structure is thus generated by the repeated action of a screw transformation acting on a subunit. A plane hexagonal lattice wrapped round a cylinder provides a useful starting point for describing the helical conformations of protein molecules, for investigating the geometrical properties of carbon nanotubes, and for certain types of dense packings of equal spheres.

Keywords

Helices • Hexagonal lattice • Nanotubes • Polyhedra

Introduction

An infinite strip of a tiling of the Euclidean plane by equilateral triangles, bounded by two parallel lines, can be wrapped around a circular cylinder so that the two strip edges meet. We shall refer to the resulting structure as a *cylindrical hexagonal lattice* or, briefly, a *CHL*. Alternatively, instead of rolling the strip around a cylinder, corresponding points on the edges may be brought into coincidence by folding along the fundamental lattice lines, keeping the triangular facets flat. We shall refer to the resulting structure as a *triangulated helical polyhedron* or, briefly, a *THP*. A THP is an "almost regular" polyhedron, in that its symmetry group, a rod group, acts transitively on the vertices and faces, although not on the edges.

The geometrical properties of the THPs are of relevance in structural chemistry for several reasons. As Sadoc and Rivier [1] have shown, the helical structures commonly occurring in proteins are metrically quite close to polygonal helices consisting of edges of THPs. The rodlike sphere packings investigated by Boerdijk [2] are derived from the Coxeter helix [3–5], which is the simplest THP. In a nanotube, the atomic positions correspond to a subset of the vertices of a THP.

The purpose of this work is twofold: to derive expressions for the metrical and topological parameters of the triangulated helical polyhedra and to indicate their importance by means of examples from the literature on protein helices, sphere packings, and nanotubes.

Nomenclature

A THP, or a CHL, is determined by the translation vector separating pairs of points on the strip boundaries that are to be identified. Let the components of this vector, referred to the underlying planar hexagonal coordinate system, be (m, n). The 6m symmetry of the plane hexagonal lattice implies that the integer pairs (m, n), $(-l, m)$, $(-n, -l)$, $(-m, -n)$, $(l, -m)$, and (n, l) all give rise to the *same* structure and (n, m), (l, n), $(-m, l)$, $(-n, -m)$, $(-l, n)$, and $(m, -l)$ refer to its mirror

[a] *Structural Chemistry* 2002, 13(3/4):305–314.

[b] This contribution is part of a collection titled Generalized Crystallography and dedicated to the 75th anniversary of Professor Alan L. Mackay, FRS.

[c] Department of Metallurgy, Indian Institute of Science, Bangalore 560012, India

I. Hargittai, B. Hargittai (eds.), *Science of Crystal Structures: Highlights in Crystallography*,
DOI 10.1007/978-3-319-19827-9_8

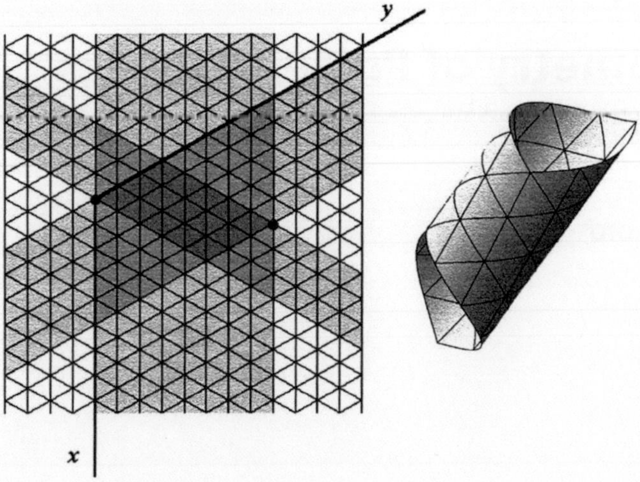

Fig. 1 Construction of the CHL (3, 5, 8).

image (where $l=n-m$.) The redundancy in the notation is eliminated by the requirement

$$0 \le l \le m \le n = l + m. \tag{1}$$

Thus each kind of THP (or CHL) may be denoted by a *unique* symbol (l, m, n). (See, for example, Sadoc and Rivier [1].)

Except for the special cases $(0, m, m)$ and $(m, m\ 2m)$, the CHLs are chiral: $(l, m, n)R$ has, by convention, *right-handed* $\{l\}$ and $\{n\}$ type helices and the $\{m\}$ type is left-handed. The mirror image of $(l, m, n)R$ is, of course, denoted by $(l, m, n)L$. Figure 1 illustrates the case (3, 5, 8). The direction of the strip edges can be arbitrary. Three special choices are indicated on the figure: strips with widths 3, 5, and 8 of the elementary equilateral triangles, along the directions [1, 1], [1, 0], and [0, 1], respectively. The corresponding lines of lattice points become helices on the cylinder: Three helices of type $\{3\}$, five of type $\{5\}$, and eight of type $\{8\}$. Figure 2 illustrates a portion of the THP of type $(3, 5, 8)L$.

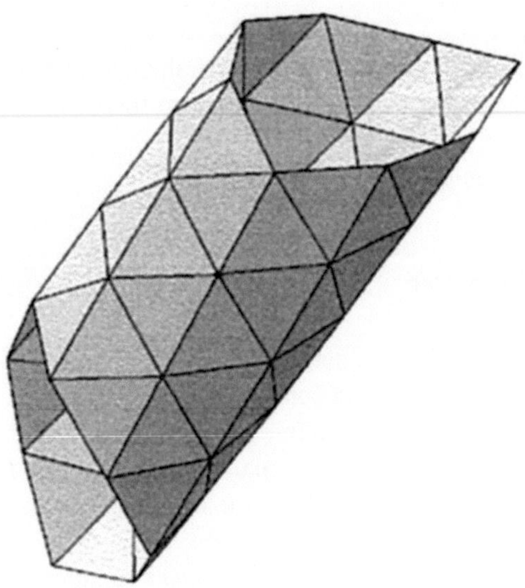

Fig. 2 The triangulated helical polyhedron $(3, 5, 8)L$.

Metrical Parameters of a CHL

A vertex of a plane hexagonal lattice can be identified by two integers (μ, ν)—its coordinates in terms of the hexagonal coordinates. In a CHL, the point (μ, ν) acquires a position on the cylinder, with three-dimensional (3-D) cylindrical coordinates (ρ, φ, z), which can be found as follows. On the plane hexagonal lattice the angle α between the directions (μ, ν) and (m, n) is given by

$$\begin{aligned} \cos\alpha &= \{(m-n/2)\mu + (n-m/2)\nu\}/\lambda k, \\ \sin\alpha &= \sqrt{3}(n\mu - m\nu)/2\lambda k \end{aligned} \tag{2}$$

where $\lambda=\sqrt{(\mu^2+\nu^2-\mu\nu)}$, $k=\sqrt{(m^2+n^2-mn)}$. Then the coordinates of the point (μ, ν) in 3-D Euclidean space are

$$\begin{pmatrix} \rho \\ \varphi \\ z \end{pmatrix} = \begin{pmatrix} k/2\pi \\ (\lambda cos\alpha)/\rho \\ \lambda\sin\alpha \end{pmatrix} = \begin{pmatrix} k/2\pi \\ 2\pi\{(m-n/2)\mu+(n-m/2)\nu\}/k^2 \\ \sqrt{3}(n\mu - m\nu)/k \end{pmatrix} \tag{3}$$

The $\{l\}$, $\{m\}$, and $\{n\}$ type helices are, respectively, along $[-1, -1]$, $[0, 1]$, and $[-1, 0]$, so that their *rotational advances per edge* are

$$\pi(-m-n)/k^2, \quad \pi(n+1)/k^2, \quad \pi(1-m)/k^2 \tag{4}$$

and the *translational advances* are

$$\sqrt{3}(l/k), \quad \sqrt{3}(m/k), \quad \sqrt{3}(n/k) \tag{5}$$

Metrical Parameters of a THP

Sadoc and Rivier [1] have identified several wellknown helices occurring in protein structure with type $\{1\}$ helices of THPs of the form $(1, m, m+1)$:

$$\begin{aligned} m &= 2: \text{the } 3_{10} \text{ helix} \\ m &= 3: \text{the } \alpha \text{ helix} \\ m &= 4: \text{the } \pi \text{ helix} \\ m &= 5: \text{the } \gamma \text{ helix.} \end{aligned}$$

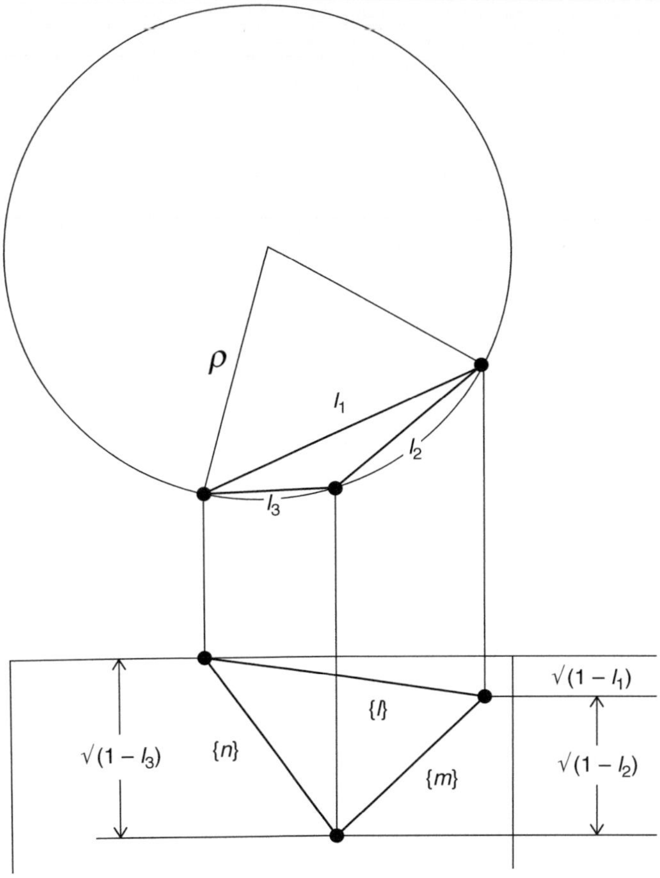

Fig. 3 Projection of a triangular facet of a THP, along the axis, and perpendicular to the axis.

See, for example, Lehninger et al. [6] for the structure and nomenclature of the corresponding protein configurations. In these models, the type {1} polygonal helix represents the polypeptide chain. Other edges of the TPH correspond to the linking hydrogen bonds that are responsible for the helical configuration.

The method of Sadoc and Rivier for computing the metrical parameters of TPHs of the special kind $(1, m, m+1)$ is as follows. Let … A_{-2} A_{-1} A_0 A_1 A_2 A_3 … denote the sequence of successive vertices of the type {1} helix of the TPH $(1, m, m+1)$ and observe that $A_0A_mA_{m+1}$ is a face—an equilateral triangle. Choose Cartesian coordinates so that A_m has coordinates ρ (cos $m\varphi$, sin $m\varphi$, mc) and equate the three edge lengths A_0A_m, A_0 A_{m+1} and $A_mA_{m+1}=A_0A_1$. This gives

$$\begin{aligned} c^2 - 2\cos\varphi &= m^2c^2 - 2\cos m\varphi \\ &= (m+1)^2 c^2 - 2\cos(m+1)\varphi. \end{aligned} \tag{6}$$

Eliminating c^2 and expressing cos$m\varphi$ and cos$(m+1)\varphi$ as polynomials in $\chi=\cos\varphi$ gives a polynomial in χ of degree $m+1$. It can be shown that this polynomial has a factor $(\chi-1)^2$, so we finally arrive at a polynomial in χ of order $m-1$. Only one root (in fact, the largest root) is relevant; the other roots correspond to triangulated surfaces with self-intersections.

Consider now more general cases (l, m, n). For large vectors (m, n) it is obvious that the formulas for the parameters of a CHL will give reasonable approximations to those for the corresponding THP. For smaller values we resort to Fig. 3, which represents the projection of a THP on planes perpendicular and parallel to its axis, of the edges of a constituent equilateral triangle. To find the radius ρ, we start from the formula for the circumradius of a triangle with edges l_1, l_2, and l_3

$$\frac{1}{\rho^2} = 2\left(\frac{1}{l_1^2}+\frac{1}{l_2^2}+\frac{1}{l_3^2}\right)-\left(\frac{l_1^2}{l_2^2l_3^2}+\frac{l_2^2}{l_3^2l_1^2}+\frac{l_3^2}{l_1^2l_2^2}\right) \tag{7}$$

From Fig. 3, we see that

$$\sqrt{\left(1-l_1^2\right)}+\sqrt{\left(1-l_2^2\right)}=\sqrt{\left(1-l_3^2\right)} \tag{8}$$

and

$$m\sqrt{\left(1-l_3^2\right)}=n\sqrt{\left(1-l_2^2\right)} \tag{9}$$

so that l_2 and l_3 can be expressed as functions of l_1:

$$\begin{aligned} 1-l_2^2 &= \left(1-l_1^2\right)(\mathrm{n}/l)^2, \\ 1-l_3^2 &= \left(1-l_1^2\right)(\mathrm{m}/l)^2 \end{aligned} \tag{10}$$

Thus, we get the radius ρ as a function of l_1^2. Now observe that m steps of an $\{n\}$ helix followed by n steps of an $\{m\}$ helix brings one back to the starting vertex. The path has rotational advance 2π. From the figure, we see that the rotational advance per edge length is $2\sin^{-1}(l_2/2\rho)$ for the type $\{m\}$ helices and $2\sin^{-1}(l_3/2\rho)$ for the type $\{n\}$ helices. This gives

$$m\sin^{-1}(l_3/2\rho)+n\sin^{-1}(l_2/2\rho)=\pi \tag{11}$$

leading to a quite formidable transcendental equation in l_1^2. The relevant root is the smallest root. A suitable starting value for finding this root by successive approximation can be taken to be the l_1^2 for the corresponding CHL, $(m+n)^2/4(m^2+n^2-mn)$. Table 1 gives the metrical parameters obtained in this way, using Mathematica, for a few small values of (m, n). The final column is the radius ρ_{CHL} of the corresponding CHL, for comparison:

$$\rho_{\mathrm{CHL}} = \frac{k}{2\pi} = \frac{\sqrt{m^2+n^2-mn}}{2\pi}. \tag{12}$$

Table 1 Parameters for THPs for Small Values of (l, m, n)

$l\ m\ n$	l_1	ρ	ρ_{CHL}
1 2 3	0.94868	0.51962	0.42108
2 2 4	0.86603	0.61237	0.55133
1 3 4	0.97207	0.64526	0.57384
1 3 5	0.91821	0.74313	0.69374
1 4 5	0.98255	0.78561	0.72934
2 4 6	0.94547	0.88462	0.84217
3 4 7	0.90435	1.00188	0.96810
1 5 6	0.98811	0.93258	0.88614
2 5 7	0.96118	1.03166	0.99392
3 5 8	0.92881	1.14441	1.11408
4 5 9	0.89632	1.26887	1.24304
1 6 7	0.99139	1.08319	1.04365
2 6 8	0.97100	1.18076	1.14768

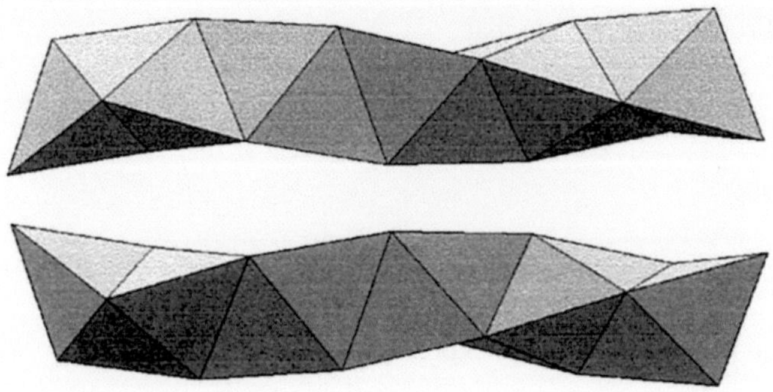

Fig. 4 The Boerdijk–Coxeter structures (1, 2, 3) *R* and (1, 2, 3)*L*.

The corresponding values for the *rotational* and *translational advance* per edge of the polygonal helices of types $\{l\}$, $\{m\}$ and $\{n\}$ are

$$\varphi_1, \varphi_2, \varphi_3 = 2\sin^{-1}(l_1/2\rho),\ \ 2\sin^{-1}(l_2/2\rho),\ \ 2\sin^{-1}(l_3/2\rho) \tag{13}$$

$$z_1, z_2, z_3, = \sqrt{(1-l_1^2)}, \sqrt{(1-l_2^2)}, \sqrt{(1-l_3^2)} \tag{14}$$

Cases (0, *m*, *m*) and (*m*, *m*, 2*m*) have not been listed in the table because they are given, respectively, by the simple formulas

$$\rho = 1/2\sin(\pi/m) \quad \text{and} \quad \rho = \sqrt{3/4}\sin(\pi/2m) \tag{15}$$

The Boerdijk–Coxeter Structure

The simplest nontrivial THP is (1, 2, 3) shown in Fig. 4. It is the surface of a stack of regular tetrahedra. We shall refer to it as the Boerdijk–Coxeter structure, or, simply, the "B-C structure." The vertices, labeled … $A_{-2}\ A_{-1}\ A_0\ A_1\ A_2\ A_3$ … consecutively along the type {1} helix are such that *any four consecutive points* are the vertices of *a regular tetrahedron* [3–5]. Buckminster Fuller [7] named this helical structure "the tetrahelix." With unit edge length, five consecutive points on the type {1} polygonal helix can be taken to be

$$(-1,-1,-1)/\sqrt{8},\ \ (-1,1,1)/\sqrt{8},$$
$$(1,-1,1)/\sqrt{8},\ \ (1,1,-1)/\sqrt{8},$$
$$(5/3,5/3,5/3)\sqrt{8}.$$

The action of a *screw transformation* $\mathbf{x} \to \mathrm{R}\mathbf{x} + \mathbf{a}$ generates the whole structure, where

$$R = \frac{1}{3}\begin{pmatrix} 2 & 2 & 1 \\ 2 & -1 & -2 \\ -1 & 2 & -2 \end{pmatrix}, \quad \boldsymbol{a} = \begin{pmatrix} 1/3\sqrt{2} \\ 1/3\sqrt{2} \\ 1/3\sqrt{2} \end{pmatrix} \tag{16}$$

The *axis direction* **n** and the *rotational advance per edge* φ (of the type {1} polygonal helix) can be extracted from the formula for rotation through an angle φ about an axis along a unit vector **n**:

$$R = I + N\sin\varphi + N^2(1-cos\varphi)$$
$$N = \begin{pmatrix} 0 & -n_3 & n_2 \\ n_3 & 0 & -n_1 \\ -n_2 & n_1 & 0 \end{pmatrix} \tag{17}$$

The translational advance **d** is the scalar product of **n** with a tetrahedron edge. The result is

$$\mathrm{n} = (2,1,0)/\sqrt{5}, \qquad \varphi = \cos^{-1}(-2/3),$$
$$d = 1/\sqrt{10}. \tag{18}$$

The radius is got by substituting $l_1 = \sqrt{(9/l_0)}$, $l_2 = \sqrt{(6/l_0)}$, $l_3 = \sqrt{(1/l_0)}$ into the formula for ρ:

$$\rho = (3\sqrt{3})/10. \tag{19}$$

Sphere Packings

Boerdijk [2] investigated the Coxeter structure [the THP (1, 2, 3)] in connection with dense packings of equal spheres. The configuration of four spheres in a tetrahedral configuration, each touching other three, gives the Rogers upper bound for the upper limit of any possible packing fraction for equal spheres. The bound can never be achieved because regular tetrahedra will not pack together in Euclidean. However, sphere packings that fill only a portion of space can come much closer to the bound than hexagonal close packing—the

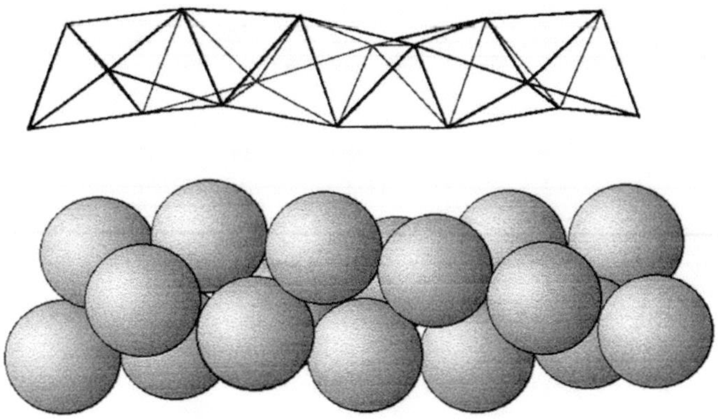

Fig. 5 The dense packing of spheres centered at the vertices of a B-C helix.

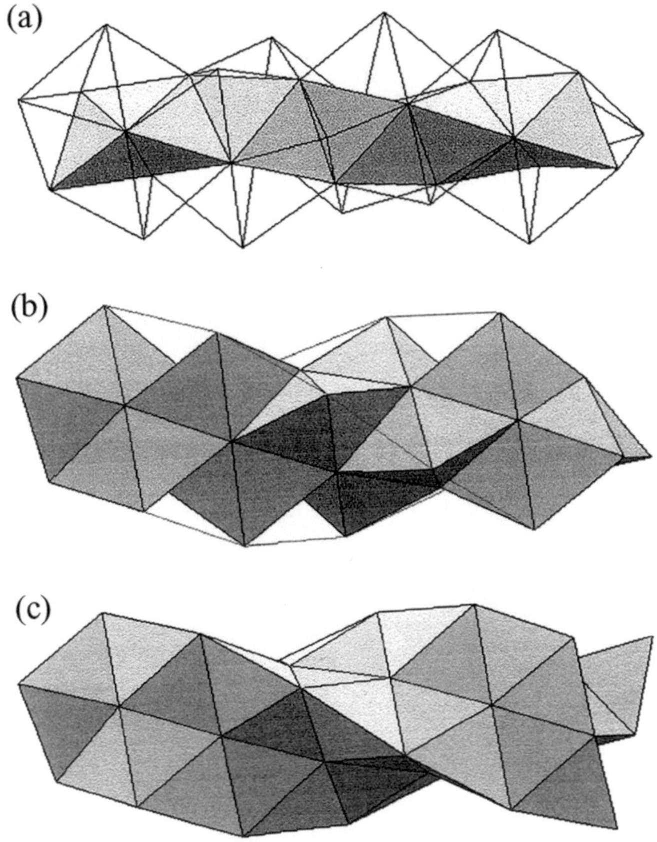

Fig. 6 Extension of the sphere packing. Sphere centers are at vertices of the tetrahedra. The additional tetrahedra are very nearly regular.

densest lattice packing. Boerdijk considered the dense rod-shaped packing in which the sphere centers lie on the vertices of a (1, 2, 3) (Fig. 5).

Extension of the Helical Sphere Packing

As pointed out by Boerdijk, the packing of spheres centred at the vertices of (1, 2, 3) can be extended by adding more spheres over the midpoints of edges of type {1} helices.

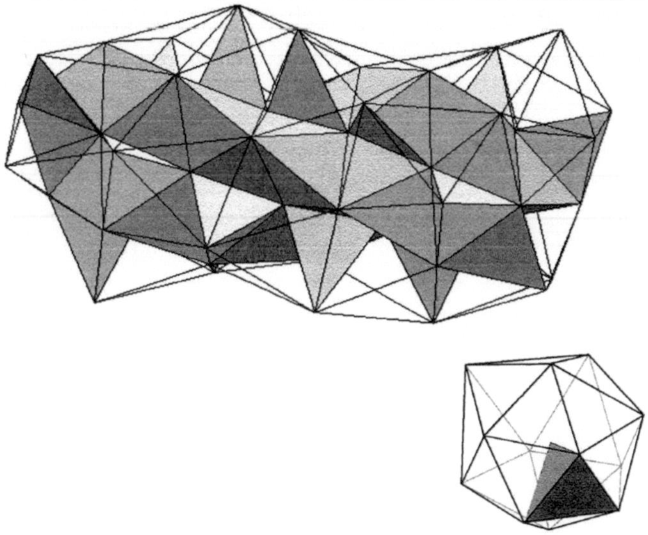

Fig. 7 Further extension of Boerdijk's sphere packing, represented by tetrahedra whose vertices are the sphere centers. Every sphere of the original configuration (Fig. 5) is now the center of a 13-sphere cluster.

Specifically, an extra vertex is placed over the midpoint of each edge of the type {1} helix, forming an equilateral triangle with the two vertices of that edge. This determines additional, only slightly irregular, tetrahedra, so that every edge of the type {1} helix is shared by five tetrahedra (Fig. 6)

Further extensions of the Boerdijk–Coxeter structure are possible. The next stage of adding spheres gives a rodlike structure in which every vertex of the original (1, 2, 3) is surrounded by 12 others, configured as a somewhat distorted *icosahedron*, as shown in Fig. 7. Thus each tetrahedron of the initial (1, 2, 3) is now shared by four icosahedra. This 26-sphere cluster is a slightly distorted form of the 26-atom γ-brass cluster. Another interesting subset of the tetrahedra in this structure is the triplet of distorted B-C helices twisted around each other (Fig. 8). One could go on adding more spheres, but the deviation of the tetrahedra from regularity (corresponding to lower sphere packing fraction) becomes severe.

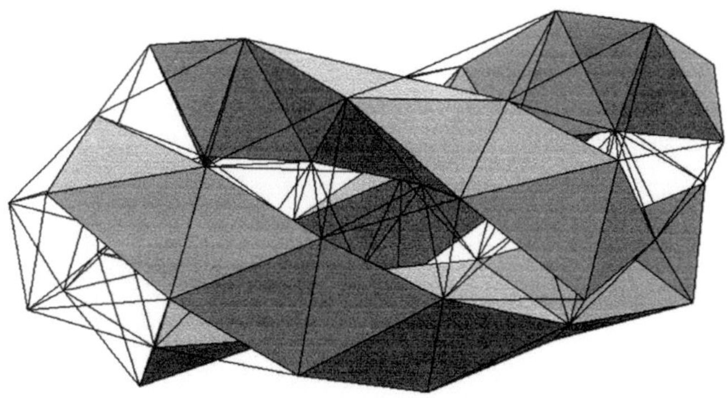

Fig. 8 Configuration of three helically coiled B-C structures.

Nanotubes

When $m+n$ is divisible by three, the THP of type (l, m, n) can be converted to a model for a net of equilateral hexagons with all vertices lying on a cylinder, simply by omitting one in three of the vertices. Figure 9 illustrates this for the case (4, 7, 11). The algorithm described in the fourth section for determining the metrical properties of the THPs is thus applicable to the structure of carbon nanotubes.

Nanotubes [8, 9] can be classified by a pairs of integers $[M, N]$, corresponding to the numbers of helices of edge-connected hexagons that wind around the nanotube to left and right. The relation between the (m, n) of a THP and the $[M, N]$ describing this subset of its vertices that corresponds to the atoms of a nanotube is

$$\begin{aligned} M &= (m+n)/3, \quad N = (2n-m)/3 \\ m &= 2M - N, \qquad n = M + N \end{aligned} \tag{20}$$

For example, starting with a (4, 7, 11), the formulas give $(7, 11) \rightarrow [6, 5]$. This corresponds to the *standard* triplet [1, 5, 6]. That is, the THP (4, 7, 11) gives rise to a metrical model for the nanotube [1, 5, 6]. More specifically, since a *reflection* is involved in the coordinate resetting of the final step, (4, 7, 11) $L \rightarrow$ [1, 5, 6] R.

The "chair" and "zigzag" type nanotubes (Fig. 10) are given by

$$(0,3\mu,3\mu) \rightarrow [\mu,\mu,2\mu] \text{ and } (\mu,\mu,2\mu) \rightarrow [0,\mu,\mu]. \tag{21}$$

The density of a carbon nanotube of a given structural type (l, m, n) is readily calculated. The type $\{n\}$ polygonal helices have a rise per edge of $\sqrt{3}(n/k)$ (using the approximation given by the CHL formula). The number density of vertices per unit advance along the cylinder axis is, therefore, $k/n\sqrt{3}$. There are n type $\{n\}$ helices. In the carbon nanotube model, one-third of them are hexagon centers, so the number of carbon atoms per unit length of the nanotube is $2k/3\sqrt{3}$. The C—C bond length B is the unit of length. Denoting the mass of a carbon atom by M, we get, for the mass per unit length of the nanotube,

$$2k\,M/3B\sqrt{3} = 4\pi RM/3B^2\sqrt{3} \tag{22}$$

where R is the radius out to an atom center—the average of internal and external radius. For example, taking $M = 10^{-3}$ gm and $B = 1.42$ Å (from graphite), gives 2.52×10^{-14} gm/cm for the case (4, 7, 11), corresponding to the [1, 5, 6] nanotube.

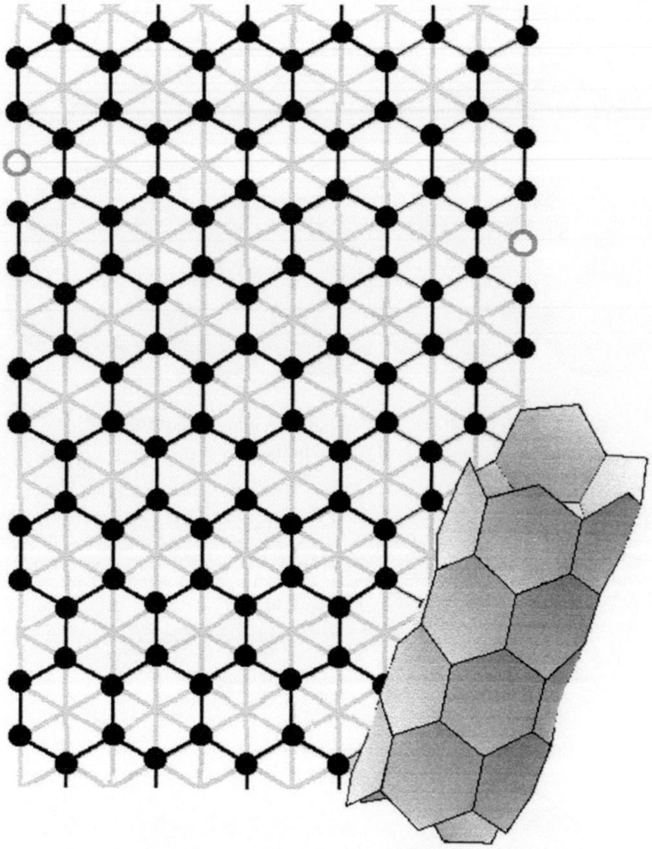

Fig. 9 The [1, 5, 6] nanotube derived from the (4, 7, 11) TPH.

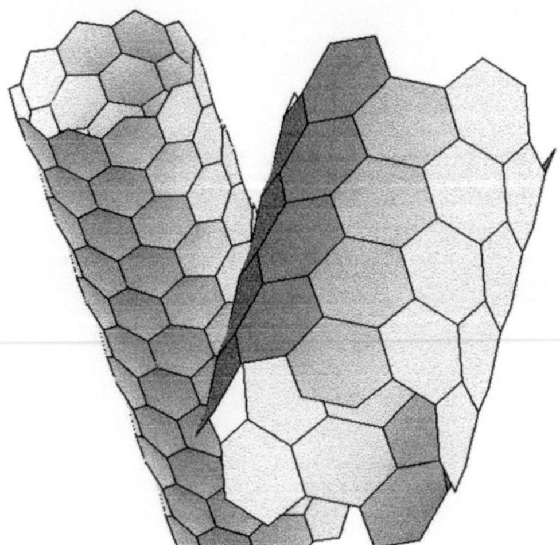

Fig. 10 The [0, 10, 10] "zigzag type" nanotube and the [5, 5, 10] "chair type".

Collagen and the Polytope {3, 3, 5}

The regular four-dimensional polytope $\{3, 3, 5\}$ has 120 vertices, 1200 edges, 720 equilateral triangle faces, and 600 *regular tetrahedral* cells. Five cells surround each edge and 20 surround each vertex—forming a *regular icosahedron* [4, 5, 10].

Circuits of 30 face-sharing tetrahedra occur in $\{3, 3, 5\}$. They are each metrically identical to the Boerdijk–Coxeter structure in 3-D Euclidean space. One can select four of them forming a configuration, which, when deformed to fit

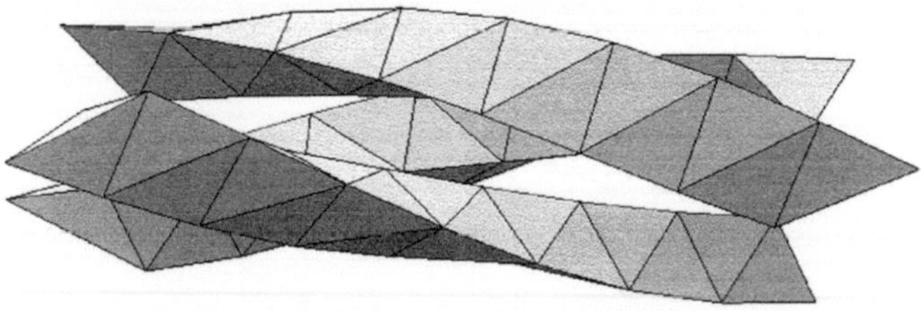

Fig. 11 General view of the three outer B-C structures of the collagen model proposed by Sadoc and Rivier. The polypeptide chains correspond to the type {2} helices of these structures. Other edges and edges of the central undeformed B-C structure (omitted for clarity) correspond to hydrogen bonds.

into 3-D Euclidean space, can give a fairly accurate structure of the *collagen molecule*. Collagen consists of three left-handed helical polypeptide chains twisted around each other with a right-handed helical twist [6]. In the model of Sadoc and Rivier [1], derived from the corresponding structure in the polytope {3, 3, 5}, a central B-C structure (1, 2, 3) R is surrounded by three somewhat distorted B-C structures (1, 2, 3) R each sharing a type {3} helix with it. The polypeptide chains correspond to the *left*-handed type {2} helices of the three outer structures. Other tetrahedron edges represent hydrogen bonds.

We have already noticed the geometrical structure in the extension of the Boerdijk's sphere packing (Fig. 8). The pictures displayed in Fig. 11 was obtained by projecting from {3, 3, 5} [11]. Observe that the structures differ in pitch; in Fig. 8, the central core B-C structure is undistorted, whereas in Fig. 11, a slight untwisting has been applied to conform to give the pitch observed in the collagen structure.

Defects in THPS and Nanotubes

If a net on a curved surfaces is adjusted so that all (straight) edge lengths are equal, it is obvious that the curvature properties of the surface and the connectivity properties of the net will be intimately related. Although no general theory exists, various fascinating structures have been explored, corresponding to the decoration of curved surfaces by graphite sheets containing defects; five-rings among the hexagons correspond to regions of positive Gaussian curvature, while seven- and eight-rings correspond to regions of negative Gaussian curvature [12–15]. Of special interest are the triply periodic minimal surfaces, tiled by hexagons and octagons.

A well-known topological relation satisfied by a triangulation of a surface of genus g (Euler characteristic χ) is

$$3n_3 + 2n_4 + n_5 - n_7 - 2n_8 - 3n_9 - \ldots = 6(2-2g) = 6\chi \tag{23}$$

where n_i is the number of i-connected vertices. The same relation is satisfied by the dual configuration: a tiling of the surface, with n_i i-gonal tiles. This purely topological relation is linked to curvature properties of the surface through the Gauss–Bonnet theorem,

$$\oint K d\ S = 2\pi\chi \tag{24}$$

where K is Gaussian curvature.

Thus, for example, a tiling of the sphere ($g=0$) by hexagons and pentagons must contain exactly 12 pentagons—a topological fact demonstrated in nature by the structure of the fullerenes. An equilateral net of hexagons, heptagons, and octagons may be associated with surfaces of negative Gaussian curvature. Mackay and Terrones [12] demonstrated the decoration of the triply periodic minimal surfaces P and D surfaces ($g=3$, per unit cell) by graphite sheets that contain 12 octagons per unit cell. In the present work, we have chosen to restrict attention to the Euclidean geometry of various structures. Mention must be made at this point, however, of the geometry of the hyperbolic plane as a valuable tool in the study of networks on surfaces of negative Gaussian curvature. For recent important results, the interested reader is referred to the work of Hyde and co-workers [16–18].

Observe that in a plane (or cylindrical) hexagonal lattice, a composite "5–7 defect" consisting of a five-coordinated vertex and a seven-coordinated vertex does not affect the genus. It gives rise to a *dislocation* of the lattice. Defects of this kind may be classified in terms of a pair of integers (p, q), which specify, in a hexagonal coordinate system associated with the lattice, the displacement of the five-vertex from the seven-vertex. The Burger's vector associated with a 5–7 defect of type (p, q) is ($p-q$, p). The *dual* of the above statements concern nanotubes, which can have analogous 5–7 defects consisting of a five-ring and a seven-ring. The Burger's vectors of a pair of such composite defects may cancel. This kind of defect has been discussed by Stone and Wales [19] in the context of fullerene structure.

The incorporation of five-rings and seven-rings (in equal numbers) into a nanotube can cause it tube to be bent, by producing regions of positive and negative Gaussian curvature (helically coiled nanotubes have been observed). The exact relationship between the location of five-rings and seven-rings, and the resulting bending effect, is a difficult problem, which we shall not enter into here. The aim of this section will be simply to demonstrate how two TPHs of different types can be joined through a region containing a (p, q) defect. In particular, the radius of a nanotube may be varied along the length of the tube, by the introduction of five-rings and seven-rings (in equal numbers).

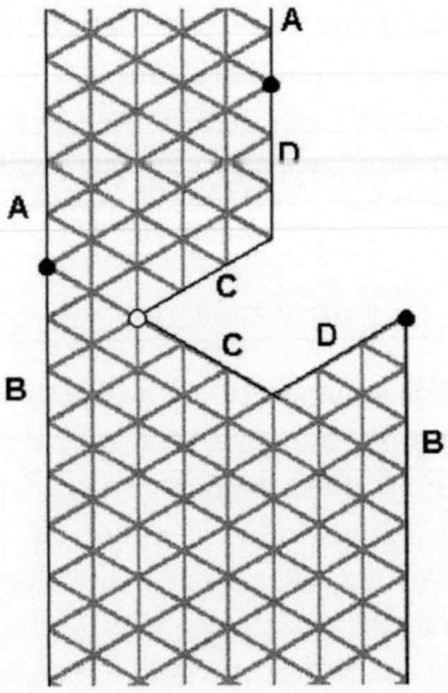

Fig. 12 Net for the construction of a TPH with a type (2, 2) defect. Observe that the lower half gives rise to a (3, 5, 8)*L*; the upper half produces a (1, 5, 6)*L*.

Figure 12 shows a net for constructing a THP with a (2, 2) defect. The net is to be folded along the grid lines to bring the edges marked with the same letter together, as in making card models of Platonic or Archimedean polyhedra. The three points indicated by the black circles becomes a single seven-connected vertex. The point indicated by the white circle becomes a five-connected vertex. Observe that the lower half of the THP will be a (3, 5, 8)*L*, while the upper half is a (1, 5, 6)*L*. We may write this in the form

$$(2,2):(3,5,8)L \rightarrow (1,5,6)L \tag{25}$$

Figure 13 demonstrates similarly the case

$$(2,4):(2,4,6)L \rightarrow (2,4,6)R \tag{26}$$

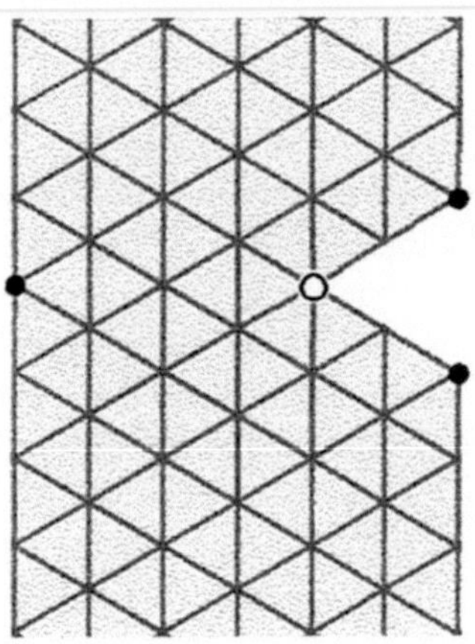

Fig. 13 Net for the construction of a "mirror twin" (2, 4): (2, 4, 6)*L*→(2, 4, 6) *R*.

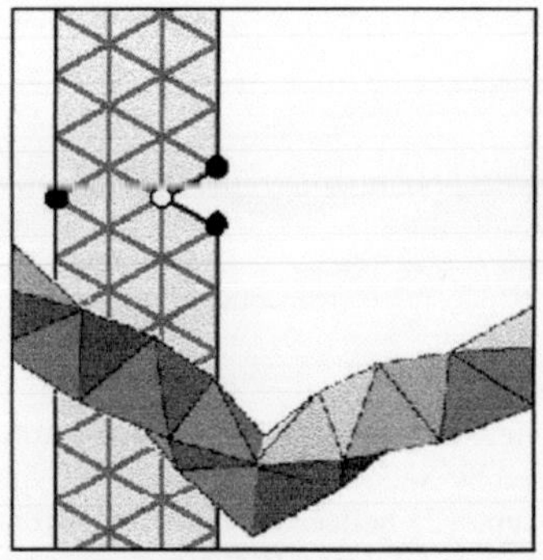

Fig. 14 A mirror twin of two B-C structures and the net for constructing a model of it. Observe the 5–7 defect of type (1, 2).

A CHL or a THP of type (m, n) can be joined to one of type (m', n'), through a region containing a (p, q) defect. We find that

$$\begin{pmatrix} m' \\ n' \end{pmatrix} = \begin{pmatrix} m-n+q \\ m+q-p \end{pmatrix} \tag{27}$$

for example, $(p, q)=(1, 2)$, $(m, n)=(2, 3)$. Then $(m', n')=(1, 3)$. Therefore

$$(1,2):(1,2,3)L \rightarrow (1,2,3)R \tag{28}$$

The handedness changes because converting (1, 3) to the three-index symbol (1, 2, 3) satisfying the inequalities (2.1) involves a *reflection*. This case corresponds simply to two mirror-twinned B-C structures; the situation is illustrated in Fig. 14.

Miscellany

Any Euclidean transformation

$$x' = R\mathbf{x} + \mathbf{a} \tag{29}$$

is uniquely determined if the four image points of four given nonplanar points are given. Writing the position vectors, referred to a cartesian coordinate system, of four given points as the columns of a 3×3 matrix A, and their four images as the corresponding columns of a matrix B, the transformation is given by

$$S = \begin{pmatrix} R & \mathbf{a} \\ 0 & 1 \end{pmatrix} = \begin{pmatrix} B & 0 \\ \mathbf{j} & 1 \end{pmatrix} \begin{pmatrix} A & 0 \\ \mathbf{j} & 1 \end{pmatrix}^{-1} \tag{30}$$

where **j** denotes the row 111. The Eq. (16), from which the metrical parameters of the B-C structure were deduced, are obtained from this prescription. The method is, clearly, readily applicable to the generation of more general structures from a given subunit once the position and orientation of a

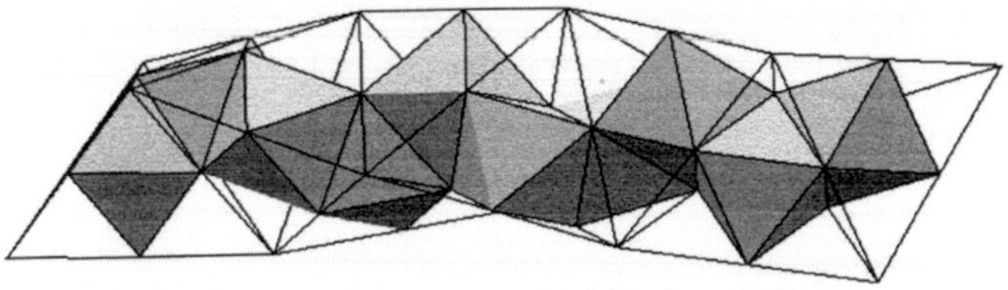

Fig. 15 Pearce's octahelix inscribed in a B-C structure.

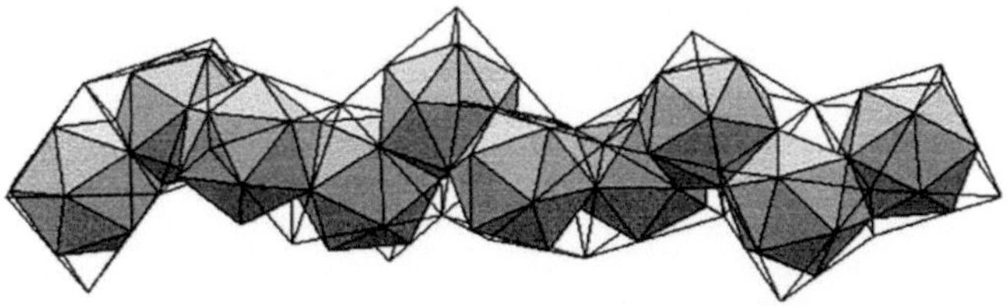

Fig. 16 Pearce's icosahelix inscribed in an octahelix.

contiguous subunit is chosen. The tiling of three-space by face-sharing polyhedra is a concept fundamental to the understanding of many complex material structures. Structures built from identical face-sharing polyhedral subunits are readily deduced from the prescription (30). A transformation S can be computed from two chosen congruent faces of a polyhedron. Repeated application of S will then produce a structure with helical, zigzag, or ring form according to the nature of S (some possibilities lead to steric hindrance). The purpose of this final section is simply to illustrate this by means of a few curious examples.

Two examples of helical structures built from polyhedral subunits are the "octahelix" and the "icosahelix" of Pearce [20]. Figure 15 indicates that the vertices of the octahelix are the midpoints of the edges of a B-C structure and Fig. 16 indicates that the vertices of the icosahelix are at golden mean positions on the edges of an octahelix. Other helical towers of octahedra, or of icosahedra, with different pitches, can be generated by choosing different orientation relationships for the initial pair of subunits, giving rise to different screw transformations S. Figure 17, for example, is another "octahelix," with a steeper pitch than the Pearce octahelix and Fig. 18 is a ring of 12 octahedra slightly distorted icosahedra generated by a roto-reflection. The helix of interpenetrating icosahedra in Fig. 19 is generated from a screw transformation S that relates two pentagonal sections of the icosahedral subunit.

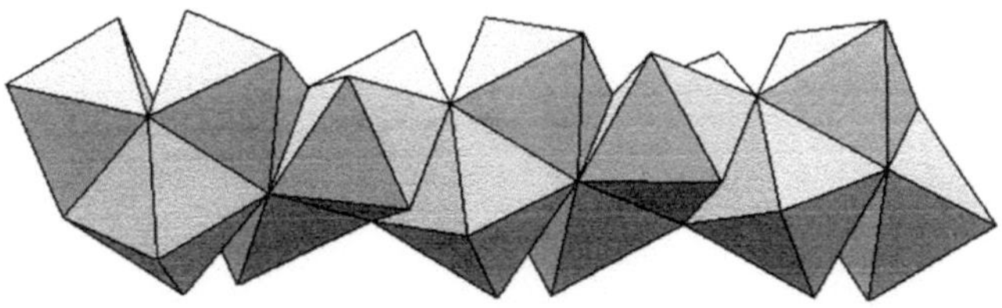

Fig. 17 An octahelix with steeper pitch.

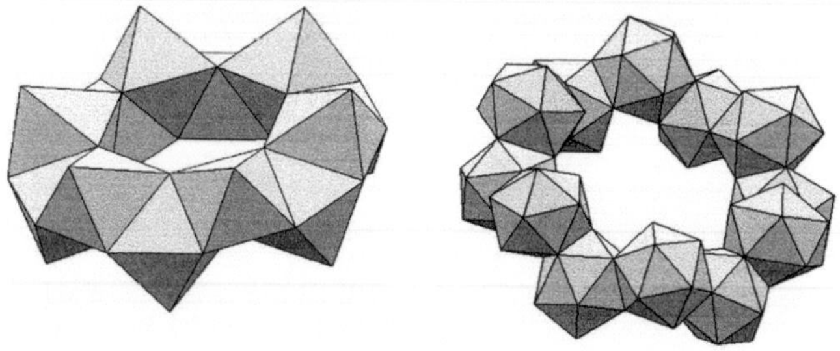

Fig. 18 A ring of 12 (slightly distorted) octahedra, generated by a roto-reflection; and the 12 "inscribed" icosahedra.

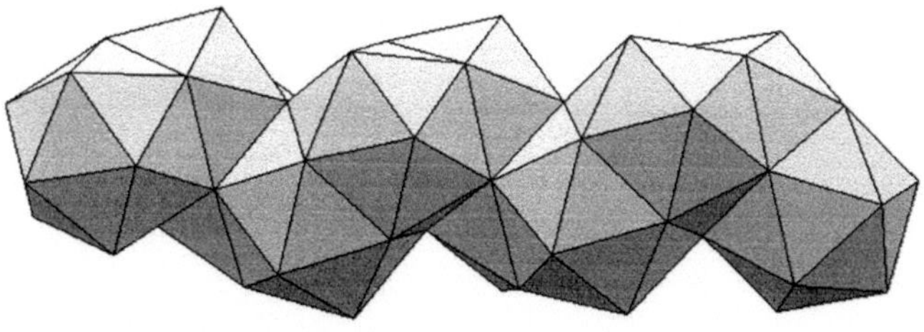

Fig. 19 A helical structure of interpenetrating icosahedra. Observe that every pair of consecutive icosahedra is a "mirror twin" with the mirror perpendicular to the fivefold axis.

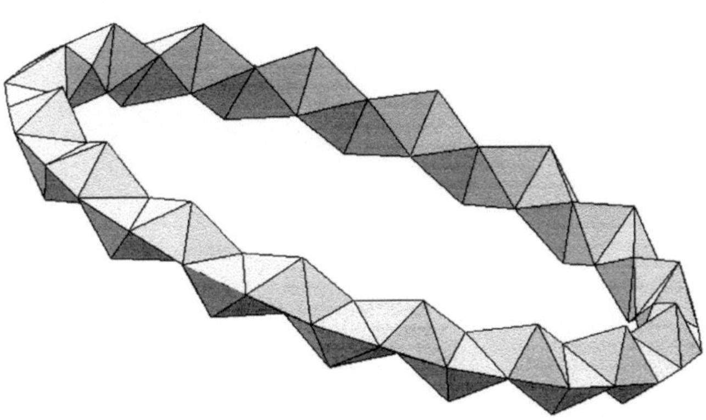

Fig. 20 The ring of 96 tetrahedra discovered by Antoine Walter. The subunit is a sequence of six of the tetrahedra of a B-C structure.

A polygonal subunit for the generation of helical structures can be chosen to be a portion of the B-C structure. This corresponds to the systematic introduction of defects along the length of the TPH (1, 2, 3). The remarkable ring structure (Fig. 20) consisting of 96 tetrahedra was discovered by Antoine Walter [21]. The subunit is a portion of the B-C structure consisting of six tetrahedra. The easiest way to understand the structure is to consider the smaller subunit of three tetrahedra (Fig. 21). The mirror symmetry of this subunit implies that the two face medians marked on the figure intersect. A simple calculation gives the angle θ between them, $\cos\theta = 53/54$, $\theta \sim 11°$. A twofold rotation about one of

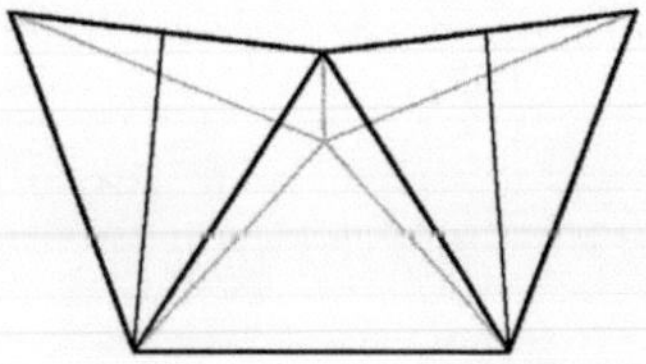

Fig. 21 Three tetrahedra. The ring of 96 tetrahedra is generated by twofold rotations about the face medians marked in the figure. Very slight deformation of the tetrahedra is necessary for exact closure of the ring, as explained in the text.

them produces a set of six tetrahedra of a B-C structure. The product of the two twofold rotations is a rotation through about 22°, whose repeated action gives a ring of 32 of the six tetrahedron units, which does not quite close. A slight deformation increasing the angle θ to $360°/32 = 11\frac{1}{4}°$ produces the 96-tetrahedron ring.

Conclusions

Nanotubes, protein helices, and the sphere packings investigated by Boerdijk, have this in common: their structures are underpinned by the Euclidean geometry of triangulated helical polyhedra. The geometrical properties of the THPs provide a simple introduction to these important structures and methods of computing their metrical and topological properties can be employed in the production of graphic representations.

Acknowledgments The figures were produced using Surface Evolver [22], which I have found to be very convenient for the rapid production of pictures of polyhedra, quite apart from its intended use for the study of the kinematics and dynamics of curved surfaces. Alan Mackay introduced me to this software and we have enjoyed exploring minimal surfaces together. Alan's enthusiasm for geometrical problems of all kinds is infectious; I have gained a great deal from my association with him. S. Ranganathan's friendly interest, support, and encouragement have also been invaluable. I also thank Anandh Subramaniam for helping me with Mathematica.

References

1. Sadoc J. F.; Rivier, N. *Eur. Phys. J.* **1999,** *B12*, 309.
2. Boerdijk, A. H. *Philips Res. Rept.* **1952,** *7*, 303.
3. Coxeter, H. S. M. *Introduction to Geometry* 1st edn.; Wiley: New York, 1961.
4. Coxeter, H. S. M. *Regular Polytopes*; Macmillan: New York, 1948 (Dover: 1973).
5. Coxeter, H. S. M. *Can. Math. Bull.* **1985,** *28*, 385.
6. Lehninger, L.; Nelson D. L.; Cox, M. M. *Principles of Biochemistry*; Worth: New York, 1993.
7. Buckminster Fuller, R. *Synergetics*; Macmillan: New York, 1975.
8. Iijima, S. *Nature (London)* **1991,** *354*, 56.
9. Tanaka, K.; Yamabe, T.; Fukui K., Eds., *The Science and Technology of Carbon Nanotubes*; Elsevier: New York, 2000.
10. Sadoc, J. F.; Mosseri, R. *Frustration Géometrique*; Eyroles: Paris, 1997; *Geometrical Frustration*; Cambridge University Press: Cambridge, 1999.
11. Lord, E. A.; Ranganathan, S. *Euro. Phys. J.* **2002**, *15,* 335.
12. Mackay, A. L.; Terrones, H. *Phil. Trans. Roy. Soc. London A* **1993**, *343*, 113.
13. Terrones, H.; Mackay, A. L. *Chem. Phys. Lett.* **1993**, *207*, 45.
14. Terrones, H.; Terrones, M. *Fullerene Sci. Technol.* **1996,** *4*, 517.
15. Terrones, H.; Terrones, M. *Phys. Rev. B* **1997,** *55*, 9969.
16. Hyde, S. T.; Andersson, S.; Blum, Z.; Lidin, S.; Larsson, K.; Landt, T.; Ninham, B. W. *The Language of Shape*; Elsevier Science: Amsterdam, 1997.
17. Hyde S. T.; Ramsden, S. In *Mathematical Chemistry* D. Bonchev and D. Rouvray, Eds.; Gordon and Breach, New York, 2000; Vol. 6, Chap. 2.
18. Hyde, S. T.; Oguey, C. *Eur. Phys. J.* **2000,** *B16*, 613.
19. Stone, A. J.; Wales, D. J. *Chem. Phys. Lett.* **1986,** *128*, 501.
20. Pearce, P. *Structure in Nature is a Strategy for Design*; MIT Press: Cambridge, MA, 1978.
21. Walter, A. *Hyperspace* **2000**, *9*, 22.
22. Brakke, K. A. *Exp. Math.* **1992**, *1*, 141.

From Waxes to Polymers—Crystallography of Polydisperse Chain Assemblies[a,b]

Douglas L. Dorset[c]

Abstract

Exploiting available single-crystal diffraction patterns from individual microareas, electron crystallography has characterized rectangular layer chain packing for polydisperse extended chain polymethylene chain assemblies. From this work, one can appreciate how these chain assemblies form stable solid solutions and also how they may phase separate. It is clear that two types of stable solid solution are formed—one that is lamellar and one that is "nemato-crystalline." Respective examples are wax fractions with narrow chain length distributions and chain-extended, low molecular weight polyethylene.

Keywords

Electron crystallography • Solid solutions • *n*-Paraffins • Petroleum wax • Polyethylene

Introduction

Wax is perhaps the first thermoplastic material used by Man. As a material of natural or synthetic origin, it can manifest a variety of physical forms from soft malleable to hard brittle solids. Although the production of candles from beeswax is a quaint reminder of earlier times, the continued economic importance of waxes worldwide in a large number of applications [1, 2] (e.g., polishes, building materials, high temperature glues, electrical insulators, raw material for fuel and lubricant production, etc.) should not be underestimated. The major natural source of wax is the paraffin fraction of crude petroleum. Nevertheless, there is sufficient additional demand for waxes worldwide to justify their synthetic production, e.g., by the Fischer–Tropsch process from "water gas," or even for the parallel production of polyethylenes from ethylene via various catalysts.

Generally speaking, waxes can be described as stable multicomponent solid solutions of paraffins or paraffin-line (e.g., fatty acid ester) chains. Although diffraction has been long used to characterize these chain assemblies [3], it is only recently that quantitative crystallographic techniques have uncovered the most important principles dictating chain assembly in the comixed solid state. Actual single-crystal structure determinations have been made possible by electron diffraction studies of epitaxially oriented microcrystals, exploiting the most informative view onto the chain axes [4]. Electron crystallography has also determined how paraffinic solid solutions can become destablilized to induce phase separation, extending pioneering crystallographic work in the field by A. I. Kitaigorodskii [5].

Why Electron Diffraction?

Although the concept of a quantitative electron crystallography has been accepted only in recent years, despite the pioneering work in the field by Pinsker [6] and Vainshtein [7], it is still safe to say that this is not the preferred method for determining material crystal structures. Nor will it ever be, since the number of available diffraction data is always less than accessible to single-crystal X-ray diffraction (XRD)

[a] *Structural Chemistry* 2002, 13(3/4):329–337.

[b] This contribution is part of a collection entitled Generalized Crystallography and is dedicated to the 75th anniversary of Professor Alan L. Mackay, FRS.

[c] Corporate Strategic Research, ExxonMobil Research and Engineering Co., 1545 Route 22 East, Annandale, NJ 08801, USA
e-mail: d.l.dorset@exxonmobil.com

I. Hargittai, B. Hargittai (eds.), *Science of Crystal Structures: Highlights in Crystallography*,
DOI 10.1007/978-3-319-19827-9_9

studies [4]. Nevertheless, there are a number of structural problems where the crystallographer is faced with a material that preferentially forms microcrystals. Linear chain assemblies are perfect examples, because most of the Van der Waals interaction in a crystal (98 % for the paraffin *n*-hexatriacontane) is expressed by the lateral packing of the polymethylene chains [8]. While individual layers will crystallize rapidly, their stacking to form a three-dimensional crystal is a much slower process, favoring the formation of thin microcrystalline lozenges. The much greater scattering cross section of matter for electrons versus X rays facilitates the use of electron beams to study individual microareas as single crystals [7]. Otherwise, powder XRD patterns from a typical bulk assembly of microcrystals [9] will provide somewhat limited (albeit valuable) information about the lamellar stacking of the chain layers (low angle region) and the preferred subcell packing of the polymethylene repeat (wide angle region), the latter as a characteristic "fingerprint" of strong intensity positions [9].

When individual lamellae, crystallized from solution, are studied by selected area electron diffraction, the projection along the molecular axes provides a somewhat limited view of the long chain packing [10], since all atoms are eclipsed. It would be far more advantageous to examine the orthogonal view onto the chain axes. Fortunately, such a chain orientation can be nucleated by certain organic and inorganic epitaxial substrates [11]. Using the eutectic phase behavior of the linear chain material with a suitable diluent [12], the first-solidified diluent crystal surface may specifically (i.e., by lattice matching) nucleate growth of the paraffin chain assembly so that the chain axes align parallel to this nucleating agent. Alternatively, the long-chain substance can be deposited from the vapor phase onto a nucleating surface that will similarly orient the chain packing [11]. If annealing is needed to improve crystal perfection, it can be carried out in the presence of the substrate with a suitably controlled furnace. Subsequently, the nucleating material can be removed to provide an array of thin microcrystals for study, again with chain axes parallel to the major crystal face.

The electron diffraction pattern obtained from an epitaxially oriented orthorhombic *n*-paraffin is shown schematically in Fig. 1. Using the goniometer tilt stage in the electron microscope (aligning the specimen rod tilt axis to the c∗ axis of the reciprocal lattice) it is possible to collect a three-dimensional (3-D) intensity data set [13]. The crystal structure can then be determined, e.g., by direct methods [4]. In Fig. 2, the orthorhombic crystal structure of the B-form of the odd-chain paraffin *n*-tritriacontane, solved from 3-D electron diffraction data [13], is shown. The structure of orthorhombic *n*-hexatriacontane was also solved earlier [14] from electron diffraction intensities obtained from oriented specimens, as were the structures of other homologs. In general, strong reflections the O*kl* patterns from even- or odd-chain paraffins grown from the melt (essentially the condition for epitaxial growth on a diluent), can be simply indexed (Fig. 1) in terms of their chain lengths.

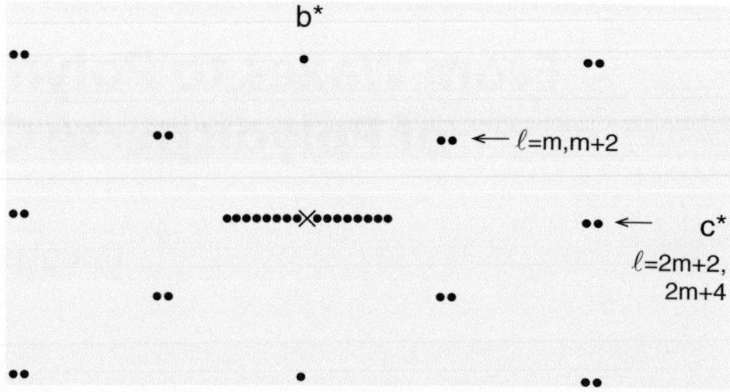

Fig. 1 Schematic 0*kl* electron diffraction pattern from epitaxially oriented orthorhombic paraffin, *n*-C_mH_{2m+2}. The *l* indices of the most intense reflections can be simply related to the chain length. For paraffin solid solutions, they correspond to the mean chain length that determines the lamellar thickness.

n-Paraffin Binary Solid Solutions

Based on available phase diagrams and powder XRD data, *n*-paraffin crystal structures were used by Kitaigorodskii [5] to predict the stability of binary solid solutions between two species. Both volumetric and symmetric factors were seen by him to be important. First, the nonoverlap volume between two molecules, Δ, should not be too large so that the quantity $\varepsilon = 1 - \Delta/r$ (where *r* is the overlap volume) is not less than, say, 0.8. Second, if a molecule of crystal symmetry A were added to a molecule with crystal symmetry B, then the symmetry of the solid solution should either remain the same or be less than that of crystal B. This symmetry argument was based on the most thermodynamically stable polymorphs of any component substances. It was, therefore, claimed [5] that an even-chain *n*-paraffin would not form a stable solid solution with an odd-chain paraffin, since the former prefers a monoclinic chain packing, whereas the latter prefers an orthorhombic structure. Experimental evidence, however, contradicted this prediction [15]. Further, it was well-known that two even-chain paraffins, each crystallizing in a triclinic structure, would form orthorhombic, not triclinic, solid solutions beyond a certain concentration of the second component [16], i.e., a higher symmetry than the original guest lattice. It is apparent, therefore, that long-lived metastable crystal forms are very important for polydisperse linear chain substances. Practically speaking, these violate the reasonable thermodynamic rules for cosolubility established from principles of lowest-energy chain packing.

Electron diffraction studies have facilitated the characterization of oriented binary solid solutions. For example, either chain combination: *n*-$C_{33}H_{68}$/*n*-$C_{36}H_{74}$ or *n*-$C_{32}H_{66}$/*n*-$C_{36}H_{74}$, forms stable solid solutions over all concentrations [17], so that odd-chain paraffins can freely comix with even paraffins in their respective orthorhombic polymorphs (B polymorph, space group $A2_1am$, for the former and the $Pca2_1$ packing for the latter). There is no observed continuity of space group symmetry, however, only the rectangular layer packing is maintained. In addition, for any nominal concentration, there may be two or three local crystal structures (on a micrometer scale), each mimicking an archetypical pure odd- or even-chain n-paraffin crystal structure (Fig. 2c), where the lamellar spacing matches the theoretical value predicted for the average chain length [18] derived from the index rule in Fig. 1. (These multiple solid solution crystal structures do not appear to result from "coring" during the solid solution growth. Neighboring contiguous microareas appear to retain a single crystal structure.) Extending the maximum chain length investigated to n-$C_{60}H_{122}$, the volumetric (ε) conditions for cosolubility were found to be similar to those determined earlier by Mathiesen and Smith [19].

The 2-D crystal structures of representative n-paraffin solid solutions [20] indicate a stable lamellar chain packing with concentration of defects at the crystal surface (Fig. 3). Crystallography depicts the average structure as one with decreasing occupancy of methylene groups as one approaches the lamellar surface. Vibrational spectroscopy detects an increasing number of nonplanar chain defects as this surface is approached [21, 22].

A subtle, but significant, variant of this packing scheme was proposed [16, 23] from two single-crystal X-ray structures of paraffin binary solids. Instead of the average flat lamellar surfaces, protruding chain ciliae were included in the structural model and even a mean lamellar thickness that was larger than the length of the largest paraffin component. Such a situation has never been observed by electron diffraction.

Electron diffraction measurements were carried out [24] on the n-$C_{24}H_{50}$/n-$C_{26}H_{54}$ solid solutions used for the X-ray crystal structure determination. Three-dimensional data from a 3:1 molar combination were used for a quantitative crystal structure analysis and a similar model was obtained as seen before, i.e., with flat lamellar ends. Reexamination of the published X-ray data [23] concluded that a flat lamellar surface, in agreement with the electron crystallographic model, would fit the diffraction intensities as well as the ciliated one, and not involve as many variable parameters as used for the ciliated model.

There are other indicators that lamellar solid solution surfaces are flat, on average, e.g., AFM measurements [25]. Also decoration experiments [26], where polymethylene chain segments are brought down onto the (disordered) solid solution (001) face from the vapor phase, demonstrate that the solid solution surface can still nucleate epitaxial growth [24], i.e., this surface has, on average, crystallographic regularity. Finally, the crystal coherence for solid solution growth along [001] extends for several micrometers, indicating that one lamellar surface serves as an effective nucleator for growth of the next, despite the accumulation of conformational disorder.

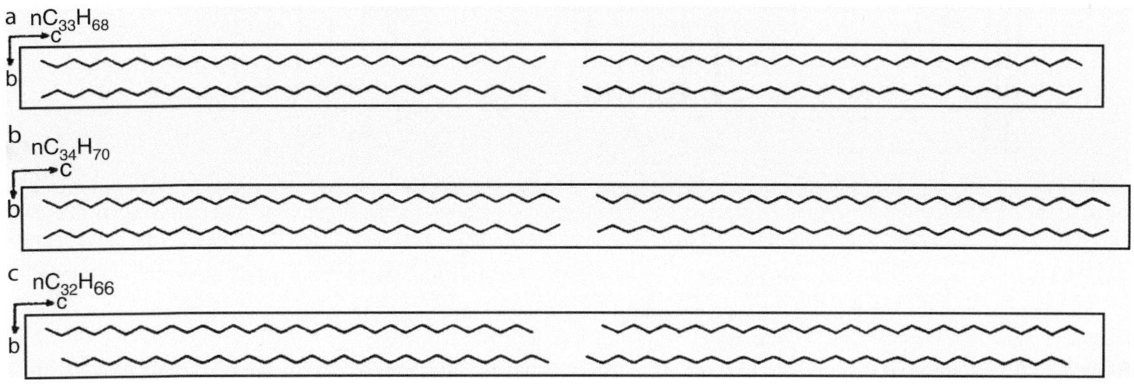

Fig. 2 Representative orthorhombic n-paraffin structures. (**a**) The odd-chain n-$C_{33}H_{68}$ crystallizing in $A2_1am$: (**b**) the even-chain n-$C_{34}H_{70}$ crystallizing in $Pca2_1$. Homologs would have the same unit cell symmetry following systematic rules for variation of the lamellar thickness (see Ref. [18]); (**c**) demonstration of how an even-type crystal structure can result from and odd chain. This is important for the formation of solid solutions.

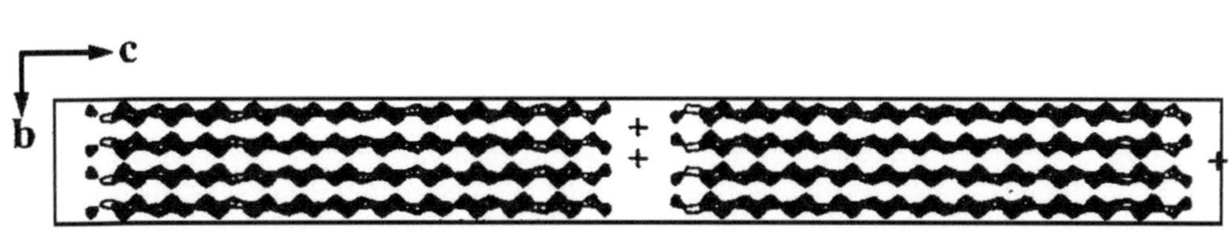

Fig. 3 Crystal structure of a typical n-paraffin lamellar solid solution. The smaller diameters of groups toward the chain ends indicate fractional methylene group occupancies. Spectroscopically an increased concentration of nonplanar conformational defects is detected in this region.

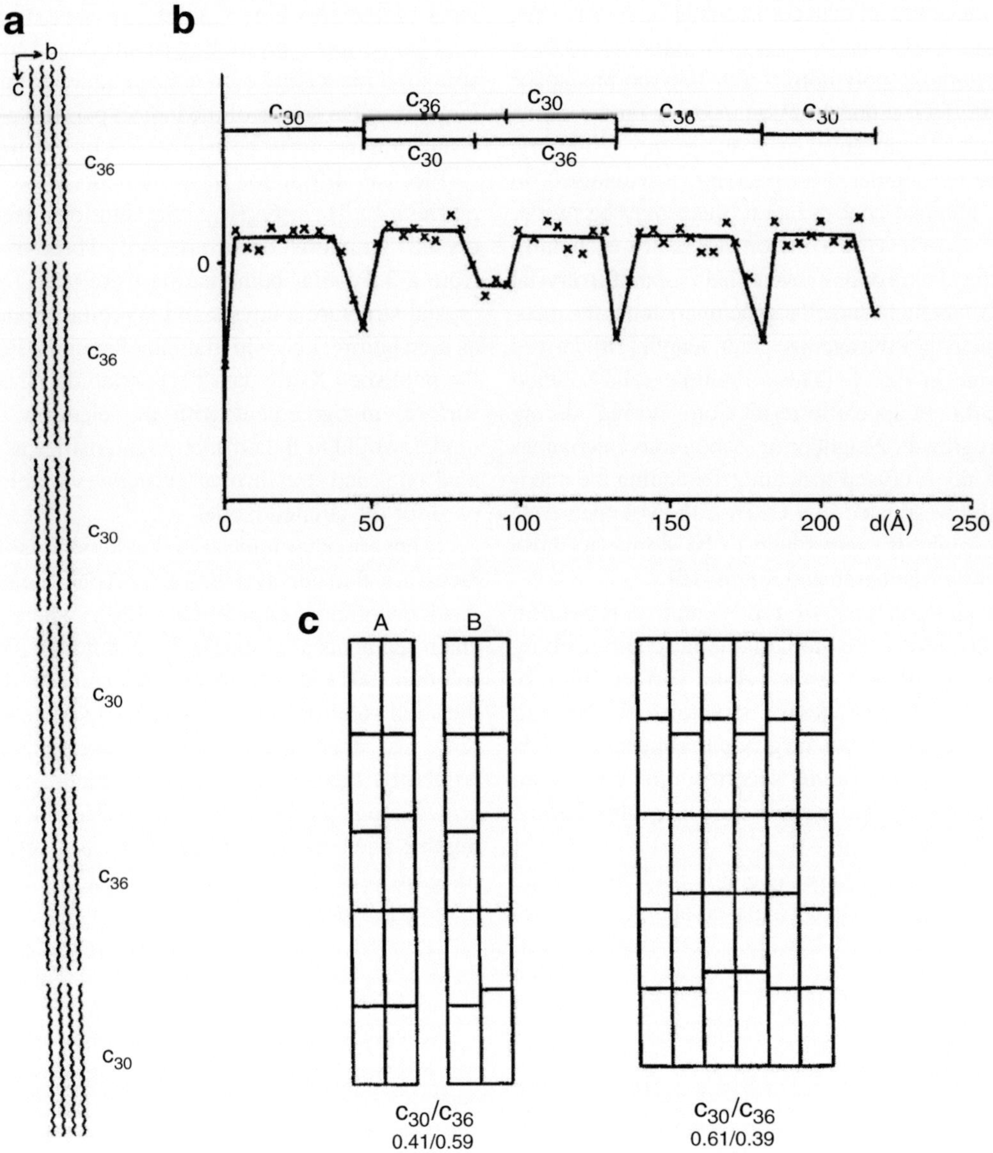

Fig. 4 Growth of superlattices within a miscibility gap. (**a**) The longitudinal separation accounting for the superlattice spots in the 00*l* row; (**b**) an analysis of the 00*l* structure factor amplitudes by direct methods, suggesting a packing sequence of chain layers; (**c**) a block representation of the chain domains for two concentrations of *n*-$C_{30}H_{62}$/*n*-$C_{36}H_{74}$, accounting for both longitudinal and lateral separation of components.

Fractionation of *n*-Paraffin Binaries—Miscibility Gap

The boundary line for solid solution stability, in terms of Cn(min) versus Cn(max) is very sharp. When it is traversed, e.g., the combination *n*-$C_{30}H_{62}$/*n*-$C_{35}H_{72}$, a metastable solid solution is initially grown from the melt that dissociates over a year [15]. This phase separation was rediscovered in our study [27, 28] of *n*-C_{30} H_{62}/*n*-$C_{36}H_{72}$, that takes only 2 days to form a superlattice. (Other paraffin binaries, *n*-$C_{28}H_{60}$/*n*-$C_{34}H_{70}$ and *n*-$C_{32}H_{66}$/*n*-$C_{37}H_{76}$, exhibit [28] similar behavior.) When these solids are crystallized from the melt, initial DSC traces resemble those from a stable *n*-paraffin solid solution. After equilibration at room temperature, a small, low enthalpy "mixing" endotherm appears that is clearly associated with the dissolution. Vibrational spectroscopic measurements with perdeuterated probes indicate the growth of lateral islands [29] from the original solid solution, whereas initial electron diffraction measurements detected a longitudinal change [27], i.e., the emergence of superlattice reflections. High-resolution, low-dose electron micrographs depict the growth of the new solid within the original solid solution, related to electron diffraction measurements *via* their optical transforms [28]. This phase separation behavior is characteristic of a miscibility gap. By analogy to metal alloys [30], the growth of fractionated solid is expected to be somewhat sluggish.

The discrepancy between lateral or longitudinal changes in the solid was resolved by quantitative single crystal electron diffraction determinations [31] of two molar combinations (ca. 3:2, 2:3) of *n*-$C_{30}H_{62}$/*n*$C_{36}H_{72}$. The block arrangements of nearly pure chain domains (Fig. 4) argue in favor of both lateral and longitudinal arrangements of nearly pure chain layers. When three-components are considered [32], e.g., n-$C_{28}H_{60}$/$nC_{32}H_{66}$/n-$C_{36}H_{74}$, the behavior is even more complicated. The first paraffin pair can behave as a "pseudocomponent" in the presence of the longest chain to bring about the initial, rapid phase separation, detected by superlattice reflections appearing along 00*l*. After a longer incubation period, layers of nearly pure components appear as the molecules rearrange into a very complicated sequence involving both longitudinal and lateral arrays of chain blocks.

The boundary between the miscibility gap behavior and eutectic phase separation is also very sharp. This has been tested with perdeuterated *n*-paraffin probes [33, 34]. For example, when either the longer chain component or both chains are perdeuterated in the n-C_{30}/n-C_{36} binaries, the phase behavior and crystallographic changes are exactly those found for the all-hydrogenated n-$C_{30}H_{62}$/$nC_{36}H_{72}$ series. However, for n-$C_{30}D_{62}$/n-$C_{36}H_{72}$, a eutectic solid forms directly from the melt. Perdeuterated chains have a slightly smaller volume than analogous hydrocarbons, so the volumetric difference is greatest for this binary pair. The subtle difference in volume is enough that the stability boundary is traversed.

Fractionation of *n*-Paraffin Binaries—Eutectic Interactions

Electron diffraction patterns from eutectic solids, e.g., of n-$C_{30}H_{62}$/n-$C_{40}H_{82}$, include contributions from a nearly pure paraffin component (depending on the binary concentration) and a coexistent superlattice of the two chains [27]. Unlike the solids grown within the miscibility gap, the positions of the 00*l* superlattice reflections do not depend very much on relative chain concentrations. A high-resolution, low-dose electron micrograph clearly depicts the sharp interfacial boundary between the pure paraffin and superlattice domains [35], an exact epitaxial nucleation of one methyl end plane by another.

The crystal structure of 1:1 n-$C_{30}H_{62}$/n-$C_{40}H_{82}$ was determined from electron diffraction intensities [35]. First, the structure of the nearly pure *n*-C40 component was verified with the expected [36] $Pca2_1$ homolog chain packing model. The associated superlattice was shown (Fig. 5) to contain regions of nearly pure chain lengths, arranged in both lateral and longitudinal blocks.

Eventually, with great enough volume differences, the component chains would be completely noncosoluble for any concentration. It is expected that two pure *n*-paraffin blocks would be adjoined across (001) planes, with appropriate registry of the methyl end planes, where chain end protrusions from one surface would project into hollows of the apposing surface, as explained by Kitaigorodskii [5].

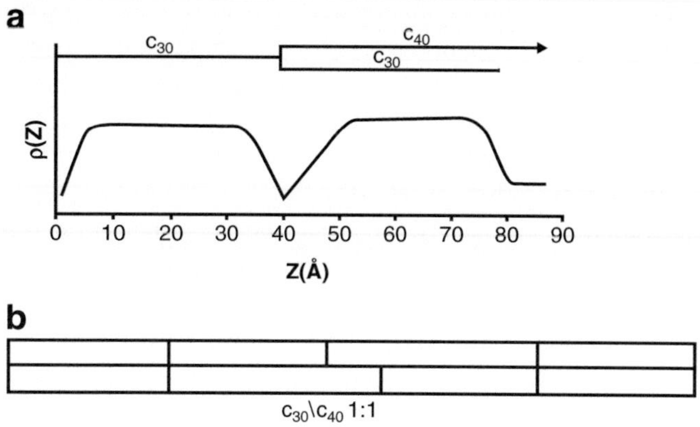

Fig. 5 Analysis of superlattice component of eutectic separation in 1:1 n-$C_{30}H_{62}$/n-$C_{40}H_{82}$. (**a**) Direct phase analysis of 00*l* amplitudes; (**b**) block representation of chain domains.

"Typical" Lamellar Paraffin Wax Structures

In an initial investigation of paraffinic wax crystallinity, a sample of a commercial petroleum distillate wax was epitaxially oriented on benzoic acid [17]. The electron diffraction pattern revealed this material to be a stable solid solution. Again, there was an observed distribution of local crystal structures, as is the case for *n*-paraffin binaries. Later, the wax crystal structure was determined [37] from electron diffraction intensity data; again there was no structural difference from that of the binary solid solutions. A similar result was obtained [38] for a wax taken from a child's birthday candle and also an artificial wax made up of a flat distribution of all even *n*-paraffins from n-$C_{26}H_{54}$ to n-$C_{36}H_{74}$. It is important to note here that the electron diffraction patterns from these waxes contain very sharp reflections and extend to very high resolution. By contrast, earlier characterizations of petroleum waxes by light microscopy and XRD [39] had indicated the crystallinity to be very poor. In other words, although vibrational spectroscopy also reveals [40] a significant defect distribution at lamellar interfaces, wax microcrystals are very highly ordered in a crystallographic sense. Imposing a Gaussian model for chain-end occupancy, revising the previous analyses made by Strobl et al. [41], the defect distribution can be shown [42] to penetrate further into the lamellar surface for complicated wax chain distributions than for simple paraffin binary solid solutions. There also seems to be a correlation of this surface defect thickness with the breadth of the DSC melting endotherm.

There is no difference between the crystal structures of waxes taken from petroleum paraffin distillate cuts and the fractionated cuts made of waxes synthesized by the Fischer–Tropsch process. A somewhat complicated, four-domain, model had been originally proposed for Fisher–Tropsch waxes [43, 44] based on solid-state NMR measurements made on rather narrow fractions of a synthetic product with a larger chain distribution. Electron crystallographic analyses [45] of narrower Fischer–Tropsh wax distillate fractions again reveal the structures to be lamellar arrays, identical to the petroleum waxes.

If the actual chain length distribution was used to model the (phenomenological) site occupancy distribution at the lamellar surface, e.g., a wax fraction characterized by Stokhuyzen and Pistorius [46] with poly-dispersity index $M_w/M_n = 1.003$, a very good fit was made to experimental electron diffraction intensity data, where the model-generated occupancy parameters accurately accounted for the chain-packing scheme [47]. This result, however, is fortuitous because a wax with a slightly broader distribution ($M_w/M_n = 1.009$), mimicking a residual petroleum wax [48], did not accurately model the fractional occupancies needed to fit experimental electron diffraction intensity data. Somewhat smaller occupancy parameters near the chain ends fit the data better [49]. This observation is significant since it suggests that chain end defects need not lie on Bragg sites. The fractional occupancies are, in themselves, meaningless if taken literally to indicate an actual distribution of "void" spaces. Nonplanar conformational defects, detected by vibrational spectroscopy [40], can also have a "noncrystallographic" distribution, detected only as a continuous diffuse signal in diffraction experiments.

Lamellar structure can also exist for some natural waxes with functionalized chains, e.g., in montan wax from brown coal [37]. Also, artificial assemblies of long-chain esters of behenic acid can crystallize in rectangular lamellar layers [38], as can the chain assembly in a synthetic substitute for spermaceti. (Natural spermaceti, however, crystallizes in oblique chain layers.)

Thus far, the crystal structure of waxes appears to be self-consistent. This viewpoint changes quickly when branched-chain waxes, such as the "microcrystalline" wax from heavier petroleum distillates [50], or natural waxes [51], such as bee honeycomb or carnauba, or intact Fischer–Tropsch synthetic products [52], are examined. After orientation of the wax on a substrate, there may be no initial evidence of a layered structure from electron diffraction patterns, which often strongly resemble the patterns from chain-folded polyethylene [53]. Only after the samples are annealed are characteristic low-angle reflections frequently observed by electron diffraction.

From Wax to Polymer Crystal—Interlayer Bridges

A model for the structural change that occurs when broader chain distributions are crystallized from the melt was accidentally discovered when very long-chain n-paraffins (e.g., n-$C_{60}H_{122}$) were deposited from the vapor phase onto an epitaxial substrate [54]. The sequence of structures is outlined in Fig. 6, associated with the corresponding 0*kl* electron diffraction patterns. Upon initial deposition, the chains retain only a common axial direction packing in the $O_\perp$ polymethylene subcell with no longitudinal registry. The term "nematocrystalline" has been assigned to this packing scheme by analogy to nematic liquid crystalline phases. Upon annealing, the chains can reptate toward a lamellar endpoint, whereupon an irreversible change is made to the system, however, via intermediate stages. If there are enough long chains to bridge nascent layers, so that the interlamellar gap is still an integral multiple of the polymethylene repeat along the chain, then the strong (01*l*) "polyethylene" reflections remain as strong singlets, although weak satellites can occur for more ordered systems. Only when true lamellae form, i.e., when there are no bridging chain segments, can these intense singlet reflections split into strong doublets, because there is then a disparity between the interlayer gap spacing and the nearest integral multiple of the polymethylene repeat.

For low molecular weight linear polyethylene that cannot undergo chain folding, e.g., $M_w = 595$, $M_n = 535$ (polydispersity 1.11), the first solid crystallized from the melt onto benzoic acid will diffract very much like that of the infinite chain polymer [55]. In other words, this assembly is "nematocrystalline." Upon annealing, some low-angle reflections appear, but the intense (01*l*) reflections never split into doublets. A quantitative crystal structure analysis from 3-D electron diffraction data reveals (Fig. 7) the presence of bridging segments across nascent layers. Considering the reflection indices of the 3-D patterns, however, it is clear that none of the orthorhombic space groups utilized by n-paraffins satisfies the observed extinction rules. The patterns represent a superlattice of the Pnam cell employed by the infinite chain polymer [56].

Why lamellae never form at some endpoint can only be speculated for now. It is known [57] that, in this narrow polyethylene fraction, that a small, but significant, distribution of chains are included in a higher molecular weight tail of chain length distribution near MW 800 (mean chain length of 57 methylene groups). These extended longer chains would never pack in lamellae with the observed layer spacing as small as 53.5 Å (mean chain length of ca. 41 methylenes), so they must bridge the layers to preserve the "nematocrystalline" order. In the projection down the chain axes, the chain axial direction through the crystal is continuous [56].

This nematocrystalline chain packing, it turns out, is ubiquitous. For example, it is also the chain packing adapted by some natural waxes [51], such as bee honeycomb wax and carnauba wax. Beeswax is known to contain fatty acid diesters of long-chain diols, which could serve as the longer bridging entities, in this case. The crystal structures of two intact

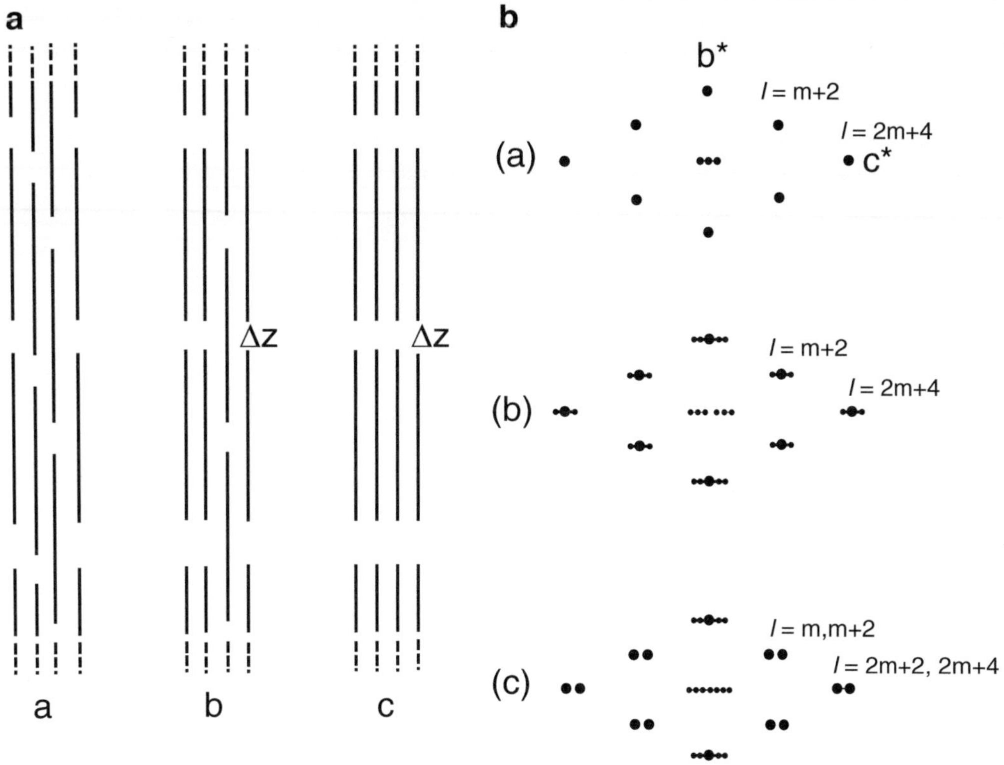

Fig. 6 Sequence of chain packing and effect on the 0*kl* electron diffraction pattern. The packing motifs in (**A**) correspond to the respective patterns in (**B**). (*a*) Nematocrystalline chain assembly—there is no longitudinal registry, although the polymethylene subcell is retained. The lowest angle 00*l* reflection would represent the case when an average layer begins to emerge; (*b*) when there are enough bridging molecules to constrain the interlamellar gap Δz to some simple multiple of the chain polymethylene repeat, the strong 01*l* peak remains a singlet, although there may be multiple orders of a layer repeat expressed by the low-angle 00*l* reflections; (*c*) true lamellar growth results in Δz being different from a simple multiple of the polymethylene repeat. Hence the strong 01*l* peaks are split into two strong reflections.

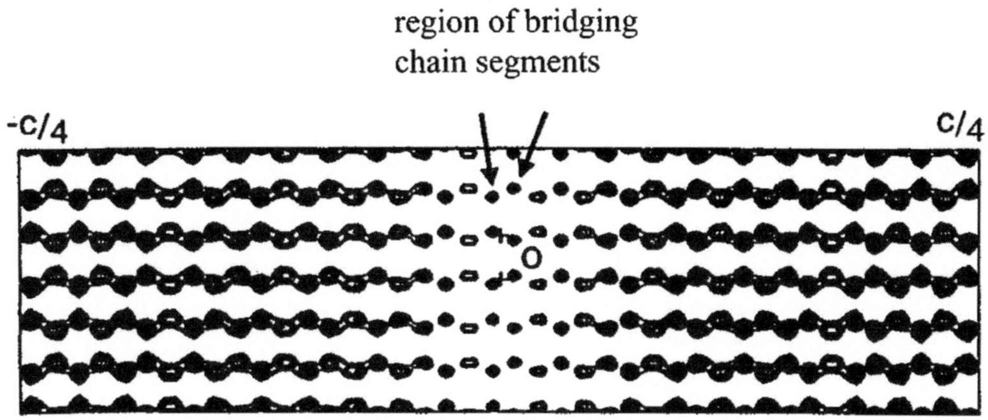

Fig. 7 Layer interface of linear low molecular weight polyethylene after structure determination. The bridging oligomethylene segments correspond to Fig. 6b.

Fischer–Tropsch waxes have also been determined, one at a relatively short [58] mean chain length (C33) and the other [52], a hard wax, with a longer mean chain length value (C68).

Another deviation from true lamellar packing is found for methyl-branched "microcrystalline" waxes. Electron diffraction patterns [50] resemble those from the polyethylene except that the intense (011) singlet is also somewhat broadened. Structure analysis reveals the presence of bridging chain segments. The same patterns can be obtained from pure 3-, 4-, or 5-methyl branched *n*-tetratriacontanes, for example, or from 3-methyl tritriacontane, when they are just crystallized from the melt. From the X-ray structures of

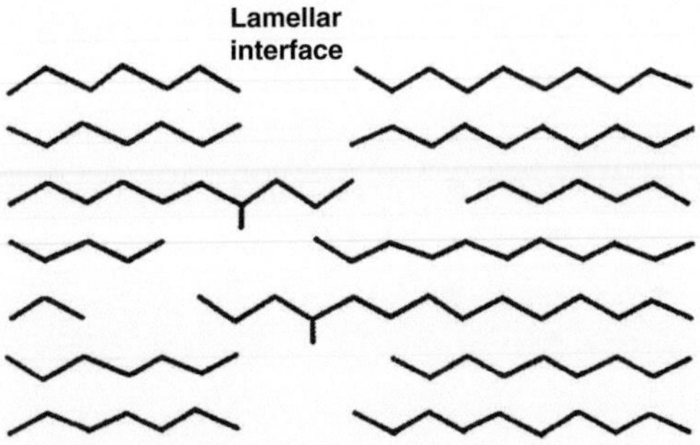

Fig. 8 Possible interpenetrating layer region for a methyl-branched *n*-paraffin or "microcrystalline" wax.

branched-chain fatty acids [59], it is well-known that methyl groups must always be sequestered into an interlamellar space. The packing model that would satisfy this requirement and lead to an interdigitated chain layer is depicted in Fig. 8.

What Is Wax and What Is Polyethylene?

From the above overview, it is clear that mere terminology for a polymethylene solid may be of little value for distinguishing actual crystal structures. Polymerization of methylene or ethylene groups to form respective linear chain distributions can end up with materials that cannot be otherwise distinguished. What is important is whether the chain distribution favors the formation of true lamellae or whether a significant fraction of bridging molecules is present to retain a nematocrystalline array. Obviously, any broad-chain distribution can change its original physical properties if it is distilled or fractionated into a narrower distribution, as evidenced by Fischer–Tropsch products [45, 52, 58].

The distinct physical properties imparted by a preferred crystal structure can be striking. Lamellar systems concentrate defects at the chain ends. Since the Van der Waals energy between layers is very small [8], certainly minimized further by inclusion of noncrystallographic disorder in this region, the resulting material can be easily plastically deformed by mechanical forces. The presence of bridging molecules spanning nascent layers, on the other hand, interpose covalent "reinforcing rods" within an emergent interlayer gap so that the layers cannot be deformed easily by lateral forces. The resultant material, then, will tend to be hard and brittle, although the chain polydispersity can be actually greater than that for a narrow "wax" distillate cut. Extension of materials properties by blending inexpensive adducts (e.g., petroleum wax) into an expensive natural product (e.g., carnauba wax from Brazilian palm, an excellent hard polish) requires that the characteristic chain-packing features are well understood. In other words, the structural motif must be preserved to retain desired properties.

Prediction of the crystallographic behavior for polydisperse linear chain arrays becomes even more complicated when chain folding is permitted in layers. Without discussing the controversial nature of random switchboard versus adjacent folds, it is clear that many of the volumetric strictures for stabilizing solid solutions are removed, at least kinetically, when admixtures are made of chains that can also fold. Progress has been made in understanding the underlying principles via the use of pure very long *n*-paraffins [60]. Hopefully accurate crystal structure analyses on well-characterized systems can be carried out in the future to complete the overview already obtained from the chain-extended arrays.

Acknowledgments Much of the work described in this review was supported by the NSF when the author was employed at the Hauptman-Woodward Medical Research Institute in Buffalo, NY. Many thanks are due to Robert G. Snyder, Jean-Claude Wittmann, Bernard Lotz, Barbara Moss, Weiping Zhang, Ilsa Basson, and Melanie Rademeyer for productive collaborations.

References

1. Warth, A. H. *The Chemistry and Technology of Waxes*; Reinhold: NY, 1947.
2. Hamilton, R. J. *Waxes: Chemistry, Molecular Biology and Function*; Oily Press: Dundee: 1995.
3. Kreger, D. R. In *Selected Topics in X-ray Crystallography from the Delft Institute*; Bouman, J., Ed.; North-Holland: Amsterdam, 1951; p. 303.
4. Dorset, D. L. *Structural Electron Crystallography*; Plenum Press: NY, 1995.
5. Kitaigorodskii, A. I. *Organic Chemical Crystallography*; Consultants Bureau: NY, 1961; p. 230.
6. Pinsker, Z. G. *Electron Diffraction*; Butterworths: London, 1951.
7. Vainshtein, B. K. *Structure Analysis by Electron Diffraction*; Pergamon; Oxford, 1964.
8. Mnyukh, Yu. V. *J. Phys. Chem. Solids* **1963,** *24*, 631.
9. Chapman, D. *The Structure of Lipids*; Wiley: NY, 1965.
10. Vainshtein, B. K.; Lobachev, A. N.; Stasova, M. M. *Sov. Phys. Crystallogr.* **1958,** *3*, 452.
11. Wittmann, J. C.; Lotz, B. *Progr. Polymer Sci.* **1990,** *15*, 909.
12. Dorset, D. L.; Hanlon, J.; Karet, G. *Macromolecules* **1989,** *22*, 2169.
13. Dorset, D. L. *Z. Kristallogr.* **1999,** *214*, 223.
14. Moss, B.; Dorset, D. L.; Wittmann, J. C.; Lotz, B. *J. Polymer Sci. Polymer Phys. Ed.* **1984,** *22*, 1919.
15. Mazee, W. M. *Amer. Chem. Soc. Div. Petro. Chem.* **1958,** 3(4), 35 (Preprints).
16. Lüth, H.; Nyburg, S. C.; Robinson, P. M.; Scott, H. G. *Mol. Cryst. Liq. Cryst.* **1974,** *27*, 337.
17. Dorset, D. L. *Macromolecules* **1987,** *20*, 2782.
18. Nyburg, S. C.; Potworowski, J. A. *Acta Crystallogr.* **1973,** *B29*, 347.
19. Mathiesen, R. R., Jr.; Smith, P. *Polymer* **1985,** *26*, 288.
20. Dorset, D. L. *Proc. Natl. Acad. Sci. USA* **1990,** *87*, 8541.
21. Maroncelli, M.; Strauss, H. L.; Snyder, R. G. *J. Phys. Chem.* **1985,** *89*, 5260.
22. Kim, Y. S.; Strauss, H. L.; Snyder, R. G. *J. Phys. Chem.* **1989,** *93*, 485.

23. Gerson, A. R.; Nyburg, S. C. *Acta Crystallogr.* **1994,** *B50*, 252.
24. Dorset, D. L. *Z. Kristallogr.* **1999,** *214*, 229.
25. Dorset, D. L.; Annis, B. K. *Macromolecules* **1996,** *29*, 2969.
26. Wittmann, J. C.; Lotz, B. *J. Polymer Sci. Polymer Phys. Ed.* **1985,** *23*, 205.
27. Dorset, D. L. *Macromolecules* **1986,** *19*, 2965.
28. Dorset D. L. *Macromolecules* **1990,** *23*, 623.
29. Snyder, R. G.; Goh, M. C.; Srivatsavoy, V. J. P.; Strauss, H. L; Dorset, D. L. *J. Phys. Chem.* **1992,** *96*, 10008.
30. Amelinckx, S.; Van Dyck, D. In *Electron Diffraction Techniques*; Cowley, J. M., Ed.; Oxford Univ. Press: NY, 1993; Vol. 2, p. 309.
31. Dorset, D. L.; Snyder, R. G. *J. Phys. Chem.* **1996,** *100*, 9848.
32. Dorset, D. L.; Snyder, R. G. *J. Phys. Chem. B* **1999,** *103*, 3282.
33. Snyder, R. G.; Srivatsavoy, V. J. P.; Cates, D. A.; Strauss, H. L.; White, J. W.; Dorset, D. L. *J. Phys. Chem.* **1994,** *98*, 674.
34. Dorset, D. L.; Snyder, R. G. *Macromolecules* **1995,** *28*, 8412.
35. Dorset, D. L. *J. Phys. Chem. B* **1997,** *101*, 4870.
36. Teare, P. W. *Acta Crystallogr.* **1959,** *12*, 294.
37. Dorset, D. L. *J. Phys. D. Appl. Phys.* **1997,** *30*, 451.
38. Dorset, D. L. *Acta Crystallogr.* **1995,** *B51*, 1021.
39. Chichakli, M.; Jessen, F. W. *Ind. Eng. Chem.* **1967,** *59*, 86.
40. Clavell-Grunbaum, D.; Strauss, H. L.; Snyder, R. G. *J. Phys. Chem. B* **1997,** *101*, 335.
41. Strobl, G.; Ewen, B.; Fischer, E. W.; Piesczek, W. *J. Chem. Phys.* **1974,** *61*, 5257.
42. Dorset, D. L. *Z. Kristallogr.* **2000,** *215*, 190.
43. Le Roux, J. A.; Loubser, J. H. *S. Afr. J. Sci.* **1983,** *76*, 157.
44. Lourens, J. A. J.; Reynhardt, E. C. *J. Phys. D. Appl. Phys.* **1979,** *12*, 1963.
45. Dorset, D. L.; Basson, I. *J. Phys. D. Appl. Phys.* **2000,** *33*, 2657.
46. Stokhuyzen, R.; Pistorius, C. W. F. T. *J. Appl. Chem.* **1979,** *20*, 1.
47. Dorset, D. L. *J. Phys. Chem. B* **2000,** *104*, 8346.
48. Gupta, A. K.; Severin, D. *Petrol. Sci. Technol.* **1997,** *15*, 943.
49. Rademeyer, M.; Dorset, D. L. *J. Phys. Chem. B* **2001,** *105*, 5139.
50. Dorset, D. L. *Energy Fuels* **2000,** *14*, 685.
51. Dorset, D. L. *J. Phys. D. Appl. Phys.* **1999,** *32*, 1276.
52. Dorset, D. L. *J. Phys. Chem. B* **2000,** *104*, 4613.
53. Hu, H. L.; Dorset, D. L. *Acta Crystallogr.* **1989,** *B45*, 283.
54. Zhang, W. P.; Dorset, D. L. *J. Polymer Sci. B. Polymer Phys.* **1990,** *28*, 1223.
55. Dorset, D. L. *Macromolecules* **1999,** *32*, 162.
56. Dorset, D. L. *J. Phys. Chem. B* **2000,** *104*, 10543.
57. Prasad, A.; Mandelkern, L. *Macromolecules* **1989,** *22*, 914.
58. Dorset, D. L. *Z. Kristallogr.* **2001,** *216*, 234.
59. Abrahamsson, S. *Ark Kemi* **1959,** *14*, 65.
60. Ungar, G.; Stejny, J.; Keller, A.; Bidd, I.; Whiting, M. C. *Science* **1985,** *229*, 386.

Structure, Chirality, and Formation of Giant Icosahedral Fullerenes and Spherical Graphitic Onions[a,b]

Mauricio Terrones[c], Guillermo Terrones[d], and Humberto Terrones[c]

Abstract

We describe the topology, structure, and stability of giant fullerenes exhibiting various symmetries (I, I_h, D_{2h}, T). Our results demonstrate that it is possible to create two new families of nested "chiral" icosahedral (I) fullerenes namely $C_{260}@C_{560}@C_{980}@C_{1520}@$ … and $C_{140}@C_{380}@C_{740}@C_{1220}@$ …, which exhibit interlayer separations of ca. 3.4 Å. These chiral fullerenes are thought to possess metalliclike conduction properties. We discuss in detail the transformation of polyhedral graphitic particles into quasispherical nested giant fullerenes by reorganization of carbon atoms, which result in the formation of additional pentagonal and heptagonal carbon rings. These "spherical" structures are metastable and we believe they could be formed under extreme conditions, such as those produced by high-energy electron irradiation. There is circumstantial experimental evidence for the presence of heptagonal rings within these spherical fullerenes.

Keywords

Fullerenes • Spherical onions • Nested fullerenes • Stability and symmetry

Introduction

Graphene sheets, which consist of honeycomb lattices of sp^2 hybridized orbitals of carbon atoms, have proved to be extremely flexible and, as a result, novel nanostructures have been generated. For example, it is possible to curve these sheets by introducing pentagonal rings so as to form closed cages known as fullerenes, e.g., C_{60} and C_{70} [1–3], polyhedral and onionlike particles [4, 5], and elongated fullerenes termed nanotubes [6]. Combinations of five, six, and seven carbon rings (an arrangement of positive and negative curvatures) [7–11] can also lead to other fascinating structures, such as corkscrewlike tubules [7], toroids [12, 13], bent tubular structures [8, 9], nanoboxes [14], or cones [15–17].

Sumio Iijima was the first to report the existence of nested carbon nanocages (known as graphitic onions) using high-resolution transmission electron microscopy (HRTEM), when studying the structure of amorphous carbon films prepared by thermal vacuum deposition (Fig. 1) [4]. In 1988, Kroto and McKay proposed, for the first time, that graphitic onions consisted of nested icosahedral fullerenes ($C_{60}@C_{240}@C_{540}@C_{960}$…) containing only pentagonal and hexagonal carbon rings (note that C_{60} has icosahedral symmetry) [18, 19]. One reason for considering these icosahedral structures is that the average separation between shells is close to that of graphite (3.35 Å). Four years later, Ugarte widened our appreciation of the curved world of carbon structures by observing the transformation of polyhedral graphitic particles into almost spherical carbon onions [5], when irradiating the

[a] *Structural Chemistry* 2002, 13(3/4):373–384.

[b] This contribution is part of a collection entitled Generalized Crystallography and dedicated to the 75th anniversary of Professor Alan L. Mackay, FRS.

[c] Advanced Materials Department, IPICYT Potosí Institute of Scientific and Technological Research, Venustiao Carranza 2425-A, San Luis Potosí 78210, Mexico

[d] Process Science and Engineering Department, Pacific Northwest National Laboratory, P.O. Box 999, Richland, WA 99352, USA

I. Hargittai, B. Hargittai (eds.), *Science of Crystal Structures: Highlights in Crystallography*,
DOI 10.1007/978-3-319-19827-9_10

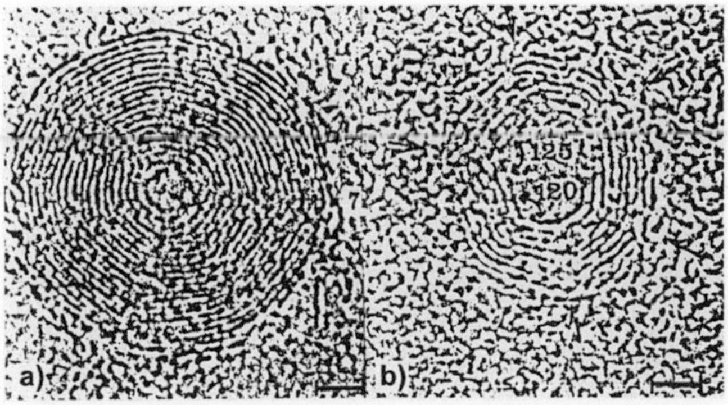

Fig. 1 Two examples (a and b) of HRTEM micrographs of spheroidal carbon particles, first observed by Iijima, exhibiting interlayer spacing close to 3.4 Å (Ref. [4]).

specimens with fast electrons inside an electron microscope. In this context, two of several questions that remain to be answered are: (1) why are onionlike structures quasispherical? and (2) how do graphitic carbon atoms rearrange in order to attain sphericity?

Since the 1970s, it has been known that the structure of carbon blacks (consisting of spherical arrangements of disordered sp^2 hybridized carbon and a few percentage of sp^3 hybridized carbon atoms) resembles that of graphitic onions. At that time, models consisting of platelets of graphite packed into spherical arrangements (Fig. 2) [20, 21] were proposed. However, these models did not account for the presence of pentagonal carbon rings embedded within the hexagonal network. Carbon blacks are now widely used as fillers in order to modify mechanical, electrical or optical properties of various materials.

In 1994, Sarkar and Kroto [22], proposed that the round structures reported by Ugarte could be generated theoretically by removing the cusps (pentagons and neighboring atoms) from the icosahedral fullerenes. The study then suggested that hydrogen or other atoms, such as oxygen or sulfur, stabilized the open-shell arrangement (eliminating dangling bonds). Nowadays, it is possible to generate larger amounts of "graphitic" onions using carbon ion bombardment

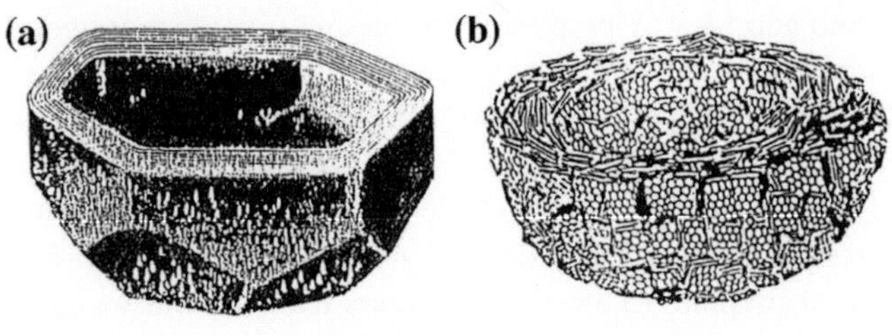

Fig. 2 Models of two possible structures of partially graphitized carbon black particles, which could be considered as concentric giant fullerenes. However, these models never considered the presence of pentagonal rings within the hexagonal network (Refs. [20, 21]).

of silver and copper targets and by arcing graphite rods in distilled water [23, 24]. However, the isolation and the bulk generation of this material have yet to be achieved. More recently, other novel methods for producing carbon onions have been developed. For instance, it is possible to produce bulk amounts of carbon nano-onions by radio-frequency plasmaenhanced chemical vapor deposition over Co and Co-SiO_2 catalysts [25].

In this paper, we demonstrate that energy minimization of spherical giant fullerenes is associated with the introduction of additional heptagons and pentagons, a result which may shed light on the mechanism whereby polyhedral graphitic particles transform into round onions. Finally, an energetic study of novel fullerene chiral families exhibiting icosahedral and tetrahedral symmetries is discussed in detail.

Euler's Law in Graphitic Structures

A clear understanding of the geometry and morphology of carbon structures, which consist of hexagonal sp^2 hybridized carbon (e.g., graphite), can be attained by applying Euler's law. The general form of this law can be expressed as $F-E+V=2$, where F is the number of faces, E is the number of edges, V is the number of vertices, and 2 is the Euler characteristic, which is a geometric invariant that characterizes topologically equivalent structures. In the case of all closed cages (polyhedra) = 2, for example in fullerene C_{60}, $F=32$ (12 pentagons and 20 hexagons), $E=90$, $V=60$, and = 2. All closed polyhedra display the same topology as the sphere.

As discussed elsewhere [26], it can be shown that Euler's geometric invariant is related to the genus g (number of handles in a structure) by

$$\chi = 2(1-g) \tag{1}$$

It can be shown that for a sphere = 2 and $g=0$, whereas for a torus = 0 and $g=1$ (one handle). For sp^2 hybridized carbon, such as graphene, and considering pentagonal, hexagonal, heptagonal, and octagonal rings of carbon, we can express

$$F = N_5 + N_6 + N_7 + N_8 \tag{2}$$

and

$$2E = 3V = 5N_5 + 6N_6 + 7N_7 + 8N_8 \tag{3}$$

where N_n is the number of polygons with n sides. Substituting (2) and the constraint (3) into Euler's law yields

$$N_5 - N_7 - 2N_8 = 6\chi = 12(1-g) \tag{4}$$

Within the graphitic structure, the pentagons, heptagons, and hexagons contribute to the curvature and topology of the arrangement and as such can be regarded as "defects." Notice that the number of hexagons is not present in Eq. (4) because they do not contribute to the curvature of the system.

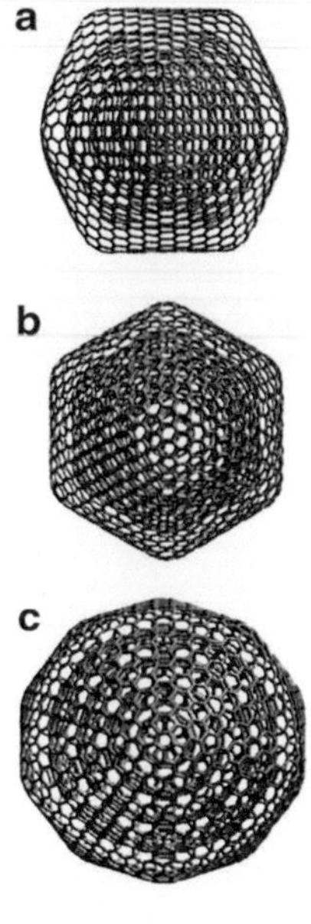

Fig. 4 Molecular models of the icosahedral $C_{60}@C_{240}@C_{540}@C_{960}@C_{1500}$ family proposed by Kroto and McKay showing: (**a**) The 2-fold axis of symmetry; (**b**) The 3-fold axis of symmetry; (**c**) The 5-fold axis of symmetry.

Stability of Giant Icosahedral Fullerenes and Geometrical Rules

Due to the large number of carbon atoms in most of the giant fullerene systems (>250–500 atoms), we have carried out molecular mechanics calculations using the DREIDING force field [27–29] in conjunction with the Tersoff potential [30]. The latter accurately reproduces cohesive energies and elastic properties of graphite and diamond (sp^2 and sp^3 carbon systems) [31, 32]. A comparison of this method with *ab initio* techniques for these giant molecules [33] validated the suitability of the force-field/Tersoff method for studying the systems described herein. Spatial coordinates, generated by the DREIDING force-field method for giant fullerenes were introduced [34] and elastic energies were computed for several graphitelike structures.

An icosahedral carbon fullerene can be constructed by fusing 20 triangular graphene segments, joined by 12 pentagons, one per vertex [35] (Fig. 3). As mentioned earlier, the first proposed family of concentric giant fullerenes ($C_{60}@C_{240}@C_{540}@C_{960}@C_{1500}@\ldots$) [18, 36] (insert, Fig. 4), exhibiting the same symmetry as C_{60} (I_h), revealed an intershell graphitic spacing of ca. 3.35 Å. The single-shell fullerenes belonging to this family can be constructed using a triangular template, consisting of a three sided graphene sheet delimited by pentagonal rings at the corners. Other giant fullerenes with I_h symmetry have also been generated, using the C_{20} cage [37–39] as a template. In this case, the sequence of consecutive fullerenes ($C_{20}@C_{80}@C_{180}@C_{320}@C_{500}@C_{720}@C_{980}@\ldots$) is not energetically favorable for graphitic onions due to a discrepancy within the intershell spacing (ca. 2 Å) as compared to that in graphite (ca. 3.35 Å). These families of giant fullerenes consist of nonhelical arrangements within the triangular templates, due to the orientation of pentagonal rings with respect to the hexagonal graphene network (Fig. 5). In other words, if the pentagons are not connected in a straight line (parallel or perpendicular) through hexagonal linear rings, the arrangement is termed helical or chiral (Fig. 5).

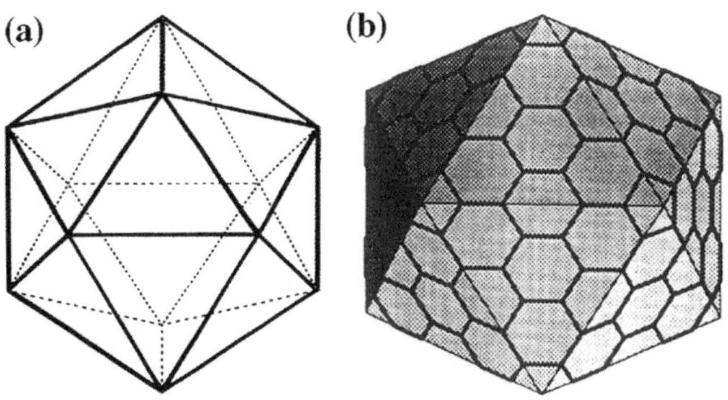

Fig. 3 (**a**) Icosahedron and (**b**) decorated icosahedron with hexagonal and pentagonal carbon rings, showing that the template necessary to generate such structure is an equilateral triangle.

A rule for creating icosahedral fullerenes can be generalized using an equilateral triangle as a building block [35]. In order to create this triangle on a graphene sheet, we first choose an origin [*(0,0)* within the sheet], attaching a pentagonal ring as shown in Fig. 6a. Subsequently, a lattice point, denoted (m, n), is placed in the sheet, and another pentagonal ring is assigned next to it (Fig 6b). At this stage, a third pentagon is located equilaterally with respect to the other two, thus generating a perfect triangle (Fig. 6c), which can be used as the template for constructing an icosahedron. The lattice point (m, n) selected to generate the fullerene determines the number of carbon atoms N_C within the icosahedral shell

$$N_C = 20\left(m^2 + mn + n^2\right) \qquad (5)$$

and the cage diameter d

$$d = \frac{5\sqrt{3}}{\pi} a_{C-C} \left(m^2 + mn + n^2\right)^{1/2} \qquad (6)$$

where a_{C-C} is the average carbon–carbon distance (ca. 1.44 Å) within the hexagonal network or fullerene cage.

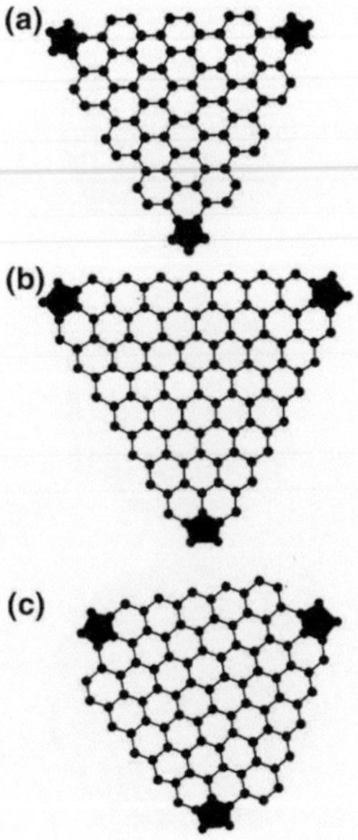

Fig. 5 (**a**, **b**) Nonhelical configurations within the triangular building template necessary for building icosahedral fullerenes; (**c**) helical arrangement, in which pentagons are not aligned through the hexagonal rings.

In particular, it is possible to construct three different families of icosahedral graphitic onions in which their interlayer spacing among shells is ca. 3.4 Å. The *n*th term and the first four elements of the sequence of these families of graphitic onions are shown below.

(i) I_h family (armchair) when $m=n$

$$\begin{aligned} N_C &= 60n^2 \\ (1,1) &= 60 \\ (2,2) &= 240 \\ (3,3) &= 540 \\ (4,4) &= 960 \end{aligned}$$

(ii) I family (helical arrangement) when $m=n+2$

$$\begin{aligned} N_C &= 20\left(3n^2+6n+4\right) \\ (2,0) &= 80 \\ (3,1) &= 260 \\ (4,2) &= 560 \\ (5,3) &= 980 \end{aligned}$$

We should note that C_{80} does not belong to the family, but it could be added in order to have smaller shells in the interior of the onion.

(iii) I family (helical arrangement) when $m=n+1$

$$\begin{aligned} N_C &= 20\left(3n^2+3n+1\right) \\ (2,1) &= 140 \\ (3,2) &= 380 \\ (4,3) &= 740 \\ (5,4) &= 1220 \end{aligned}$$

Using the above procedure, any fullerene or giant fullerene exhibiting icosahedral symmetries (I or I_h) can be generated. Furthermore, structures with other symmetries such as octahedral and tetrahedral can also be created using this method, in addition to other restrictions [39]. In particular, two other new families with icosahedral (I) symmetry, exhibiting intershell spacings close to that of graphite can be generated: (1) C_{260}@C_{560}@C_{980}@C_{1520}@… and (2) C_{140}@C_{380}@C_{740}@C_{1220}@ … These nested fullerenes are helical and possess chiral images with two possible isomers per icosahedral cage (Fig. 7). In any icosahedral cage, as the number of carbon atoms increase, the more faceted the structure becomes. This is due to the fact that only 12 pentagonal rings are necessary to close any cage. Therefore, for systems greater than 540 carbon atoms (which contain more hexagonal rings), sphericity decreases dramatically (Fig. 8).

Laser vaporization of graphite has resulted in the formation of high mass fullerenelike clusters in the 300–350 and 600–700 carbon atom range [40]. This observation provides circumstantial evidence for the formation of icosahedral concentric fullerenes as stable forms of carbon. Specifically, the masses observed in the 300–350 atom range may correspond to single or multiple shells: C_{60}@C_{240}; C_{80}@C_{260}; or to any single helical shell within the same mass range. Furthermore, C_{240}@C_{540}, or single clusters containing roughly the same number of atoms, may match with higher fullerenes reported in the 600–700 carbon atom range.

HRTEM, in conjunction with electron irradiation studies of particles generated by arcing graphite electrodes in inert atmospheres, revealed the presence of graphitelike onions corresponding to: single-shell C_{240} [41], C_{50}@C_{230} [42], C_{120}@C_{390}@C_{760} [42], C_{60}@C_{240}@C_{540}@C_{930} [41, 42], C_{70}@C_{290}@C_{680}@C_{1200}@C_{2000} [42]. These observations suggest that it may be possible to produce helical as well as nonhelical concentric fullerene cages. Moreover, electron diffraction studies of single polyhedral particles reveal *hk0* reflections indicating that the stacking of the graphitic shells is disordered (turbostratic) [43]. This may also be due to helical (I symmetry) concentric cages, which generate turbostraticlike arrangements within the onions.

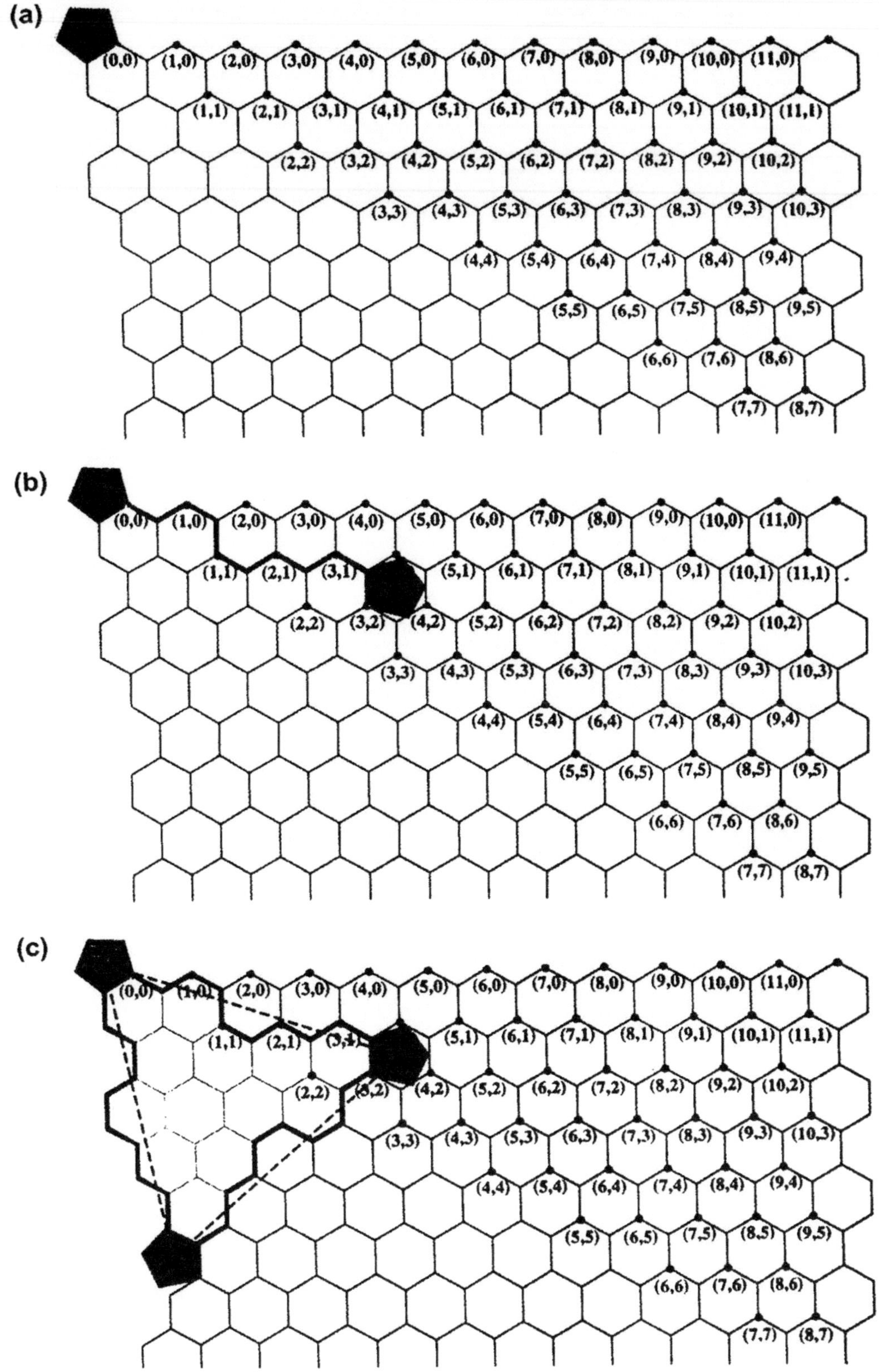

Fig. 6 Procedure followed to generate an icosahedral fullerene: (**a**) select an origin *(0, 0)* within the graphene sheet and attach a pentagonal ring next to it; (**b**) choose a lattice point *(m, n)* and place another pentagon next to it; (**c**) locate a third pentagon equilaterally to the other two pentagonal rings, thus generating the triangular building template. Note that *(m, n)* determine the number of atoms and the average diameter of the fullerene.

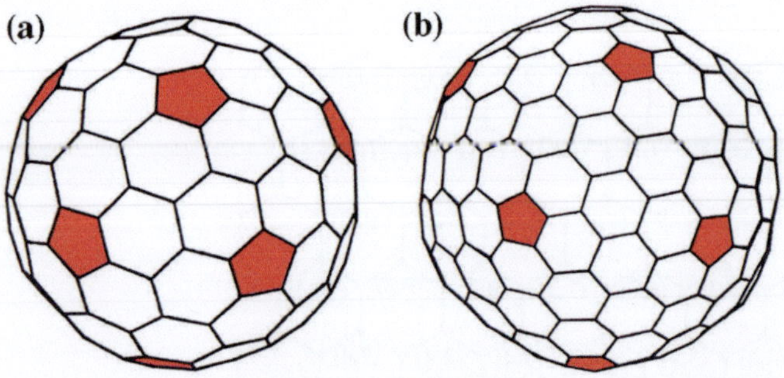

Fig. 7 Generated icosahedral (I) fullerenes with a helical arrangement; (**a**) C_{140}; (**b**) C_{260}. The families ($C_{80}@C_{260}@C_{560}@C_{980}@C_{1520}@\ldots$ and $C_{140}@C_{380}@C_{740}@C_{1220}@\ldots$) are predicted to be metallic (M. I. Heggie, private communication), in contrast to the I_h family ($C_{60}@C_{240}@C_{540}@C_{960}@C_{1500}$), which is considered a semiconductor.

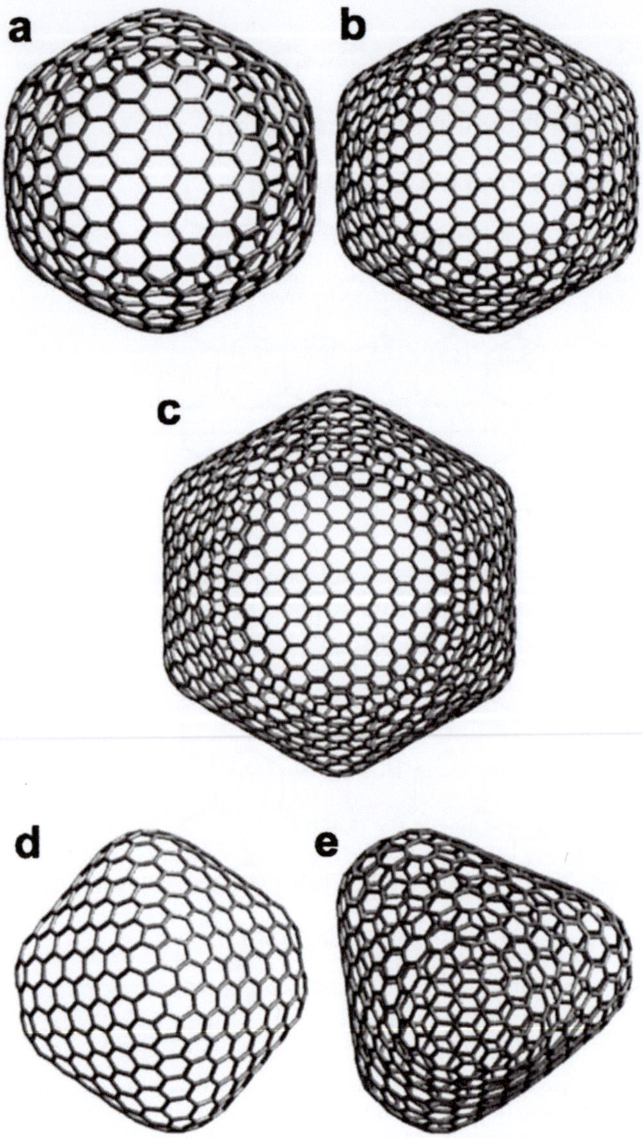

Fig. 8 Fullerenes with 12 pentagonal rings. The faceting increases in icosahedral fullerenes as the structure becomes larger. (**a**) C_{540} (I_h); (**b**) C_{960} (I_h); (**c**) C_{1500} (I_h). Other symmetries: (**d**) D_{2h} with 600 atoms; (**e**) T_d with 660 atoms.

It has been demonstrated experimentally that smallsized and multishell "bucky onions" behave like semiconductors [44]. It has been also noted that as their size increases they exhibit metalliclike behavior because of their nonlinear conducting properties. It is conceivable that single-shell giant fullerenes may be metallic [33], thus propelling viable candidates for the construction of single electron devices. However, the electronic properties of chiral single-shell onions have not yet been reported.

Polyhedral Particles and Graphitic Onion Formation

In 1980, when Iijima first observed (by HRTEM) onionlike graphitized nanoparticles [4] with an interlayer spacing of ca. 3.44 Å, he noted that the angles intersecting the graphitic planes corresponded to ca. 120 ± 20° He then concluded that these angles were fairly close to the ideal angle (109°) of a tetrahedral sp^3 carbon attached to graphite. He explained the latter in terms of tetrahedral bonds or bent trigonal graphitic systems, possibly due to the insertion of pentagonal carbon rings or to the presence of dangling bonds into the graphitic system [4]. Three years after the discovery of C_{60}, Kroto and McKay [18] also suggested that Iijima's structures might be explained as quasi-icosahedral concentric carbon shells that grew through a spiraling mechanism as a result of introducing pentagonal rings into the hexagonal network (Fig. 9).

Ugarte's experiments demonstrated that high electron irradiation over polyhedral particles (generated in the fullerene Krätschmer–Huffman reactor) was able to produce quasispherical graphitic arrangements that appeared to be stable [5]. In 1996, Banhart and Ajayan [45] also showed that high electron doses at high temperatures (700–800 °C) resulted in the formation of perfect spherical onions. Continued irradiation at this temperature was responsible for a contraction of the intershell separation, which resulted in the formation of diamond (self-compression phenomenon).

High-energy electron irradiation is capable of producing transformations and atomic rearrangements within the irradiated specimen. These transformations may be due to: (1) bond rupture due to electron excitations; (2) momentum transfer to the atomic nuclei, which may result in the atom displacement to the interstitial sites (*knock-on* effect); (3) excitation of vibrational or rotational modes of the molecules; and (4) energy transfer (loss) from the fast electrons to the specimen, thus raising the average temperature of the irradiated sample [5, 39].

In particular, Ugarte's experiments consisted of irradiating carbon polyhedral particles with a very intense electron beam in a HRTEM, using electron densities 10–20 times higher than those found under normal operating conditions (ca. 10 Å/cm^2) [5]. He noticed that the polyhedral particles transformed gradually into quasispherical graphitic onions, a

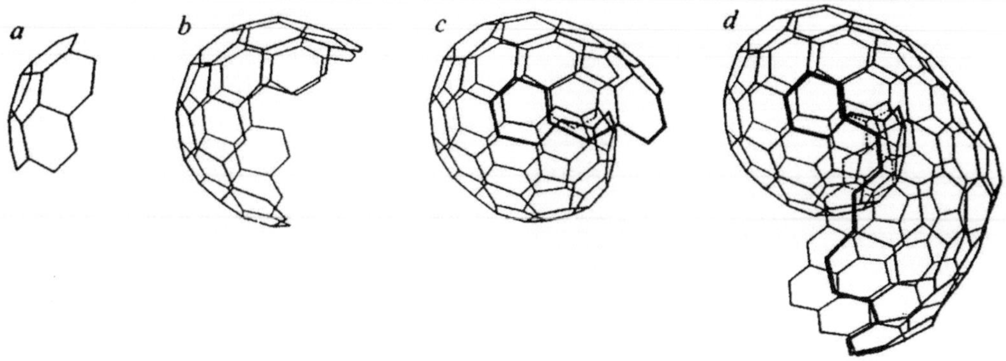

Fig. 9 Hypothetical initial growth sequence (proposed by Kroto and McKay) evolving from: (**a**) a corannulene molecule; (**b**) and (**c**) species in which edge by-pass occurs to (**d**), an embryo in which the second shell is forming. The sequence leads in a natural way to both carbon particles and closed cages, fullerenes (Courtesy of H. W. Kroto).

transformation which can be divided into several steps [46]. Figure 10 shows this sequence. Initially, the polyhedral particle collapses occupying the inner hollow space, thus causing the generation of a disordered particle (Fig 10b). Subsequently, reorganization of the outer layers begins (Fig. 10c), progressing by a kind of internal epitaxy to a final structure, which contains small inner shells, usually very close to the diameter of C_{60} (ca. 1 nm; Fig. 10d).

Ugarte proposed that their transformation under irradiation was caused by the electron beam (resembling high-temperature regimes), thus promoting bond rupture and allowing a degree of structural fluidity [5, 42] responsible for a gradual reorganization into quasispherical arrangements. Notwithstanding this suggestion, Banhart and co-workers discussed [41, 47, 48] onion formation primarily based upon *knock-on* collisions from electrons and carbon atoms, concluding that heating or electronic excitation do not dominate the process [47, 48]. These authors used much higher-energy electron beams (1250 KeV) in order to confirm the latter. It is important to note that high temperature regimes within the process are localized for a few picoseconds. Therefore, it is unlikely that this could lead to an explanation for bucky-onion formation.

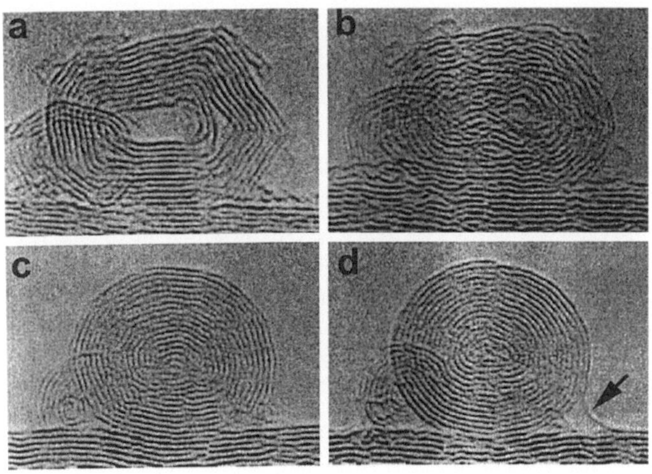

Fig. 10 Formation of quasispherical onionlike particles under electron irradiation. (**a**) polyhedral particle as obtained from the electric arc; (**b**) after 20 s of irradiation the particle collapses and the inner central void disappears; (**c**) after 40 s reordering continues at the surface toward the center progressively; (**d**) 180 s further irradiation yields to quasispherical onionlike particle with a very small central shell (interlayer spacing 0.34 nm; Courtesy of Daniel Ugarte; see also Ref. [46]).

The quasispherical arrangements (onions formed by two to four shells) were reported to be very stable under intense electron bombardment [5, 41]. However, further investigations [47] regarding their stability after irradiation revealed that (when deflecting the electron beam from the irradiated area), the outermost shells dissolved into noncrystalline material, losing their spherical shape and decaying completely into disordered graphitic carbon. Therefore, spherical onions produced by electron irradiation are, indeed, *metastable forms* of carbon.

Sphericity of Graphitic Onions

For giant icosahedral fullerenes containing only 12 pentagons, faceting (i.e., progressive loss of sphericity; Fig. 11) is enhanced as the number of hexagons increases. Transformations of hexagons by heptagon/pentagon pairs will thus increase sphericity at the expense of faceting (discussed by Hare and Fowler) [49]. According to Euler's Law, it is possible to introduce heptagons into a fullerene provided that each heptagon is matched by an additional pentagon, in order to compensate for the negative curvature associated with the heptagon (i.e., $N_5-N_7=12$).

In this section, through energy minimization calculations using the DREIDING force field [27–29] and the Tersoff potential [30], we demonstrate that quasispherical structures containing pentagons and heptagons are metastable, and could be favored under high-energy electron irradiation. From the experimental point of view (Fig. 10), it is possible that the strained regions of polyhedral particles

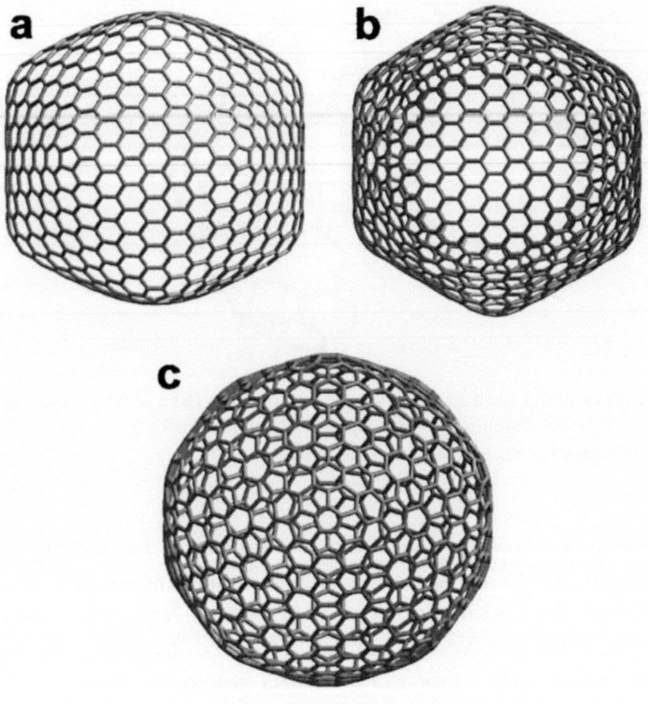

Fig. 11 Molecular model of C_{940} oriented along various symmetry axes: (**a**) C_2; (**b**) C_3; (**c**) C_5. It is clear that as the number of hexagons increases the more faceted the structure becomes.

(Fig. 10a) are selectively disrupted by electron irradiation with ejection of carbon fragments by knock-on effects centered upon the pentagonal cusps, thus generating graphitic domains containing holes (dangling bonds). Such perforated structures are likely to be flexible and could be easily deformed.

In order to generate quasispherical fullerenes theoretically, the following steps should be followed:

1. A polyhedral fullerene (generally exhibiting icosahedral symmetry) possessing hexagons and 12 pentagonal rings is relaxed using the DREIDING force field with the conjugate gradient method in order to find the local minimum of the structure. Subsequently, the Tersoff potential is used to obtain the elastic energy of the structures (For graphite this potential prefers bond distances of 1.46 instead of 1.42 Å) [39].
2. The pentagonal carbon rings and adjacent atoms of the icosahedral fullerene are removed, resulting in perforated structures containing dangling bonds (for larger fullerenes, several carbon atoms surrounding the pentagons need to be removed). It should be noted that the perforations created in the faceted structure emulate experimental observations witnessed under high-electron irradiation (Fig. 10a, b), in which *strained* regions of the structure break and/or fragment, leading to less rigid arrangements [22]. After relaxing using the DREIDING force field in conjunction with the Tersoff potential, these *holey* structures become rounder containing dangling bonds at the edges (Fig. 12). From the experimental point of view, it must be borne in mind that such dangling bonds are energetically costly. Therefore, these voids are likely to be quickly occupied with carbon atoms, possibly by the introduction of heptagon/pentagon arrays (see point 3).
3. The perforations within the *holey* fullerene are subsequently closed by introducing more pentagons (more than 12) and additional heptagons, responsible for maximizing sphericity (Fig. 13).

We believe that, under the electron beam, the rearrangements are caused by *rupture* and/or *knock-on* with subsequent reaggregation of strained regions (pentagonal cusps and defects), in which broken carbon chains, delocalized atoms, interstitials or graphitic fragments arrange in more stable spherical-concentric configurations. Van der Waals interactions should also play a positive role in this context, because a constant interlayer distance between nesting spheres must be maintained.

According to our model, some random reaggregation takes place in order to satisfy dangling bonds once bond rupture occurs. In addition, the initial symmetry of the precursor particle (parent-fullerene) needs to be broken after heptagons and pentagons are introduced in the cage, thus resulting in a more spherical structure (Fig. 13).

The calculations begin with symmetrical polyhedral particles (icosahedral, D_{2h} and tetrahedral), possessing only hexagonal and 12 pentagonal rings of carbon (Tables 1 and 2).

For the D_{2h} system for icosahedral fullerenes with 600 and 540 atoms, respectively (Fig. 13), it is found that the most stable and spherical structure contained 540 atoms, with was obtained from a D_{2h} fullerence with 600 atoms (D_{2h} C_{540}-A in Table 3 and Fig. 13). In this case the structure has lost 60 atoms during rearrangement.

However, starting from icosahedral (I_h) C_{540}, we found highly spherical structures (ico C_{540}-A, B, C, Table 3), which were not as stable as the D_{2h} C_{540}-A. The data from Tables 1, 2, and 3 are summarized in Fig. 14. For quasispherical fullerenes with additional pentagons and heptagons, the stability decreases due to a reduction in the number of hexagonal rings, confirming that these spherical isomers are *metastable arrangements*.

Another important result is provided by a Stone–Wales-type transformation [50] (Fig. 15), which converts icosahedral fullerenes into quasispherical fullerenes [39, 51–53], with minimal expenditure of energy. This transformation has been applied in the following cases: (1) to convert four adjacent hexagons into two pentagons and two hexagons; (2) three hexagons and one pentagon into two pentagons, one hexagon, and one heptagon; (3) two hexagons and two

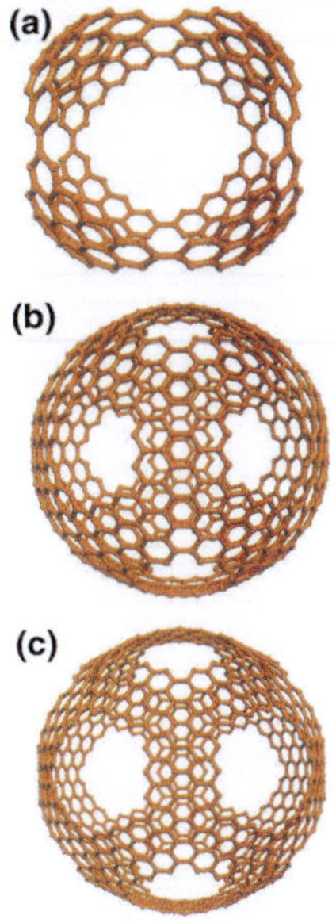

Fig. 12 Quasispherical fullerenes with holes obtained via energy minimization. (**a**) Precursor of D_{2h} C_{540}; (**b**) precursor of ico-C_{888}-A; (**c**) precursor of ico-C_{1500}-A.

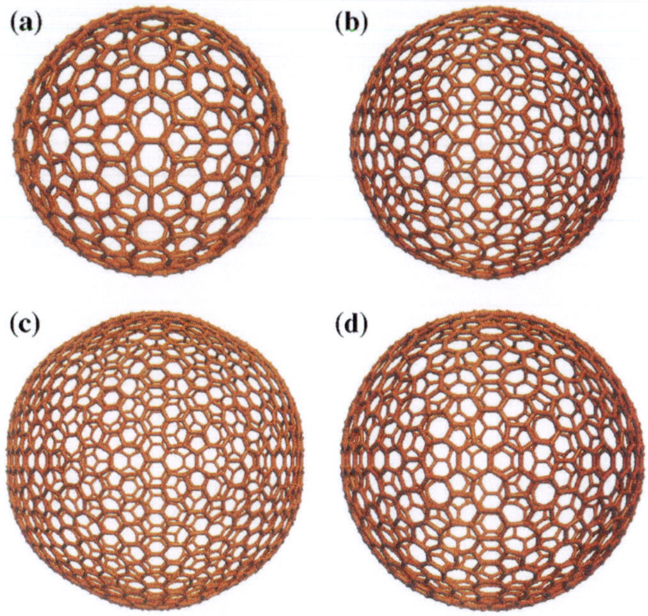

Fig. 13 Quasispherical fullerenes in which the holes have been filled with heptagonal and pentagonal rings. (**a**) D_{2h} C_{540} with 76 pentagons and 64 heptagons; (**b**) ico-C_{888}-A with 60 pentagons and 72 heptagons; (**c**) ico-C_{1500} with 132 pentagons and 120 heptagons (I_h symmetry); (**d**) C_{1020} obtained by adding 132 atoms to the ico-C_{888}, thus generating more heptagon–pentagon pairs responsible for a more spherical isomer.

pentagons into two pentagons and two hexagons. In the D_{2h} C_{540}, one Stone–Walestype transformation involving two pentagons and two hexagons needs 0.004 eV; on three hexagons and one pentagon requires 0.086 eV; four hexagons require 11.349 eV (Fig. 15). Starting with an icosahedral C_{540}, with heptagons and pentagons, a more spherical structure is obtained via several random Stone–Wales-type transformations, requiring 23.936 eV to generate the ico C_{540}-A fullerene (see Tables 1, 2, and 3).

It is noteworthy that quasispherical onions are not more stable than the symmetrical faceted giant fullerenes on which we based our calculations. However, under the conditions of formation (high-energy electron irradiation), quasispherical structures possessing negative curvature are preferred. Experimentally, it has been found that the maximum energy transfer to the carbon atoms during the high-energy irradiation of 1250 keV is ca. 500 eV [47]. Therefore, the possibility to promote Stone–Wales-type defects is likely. In all cases, except for the D_{2h} C_{228}-A, the quasispherical fullerenes are more stable than C_{60} (see Table 3), provided the isolated pentagon rule is not violated.

To quantify the degree of sphericity in our structures, the distances r_i between each atom and the center of mass were computed. For a perfect spherical arrangement, r_i corresponds to the radius of a sphere. In the quasispherical structures studied, r_i fluctuates, however, all r_i lie within a small range Δr_i, calculated from the difference between the largest and shortest r_i (Table 3). It is noteworthy that r_i for icosahedral arrangements with just 12 pentagons increases as the fullerene grows. For example, C_{60} is a quasispherical icosahedron $\Delta r_i = 6.4\times10^{-4}$ (almost spherical), and can, therefore, be used as the first shell in an onion structure. In the quasispherical particles, the distance between layers is close to that observed experimentally (Fig. 16).

Table 1 Energy/atom of icosahedral (I_h) and tetrahedral fullerenes relative to graphite, mean radius of fullerene, and difference Δr_i of the largest and shortest distance to the center of mass of fullerene[a]

Structure	Energy/atom (eV)	Mean radius (Å)	Δr_i (Å)
C_{20} (I_h)	−1.7449	2.045	1.09×10^{-4}
C_{60} (I_h)	−0.6600	3.643	6.4×10^{-4}
C_{70} (D_{5h})	−0.6071	3.912	0.647
C_{80} (I_h)	−0.5576	4.182	8.58×10^{-4}
C_{180} (I_h)	−0.3597	6.282	0.1349
C_{240} (I_h)	−0.3136	7.253	0.198
C_{264} (T)	−0.2874	7.591	1.210
C_{320} (I_h)	−0.2586	8.372	0.388
C_{500} (I_h)	−0.1836	10.452	0.753
C_{540} (I_h)	−0.1748	10.860	0.774
C_{660} (I_h)	−0.1305	11.904	4.071
C_{720} (I_h)	−0.1308	12.522	1.144
C_{960} (I_h)	−0.0987	14.440	1.570
C_{980} (I_h)	−0.0961	14.588	1.607
C_{1248} (T)	−0.0753	16.297	8.257
C_{1500} (I_h)	−0.0631	18.014	2.415

[a]All structures possess 1.46 Å C-C bond lengths, the parameter needed for the Tersoff potential

Table 2 Energy/atom of icosahedral/helical (*I*) fullerenes relative to graphite, mean radius of fullerene and difference Δr_i of largest and shortest distance to center of mass of fullerene[a]

Structure	Energy/atom (eV)	Mean radius (Å)	Δr_i (Å)
C_{140} (*I*)	–0.4148	5.539	0.090
C_{260} (*I*)	–0.2789	7.662	0.275
C_{380} (*I*)	–0.2317	9.121	0.496
C_{420} (*I*)	–0.2020	9.7330	0.640
C_{560} (*I*)	–0.1673	11.057	0.851
C_{740} (*I*)	–0.1274	12.694	1.228
C_{980} (*I*)	–0.0961	14.588	1.608

[a]All structures possess 1.46 Å C-C bond lengths, the parameter needed for the Tersoff potential

Table 3 Energy/atom of quasispherical fullerenes relative to graphite, mean radius of fullerene and difference Δr_i of largest and shortest distance to center of mass of fullerene[a]

Structure	Energy/atom (eV)	Mean radius (Å)	Δr_i (Å)
D_{2h} C_{228}-A	–0.8095	7.103	0.399
D_{2h} C_{540}-A	–0.5591	10.962	0.492
D_{2h} C_{540}-B	–0.5588	10.967	0.803
ico C_{540}-A	–0.6287	10.958	0.539
ico C_{540}-B	–0.6002	10.956	0.656
ico C_{540}-C	–0.5987	10.955	0.693
ico C_{888}-A	–0.3648	14.008	0.699
ico C_{1020}-B	–0.5952	15.067	0.159
ico C_{1500}-A	–0.5539	18.201	0.591
ico C_{1620}-B	–0.5938	18.939	0.976

[a]All structures possess 1.46 Å C-C bond lengths, the parameter needed for the Tersoff potential

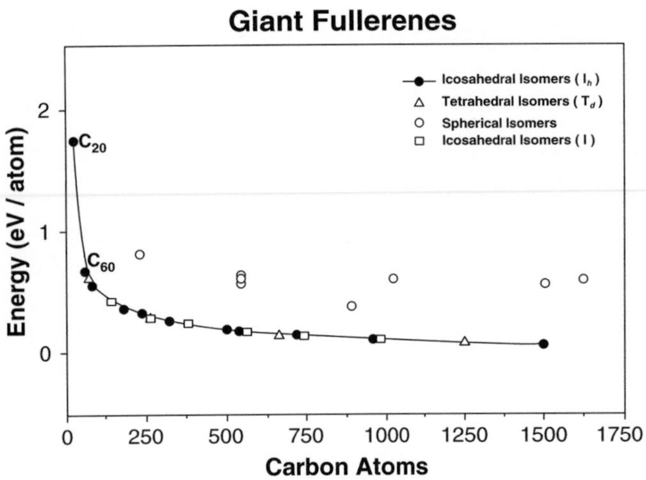

Fig. 14 Energy/atom relative to graphite versus carbon atoms within the fullerenes exhibiting different symmetries and shapes. All fullerenes with twelve pentagonal rings only lie in the same curve (e.g., I_h, I, T_d symmetries). All spherical isomers possess higher energies and, therefore, may be considered as metastable configurations. Energies were obtained using DREIDING force fields and Tersoff potential with 1.46 Å C-C distance.

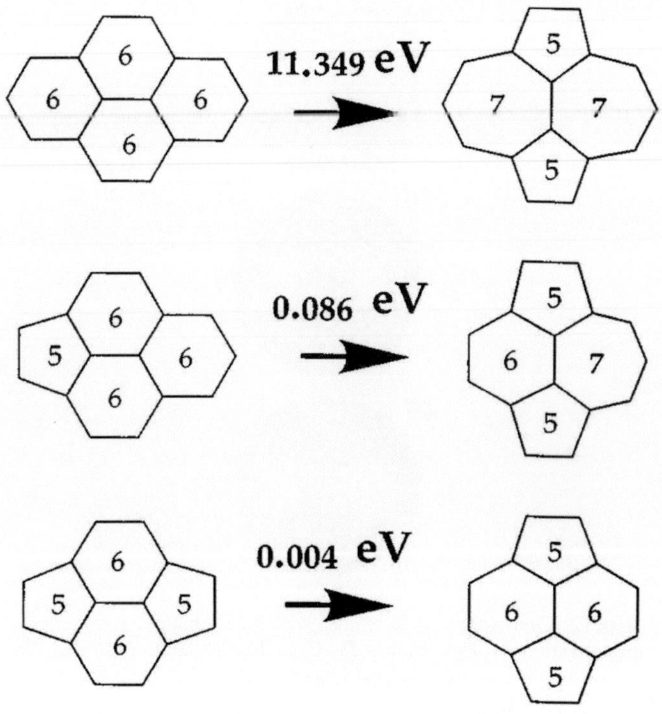

Fig. 15 Energy cost of different Stone–Wales defects for the fullerene D_{2h} C_{540}.

An important experimental observation is the presence of negative curvature [54, 55] (possibly heptagonal rings) in the exterior shells of graphitic onions (see arrow Fig. 10d) [46, 56, 57]. Therefore, it is likely that heptagonal rings are also formed within the carbon shell. Along this line, electron irradiation over diamond, using an intense flux (>10^7 electrons/nm^2 at 200 keV), yields concentric spherical graphitic shells. In the latter, a three step mechanism was followed: comprising displacements, migration, and aggregation [58]. However, further theoretical studies explaining these transformations are awaiting development.

Conclusions

It has been shown that fullerene cages, exhibiting icosahedral symmetries, can easily be generated using equilateral triangles as building templates. As a consequence, two new families of graphitic onions possessing *I* symmetry have been created: (1) C_{260}@C_{560}@C_{980}@ C_{1520}@…and (2) C_{140}@ C_{380}@C_{740}@C_{1220}@…. Experimental evidence for their possible existence has also been discussed. Molecular simulation programs based upon force-field techniques and the Tersoff potential had to be considered when dealing with large atomic systems (>250 atoms). In addition, a model which explains the transformation of faceted layered structures into quasispherical graphitic onions under high electron irradiation

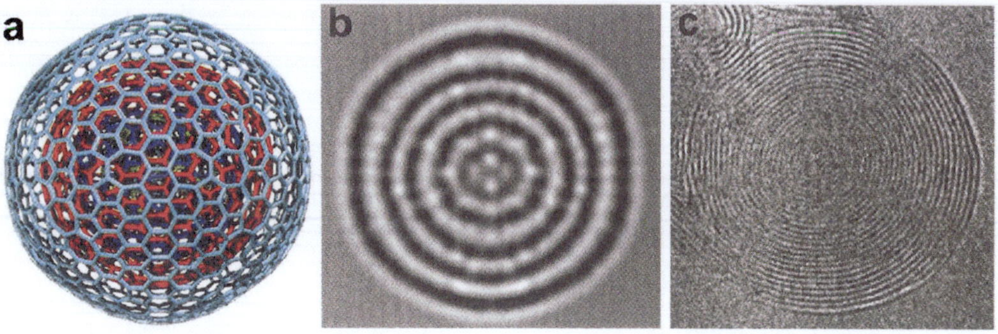

Fig. 16 Nested quasispherical fullerenes created by introducing heptagonal and pentagonal carbon rings. The onion is composed of C_{60}@C_{240}@$D_{2h}C_{540}$@C_{888}-A@C_{1500}-A. (**a**) Molecular simulation model of the minimized structure; (**b**) HRTEM simulation of (**a**); (**c**) experimental spherical carbon onion obtained under high-energy irradiation (interlayer spacing ca. 3.4 Å). Note the almost perfect sphericity of the structure.

has been proposed. This model is based upon the destruction of strained regions, which include pentagonal rings, in order to obtain fullerenes with holes that have more flexibility and which can be filled in different ways with heptagonal and pentagonal carbon rings, generating crystalline or amorphous quasispherical onions. It is believed that the aggregation process, which fills the holes, is random. The introduction and/or production of heptagonal rings via Stone–Wales-type transformations is supported by experimental (HRTEM) observations in onions with negative curvature and related systems reported in the literature. Therefore, quasispherical onions do not need necessarily to follow a predetermined sequence of sizes, such as $60n^2$, and can be considered as *metastable* carbon arrangements. We have ignored the van der Waals interaction. However, this plays an important role in stabilizing the onions, given the adequate interlayer spacing.

Acknowledgments We thank CONACYT-Mèxico grants W-8001millennium initiative (HT, MT), G-25851-E (HT, MT), and the European Union grant CNT-NET project contract G5RT-CT2001-05026 (MT), for financial support. Pacific Northwest National Laboratory is operated by Battelle for the U.S. Department of Energy under contract DE-AC06-76RLO 1830.

References

1. Kroto, H. W.; Heath, J. R.; O'Brien, S. C.; Curl, R. F.; Smalley, R. E. *Nature* (*London*) **1985,** *318*, 162.
2. Krätschmer, W.; Lamb, L. D.; Fostiropoulos, K.; Huffman, D. R. *Nature* (*London*) **1990,** *347*, 354.
3. Taylor, R.; Hare, J. P.; Abdul-Sada, A. K.; Kroto, H. W. *J. Chem. Soc. Chem. Commun.* **1990,** p. 1423.
4. Iijima, S. *J. Cryst. Growth* **1980,** *5*, 675.
5. Ugarte, D. *Nature* (*London*) **1992,** *359*, 707.
6. Iijima, S. *Nature* (*London*) **1991,** *354*, 56.
7. Amelinckx, S.; Zhang, X. B.; Bernaerts, D.; Zhang, X. F.; Ivanov, V.; Nagy, J. B. *Science* **1994,** *265*, 635.
8. Iijima, S.; Ichihashi, T.; Ando, Y. *Nature* (*London*) **1992,** *356*, 776.
9. Zhou, D.; Seraphin, S. *Chem. Phys. Lett.* **1995,** *238*, 286.
10. Mackay, A. L.; Terrones, H. *Nature* (*London*) **1991,** *352*, 762.
11. Terrones, H.; Fayos, J.; Aragón, J. L. *Acta Met. Mater.* **1994,** *42*, 2687.
12. Liu, J.; Dai, H.; Hafner, J. H.; Colbert, D. T.; Smalley, R. E.; Tans, S. J.; Dekker, C. *Nature* (*London*) **1997,** *385*, 780.
13. Martel, R.; Shea, H. R.; Avouris, P. *Nature* (*London*) **1999,** *398*, 299.
14. Saito, Y.; Matsumoto, T. *Nature* (*London*) **1998,** *392*, 237.
15. Sarkar, A.; Kroto, H. W.; Endo, M. *Carbon* **1995,** *33*, 51.
16. Krishnan, A.; Dujardin, E.; Treacy, M. M. J.; Hugdahl, J.; Lynum, S.; Ebbesen, T. W. *Nature* (*London*) **1997,** *388*, 451.
17. Terrones, H.; Hayashi, T.; Muñoz-Navia, M.; Terrones, M.; Kim, Y. A.; Grobert, N.; Kamalakaran, R.; Dorantes-Dávila. J.; Escudero, R.; Dresselhaus, M. S.; Endo, M. *Chem. Phys. Lett.* **2001,** *343*, 241.
18. Kroto, H. W.; McKay, K. G. *Nature* (*London*) **1988,** *331*, 328.
19. Maiti, A.; Brebec, C. J.; Bernholc, J. *Phys. Rev. Lett.* **1993,** *70*, 3023.
20. Heidenreich, R. D.; Hess, W. M.; Ban, L. L. *J. Appl. Crystallogr.* **1968,** *1*, 1.
21. Speck, J. S. Ph.D. Thesis, MIT (1989).
22. Sarkar, A. K. Ph.D. Thesis, University of Sussex (1994).
23. Cabioc'h, T.; Riviere, J. P.; Delafond, J. *J. Mater. Sci.* **1995,** *30*, 4787.
24. Sano, N.; Wang, H.; Chhowalla, M.; Alexandrou, I.; Amaratunga, G. A. J. *Nature* (*London*) **2001,** *414*, 506.
25. Chen, X. H.; Wu, G. T.; Deng, F. M.; Wang, J. X.; Yang, H. S.; Wang, M.; Lu, X. N.; Peng, J. C.; Li, W. Z. *Acta Phys. Sinica* **2001,** *50*, 1264.
26. Terrones, H.; Terrones, M.; Hsu, W. K. *Chem. Soc. Rev.* **1995,** *24*, 341.
27. Mayo, S. L.; Olafson, B. D.; Goddard, W. A. III, *J. Phys. Chem.* **1990,** *94*, 8897.
28. Barth, W. E.; Lawton, R. G. *J. Amer. Chem. Soc.* **1971,** *93*, 1730.
29. Fletcher, R.; Reeves, C. M. *Comput. J.* **1964,** *7*, 149.
30. Tersoff, J. *Phys. Rev. B* **1988,** *37*, 6991.
31. Maiti, A.; Brabec, C. J.; Roland, C.; Bernholc, J. *Phys. Rev. B* **1995,** *52*, 14850.
32. Terrones, M.; Terrones, H.; Charlier, J. C.; Banhart, F.; Ajayan, P. M. *Science* **2000,** *288*, 1226.
33. Heggie, M. I.; Terrones, M.; Eggen, B. R.; Jungnickel, G.; Jones, R.; Latham, C. D.; Briddon, P. R.; Terrones, H. *Phys. Rev. B.* **1998,** *57*, 13339.
34. Minimized structures using DREIDING force field where they were reparametrised to C-C bond lengths of 1.46 Å, which are suitable for the Tersoff Potential and approximates, with good accuracy, to experimental data (see Ref. 30).
35. Goldberg, M. *Tohoku Math. J.* **1937,** *43*, 104.
36. McKay, K. G.; Kroto, H. W.; Wales, D. J. *J. Chem. Soc. Faraday Trans.* **1992,** *88*, 2815.

37. Yoshida, M.; Osawa, E. *Fullerene Sci. Technol.* **1993,** *1*, 55.
38. Tang, A. C.; Huang, F. Q. *Chem. Phys. Lett.* **1995,** *245*, 561.
39. Terrones, H.; Terrones, M. *J. Phys. Chem. Solids* **1997,** *58*, 1789.
40. Smalley, R. E. *Account Chem. Res.* **1992,** *25*, 98.
41. Füller, T.; Banhart, F. *Chem. Phys. Lett.* **1996,** *254*, 372.
42. Ugarte, D. *Europhys. Lett.* **1993,** *22*, 45.
43. Saito, Y.; Yoshikawa, T.; Inagaki, M.; Tomita, M.; Hayashi, T. *Chem. Phys. Lett.* **1994,** *204*, 277.
44. Liu, H. W.; Hou, S. M.; Liu, S. J.; Tao, C. G.; Shi, Z. J.; Gu, Z. N.; Peng, L. M.; Wu, J. L.; Xue, Z. Q. *Acta Phys. Chim. Sinica* **2001,** *17*, 427.
45. Banhart, F.; Ajayan, P. M. *Nature* (*London*) **1996,** *382*, 433.
46. Ugarte, D. *Chem Phys. Lett.* **1993,** *207*, 473.
47. Zwanger, M. S.; Banhart, F.; Seeger, A. *J. Cryst. Growth* **1996,** *163*, 445.
48. Banhart, F. *Rept. Progr. Phys.* **1999,** *62*, 1181.
49. Tarnai, T. (discussion by Hare, J. P. & Fowler, P. W.). In *The Fullerenes: New Horizons for the Chemistry, Physics and Astrophysics of Carbon*, Kroto, H. W.; Walton D. R. M., Eds.; Cambridge University Press, Cambridge, 1993; p. 45.
50. Stone, A. J., Wales, D. J. *Chem. Phys. Lett.* **1986,** *128*, 501.
51. Brabec, C.; Maiti, A.; Bernholc, J. *J. Chem. Phys. Lett.* **1994,** *219*, 473.
52. Terrones, M.; Terrones, H. *Fullerene Sci. Technol.* **1996,** *4*, 517.
53. Bates, K. R.; Scuseria, G. E. *Theoret. Chem. Acc.* **1998,** *99*, 29.
54. Mackay, A. L. *Nature* (*London*) **1982,** *314*, 604.
55. Mackay, A. L.; Terrones, H. *Phil. Trans. Roy. Soc. London A* **1993,** *343*, 113.
56. Harris, P. J. F.; Tsang, S. C.; Claridge, J. B.; Green, M. L. H. *J. Chem. Soc. Faraday Trans.* **1994,** *90*, 2799.
57. Harris, P. J. F. *Carbon Nanotubes and Related Structures: New Materials for the Twenty-First* Century; Cambridge University Press: Cambridge, 1999.
58. Qin, L. C.; Iijima, S. *Chem. Phys. Lett.* **1996,** *262*, 252.

A Generalized Model for the Shell Structure of Icosahedral Viruses[a]

V.Ya. Shevchenko[b], A.E. Madison[b], and A.L. Mackay[c]

Abstract
The principles of algebraic geometry have been applied for the modelling of substances of different natures. A local packing approach provides a tool for constructing not only the special icosa-geodesic 2D tilings of a sphere but also the 3D tilings of spherical layers. The structures of the pseudo-T=7 polyoma and papilloma viral capsids can thus be modelled. This general approach is applicable to inorganic, organometallic, and biological nanoparticles.

Keywords
Quasi-equivalence • Triangulation number • Virus • Capsid • Polyoma • Papilloma

Introduction

The problem of basic differences between the structures of inorganic and of biological (organic) objects dates back to the origins of X-ray diffraction analysis in 1912. The main distinction is that, among the 230 Fedorov space groups, none contains point subgroups with fivefold and sevenfold axes, or with non-integral rotation axes. Until recently most known and well-characterized structures of inorganic substances corresponded strictly to the classical description. Nevertheless, fivefold axes frequently appeared in biological substances.

Among biological objects, the best-known examples of icosahedral symmetry are the isometric viruses, systematic investigation of which started in the 1950s [1, 2]. It appeared that the regular virus capsids are built by the self-assembly of identical subunits. The quasi-equivalence theory [2, 3], recalling the concept of the Buckminster Fuller's geodesic dome [4] and of earlier constructions, is still successful. Nature does not insist on exact mathematical identity. Exact equivalence is replaced by quasi-equivalence, for even in a very small crystal all the atoms cannot be exactly equivalent as they would be in an infinite crystal. The problem of the synthesis and structure analysis of small inorganic particles (the term *nano-structure* is now more convenient) is similar. A step in understanding the non-crystallographic regular structures was made when the multi-layer icosahedral packing, which underlies the structure of many objects of the inorganic world, was recognised [5]. Bernal's efforts to generalise orthodox crystallography towards including the ordered structures actually found in living matter and his assumption about the non-crystalline ordering of atoms in small particles [6] were further steps deriving from his earlier work with liquid crystals and tactoids.

By 2000, much experimental data on the synthesis of various small particles of different natures, such as inorganic, organic, and metal-organic particles, metal clusters, composites, etc., had accumulated. The development of techniques (HREM—high resolution electron microscopy) for the direct observation of atomic positions in small particles and crystals made it possible to describe and systematize their structures. The paradigm of classical X-ray crystal structure

[a] *Structural Chemistry* 2007, 18(3):343–346.

[b] Grebenshchikov Institute of Silicate Chemistry, Russian Academy of Sciences, Nab. Makarova 2, St., Petersburg 199034, Russia
e-mail: shevchenko@isc.nw.ru

[c] School of Crystallography, Birkbeck College, University of London, Malet Street, London WC1E 7HX, UK

I. Hargittai, B. Hargittai (eds.), *Science of Crystal Structures: Highlights in Crystallography*,
DOI 10.1007/978-3-319-19827-9_11

analysis, where the phases of the X-ray reflections were lost, had been, in many areas, overtaken by electron microscopy where the phases could be recovered experimentally. The new world of small nanoparticles was revealed, comparable in its beauty and peculiarity with the rich world of biological objects.

When studying the structure of zirconium dioxide nanoparticles (since the 1980s), we came across this unusual class of objects, whose properties are not always consistent with classical notions. Properties of the nanoparticles such as the structural inhomogeneity, fragmentariness, boundary coherence, which give rise to the structural variety, can be described within finite geometries. A trend towards a minimal volume is always clearly apparent in the structures of nanosized objects. This feature is attained in the best way by one of the highly symmetrical close packings (related to Euclidean or non-Euclidean geometry).

A basic concept that allows one to conceive, describe, and predict the nanoparticle structure and properties is the idea of the existence of the nanoworld as a separate part of the Nature with its characteristic laws [7]. The geometry of space wherein the nanoparticles are formed and exist may not be that of Euclidean geometry in the sense that the local metric, appropriate to local bonding requirements, may not be that of R^3. The geometry of S^3 was early examined by Sadoc [8, 9] and by Mackay [10] and the procedure of decurving such structures back to R^3 was developed fully by Sadoc. Structures may be in local energy minima rather than in absolute minima and compromises must take place between local and long-range order for existence in R^3. Various kinds of curved geometries and finite projective geometries may provide for local order, fragmentation, inhomogeneity, and curvilinear symmetry.

We introduce here and adapt terms necessary for the description of nano-objects. These are the spatial *coherence* as a capability of individual fragments to combine without significant distortions at an interface; *geometrical structural complexes* as "building blocks" for more complex structures of the next level of hierarchy; *locally minimal manifolds* as a generalization of planar sheets. In such ways the structure of inhomogeneous zirconia "centaur" nanoparticles, organometallic Keplerates, and giant metal clusters can be explained or visualised [11–13].

Capsids of Non-Quasi-Equivalent Viruses

In terms of such concepts, there is no basic difference between the structures of nanoparticles of the biological and inorganic types. Let us consider the structure of the family of isometric viruses and, in particular, of a polyoma virus whose structure only in a first approximation corresponds to the icosa-geodesic design with the triangulation number $T=7$ but, at the same time, basically differs from that predicted from the quasi-equivalence theory [14].

According to the quasi-equivalence theory, the capsid conforms to the triangulation of the icosahedron surface and is formed by self-assembly of 60 T quasi-equivalent subunits, where $T=(h^2+hk+k^2)$ is the triangulation number. The allowed values for T are thus 1, 3, 4, 7, 9, 13… The structure of most isometric viruses follows the existing theory [15, 16]. A specific group is singled out with the pseudo- $T=7$ viruses of the *Papovaviridae* family (which comprises the genus *Polyomavirus* and the genus *Papillomavirus*). According to the quasi-equivalence view [2, 3], the capsomers located at the fivefold axes should be pentamers, whereas all the other similar capsomers should be hexamers. This structure is characteristic of the true $T=7$ capsid of bacteriophage *HK*-97 [17]. At the same time, the viral capsids of the *Papovaviridae* family are characterized by a drastically different arrangement of capsomers. Despite the fact that there are seven non-equivalent positions for the capsomers, all the capsomers are pentamers, including those located at the positions of the quasi-sixfold axes of the $T=7$ surface grid [14]. These features of the structure are reliably corroborated for the papilloma and polyoma viruses, as well as for simian virus *SV*-40 [14, 18–20], and cannot be explained within the frameworks of the quasi-equivalence theory. One more unexpected example was the finding of the structures with the triangulation number $T=2$ [21], which is forbidden by the classical theory.

To model these features, various theoretical approaches, mostly based on methods of partition theory and the local matching concept, were proposed earlier [22–26]. A majority of the existing methods, as well as the classical theory [2], (either explicitly or implicitly) imply the geodesic design or a different tiling of some isometric surface into triangles. Recently we have set out a general principle, which can underlie all the icosahedral packings [13]. The Euclidean space may not be filled by ideal interpenetrating icosahedra but there exists the closest packing of ideal tetrahedra in the curved space—namely, the {3,3,5} polytope. Cutting various fragments from it and projecting them with slight distortions into Euclidean space one can obtain finite packings with slightly distorted icosahedral arrangement of subunits everywhere (e.g. [27]). Only the simplest icosahedral structures may be considered as cyclic multiple twins or geodesic domes. In the general case, there are rigorous assembling rules, which are strictly predefined by the true group-theoretical equivalence of corresponding blocks in the parent structure, which may be not realizable in Euclidean space but which can exist in curved (or projective) space.

The polytope {3,3,5} should be chosen as an initial parent packing. On the one hand, it is described by a finite discrete subgroup of the displacement group of the 3D Riemannian space S^3 or by a subgroup of the point group of the 4D

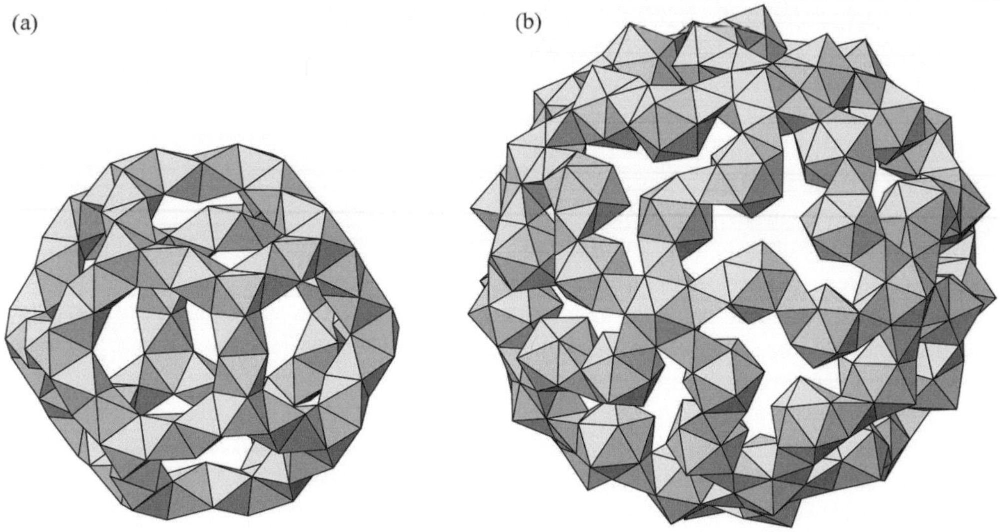

Fig. 1 Geometrical model of the structure of a papilloma virus capsid (all pictures are presented on comparable scales). (**a**) Capsid scaffold of a papilloma virus: 15 slightly distorted octahedra form a ring and 80 octahedra make up a dodecahedron-shaped frame. (**b**) Outer shell of a pseudo-$T=7$ papilloma virus capsid (only the front part is shown). The icosahedra rotated with respect to one another and joined by zigzags made of segments of the Boerdijk-Coxeter helices [27 (p. 112), 28].

Euclidean space E^4. Furthermore, it can be treated as the closest sphere packing in the positive curvature space S^3 and as an ideal icosahedral packing of regular tetrahedra in space E^4. The icosahedral packings in the 3D Euclidean space E^3 can be produced by projecting (mapping) of the fragments of a multidimensional icosahedral packing, which, in general, is accompanied by some insignificant distortions. According to the choice of the origin of coordinates or the method of projection (on a collection of concentric spheres or on Clifford surfaces), a great variety of sphere- or rod-shaped structures can be produced with the relative position of the subunits everywhere locally close to the icosahedral arrangement [28]. In contrast to the classical quasi-equivalence theory [2], we apply the geodesic design to polytope {3,3,5}, and only then the fragments of the obtained tiling should be mapped into the Euclidean space.

The geometrical model of the structure of a papilloma virus capsid derived by us is presented on the Fig. 1. Within this model, the papilloma virus capsid differs from the classical model. Capsid pentamers are represented by icosahedra rotated with respect to one another and are cross-linked by zigzags made of the fragments of the Boerdijk-Coxeter helices (Fig. 1b). All subunits are arranged almost as in the polytope {3,3,5}.

The structure and the arrangement of the scaffolding proteins are yet not known for the viruses of the *Papovaviridae* family [29–31]. We supposed that there should be no discrimination between the rules determining self-assembling of subunits within the separate shell and rules determining cross-linking of the outer shell with the scaffolding. The most likely scaffolding structure is predicted to be a dodecahedron consisting of 80 octahedra (Fig. 1a). Exactly two nonequivalent variants are available for coherent arrangement of the icosahedral pentamers on a scaffold made up of octahedra; which result in the *laevo* and *dextro* capsid forms. The geometrical model derived upon coherent joining of the scaffolding with the outer shell agrees well enough with experimental results [20] obtained by cryo-electron microscopy upon a 3D image reconstruction.

Concluding Remarks

Let us emphasize some advantages of this approach. Firstly, both the icosa-geodesic design of the surface of a sphere [2] and the multilayered icosahedral packing [5] can be produced as particular cases of the above procedure if the origin of coordinates is chosen at one of the vertices of the polytope (subjected to the geodesic design) and the mapping is limited to the nearest neighbors only. Secondly, we change over from the consideration of the tiling of the spherical surface into triangles to the tiling of a layer of the finite thickness into slightly distorted tetrahedra (or into tetrahedra and octahedra in the case of the geodesic design of the polytope). Proteins of the capsid are assigned to the locally minimal polyhedra rather than to the points on the surface grid. Owing to this procedure, whereby the tiling of the outer surface of the capsid is known, we can predict the tiling of its inner surface and the most likely structure of the scaffolding, as well as how the scaffolding could predetermine the structure of the outer capsid shell [29–31]. Thirdly, the stated concept is not limited to consideration of only the isometric particles, but may explain the polymorphism of the virus particles [32, 33], in which the self-assembly of the same protein subunits results in the appearance of a basically different (including the non-isometric) morphology without any essential disturbances in the quasi-equivalence and local conditions between the subunits. For instance, the rod-like and zigzag structures spatially compatible with interpenetrating multi-layer icosahedral packings may be designed by projecting the polytope fragments onto the Clifford surfaces. The Clifford surfaces are a special kind of manifold of zero Gaussian curvature defined in elliptical space. The corresponding surfaces in Euclidean space are cylinders. In this way the hierarchical structures built up of multi-layer icosahedral packings interconnected by the intersecting fragments of the Boerdijk-Coxeter helices may be designed.

The authors of the quasi-equivalence theory [2, 3] seemed to avoid intentionally the strict mathematical definition of the term *quasi-equivalence*. We consider the icosahedral packing as a united whole, as a fragment of the corresponding parent non-Euclidean packing. Different parts of the real structure correspond to the certain parts of the parent non-Euclidean structure and moreover, they are assembled by rules, which in turn have prototypes among the group operations of the parent structure. Strictly equivalent subunits of the non-Euclidean packings become quasi-equivalent after being subjected to unique distortions inevitable due to the mapping into Euclidean space. Straight lines and rotation axes become geodesic lines followed topologically by quasi-equivalent subunits.

Thus, the principles of algebraic geometry may be applied for the modelling of various substances. The local approach provides a tool for constructing not only the special icosageodesic 2D tilings of a sphere but also the 3D tilings of spherical layers (shells) with vertices being located onto concentric spheres. The structures of the pseudo-T=7 polyoma and papilloma viral capsids may be explained and this general approach is applicable to inorganic, organometallic, and biological nanoparticles.

Acknowledgments The authors thank the Russian Foundation for Basic Research, grant 06-03-32469, for financial support.

References

1. Crick FHC, Watson JD (1956) Nature (London) 177:473
2. Caspar DLD, Klug A (1962) Cold Spring Harbor Symp Quant Biol 27:1
3. Caspar DLD (1980) Biophys J 32:103
4. Marks RW (1960) The dymaxion world of Buckminster Fuller. Anchor, New York
5. Mackay AL (1962) Acta Cryst 15:916
6. Bernal JD, Carlisle CH (1968/1969) Kristallografiya 13:927; Soviet Phys —Crystallogr. 13:811
7. Shevchenko VYa, Samoilovich MI (2005) In: Shevchenko, VYa (ed.) Nanoparticles, nanostructures and nanocomposites, VVM Com. Ltd., St.-Petersburg p. 7.
8. Sadoc J-F (ed.) (1990) Geometry in condensed matter physics. World Scientific, Singapore
9. Sadoc J-F (1980) J Phys (Paris) Coll 41:C8–326
10. Mackay AL (1980) J Phys A 13:3373
11. Shevchenko VYa, Samoilovich MI, Talis AL, (2005) Madison AE: Glass Phys Chem 31:407
12. Shevchenko VYa, Samoilovich MI, Talis AL, Madison AE (2005) Glass Phys Chem 31:402
13. Shevchenko VYa, Samoilovich MI, Talis AL, Madison AE (2005) Glass Phys Chem 31:259
14. Rayment I, Baker TS, Caspar DLD, Murakami, WT (1982) Nature (London) 295:110
15. Baker TS, Olson NH, Fuller SD (1999) Microbiol Mol Biol Rev 63:862
16. Johnson JE, Speir JA (1997) J. Mol Biol 269:665
17. Wikoff WR, Liljas L, Duda RL, Tsuruta H, Hendrix RW, Johnson JE (2000) Science 289:2129
18. Baker TS, Drak J, Bina M (1989) Biophys J 55:243
19. Belnap DM, Olson NH, Cladel NM, Newcomb WW, Brown JC, Kreider JW, Christensen ND, Baker TS (1996) J Mol Biol 259:249
20. Modis Y, Trus BL, Harrison SC (2002) EMBO J 21:4754
21. Rossmann MG, Tao Y (1999) Structure 7:R43
22. Zandi R, Reguera D, Bruinsma RF, Gelbart WM, Rudnick J (2004) Proc Natl Acad Sci USA 101:15556
23. Berger B, Shor PW, Tucker-Kellogg L, King J (1994) Proc Natl Acad Sci USA 91:7732
24. Keef T, Taormina, A, Twarock R (2005) Phys Biol 2:175
25. Andersson S (2006) Virus Structure in Spherical Space: Polyhedra; and Beyond Polyhedra. http://www.sandforsk.se/articles
26. Damodaran KV, Reddy VS, Johnson JE, Brooks CL (2002) J Mol Biol 324:723
27. Lord EA, Mackay AL, Ranganathan S (2006) New geometries for new materials. Cambridge University Press, Cambridge
28. Lord EA, Ranganathan S (2001) Eur Phys J D 15:335
29. Ishii Y, Ozaki S, Tanaka K, Kanda T (2005) Virus Genes 31:321
30. Hagensee ME, Yaegashi N, Galloway DA (1993) J Virol 67:315
31. Barouch DH, Harrison SC (1994) J Virol 68:3982
32. Salunke DM, Caspar DLD, Garcea RL (1989) Biophys J 56:887
33. Cielens I, Ose V, Petrovskis I, Srelnikova A, Renhofa R, Kozlovska T, Pumpens P (2000) FEBS Lett 482:261

Intermetallic Compounds of the $NaCd_2$ Family Perceived as Assemblies of Nanoclusters[a]

V.Ya. Shevchenko[b], V.A. Blatov[b], and G.D. Ilyushin[b]

Abstract

In this work, we proceed with the concept of developing nanoclusters (fundamental configurations), which are the basic building blocks of certain crystal structures, allowed us to propose a universal algorithm for crystal self-assembly structure, and hence to elaborate a uniform approach to the analysis of the essentially hierarchic structure of natural objects under the nanoscale. It should be noted that we use the conception, where the configuration of points on the single sphere that minimize potential energy for broad class of potential functions (with the characteristic quadric function of potential versus distance between points) has been studied. The model is not connected both with the undefined isolated from structure nets (like diamond net) and polyhedra (like Friauf polyhedra), which are the basis of clusters, but is hierarchically constructed on local polyhedral structure's searching according to universal algorithm. Crystal structure can explicitly store the information on the structure of the particles that formed it. These particles, geometrically distorted in some extent, can remain in parts of fundamental nanoclusters, coherently grow into them, and emerge as clusters-centaurs.

Keywords

Structural chemistry • Nanoclusters • Nanoparticles • Centaurs

The geometry of nature never ceases to astound, nor does the ingenuity of human beings in discerning patterns.

D. C. Fredrickson, S. Lee, R. Hoffmann

Introduction

The term *nanocluster* is used here in the wide sense as a nanomeric set of connected atoms, stable either in isolation state or in building unit of condensed matter. Nanoclusters defined in this way can be found in substances of different chemical nature, but we will consider only intermetallic compounds, taking into account that the latter consist of only metal atoms. Metallic nanoclusters are particularly important objects in the nanoworld, and their structural chemistry is a new chapter of modern inorganic chemistry and materials science. The former is characterized by the great diversity in composition and structure, and often is the primary unit in the self-assembly of nanostructured materials. Nanoclusters existing in solids and interacting with each other (so-called *quantum dots*) are of special interest as well as the process of the self-organization of nanoclusters (nanoparticles) from isolated atoms in some environment (such as solution, melt, vacuum), which gives rise to three-periodic structures. Quantum processes are still important in large semiconducting nanoclusters (*quantum dots*). Nanoclusters can be obtained by gas aggregation, vaporization from surfaces, dispersion, laser ablation, electro-blasting, plasma methods, supersonic gas-dynamic jets, preparative chemical synthesis, etc.

[a] *Structural Chemistry* 2009, 20(6):975–982.

[b] Institute of Silicate Chemistry, St. Petersburg, Russia
e-mail: exchem@isc.nw.ru

I. Hargittai, B. Hargittai (eds.), *Science of Crystal Structures: Highlights in Crystallography*,
DOI 10.1007/978-3-319-19827-9_12

In this work, we proceed with the concept of developing nanoclusters (fundamental configurations), which are the basic building blocks of certain crystal structures, allowed us to propose a universal algorithm for crystal structure self-assembly, and hence to elaborate a uniform approach to the analysis of the essentially hierarchic structure of natural objects at the nanoscale.

In three earlier works [1–3], we have formulated the basis of this approach. We have studied Rh_7Mg_{44} and $ZrZn_{22}$, the most complicated cubic structures of intermetallic compounds with unusual stoichiometry [2, 3]. Some other cubic structures have the simpler composition, but crystallochemically the former is incredibly complicated. Moreover, they always have the homogeneity regions on their phase diagrams, i.e., their compositions are not distinct, but vary, depending on the ambient conditions.

In the 1960s, Samson determined the structures of the cubic phases $NaCd_2$ [4, 5], Mg_2Al_3 [6], and Cu_3Cd_4 [7], each of all has more than 1,000 atoms in the unit cell. In particular, the unit cell of $NaCd_2$ with $a = 30.56$ Å and space group $Fd\bar{3}m$ contains 1,192 atoms (Pearson code cF1192), which occupy 17 inequivalent positions (their Wyckoff sequence is ihg^7fe^5cb) and form 672 icosahedra, 252 Friauf polyhedra, and 254 other polyhedra.

Under constructing the model of $NaCd_2$, Samson [4] selected a 47-atom cluster of mm2 symmetry composed of five Friauf polyhedra, which were joined face to face. Then, he united six such clusters into 234-atom octahedral clusters of $\bar{4}3m$ symmetry. The centers of the clusters occupied Wyckoff positions 8*a*, while one sodium atom, not belonging to the cluster, laid in a position 8*b* with the same $\bar{4}3m$ symmetry as that of the center of a Friauf polyhedron. Other models for Samson's phases were proposed in the next years [8–12].

The authors [9–11] described the $NaCd_2$ crystal structure in terms of two clusters in positions 8*a* and 8*b*, i.e., selected a new cluster based on the Friauf polyhedron centered by a sodium atom. The centers of two nonequivalent clusters formed two interpenetrating diamond nets. Andersson [9] was the first who have noticed the atoms not belonging to the clusters were laid near the Riemann D surface separating the clusters of the diamond nets. The other authors [11] gave a quantum-mechanical substantiation of model, supported by Andersson et al. [9], and described in detail the two types of clusters, the 90-atom cluster as major and the 45-atom as the minor cluster. Bergman [10] proposed another way of atom's location as distributed on 76- and 73-atom clusters in the 8*a* and 8*b* positions, respectively.

It is clear that the crystal structure of $NaCd_2$ is a remarkable target for computer modeling with the use of self-assembly structure from nanoclusters concept, and moreover the attention of such outstanding researchers as Samson, Andersson, and Hoffmann was attracted to this problem during different years.

Principles and Methods of Analysis in Terms of Nanoclusters

When constructing the model of the $NaCd_2$ crystal structure, we have used the following principles (cf. [2]):

1. The structure is formed as a result of the self-assembly of the nanoclusters corresponding to optimal (fundamental) configurations of atoms. With smooth potentials for interaction, the configuration stability has generally been studied by the methods of mathematical modeling [13–15]. In the approximation of isotropic interactions, any fundamental configuration corresponds to a close packing of atoms. However, in crystal structures the atomic configurations can differ from ideal close packings. We consider the occurrence of an atomic configuration in different structure types as one of the important indicators of its stability. Thus, the separation of any nanocluster in a crystalline substance should be proved by finding the nanocluster in the compounds of quite another crystal structure.
2. Nanoclusters are considered as the highest point symmetry in the structure (their centers occupy the most symmetrical positions in the unit cell). At present, our algorithm assumes that their centers are occupied by atoms, but it can be easily modified by search of non-centered nanoclusters.
3. Centers of nanoclusters are occupied by high-coordinated atoms, i.e., atoms have the largest coordination numbers (CN; this is the principle of the maximal connectivity of the central atom).
4. Nanoclusters have no common internal atoms (i.e., they should not interpenetrate), but can have the shared surface atoms.
5. Ordinarily, the clusters include all atoms of the structure. In some cases, there can be a few atoms that are separated between clusters.
6. If several models follow the rules (1–5; i.e., in the case of low-symmetry crystals), they should all be enumerated and tested, to see whether they indicate different methods of self-assembly.
7. The nanocluster that determines the particular structure type is taken place; the former can be in low abundance or can never be occurred in other structures. The function of above-mentioned cluster is the definition of unique mechanism of building blocks assembling, which gives the rise of certain structural type, and therefore it can be called a 'structure determinant.' However, it can be frequently found, in the case of more conformability to other types of fundamental configurations.

The main advantages of the principles are that they have clear physical meaning and can easily be realized by a

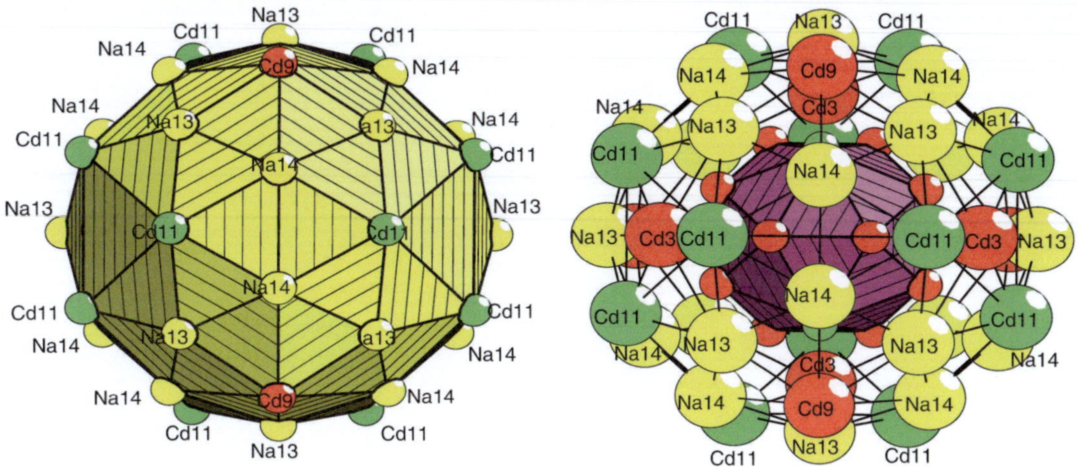

Fig. 1 61-atom microcluster A. (*left*) General view; (*right*) shell structure.

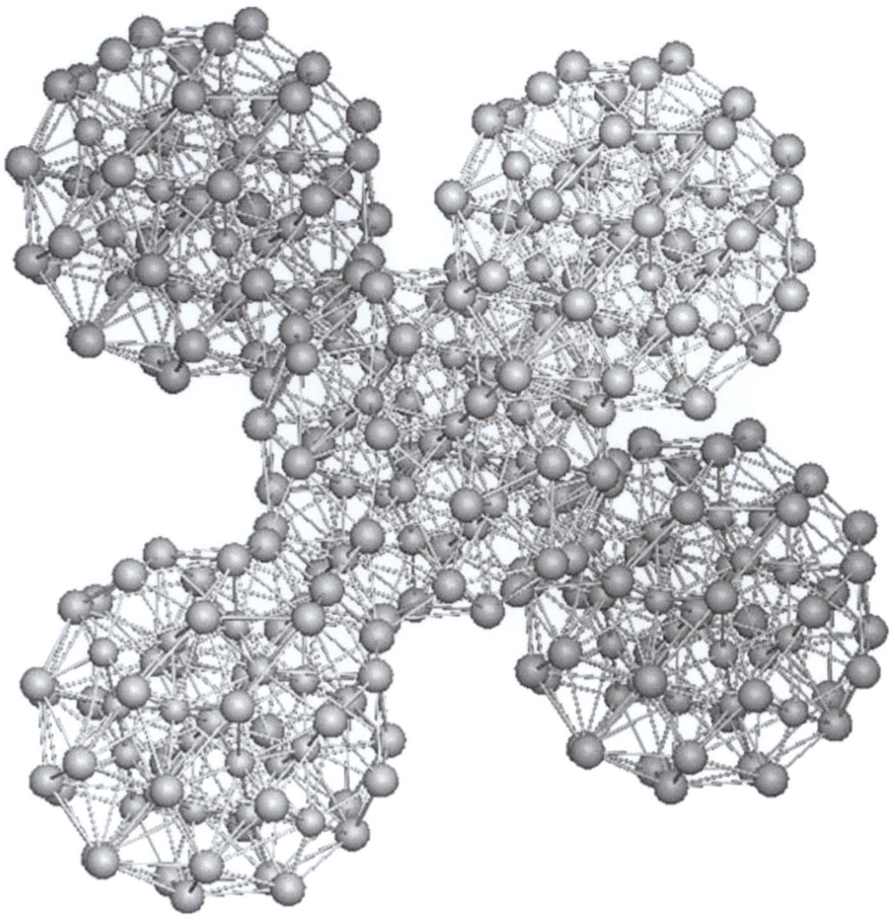

Fig. 2 The method of connecting microclusters A with each other.

computer program as a formal algorithm; we made it within TOPOS[1] program package. Note that all previous descriptive approaches in the crystal chemistry of intermetallic compounds could have not to be algorithmized. In general, the problem of the classification of structure types of intermetallic compounds includes a search for all the possible types of the nanoclusters participating in the selfassembly. Until recently, this problem could not have to be solved, since it meant a search for nanomeric nanoclusters consisting of several tens of atoms, in the crystal structures of all known intermetallic compounds. However, we have developed the

[1] http://www.topos.ssu.samara.ru.

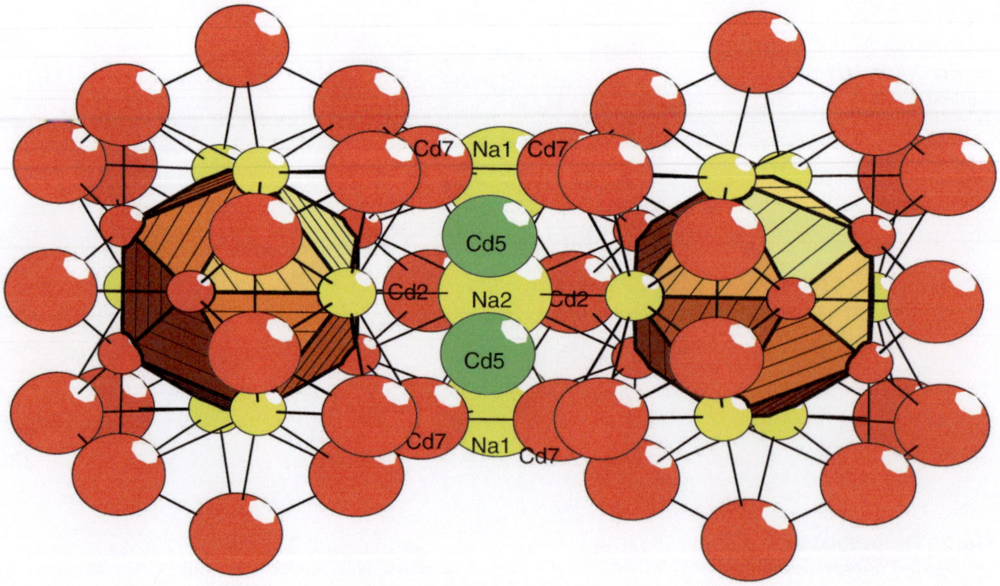

Fig. 3 Two connected microclusters ***B***, common atoms are marked. For each cluster, the shell structure with internal icosahedron is shown.

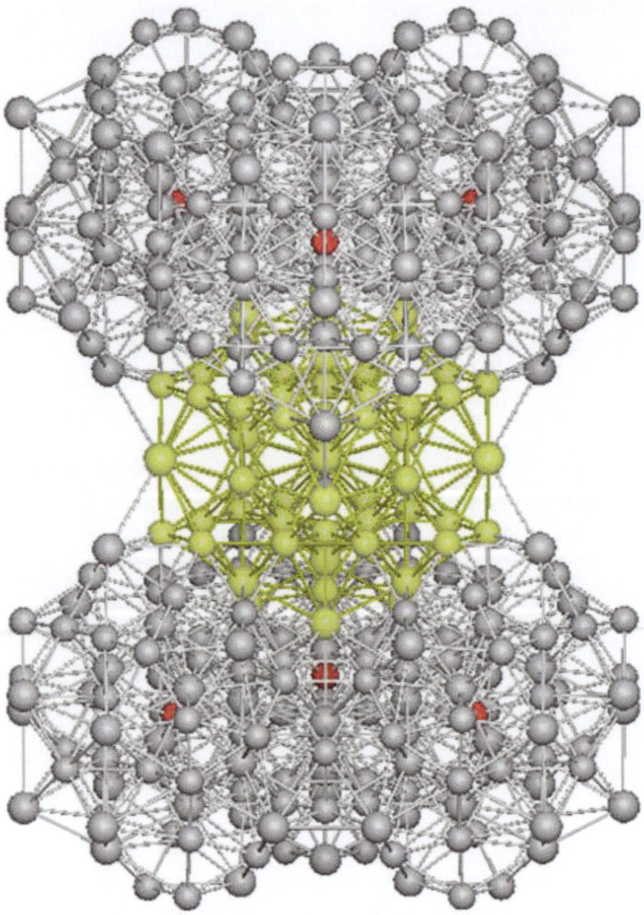

Fig. 4 The method of connecting microclusters ***B*** with each other. *Yellow balls* are the atoms of the central microcluster, *red balls* are central atoms of neighboring microclusters (Color figure online).

necessary algorithm for such search [2], and it has been implemented by TOPOS. The algorithm is based on the search of a finite subgraph (nanocluster) in the infinite graph (overall structure) and is insensible to geometrical distortions of crystal structures.

Here, the crystallochemical analysis of the $NaCd_2$ crystal structure, as well as its modeling with the concept of nanocluster self-assembly in mind, has been carried out with TOPOS. The structural information on intermetallic compounds required for occurrence of nanoclusters checking was taken from the ICSD [16] and CRYSTMET [17] databases. A group of 4,812 compounds, containing icosahedral clusters arranged by 542 Pearson classes, was selected.

The Nanocluster Model of the $NaCd_2$ Crystal Structure

When analyzing the $NaCd_2$ crystal structure, we were using the crystallographic data corresponding to a so-called basic or ordered model, i.e., all positions were assumed to be completely occupied by the atoms with the largest occupancies according to Andersson et al. [9]. The ordered model was considered only for convenience in assigning atom positions to Na or Cd; the structure of the resulting nanoclusters does not depend on disordering. The Cd position with occupancy 0.80 [9] was considered as completely occupied by Cd atoms. We also used the same numbering of atoms as done by Andersson et al. [9].

The first step in distinguishing nanoclusters is the identification of their central atoms. According to the principles mentioned earlier, such atoms should be allocated to the most symmetrical positions and should have high coordination numbers.

Only the Na(5) atoms occupying *8b* positions and having CN = 16 correspond to these conditions. To determine the size of the nanocluster, we modeled it layer by layer until the equivalent nanoclusters became intersecting. Making this

Fig. 5 Integrated microclusters *A* and *B*. (*top left*) Supercluster AB_2; the method of connecting the superclusters into (*top right*) chain and (*bottom*) layer.

twice, we obtained the 61-atom two-layer nanocluster (1+16+44=61 atoms; Fig. 1) containing the Cd(7)–Cd(10) and Na(2), Na(5)–Na(7) atoms. We will call this *nanocluster A*. *A* nanoclusters interact with each other, but have no common atoms (Fig. 2).

Then, we repeated the procedure between all other atoms not entering the *A* nanoclusters. According these, the most symmetrical position was occupied by Cd(6) with CN=12. The layer sequence of coordination is shelled around by this atom gave rise to another two-layer nanocluster (B) consisting of 1+12+50=63 atoms. This nanocluster included all remaining non-equivalent atoms as well as six atoms Na(9) from *A* nanocluster. Moreover, each *B* nanocluster had atoms shared with other *B* nanoclusters (Figs. 3 and 4).

The method of the hierarchical self-assembly of the whole crystal structure, by combining nanoclusters *A* and *B* into

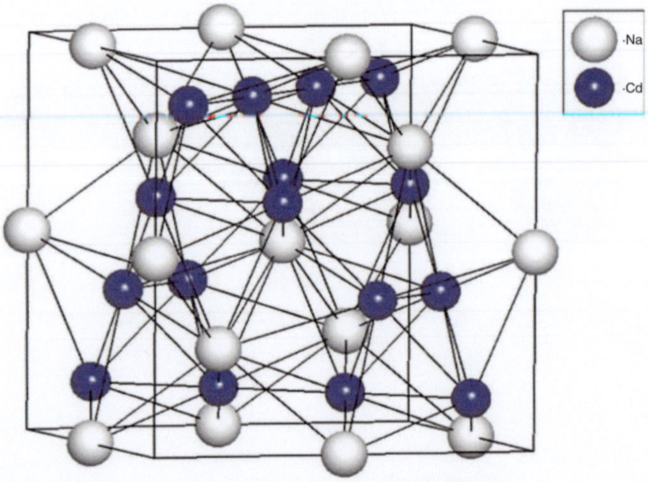

Fig. 6 Centers of clusters *A* (Na) and *B* (Cd), forming the Laves phase $MgCu_2$ network.

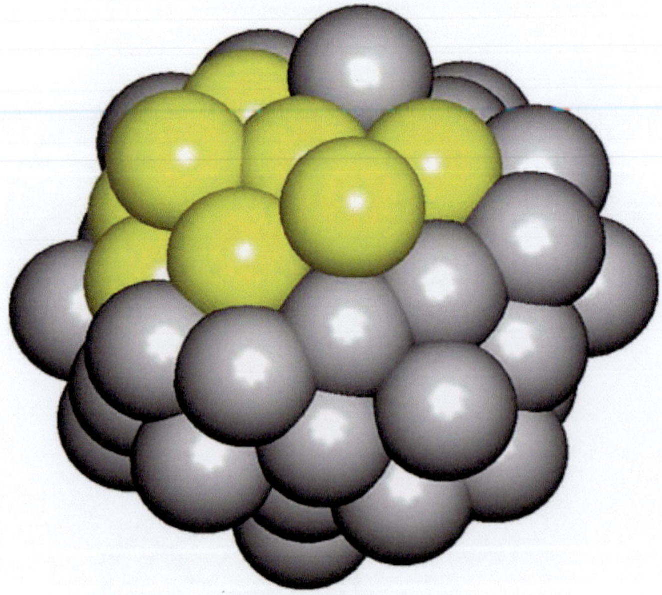

Fig. 7 21-atom model microcluster (*yellow balls*) within the microcluster ***B*** (Color figure online).

supraclusters AB_2 (i.e., building units, unambiguously determining three-periodic structure), microchains, and microlayers is shown in Fig. 5. The centers of the nanoclusters *A* and *B* form a network with the topology of the Laves phase $MgCu_2$ (Fig. 6).

The proposed model differs from previous models [9–11] both in the composition of the structural units and by their positions in space (Table 1). Nanoclusters *A* and *B* are based on the Friauf polyhedron $Na(Na_4Cd_{12})$ and the icosahedron $Cd(Cd_6)Na_6$, respectively; they include all crystallographically inequivalent atoms, have different composition, and their stoichiometric ratio is 1:2.

Our nanoclusters *A* and *B* can be seen to be related to the major 90-atom and minor 45-atom clusters, perceived in the Hoffmann model. Figure 7 shows that the minor Hoffmann cluster is completely contained in our *A* cluster. The major Hoffmann cluster includes the *B* cluster (Fig. 7).

Recently [2], when considering the Rh_7Mg_{44} crystal structure, we have revealed *model* nanoclusters (i.e., obtained by the methods of mathematical modeling [14, 15]) containing up to 62 atoms and partially embedded in the superclusters forming the structure. The cluster, being a result of the intergrowth of a model nanocluster (*M*) and a supercluster (*C*), was called a cluster-centaur $M+C$. This demonstrates the relationship between a non-periodic phase (of nanoparticles), where the nanoclusters in question should exist, and a periodic crystal formed by superclusters *C*. In the $NaCd_2$ crystal structure, the 21-atom nanoclusters of the model [14] exist, which is completely included in the *B* nanoclusters (Fig. 7). Thus, we have a special case of a cluster-centaur that coincides with one of our initial nanoclusters; in this case $M+C=C$.

A and *B* nanoclusters follow the principles (1) and (7). Indeed, *B* nanoclusters exist in the Laves phase $MgCu_2$ (Pearson class cF24) and in 37th other structures of Pearson classes (Table 2), some of which are corresponded to other Laves phases ($MgZn_2$ and $MgNi_2$) and its superstructures, while the remaining ones can be considered as containing fragments of Laves phases (cf. [11]). On the other hand,

Table 1 Comparison of different models of $NaCd_2$

	Samson [3]	Andersson [7]	Bergman [8]	Hoffmann [9]	Our data
Composition of structural units	Clusters of 47 atoms united in cluster of 234 atoms	Two types of clusters (composition is not specified)	Two types of clusters of 76 and 73 atoms	Two types of clusters of 90 and 45 atoms	Two types of microclusters of 61 and 63 atoms
Center of cluster	8*a* (0, 0, 0)	8*a* (0, 0, 0) and 8*b* (1/2, 1/2, 1/2)	8*a* (0, 0, 0) and 8*b* (1/2, 1/2, 1/2)	8*a* (0, 0, 0) and 8*b* (1/2, 1/2, 1/2)	8*b* (1/2, 1/2, 1/2) and 16*c* (1/8, 1/8, 1/8)
Atoms, not belonging to clusters	Cd4[a], Na1	Not specified	Not specified	Specified only Cd7	No
Stoichiometric ratio of clusters	–	1:1	1:1	1:1	1:2
Type of the cluster network	NaCl	NaCl and NaCl	NaCl and NaCl	NaCl and NaCl	$MgCu_2$

[a]Atomic numbers correspond to ICSD data

Table 2 Pearson classes where microclusters *B* were found

Pearson class	Number of representatives	Example	Pearson class	Number of representatives	Example
cF24	1,148	$MgCu_2$	hP84	2	$LiMg_7Zn_{13}$
hP12	212	$MgZn_2$	hP48	2	$Ce_2Co_{5.32}Cu_{2.28}$
hR36	51	$LaNi_3$	mB110	1	Mg_4Zn_7
hP24	34	$MgNi_2$	mC12	1	$FeAl_2Be_{2.5}$
cF192	20	$CeNi_2$	hR72	1	Ce_5Co_{19}
hR54	11	Y_2Ni_7	hR6	1	$TbFe_2$
hR18	11	$TbCo_2$	oC24	1	URe_2
hR39	10	NbFe	tI12	1	HfV_2
hP36	10	La_2Ni_7	aP96	1	MgZn
cF32	5	$CsTaN_2$	hR144	1	$TbNi_2$
hR13	4	Mo_6Fe_7	hP26	1	Cs_6K_7
cP24	4	$TmNi_2$	hP92	1	$Mg_{26}Zn_{60}Sm_6$
hP19	4	$Mg_4Zn_{11}Al$	hP60	1	$Ag_{0.15}MgZn_{1.85}$
hP96	3	$LiMg_2Zn_3$	hP248	1	$Sm_{10.5}Mg_{24.2}Zn_{65.3}$
oP74	2	$Nb_{28}Ni_{33.5}Sb_{12.5}$	cF1192	1	$NaCd_2$
oI12	2	UMn_2	cF1124	1	$Cd_{2.52}Cu_{3.13}$
hR276	2	$Mg_{21}Zn_{25}$	hP20	1	Ba_7Al_{13}
hR42	2	Sr_5Al_9	cF1400	1	Mg_2Al_3

nanoclusters *A* were found in merely two other Pearson classes: cF464 (including $K_{17}In_{41}$, $Li_{18}Cu_5In_4Ga_{31}$, $Na_{35}Cd_{23.77}Ga_{56.23}$, $Mg_{35.07}Cu_{24}Ga_{52.67}$, $Na_{36}Ag_7Ga_{73}$, $Na_{17}In_{12}Ga_{29}$) and hP238 ($K_{39}In_{80}$). Hence, nanocluster *A* can be considered as a 'gene' of the $NaCd_2$ crystal structure specifying the unique method of assembling *B* nanoclusters on its surface and determining the crystal structure individuality. The fact, that *A* nanocluster contains the Laves phase $MgCu_2$ fragment is once more underlined both with the cognation and distinction of the shown structures. It should be noted that $NaCd_2$ structure is kindred with $MgCu_2$ both on local level (in the framework of nanoclusters) and on global linking level of nanoclusters to the three-dimensional net.

Conclusion

The main advantages of our proposed model of the $NaCd_2$ crystal structure in comparison with other models is summarized [7, 9–11] (Table 1).

1. The model is not connected both with the undefined isolated from structure nets (like diamond net) and polyhedra (like Friauf polyhedra), which are the basis of clusters, but is hierarchically constructed on local polyhedral structure's searching according to universal algorithm.
2. The occurrence of these particular nanoclusters was sought in other intermetallic compounds. The *B* nanoclusters, described here for the first time, were found as typical building blocks determining the relationships between Laves phases, as well as between quite different, often unique structures of 1,557 binary and ternary intermetallic compounds (16 Pearson classes have the single representative shown in Table 2). *A* nanocluster can be considered as a special structural fragment, 'determinant' of the $NaCd_2$ crystal structure. The sizes of both nanoclusters are almost the same, and their forms are close to that of a sphere with diameter of about 1.2 nm. The clusters are arranged as a network that follows the motif of the Laves phase $MgCu_2$.
3. The whole structure self-organizes, since each component occupies its position in accordance to a given local order. In simple structures, the rules of selforganization are determined by the fundamental nanoclusters (configurations). We could expect that in complicated structures the local order is determined by components-'determinants,' whereas other components provide the connectivity of the 'determinants,' transmitting the information on local order through the structure. As a result, we have a composite structure that contains its own description as, for instance, bacteriophage contains its DNA. This process can be repeated on a higher hierarchical level. Local order can also be determined without using translational symmetry as, for instance, in the cellular automata formalism. One further example is that in cubic crystals the process of successive twinning along the {111} plane around the [110] axis can be replaced by icosahedral twinning. Dihedral angle between two {111} planes is equal to 71.5°, so there is a gap of about 2.5° after five twinning operations around the [110] axis. This gives rise to significant strain of the icosahedral particle and, probably, to growing a cuboctahedral particle.

These cubic icosa-twins can exist only in the nanomeric scale. On the other hand, icosa-twinning of rhombohedral crystals with $\alpha=63.43°$ will fill the space without gaps and can proceed infinitely (at least in the nanomeric scale).

4. Crystal structure can explicitly store the information on the structure of the particles that formed it. These particles, geometrically distorted in some extent, can remain parts of fundamental nanoclusters, coherently grow into them, and emerge as clusters-centaurs. Thus, clusters-centaurs can be defined as single-connected atomic ensembles formed by coherent fusing of two or more number of atomic ensembles of a smaller size. A typical feature of the clusters-centaurs is the lack of distinct boundaries between different ensembles.

Note that if there are several structure units in the crystal, they can be separated by an interface [9, 11]. The geometry of the interface may correspond to Coxeter's idea that some configurations found in finite structures (the icosahedron ahedron) are close to dense packings in a higher dimensionality (the 600-cell). Strains in the structure make these configurations elliptically or hyperbolically curved.

Acknowledgments Authors would like to thank A.L. Mackay for the great work on the article editing. A.L. Mackay replaced the word «inorganic gene», which we have taken from his article with J. Klinovski, on the «determinant» one. At the article basis, the deep principles, evolved by A.L. Mackay at his fundamental contributions on generalized crystallography and also said by him during the scientific discussions on the observed and other articles are underlain. Authors would like to thank V.V. Ivashkin for both technical and creative help during all stages of this work.

References

1. Shevchenko VYa, Mackay AL (2008) Glass Phys Chem 34:1
2. Shevchenko VYa, Blatov VA, Ilyushin GD (2009) Glass Phys Chem 35:1
3. Ilyushin GD, Blatov VA (2009) Acta Cryst B65:300
4. Samson S (1962) Nature 195:259
5. Samson S (1964) Acta Cryst 17:491
6. Samson S (1965) Acta Cryst 19:401
7. Samson S (1967) Acta Cryst 23:586
8. Kreiner G, Franzen HF (1997) J Alloys Compd 259:83
9. Yang Q-B, Andersson S, Stenberg L (1987) Acta Cryst B43:14
10. Bergman G (1996) Acta Cryst B52:54
11. Fredrickson DC, Lee S, Hoffmann R (2007) Angew Chem Int Ed 46:1958
12. Feuerbacher M et al. (2007) Z Kristallogr 222:259
13. Cohn H, Kumar A (2007) J Amer Math Soc 20:99
14. Wales DJ, Doye JPK, Dullweber A, Naumkin FY, Calvo F, Hernandez-Rojas J, Middleton TF (2004) The Cambridge Cluster Database. http://www-wales.ch.cam.ac.uk/CCD.html
15. Doye JPK, Wales DJ (1998) New J Chem 22:733
16. Belsky A, Hellenbrandt M, Karen VL, Luksch P (2002) Acta Cryst B58:364
17. CRYSTMET (2004) Structure and powder database for metals. Materials Toolkit

Linus Pauling's Quest for the Structure of Proteins[a]

István Hargittai[b]

Abstract

Linus Pauling, arguably the greatest chemist of the twentieth century, never publicly admitted that there was a race for the determination of the structure of the most important biopolymers. But according to his competitors there was a race, in fact, there were two, and Pauling won one and lost the other. He had a tremendous amount of ideas, many of them worthless, but a few were spectacular. Not only did he make seminal discoveries, he was also a master of announcing them in a most dramatic way. Eventually, Pauling shifted toward politics and controversial issues, but his science ensured him his place among the greats. Here, we follow Pauling's route to the discovery of the alpha-helix; the defeat of the star-studded British team in the same quest; and a seemingly unrelated story about the fate of the theory of resonance that assured Pauling's victory yet at the same time it was excommunicated in the Soviet Union.

Keywords

Linus Pauling • Structure of proteins • Theory of resonance • Peptide bond • Alpha-helix

For most of the first half of the twentieth century a large number of scientists were not even sure biopolymers existed. The view survived for long that the principal components of living matter were in a colloidal state, that is, conglomerates of smaller molecules. It was only in 1953 when Hermann Staudinger was awarded the Nobel Prize in Chemistry for his discoveries about macromolecules that the existence of polymers was irrevocable accepted. By then, though, a lot about the structures of the biologically important macromolecules had been discovered. Thus, during the first half of the twentieth century, the efforts to establish the nature of biological substances and to uncover their structures went in parallel.

The British father and son team, W. H. Bragg and W. L. Bragg, pioneered the technique of X-ray diffraction crystallography in 1913, with the son playing the leading role. When the two Braggs were awarded the Nobel Prize in 1915, the son became the youngest ever Nobel laureate and has stayed the youngest to this date. After a hiatus due to World War I, this field took off spectacularly in the realm of small molecular systems. As early as the 1920s, fibrous materials were subjected to X-ray diffraction for the first time by Michael Polanyi in Herman Mark's laboratory at the Kaiser Wilhelm Institute in Berlin. Polanyi's experiments on cellulose indicated the presence of crystallites in cellulose and they were oriented in the direction of the fiber axis. He could not have performed a full structure analysis at that time, but Mark and Polanyi observed characteristic changes on stretching the cellulose fibers. Mark was to become one of the century's foremost polymer chemists. When he was forced out of Germany, he moved to his native Vienna where he helped one of his students, Max Perutz to be accepted as a doctoral student in Cambridge, UK, in 1935. Perutz would

[a] *Structural Chemistry* 2010, 21(1):1–7.

[b] Department of Inorganic and Analytical Chemistry and Materials, Structure and Modeling Research Group of the Hungarian Academy of Sciences, Budapest University of Technology and Economics, P.O. Box 91, H-1521 Budapest, Hungary
e-mail: istvan.hargittai@gmail.com

I. Hargittai, B. Hargittai (eds.), *Science of Crystal Structures: Highlights in Crystallography*,
DOI 10.1007/978-3-319-19827-9_13

become a key player in the quest for the structure of proteins. But he was only slowly moving to the area of his ultimate success when Linus Pauling was already a major force in the field.

Pauling came from a humble background, but he was ambitious. He lost his father when he was 9 years old and his mother found it difficult to cope with her obligations. She certainly did not appreciate her son's intentions to stay in school even when it was no longer mandated for him to continue his studies. Pauling's schoolings were not at top places, and when in 1922 he went to the California Institute of Technology (as it was later; Caltech in short) it was far from the preeminent research-oriented institution into which it would develop. But the school was as ambitious as its new student, and there were visionary movers of it who were set to making Caltech a top-notch institution of higher education and research. They were smart enough not only looking into the distant future and only for big names to recruit from faraway places, but recognized in Pauling the potentials of a star scientist who would even challenge British preeminence in the science of chemical structures.

When Pauling started his doctoral studies with Roscoe Dickenson, a fresh home-grown PhD in X-ray crystallography at Caltech, this field was less than a decade old. Pauling became engaged in the determination of the structure of many inorganic and organic molecules and amassed a large amount of information about them during the ensuing decade. What kind of information was that? It was about the geometrical arrangement of the atoms in the molecules and the arrangement of the molecules in the crystals.

Not all the modern knowledge was to be had at Caltech at the time, and not even in other laboratories in the United States. The leading country of science was Germany and a few other places in Europe, and Pauling—like many other aspiring American scientists—paid pilgrimage to a series of European research centers in order to learn from the likes of Arnold Sommerfeld in Munich and Erwin Schrödinger in Zurich. They were both physicists, but Pauling's aim was not to transform himself into a physicist. Rather, his goal was to apply the latest discoveries in physics, and above all the new quantum mechanics, to solving a wealth of problems in chemistry in which he proved to be unique.

The most intriguing question in chemistry at that time was about the forces that keep the atoms together in a molecule, that is, about the nature of the chemical bond. If there is anything truly associated with Pauling's name, it is the understanding the nature of the chemical bond. He used the achievements of modern physics, the experimental information about the geometry of molecules and his thinking, to put together a theory. He then kept refining it in accordance with the emergence of the latest experimental information. The science of chemistry has a great deal of intuitive approach in it, very often stemming from a desire to represent on paper what the chemists experience in the laboratory. Thus, for example, they started using a straight line connecting the symbols of two elements to represent their bonding without really understanding anything about what that straight line represented. Nowadays when we know so much about what it means, we still find this straight line an excellent representation of the chemical bond. Lewis's description of the covalent bond in 1916 was not much less intuitive than this; nonetheless he made a big step forward. He introduced the idea of the shared electron pair, meaning the covalent bond between two atoms. During the late 1920s two physicists, Walter Heitler and Fritz London used the new quantum mechanics and their sophisticated mathematical apparatus to rigorously describe this covalent bond. It was so rigorous that it was too sophisticated for most chemists to understand it let alone to apply it to solving their problems that were usually more complex than the hydrogen molecule for which Heitler and London had worked out their theory.

Linus Pauling bridged this gap in a series of brilliant articles in the *Journal of the American Chemical Society*. Eventually he developed his ideas and his repository of structural information into a bestseller *The Nature of the Chemical Bond* [1]. Its last, third edition appeared in 1960 and many of the later stars of chemistry benefited from it by getting their introduction to the intricacies of this branch of science. A new book would be timely, but nobody seems brave enough to try filling Pauling's shoes in producing a new comprehensive monograph about the chemical bond.

Had Pauling produced his series of articles about the chemical bond, and nothing else, he would have already written his name into the annals of the history of chemistry. However, he did not limit his interest to theoretical studies. He utilized X-ray crystallography broadly and was constantly on the lookout for new techniques. While in Europe, he visited Herman Mark's laboratory in Ludwigshafen, Germany (where he was at the time), and Mark introduced a new experimental technique to his visitor for the determination of molecular structure, gas-phase electron diffraction. It was similar to X-ray crystallography, but there were two major differences. It used electrons rather than X-rays and the target was not a crystal but a gaseous sample in which the molecules had no well-defined order in their mutual arrangements.

One of the great advantages of using electrons was the very high intensity of the interaction between electrons and molecules. Thus, the duration of the required interaction was measured in minutes rather than many hours as with X-rays. The other important advantage was that in the gaseous sample the molecules were by themselves and their structures were not impacted by the closeness of their neighbors. For the X-ray technique, the molecules were required to be able to form a crystal in the first place, and there was no such requirement for using the electron diffraction technique.

The structures determined by the new technique depended only by the molecule itself and not by the way they were arranged relative to each other as was the case in the crystal. Other limitations of the new technique, however, have restricted it from becoming so widely used as X-ray crystallography, which truly has been the preeminent tool for uncovering the structures of biopolymers.

Mark's industrial laboratory was not the proper environment to expand the studies of molecular structures and he happily offered Pauling to take the new technique with him to Caltech. Mark even supplied him with the blueprints of his apparatus. Pauling not only introduced the gas-phase electron diffraction technique quickly in the United States, but he and his student, Lawrence Brockway further developed it. They added a mathematical step to handling the experimental data that made it possible to extract structural information in a graphically direct and attractive way from the probability density distribution of the inter-nuclear distances in the molecule (usually it is referred to as the radial distribution curve, which is a misnomer). From the experiment to reading off the curves directly the distances between atoms in simple molecules took only a few days' work.

Pauling (Fig. 1) established relationships among various experimental facts and made predictions about structures not yet investigated. He then worked out a theoretical technique based on quantum mechanics, but simple enough for a broad circle of chemists, to describe molecular structures. It was called the valence-bond or VB theory and it was one of the two major theoretical approaches developed over the decades. The other is the molecular orbital or MO theory. The VB theory builds the molecules from individual atoms linked by electron-pair bonds. For chemists, the VB theory appealed as more straightforward, alas, it did not stand well the test of time. The MO theory has proved more amenable to computations, which itself has become a major thrust in modern structural chemistry. However, for a long time the VB theory dominated the field.

Fig. 1 Ava and Linus Pauling (Photograph by and courtesy of Karl Maramorosch, Scarsdale, NY).

An important feature of the VB theory was that a molecular structure could be described by a set of "resonating" structures. This did not mean that each structure in such a set would be considered as present individually, but that the sum of these resonating structures represented the emerging structure better than any other description at the time. It needs to be stressed that what the resonance theory provides is merely a model, an approach, rather than a unique reflection of reality. There were proponents and opponents of the theory as is the case with most theories. Yet the resonance theory proved to be eminently useful for Linus Pauling—who was one of its initiators—in his quest for the protein structure. It happened so that this theory showed him the way and brought him a resounding victory over his competitors who lacked this tool and could not arrive at the right solution.

Pauling was advancing in a systematic manner in his quest for building up structural chemistry. First, he busied himself with inorganic substances and after the first 10 years he moved to organic substances. Among the organic molecules he often observed structures in which the lengths of the bonds between atoms were intermediate between single bonds and double bonds, so the theory of resonance came in handy in their understanding and description. Today, chemists no longer tend to think in terms of purely single bonds and double bonds, or triple bonds for that matter, and, accordingly, the utility of the resonance theory has largely disappeared, but in the 1930s it was considered to be of great help.

As Pauling was learning more and more about the structures of relatively simple molecules, in the mid-1930s, it occurred to him that he might as well make an attempt to learn about larger systems. He was aware of the importance of biopolymers and that the understanding of their structures might be a step toward understanding biological processes. Proteins were an obvious choice, because they were the most important biopolymers. At that time nucleic acids were already known, and their building blocks, the nucleotides, had been identified, but the nucleic acids were not considered to be of great significance. There was a hypothesis by Phoebus Levene about the tetranucleotide structure that was based on an erroneous observation that the four nucleotides in nucleic acid were present in equal amounts [2]. Hence, the nucleic acids were thought to be dull, uninteresting molecules, not capable of carrying any great amount of information.

When Pauling started thinking about protein structures, the first protein to attract his attention was hemoglobin, which is the vehicle of carrying oxygen in our organism. Incidentally, the British group engaged in protein structure studies had also selected hemoglobin for their target; their choice was independent of Pauling's interest. At the end of the 1920s, Gilbert Adair in Cambridge, UK, showed that the hemoglobin molecule consists of four units each with an iron atom, and each iron could bind an oxygen atom. Pauling formulated a theory about the oxygen uptake of hemoglobin and the structural features of this molecule related to its function of disposing of and taking up oxygen.

His interest in protein structures was further whetted when a visiting scientist and protein specialist, Alfred Mirsky of the Rockefeller Institute, spent the academic year 1935–1936 in his laboratory. They jointly studied the phenomenon of denaturation of proteins by heat or chemical substances, and formulated a theory about it. In this theory, they described the native protein as having a regularly folded structure in which hydrogen bonds provided the stability of the structure. Hydrogen bonding was a recently discovered phenomenon; it was becoming recognized as a crucial mode of interaction in chemical structures and especially in those of biological importance. In retrospect, it was a pivotal discovery, but its significance emerged only gradually over the years. For many biological molecules it is the hydrogen bonds that keep their different parts together.

Pauling postulated that the subsequent amino acid units are linked to each other in the folded protein molecule not only by the normal peptide bond but also by hydrogen bonding that is facilitated by the folding of the protein, which brings the participating atoms sufficiently close to each other for such interactions. In Pauling's and Mirsky's conclusion, when the protein molecule is denatured it undergoes complete or partial unfolding accompanied by breaking the hydrogen bonds. This was a hypothesis, because they knew practically nothing about the nature of folding; finding more about it occupied Pauling's mind for the next 15 years.

By the time Pauling became engaged in this research it had been established from rudimentary X-ray diffraction patterns that there might be two principal types of protein structure. Keratin fibers, such as hair, horn, porcupine quill, and fingernail belonged to one, and silk to the other. The foremost British crystallographer of fibers, William T. Astbury showed in the early 1930s that the diffraction pattern of hair underwent changes when it was stretched. He called the one producing the normal pattern alpha keratin and the other, which was similar to the pattern from silk, beta keratin. In 1937, Pauling set out to determine the structure of alpha keratin. He did not just want to rely on a single source of information. He planned to use all his accumulated knowledge in structural chemistry and find the best model that would make sense on this background and would be compatible with the X-ray diffraction pattern.

There was one piece of information from X-ray diffraction that seemed to be a good point of reference and that was the structural unit—whatever it would be—along the axis of the protein molecules repeated at the distance of 5.1 angstrom. He also knew the dimensions of the peptide group, that is, the characteristic sizes of the group linking the amino acids to each other in the protein chain. The C–N bond in the peptide linkage was not simply a single bond, but it was not a purely double bond either. Pauling's involvement with the resonance theory taught him that the emerging structure could be represented by two resonating structures.

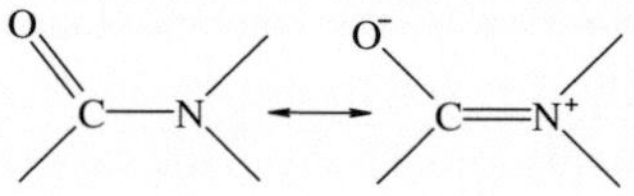

Hence, the resonance theory suggested that the C–N bond in the peptide linkage had a partial double bond character. From the accumulated structural information he also knew that the bonds around a double bond are all in the same plane. This was a very important piece of information because rather than taking into account all kinds of rotational forms with respect to the peptide bond, he could assume that it was a planar configuration. This assumption greatly reduced the number of possible models he had to consider for describing the structure of alpha keratin. Nonetheless, at this time Pauling was unable to find a model that would fit the X-ray diffraction pattern and he postponed further study on protein structures.

During the ensuing years Pauling and his newly arrived associate, Robert Corey, an expert in X-ray crystallography, carried out a large amount of experimental work determining the structures of individual amino acids and simple peptides. At some time every doctoral student in Pauling's laboratory was supposed to determine the structure of an amino acid for his PhD dissertation. The study was interrupted by World War II, but continued vigorously upon its conclusion. Pauling returned to the question of the structure of alpha keratin in 1948 while he was a visiting professor at Oxford University in England.

Not only had the amount of experimental information in the meantime expanded considerably, but Pauling could take a more detached view of the problem in his renewed efforts. When he was looking for the solution more than a decade before, he was bothered by the knowledge that his model was supposed to accommodate the possible presence of 20 different amino acids in the protein chain. At this time, in 1948, he decided to ignore their differences and assumed them to be equivalent for the purpose of his model. This was yet another example of Pauling's ability to distinguish between essential features and those that could be ignored in building his models.

Pauling remembered a theorem in mathematics he learned about at Caltech a quarter of a century before. It stated that the most general operation to convert an asymmetric object into an equivalent asymmetric object is a rotation–translation and that repeated application of this operation produces a helix. Here the asymmetric objects are the amino acids constituting the protein chain; the rotation should take place about the molecular axis of the protein; and the translation is the movement ahead along the chain. The amount of rotation was such that took the chain from one amino acid to the next while the peptide group was kept planar, and this operation was being repeated and repeated all the time. An additional restriction was keeping the adjacent peptide groups apart at a distance that corresponded to hydrogen bonding. In Pauling's model the turn of the protein chain did not involve an integral number of amino acids—he did not consider this a requirement whereas his British counterparts did. This was yet another relaxed feature of the structure that served him well in finding the best model whereas it served as an unnecessary restriction for his competitors.

Pauling—ever the model builder—sketched a protein chain on a piece of paper and folded the paper while looking for structures that would satisfy the assumptions he had made (Fig. 2). He found two and called one the alpha helix and the other the gamma helix, the latter being much less probable than the former. He determined the distance between repeating units in the protein chain and noticed a marked difference between his estimation from the model and the experimental value from the diffraction pattern. This was disappointing but the model was so attractive and so sensible that Pauling had little doubt in its correctness. Nonetheless, he decided to wait with its publication until the discrepancy would be understood. His confidence was enhanced when he visited the British group involved also in the structure elucidation of proteins and Max Perutz showed him his diffraction patters. From the X-ray diagrams it was obvious to Pauling—though not yet to Perutz—that the structure was alpha helix. Pauling did not say anything to Perutz.

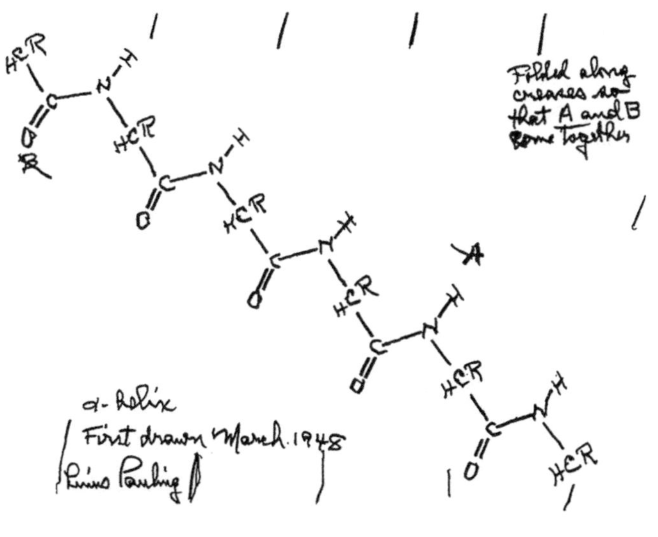

Fig. 2 Linus Pauling's sketch of the polypeptide chain in 1948. When he folded the paper along the creases, the alpha-helix appeared [3] (Fig. 3).

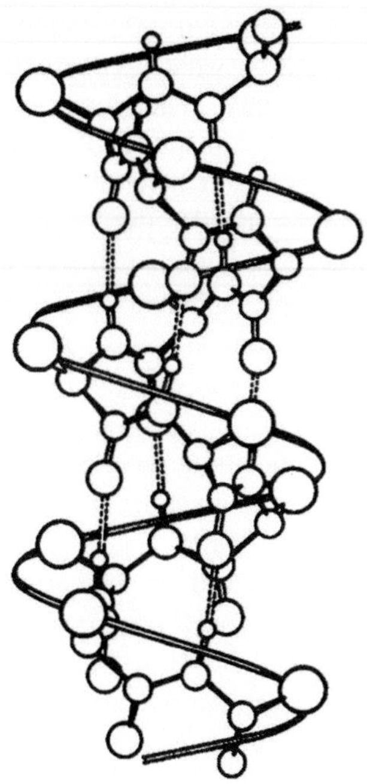

Fig. 3 Model of alpha-helix with 3.7 amino acid residues per pitch after [4].

When Pauling returned to Pasadena, he and his associates double checked all his calculations and found no errors in them. In the meantime, after about a year, Bragg, Perutz, and John Kendrew of Cambridge, UK, published a big article about protein structures and communicated about 20 models, none of which contained a planar peptide group and none of which described alpha keratin satisfactorily [5]. Finally, Pauling decided to ignore the discrepancy of the repeat distance between his model and the experimental observation and he and his associates published the alpha helix.

Eventually, the origin of the discrepancy was understood; it was caused by the alpha helices twisting together into ropes. This interaction between the chains caused a change in the experimental data as compared to what it would be for a single chain for which the model had been constructed. Thus, Pauling's alpha helix was confirmed even in this detail. The alpha helix has proved to be a great discovery because it is a conspicuously frequent structural feature of proteins.

Pauling's approach to solving this complex problem was exemplary in focusing on what was essential and ignoring what had little consequence. When it turned out that the turn about the chain did not correspond to an integer number of amino acids, hinting at less than perfect symmetry, he did not let himself bothered by this. He thus expanded the realm of crystallography toward structures that were not part of classical crystallography yet included literally vital substances. It was also noteworthy that he could skip a decade in working on this most important discovery without much danger of others scooping him.

They almost did, but only in their timing and not in knowledge, because his knowledge proved to be superior to anyone in his field at that time.

Pauling must have sensed the precarious nature of the situation and restrained himself from revealing crucial information to Perutz during his visit to Cambridge (vide supra). The Cambridge X-ray diffraction pattern showed the helical nature but Perutz did not think about it and thus did not notice it whereas for Pauling it provided additional evidence of the correctness of his model. This episode showed both his competitive spirit and his self-discipline. Finally, Pauling was sure enough in himself and his model that he went ahead with publishing the alpha helix without having yet resolved the remaining (apparent) discrepancy between his model and the available experimental evidence. First they published a short note [6], followed by a longer article [7] and soon they wrote seven more papers to report their findings.

Pauling was a master in creating publicity for his discoveries. When he prepared for announcing the discovery of the alpha helix it was to be in a big lecture hall at Caltech. The model stood on the rostrum, but it was under a cover, waiting to be unveiled, just as a sculpture would be, and it came toward the end of Pauling's lecture. When it was finally unveiled, the effect was dramatic and the audience was stunned by its beauty. I myself experienced the mesmerizing effect of Pauling's lecturing at the University of Oslo in 1982. He covered the board with complicated formulas and from time to time he looked at the audience as if checking whether we were duly impressed. Otherwise, the formulas were not at all necessary for us to understand the points he was making. He was already an octogenarian, but watching him gave an impression of a young assistant professor who came for interview and was presenting his research with the usual arrogance of such scenes. During the lunch following the lecture he was more vigorous than the rest, led the discussion, and fired away questions, mostly answering them himself.

In research publications there is no place for the human sides of the discoveries and Pauling wrote up the story of his alpha helix discovery separately, but it never appeared while he was alive. It was published 2 years after he died when I was running a chemical magazine and his former secretary of his last 20 years, Dorothy Monro, suggested to bring it out there. Research papers usually lack the human element and the blind alleys in research, so this paper by Pauling was especially valuable for our understanding how this particular discovery happened [3].

The Cambridge group suffered a defeat in this case, which was especially heavy for W. Lawrence Bragg to bear, because he was the pioneer of X-ray crystallography and the American group came out on top in their undeclared race. It was not possible to pinpoint a single reason for this defeat, but it was a crucial difference that Pauling could limit the number of possible models because of his superior knowledge of structural chemistry. The Cambridge group had no such guideline although it could have. It turned out that Lord Todd the soon to be Nobel laureate organic chemist who worked in the next building to Perutz's and Kendrew's laboratory had told Bragg that the peptide bond had some double-bond character. Bragg, however, could not from this piece of information make any conclusion about the configuration of the peptide bond, namely, that it was planar.

Years after this fiasco, Perutz complained about their lack of knowledge of the planarity of the peptide group. He blamed the Medical Research Council (MRC) for having him denied the use of a Rockefeller Fellowship for travel to America in 1948. The Secretary of the MRC thought that rather than going to learn from the Americans, the Americans should come and learn from the British. In hindsight, Perutz thought that he could have learned about the peptide bond planarity from Pauling had he been allowed to travel [8]. Of course, he could have just walked across the street to visit Lord Todd for the same information.

It is not at all sure whether had Perutz visited Pauling he would have learned from Pauling as much as he might have supposed in retrospect. We have seen Pauling withholding his observation from Perutz that he had noticed the evidence of helical structure on Perutz's X-ray diffraction diagram. During his Oxford sojourn, Pauling wrote to Corey back to Pasadena that he felt uncomfortable about the English competition. In their turn, the British considered protein crystallography their own territory. It was not only that the Braggs discovered X-ray crystallography and that Astbury was a pioneer in taking X-ray pictures of proteins. It was also J. Desmond Bernal who had prepared the first ever X-ray diffraction diagrams of a single-crystal protein—a pepsin single crystal—that clearly showed the possibility of deducing atomic positions from it. This was in 1934. In the future Nobel laureate X-ray crystallographer Dorothy Hodgkin's description, "that night, Bernal, full of excitement, wandered about the streets of Cambridge, thinking of the future and how much it might be possible to know about the structure of proteins if the photographs he had just taken could be interpreted in every detail" [9]. The British self-confidence in dominating this field reached such proportion that Astbury and Bernal divided it by a gentlemen's agreement between the two of them. They decided that Bernal would take up the investigation of the crystalline substances and Astbury the fibrous ones [10].

Perutz on his part, for their failure blamed Astbury's X-ray diffraction picture, which showed a discrepancy between the repeat distances as compared with reasonable structures, a discrepancy—as we have seen—Pauling daringly disregarded. Perutz was disheartened when he found Pauling's paper about the alpha helix model. He devised an additional X-ray experiment that gave further evidence for the correctness of Pauling's result, something that Pauling had missed. When Perutz reported his finding to Bragg, Bragg asked him, "How did you think of that?" Perutz's response was that it was because he was so angry that he hadn't thought of the structure himself. To which Bragg replied coldly, "I wish I'd made you angry earlier" [11]. Perutz told me this story in 1997, and he used this phrase as the title of his next book. Perutz might have thought that Pauling would be pleased that he provided additional evidence for alpha helix, but was disappointed by Pauling's reaction, which was clearly dismissing.

Pauling's fascination with proteins served him well in his focusing his attention to their structures at a crucial period in twentieth century science. However, he continued his protein bias even when the next big task appeared before structural chemistry that was the structure of nucleic acids. Pauling entered that race too, but there is ample evidence that Pauling did not concentrate on it with the intensity and dedication as he had done for the protein structures. In case of the quest for the structure of nucleic acids he was defeated by the British teams. Pauling published an erroneous triple helix and he was not in possession of the best X-ray diffraction patterns of nucleic acids that were available at the early 1950s either. As is well known, those patterns were produced at King's College in London and the winning double helix model came out from the Cavendish Laboratory in Cambridge, UK, but this is a different story.

We can add a footnote about Pauling's theory of chemical resonance, which served him so well in the above story. At about the same time, this theory was in the center of attack by rabid ideologists in the Soviet Union [12]. The culmination was a 4-day conference in Moscow in 1951, organized by the Soviet Academy of Sciences. Leading Soviet chemists, physicists, philosophers, and others attended the meeting. A small but vocal group of chemists attacked the theory of resonance as an ideological aberration and together with it quantum theory and the science of the West. They insisted on returning to traditional Russian values and offered their own worthless theories. Excellent scientists suffered ruthless criticism for having applied the theory of resonance in their work, and they, in turn, offered humiliating self-criticism.

The affair has been referred to as the great Soviet resonance controversy and it was a chapter in the anti-science events following World War II that touched biology even more severely. Physics was spared in the last minute due to its decisive role in producing nuclear weapons. Stalin's terror did everything to protect his empire from even the slightest influence by the West, the purest sciences included. There was irony in this story in that Pauling was a friend of the Soviet Union and suffered persecution in the McCarthy era, but this was not yet known in the Soviet Union. In 1993, I asked Pauling for his comments about this affair. He appeared as if he misunderstood it or did not want to understand it. He wrote that it took years "for the chemists in the Soviet Union to get a proper understanding of the resonance theory" [12, p 5]. In reality, they understood it well enough and applied it with great success, that is, until 1951, when the main proponents of the theory lost their jobs. If it was a consolation, their lives were spared in contrast with some of their biologist colleagues in a similar ideological controversy.

Acknowledgment Our research is being supported in part by the Hungarian Scientific Research Foundation (OTKA Nos. K60365 and T046183).

References

1. Pauling L (1960) The nature of the chemical bond and the structure of molecules and crystals: an introduction to modern structural chemistry (Third edition, First edition, 1939). Cornell University Press, Ithaca, NY
2. Hargittai I (2009) Struct Chem 20:753–756
3. Pauling L (1996) Chem Intell 2(1):32–38
4. Corradini P (1982) In: Carra S, Parisi F, Pasquon I, Pino P (eds) Giulio Natta: present significance of his scientific contribution. Editrice di Chimica, Milano, p 134
5. Bragg WL, Kendrew JC, Perutz MF (1950) Proc R Soc 203A:321–357
6. Pauling L, Corey RB (1950) J Am Chem Soc 72:5349
7. Pauling L, Corey RB, Branson HR (1951) Proc Natl Acad Sci USA 37:205–211
8. Olby R (1994) The path to the double helix: the discovery of DNA. Dover Publications, New York, p 291
9. Hodgkin DC, Riley DP (1968) In: Rich A, Davidson N (eds) Structural chemistry and molecular biology. WH Freeman, San Francisco and London, pp 15–28
10. Bernal JD (1968) Labour Monthly, pp 323–326
11. Hargittai I (2002) Candid science II: conversations with famous biomedical scientists (edited by Magdolna Hargittai). Imperial College Press, London, p 288
12. Hargittai I (2000) Candid science: conversations with famous chemists (edited by Magdolna Hargittai). Imperial College Press, London, pp 8–13

"There Is No Such Animal (אין חיה כזו)"—Lessons of a Discovery[a,b]

István Hargittai[c]

Abstract

The discovery of quasicrystals by Dan Shechtman in the early 1980s was a conspicuous event in materials science not only because it led to the production of a plethora of new materials but also because it signified the demise of a dogma in the science of condensed phase materials concerning symmetry restrictions. Having the discovery recognized was not easy and it required stamina on Shechtman's part. The story of the quasicrystal discovery offers a set of lessons that might be useful to remember in similar situations.

Keywords

Dan Shechtman • Quasicrystals • Scientific discovery

"Do not consider it proof just because it is written in books..."
Maimonides (attributed)

Quasicrystals are not only important for science; they are also visually attractive! (Fig. 1). The story of Dan Shechtman's discovery of quasicrystals gives us an opportunity to ponder about the nature of scientific discoveries. This was the subject of my after-dinner talk on January 12, 2011, at the Technion in Haifa, on the occasion of Shechtman's seventieth birthday celebrations. The title of the presentation was "Lessons of the Quasicrystals Discovery."

As the readers of Structural Chemistry know, quasicrystal structures have order but no periodicity; thus they are different from crystals having both order and periodicity and from amorphous materials having neither. Thus, quasicrystals are in between the two other kinds of materials. Crystals and amorphous bodies (glass, for example) have been known for a very long time, but what we call today quasicrystals used to

Fig. 1 Flowerlike icosahedral quasicrystals in quenched aluminum–manganese sample (By Ágnes Csanády, Budapest and Hans-Ude Nissen, Zurich; Courtesy of Ágnes Csanády) [1].

[a] *Structural Chemistry* 2011, 22(4):745–748.

[b] This Editorial is based on an invited contribution to IGGERET, the Hebrew-language publication of the Israel Academy of Sciences and Humanities.

[c] Department of Inorganic and Analytical Chemistry and Materials Structure and Modeling Research Group of the Hungarian Academy of Sciences, Budapest University of Technology and Economics, PO Box 91, Budapest 1521, Hungary
e-mail: istvan.hargittai@gmail.com

I. Hargittai, B. Hargittai (eds.), *Science of Crystal Structures: Highlights in Crystallography*,
DOI 10.1007/978-3-319-19827-9_14

be considered impossible until Shechtman's experiment. Shechtman discovered the quasicrystals; he did not invent them, because they have always been around without our knowing about them.

The official date of the quasicrystals discovery was marked by the appearance of Shechtman's and his co-authors' report in Physical Review Letters at the end of 1984 [2]. However, the actual date of discovery was April 8, 1982, when Shechtman noticed that one of his electron diffraction patterns of a manganese–aluminum alloy displayed tenfold symmetry. He knew that it was "impossible," if from nothing else, because many years before he had passed a university examination proving that there was no such thing. His own disbelief was marked by his words (in Hebrew), "There is no such animal," hence the title of this contribution. According to the rules of classical crystallography, fivefold, sevenfold, eightfold, etc., symmetries are impossible in the condensed phase.

What happened was an example *par excellence* of what Louis Pasteur expressed as "Chance favors the prepared mind." Danny could have dismissed his serendipitous experiment as some artifact, but he did not. Also, a weaker character might have sooner or later succumbed to the pressure of friendly advice and not so friendly ridicule that he had been exposed to when he insisted on the validity of his discovery. Shechtman's going against the dogma of symmetry restrictions in the condensed phase reminded me of the statement attributed to Maimonides: "Do not consider it proof just because it is written in books…"

For a long time illustrious scientists and artists such as Johannes Kepler and Albrecht Dürer, tried to create patterns in which regular pentagons covered a surface without gaps or overlaps, but they did not succeed. Then, the British mathematician Roger Penrose created a pattern in which a pentagon was surrounded by five other pentagons within a larger pentagon, and the pattern continued by including larger and larger pentagons. He used not only the whole pentagons, but parts of them, too, and produced an appealing pattern, which covered the surface with regular pentagons, but of gradually changing sizes (Fig. 2). Another British scientist, the crystallographer Alan Mackay was intrigued by Penrose's pattern, and simulated on this pattern a diffraction experiment. He figured that if he could produce a diffraction pattern from this planar pattern, the same should be possible even from three-dimensional structures.

When in the fall of 1982 Mackay told me about his simulation, Shechtman had already performed his crucial experiment, but we did not know about it. Neither did Shechtman know about Mackay's attempts. Shechtman, who earned his BS, MSc, and PhD degrees at the Technion, was a visiting scientist at the then U.S. National Bureau of Standards (NBS) where he was experimenting with aluminum–manganese alloys. That is when he ran his experiment that shocked him and that later would shock the world of science of condensed materials. But it took another 2 years before this happened.

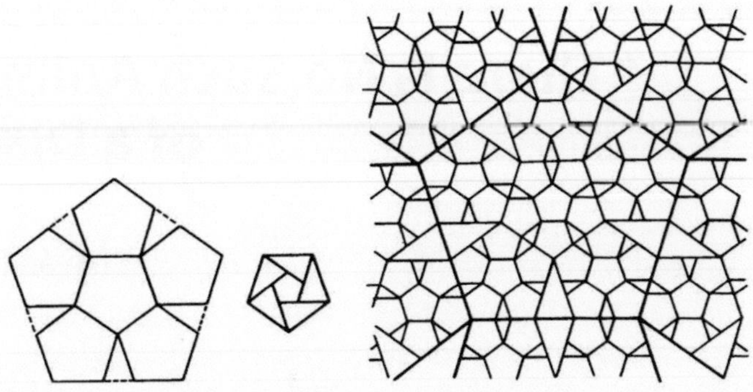

Fig. 2 Tiling with regular pentagons of changing sizes after Alan Mackay [3].

Once Shechtman had completed his experiment, he became a very lonely person as every scientific discoverer does: the discoverer knows something nobody else does. At the time of his crucial experiment Shechtman was alone in the laboratory, but felt the urge to share his excitement with somebody else, and he ran out into the corridor, but nobody was there, so he returned to his electron microscope and performed a series of additional experiments. Other scientists, including Louis Pasteur have described similar experience of having the urge to share the new knowledge at the moment of discovery.

Right after his experiment Shechtman started inquiring what others knew about the possibility of whether or not what he had observed might be real. Sadly, however, scientists, who were authorities in his field—where he had not made his name yet—dismissed his claims. In one aspect this was also fortunate as nobody tried to expropriate his discovery. Finally, he found an expert in X-ray diffraction, Ilan Blech, who was willing to listen to him, and the two started making models that might produce the diffraction pattern that Shechtman had observed.

There was yet another direction of related investigation, at the physics department at the University of Pennsylvania in Dov Levine's PhD work under Paul Steinhardt's supervision. Levine's dissertational work produced a theoretical model that in time would prove very suitable to interpret Shechtman's experimental observation. Levine was anxious to publish his findings, but his advisor was reluctant, probably fearing the reaction that a suggestion showing an established dogma invalid might confront them. Both Dan Shechtman and Dov Levine are currently professors at the Technion (Fig. 3), Shechtman at the Department of Materials Engineering and Levine at the Department of Physics.

For Shechtman it was not easy to bring out his discovery in a scientific paper. A more established scientist might have

Fig. 3 Dan Shechtman (on the *right*) and Dov Levine in Shechtman's office at the Technion in 1996 (Photograph by I. Hargittai).

rushed to publish such a new finding to avoid losing the claim of priority. On the other hand, a more established scientist might have been very cautious, too, in coming out with a discovery that others thought impossible. Shechtman was convinced that he had a novel observation, but it bothered him that he could not offer a credible explanation for it, and this restrained him from trying to publish it.

When Shechtman finally decided to submit a manuscript in which he and Blech described his experiment, it was done in a tentative manner. The information was there but an unsuspecting reader might have not noticed it, since it was buried under a mountain of information about alloys. It read like a report in metallurgy yet they submitted it to a physics publication, the Journal of Applied Physics. However, the editor of the journal wrote back by return of mail that the material of the manuscript was not suitable for the journal and—incredibly—that it would not interest physicists. It is not that rare that it is difficult to publish a seminal discovery. Of course, bad papers are rejected. But often this is the fate of manuscripts that report radically new results. As the Shechtman–Blech manuscript was more a report in metallurgy than in physics, they re-submitted it to a metallurgical periodical, and it appeared there, eventually [4].

Shechtman knew that he should considerably improve the presentation of his discovery. He turned for help to John Cahn, a veteran scientist in the section where he worked at NBS. Cahn for two long years did not believe in the discovery, but now he changed his mind and the grateful Shechtman invited Cahn to be a co-author of the report. Then, Cahn added yet another name, Denis Gratias, a young French mathematical crystallographer who helped in shaping the mathematical format of the article, and Shechtman added Blech's name out of loyalty, hence the four authors on the paper [2].

They sent the new manuscript to the prestigious periodical, *Physical Review Letters*, where it was received on October 9, 1984. The manuscript was accepted for publication without delay, and appeared in the issue dated November 12, 1984. It had a somewhat innocuous title, "Metallic Phase with Long-Range Orientational Order and No Translational Symmetry" [2]. The impact was tremendous, as if a floodgate had been lifted. Scientists, especially theoreticians, had been working on related problems and those who were versed in the literature could easily connect their work to the Penrose pattern, Mackay's simulated diffraction diagram, and Shechtman's discovery.

Dov Levine's and Paul Steinhardt's purely theoretical paper quickly followed in the same journal. Their manuscript arrived on November 2, 1984, that is less than 2 weeks before the Shechtman paper appeared, and this was no coincidence. They learned about the Shechtman manuscript back in October from a Harvard professor to whom Cahn had sent a copy. The Levine–Steinhardt report had an elegant title, "Quasicrystals: A New Class of Ordered Structures," thus it gave a name to the new structures [5]. The name was easy to remember, it well expressed the essence of the new substance, and it stuck. There are some who argue that giving a name to a discovery may be as important as making the discovery itself. A scientific discovery is made sooner or later and if not by this scientist, then by another scientist. In contrast, giving it a name is a unique act that belongs to an individual. Still, it is the discovery rather than its name that impacts science and the application of its fruits.

It is a question of contention to what extent should Shechtman's finding be considered a real discovery. To some, the identification of quasicrystals merely pointed to the inadequacy of how crystals had been defined. To this we might ask, if it was merely an easy correction, why nobody had bothered to make it before Shechtman. In reality, this discovery expanded our concept of materials by adding a new class to them. It brought about a paradigm change. There had been earlier attempts to challenge the dogma, but any prior "disturbing" experiment could be brought into line and the dogma could be left in place.

The main critique of the quasicrystal concept came from the great chemist Linus Pauling. He had distinguished himself not only with exceptional knowledge of structures of materials, but also with fundamental discoveries that broke with previous dogmas! His refusal of Shechtman's discovery made Shechtman's case even harder because this came from a scientist of Pauling's reputation. But Shechtman persevered. Pauling died in 1994 without having accepted Shechtman's discovery. Pauling's many followers found it easier to accept the new concept afterwards. The way it happened strengthened what Max Planck wrote in his scientific biography: "A new scientific truth does not triumph by convincing its opponents and making them see the light, but rather because its opponents eventually die, and a new generation grows up that is familiar with it" [6].

There are a few general lessons from the quasicrystal discovery that include the following:

- Start with a practical problem. In this case it was the quest for alloys with special properties.
- Play with it. In this case Shechtman did not limit his testing of alloy compositions that might have been expected to become useful.
- Do not discard the unexpected. When we get what we expect it is not a discovery; it happens when we get something we did not expect. Shechtman certainly did not discard the unexpected.
- Expose what you found to wide scrutiny. This Shechtman did impeccably.
- Coin a name for it. This happened here, too, though by others than the discoverer.
- Publish it conspicuously. This did not happen right away, but eventually it did.
- Talk about it to all that will listen, and try in every way easing your loneliness. Public Relations is also important in science.
- Stay with your discovery while it brings new knowledge and leave it only when it becomes routine. In this, Shechtman might have stayed with it a little longer although after a while he returned to it.

The quasicrystal story is yet another example of the unique role symmetry considerations play in many scientific discoveries [7]. This is part of the visual effects quasicrystals exert that led to their impact not only in science but in the visual arts and design as well [8]. The discovery opened the door to a world of "structures beyond crystals" [9]. Today Shechtman is the undisputed discoverer of quasicrystals, which is the recognition not only of his ingenuity, but also of his stamina and perseverance [10]. The discovery caused a paradigm change in physics, chemistry, crystallography, and materials science. It deserves the highest recognition. Dan Shechtman is a most decorated scientist, but whether or not such recognition will further expand, it is difficult to predict. This author certainly feels that it should.

References

1. Csanády A, Papp K, Dobosy M, Bauer M (1990) Symmetry 1:75–79
2. Shechtman D, Blech I, Gratias D, Cahn JW (1984) Phys Rev Lett 53:1951–1953
3. Mackay AL (1981) Kristallografiya 26:910–919
4. Shechtman D, Blech I (1985) Metall Trans 16A:1005–1012
5. Levine D, Steinhardt PJ (1984) Phys Rev Lett 53:2477–2480
6. From Max Planck (1948) A Scientific Autobiography, quoted in W. F. Bynum, Roy Porter, eds, Oxford Dictionary of Scientific Quotations. OUP, Oxford, UK, 2005, p 494
7. Hargittai M, Hargittai I (2009; 2010) Symmetry through the eyes of a chemist, 3rd edn. Springer
8. Hargittai I, Hargittai M (2000) In Our Own Image: Personal Symmetry in Discovery. Kluwer, NY
9. Hargittai I (2010) J Mol Struct 976:81–86
10. Hargittai I (2011) Drive and curiosity: What fuels the passion for science. Prometheus, Amherst, NY

Quasicrystal Discovery—From NBS/NIST to Stockholm[a]

Balazs Hargittai[b] and István Hargittai[c]

Abstract

There were ideal conditions in 1982 at the NBS/NIST for the kind of exploratory research that led Dan Shechtman to the quasicrystal discovery. Almost 30 years later, in 2011, Shechtman received the Nobel Prize for this achievement. What happened in between these two dates has been covered quite extensively elsewhere. But it is intriguing to learn more about the very beginning. Then, the festivities in Stockholm added to our seeing the story in a fuller perspective.

Keywords

Quasicrystals • Dan Shechtman • Quasiperiodicity • NIST • Nobel prize

Many aspects of the quasicrystal story have been covered (see, e.g., [1–4]), but very little has been described of the circumstances of the original discovery, which happened at the (then) National Bureau of Standards, NBS—today's National Institute of Standards and Technology, NIST—in Gaithersburg, Maryland.

In 1981, Dan Shechtman of the Technion—The Israel Institute of Technology—went for his first sabbatical at NBS. His stay at NBS had been initiated by a senior scientist there, John Cahn, because of his interest in a technique Shechtman had developed for studying metallic powders by transmission electron microscopy. Shechtman's work at NBS was sponsored by the then U.S. Defense Advanced Research Project Agency (DARPA). Shechtman remembered that Jake Jacobson of DARPA told him specifically not to limit his studies to his originally proposed plan but rather to expand in any direction he felt was interesting. Shechtman further remembered in a conversation in 1995 during an international school on quasicrystals: "I started by studying rapidly solidified aluminum-iron alloys. I analyzed the phases present and the solidification patterns. I collaborated mainly with members of the metallurgy group, Bill Boettinger, Bob Schaefer and Frank Biancaniello. We wrote a series of papers together and understood rapid solidification better. It was in April 1982, half a year after I had arrived, that I discovered the icosahedral phase" [5].

This conversation was part of an extensive project, which included interviews of some other scientists who figured in the quasicrystal story (Fig. 1).

Frank Biancaniello's role was especially important in Shechtman's studies. Biancaniello was not yet a full-fledged researcher at the time (as he has become soon afterwards); he was, rather, a technician whose task included the preparation of the alloy samples of various compositions for the electron microscope–electron diffraction experiments. Biancaniello was—according to the testimony of the other members of the group—most enthusiastic about this investigation. He was happy to work many extra hours, including nights and weekends, and he was as skilled as he was devoted to this project. This is how the tests continued outside the range of reasonable compositions—reasonable, that is, for expected practical

[a] *Structural Chemistry* 2012, 23(2):301–306.

[b] Department of Chemistry, Saint Francis University, P.O. Box 600, Loretto, PA, USA
e-mail: bhargittai@francis.edu

[c] Department of Inorganic and Analytical Chemistry, Budapest University of Technology and Economics, P.O. Box 91, Budapest 1521, Hungary
e-mail: istvan.hargittai@gmail.com

I. Hargittai, B. Hargittai (eds.), *Science of Crystal Structures: Highlights in Crystallography*,
DOI 10.1007/978-3-319-19827-9_15

Fig. 1 A collection of interviews containing conversations with a few of the principal players of the quasicrystal story. Portraits of Roger Penrose, Alan Mackay, and Dan Shechtman are highlighted.

applications. Biancaniello remembers, "Obviously, the quasicrystal discovery was the most important one culminating in Danny's Nobel. But there were many other exciting projects we worked on that came to fruition over the years. I'm sure you have heard of some of the bulk metallic glass research being carried out over the years, with yield strengths on the order of 1 to 2 gigapascals with no ductility. Steve Ridder and I developed a High Nitrogen Stainless Steel with a yield strength of 1.1 gigapascals with an elongation of 60% using so-called rapid solidification technology. I only mention this in an attempt to demonstrate how intellectually stimulating these thirty years were starting with working with Danny Shechtman and continuing on to recent times. As Bill Boettinger mentioned, we worked on many projects with Danny. Steve Ridder and I did more recent research with Danny using a quasicrystal for a wear-resistant coating while doing thermal spray studies" [6].

Shechtman narrated in the 1995 conversation about some of the details of the seminal experiment [7]:

> At first, I was studying rapidly solidified aluminum-iron alloy, which we thought had some commercial future. Eventually, it turned out that although rapid solidification research resulted in several useful products, it did not develop into a widespread technology. This, however, is not important for our story. In the aluminum-iron binary system there was one metastable phase Al_6Fe, which I studied. The equivalent Al_6Mn in the aluminum-manganese system is a stable phase, and I wanted to compare some crystallographic features of the two. We started therefore to produce a series of aluminum-manganese alloys with increasing amounts of Mn in them. Eventually I ran wild, from a practical point of view, since beyond several percents of manganese the rapidly solidified alloy becomes brittle and therefore useless. Among the alloy ribbons which I have prepared with Frank Biancaniello by melt spinning, there were alloys which contained over 25 weight percent manganese. On April 8, 1982, as I was studying by electron microscopy rapidly solidified aluminum alloy which contained 25% manganese, something very strange and unexpected happened. It is worthwhile to look at my TEM [transmission electron microscope] logbook records of that day. For plate number 1725 (Al-25% Mn) I wrote: "10 Fold ???"
>
> There were ten bright spots in the selected area diffraction pattern, equally spaced from the center and from one another. I counted them and repeated the count in the other direction and said to my self: "There is no such animal". In Hebrew: "Ein Chaya Kazo". I then walked out to the corridor to share it with somebody, but there was nobody there, so I returned to the microscope and in the next couple of hours performed a series of experiments. Most of the needed experiments were performed at that time. A few days later all my work was complete, and everything was ready for the announcement. Then it took two years to publish it.

The diffraction pattern with bright spots indicating tenfold symmetry in the solid structure is reproduced in Fig. 2 and the logbook page where the observation of the conspicuously unexpected pattern is recorded is reproduced in Fig. 3.

Shechtman almost immediately excluded the possibility that the pattern originated from crystal twin formation—something that many crystallographers first suspected upon becoming acquainted with Shechtman's observations. Of course, Shechtman could not dismiss such a supposition off hand. Rather, he had prepared himself for such a question, the more so, because this possibility had also occurred to himself. In his words, [referring to the fact that the sequence of distances of the bright spots from the central spot was not periodic in his observation] "When an electron microscopist sees such a sequence the first thought is that these must be

Fig. 2 Shechtman's historic electron diffraction image of the aluminium/25%-manganese alloy (Courtesy of Dan Shechtman).

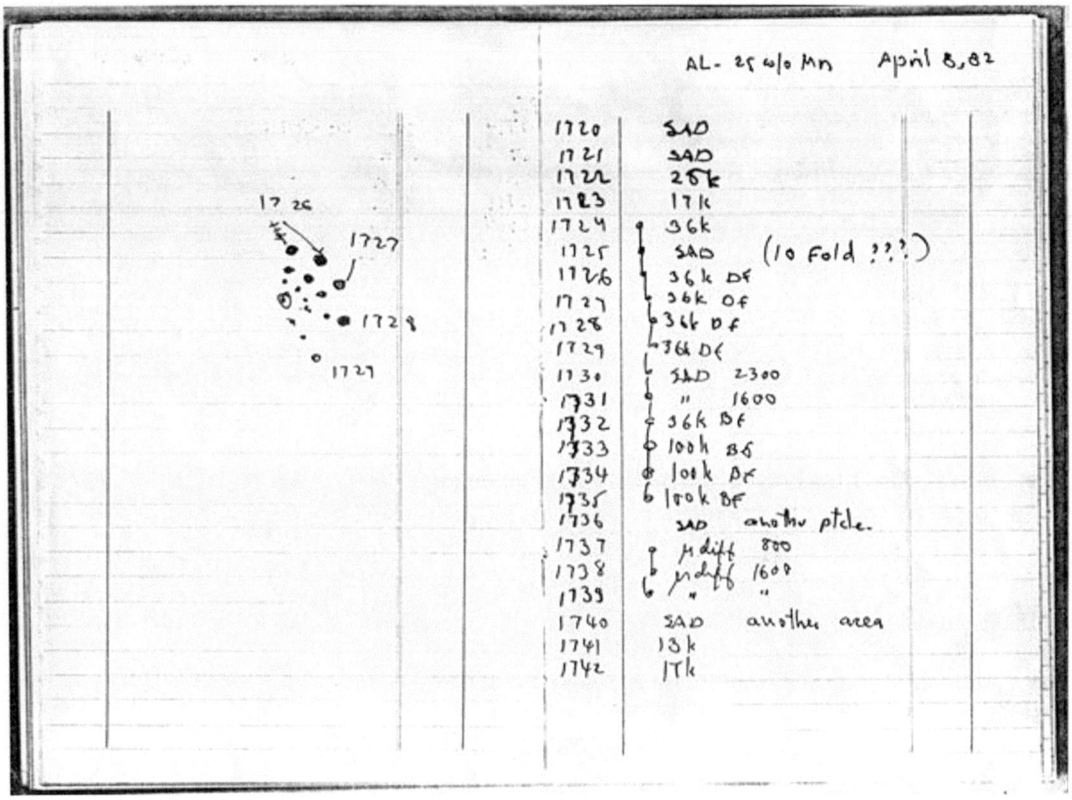

Fig. 3 Page in Shechtman's logbook where he recorded his observation of tenfold symmetry of plate 1725 on April 8, 1982, with highlighted information (Courtesy of Dan Shechtman).

twins. The alternative explanation is single crystal icosahedral symmetry, which was assumed to be forbidden symmetry for crystals although for molecules this symmetry has been known to exist. In order to prove that my diffraction patterns did not originate from twins, I generated a series of dark field images. In these experiments you take an image from a diffracted point such that all the information that passes to the plate is contained in the beam only. Examining the information I have found that the dark field images were almost identical which means that the same part of the crystal produced all the diffracted beams. It was clear to me that the crystal, the origin of diffraction was not twinned" [8].

When Shechtman mentioned that "it took two years to publish" his observation, this was an understatement, because it took more than two and a half years. But the delay was not only due to the disbelief of the scientific community that the dogma about the impossibility of five-fold symmetry (and other "forbidden" symmetries) should fall. It was also a delay of Shechtman's own making. Of course, the resistance of the scientific community bothered him and he would have liked to come out with a suitable model along with the publication of his observation. By mid-1984, however, he realized that he should have his observation printed in the scientific literature and produced a manuscript in co-authorship with Ilan Blech. They submitted their manuscript to the Journal of Applied Physics in the summer of 1984. The manuscript was more of the nature of a report in metallurgy than in physics and this may be judged as being so because eventually about the same manuscript appeared in a metallurgy journal [9]. The manuscript to Journal of Applied Physics was submitted in the summer of 1984 and the editorial decision came back soon. According to the letter two items were enclosed with it, the original manuscript (Ms. #R-6168) and a Reviewer's report. The Reviewer's report could not be found and Shechtman does not remember having seen such a report [10]. The Editor of Journal of Applied Physics, Lester Guttman said, "Our reviewer and we both believe that your paper will not reach the most appropriate audience through this journal, and we are therefore returning your manuscript. We wish to make it clear that this is not a comment on the technical quality of your work; rather it is an attempt on our part to try to place papers where they will be of the greatest use to the greatest number of people."

Even before Shechtman received the letter of rejection, he showed the manuscript to John Cahn who described his reaction to one of us, in a conversation in 1995 in Gaithersburg [11]:

> I read this paper immediately, and got very excited, because I saw the data for the first time. I wasn't convinced by the Blech model, but I was very much convinced by the experiments. It was obvious that it was not twinning, that it was something outside of what was known. I'm sure that had I seen the data in '82 the same day when they were taken, we wouldn't have wasted

2 years. On Friday I met with Danny and said, this is a poorly written paper, you're hiding this new phase. The entire paper was about experiments on aluminum–manganese. It started out by saying that we did this, we found this, and gave a list of the known periodic phases for each composition range, and buried, in half a paragraph, was a new phase, in the middle of an otherwise very conventional paper. My impression was that it was written that way to sneak it pass the referee, because everybody, except Blech, had told him that he had nothing.

I remember that Friday going to the bosses and telling them that this is really something and it's going to be big. I more or less said that we should mount a major effort; lots of people should shift because this has got to be investigated. At this stage I was not a co-author and I did not expect to be a co-author. I told Danny that this is the wrong paper but Danny told me that the paper had been submitted.

By the time Shechtman and Cahn talked again, the manuscript had been rejected by the Journal of Applied Physics and submitted to Metallurgical Transactions. Cahn still thought that this was the wrong manuscript and said so to Shechtman. At this point, Shechtman told Cahn if he thought so strongly about it he should write the right manuscript. This is how the paper that finally made the breakthrough was initiated [12]. When the manuscript was ready there was still another hurdle to overcome before it could be submitted for publication. This is how Cahn remembered [13]:

At NIST we have an Editorial Review Board. Every paper from the Institute has to be reviewed internally before it can be sent out. This Review Board was afraid of another polywater. The Bureau of Standards had been burned by polywater publications. The Board wanted to be on the right side this time and had asked local crystallographers to give them lectures about fivefold symmetry, why it can't be. They were spending a great deal of time. The Board was split. My Division Chief of Metallurgy, also an old friend, said to me, "John, you have a wonderful reputation. Why ruin it by putting your name on something like such a paper." I said, "please, read this paper, it is very important and very exciting, and it's not a time to be conservative." The Board took a very long time, they had lots of meetings, sometimes they came to me for advice but I was never present at these meetings. They carried an inquiry very carefully, and eventually they approved the paper. Finally, in mid-October [1984], we could send off the manuscript. It came out 3 weeks after it was submitted.

NIST followed up Cahn's recommendation in forming a quasicrystal group to conduct experimental as well as theoretical work. For a while, the outside world considered NIST to be in the focal point of this new field and researchers kept sending the NIST group their manuscripts, about 300 preprints already in the first year (Fig. 4).

On November 3, 2011, the authors of this Editorial visited the Metallurgy Division at NIST and at their request met with the associates of NIST who had been Shechtman's colleagues at the time of the quasicrystal discovery. At this informal meeting, present was the current Chief of the Metallurgy Division, Frank Gayle. The very friendly sentiments toward Dan Shechtman shone through this gathering; the reminiscences made it obvious that the helpful and creative atmosphere in the group must have greatly facilitated the birth of new experiments and new ideas. One of us has written about the loneliness of the scientific discoverer, which their supportive collective spirit which must have at least eased Shechtman's scientific loneliness (Fig. 5). stems from the fact that the discoverer knows something—for a while anyway—that nobody else does and sometimes has to go against the mainstream of science; this was the case of the quasicrystal discovery. However, in a human sense, socially, Dan Shechtman was never alone in this community and his former colleagues take pride in the strength of their supportive collective spirit which must have at least eased Shechtman's scientific loneliness (Fig. 5).

Fig. 4 In 1984, at NBS, Dan Shechtman, Frank Biancaniello, Denis Gratias, John Cahn, Leonid Bendersky, and Robert Schaefer (Photograph: H. Mark Helfer/NIST; Courtesy of NIST).

Eventually, Dan Shechtman being primarily at the Technion, his principal research activities moved there. He did not immediately focus his further studies on quasicrystals, but he never abandoned the area entirely. In the press conference during the Nobel festivities in December 2011 he was relaxed about the period of frustration when he had been waging an uphill battle in the scientific community for the recognition of his discovery. Incidentally, among the many awards he received for his achievement, the Georgi Aminoff Prize in 2000 was very special, because it is given by the same institution as the chemistry Nobel Prize, the Royal Swedish Academy of Sciences. The motivation was "for your discovery of quasicrystals." Those of us who were present at his Aminoff lecture in 2000 and at his Nobel lecture on

Fig. 5 From left to right, Istvan Hargittai, William Boettinger, Frank Gayle, Francis Biancaniello, Stephen Ridder, Leonid Bendersky, Robert Schaefer, and Balazs Hargittai at NIST on November 3, 2011.

Fig. 6 From among the quasicrystal images produced at NIST: triacontahedral Al-Cu-Li quasicrystal (facet edges about 100 μm long) by Frank Gayle.

Fig. 7 From among the quasicrystal images produced at NIST: dodecahedral Al-Cu-Fe quasicrystal (facet edges about 100 μm long) by Frank Gayle.

December 8, 2011, could notice the similarities between the two presentations. This was no accident; the motivation of his Nobel Prize was the same by the same institution, "for the discovery of quasicrystals." In his Nobel lecture, Shechtman made an effort to avoid using the name "quasicrystal" and preferred "quasiperiodic crystal" instead, though the term "quasicrystal" slipped through several times (Figs. 6 and 7). In 1995, he explained the difference: "At least some of the materials we call quasiperiodic crystals can be explained by quasiperiodic tiling of space. Others can be explained by the icosahedral glass model. The general term quasicrystal, coined by Dov Levine and Paul Steinhardt, is a nice popular term but it does not say scientifically what it is. In addition, the term quasicrystallographer, for example, is not acceptable; quasiperiodic crystallographer may sound better" [14].

At the award ceremony, Sven Lidin, Member of the Royal Swedish Academy of Sciences and Member of the Nobel Committee for Chemistry in his presentation speech pointed out the importance of observations in making and breaking theories. The discovery of quasicrystals did away with the dogma that periodicity was a prerequisite for crystallinity. Much of his presentation speech was about a metaphor of the dwarf and the giant that has been known from antiquity but made most famous by Isaac Newton, according to which if the dwarfs sees farther it is because he stands on the shoulder of giants. The giant is the accumulated knowledge and the dwarf is the discoverer. As Lidin put it, "The giant provides established truths. The dwarf strives for new insight." The connection to the 2011 Nobel Prize was obvious in that Shechtman was the discoverer and he was standing on the foundation of accumulated knowledge, but saw farther (Figs. 8 and 9). Lidin stressed that the metaphor also hinted at the dangers of this arrangement [15]:

> The relation between the dwarf and the giant is fundamentally asymmetric. The dwarf can see, but the giant decides on which road the two shall take. The dilemma of the giant is that he is at the mercy of the dwarf, but he cannot trust him blindly. The paradigms of science are challenged daily on more or less solid grounds and the difficulty is to know when to take these challenges seriously. The dwarf faces the reverse problem. He depends on the giant, and without him he gets nowhere despite the clarity of his vision. In order to make his own choices he is forced down on the ground, to walk alone without the support he enjoyed on the shoulders of the giant. … The dwarf doesn't serve the giant by subservience but through independence.

Lidin called the disbelief with which the scientific community initially met the scientific discovery healthy, but the ridicule to which Shechtman was subjected, deeply unfair. In addition to having created a new branch of science, Lidin found great achievement in that the quasicrystal discovery "has given us a reminder off how little we really know and perhaps even taught us some humility."

Fig. 8 Dan Shechtman at the press conference on December 7, 2011, Royal Swedish Academy of Sciences (Photograph by Istvan Hargittai).

Fig. 9 Dan Shechtman during his Nobel lecture on December 8, 2011, in Stockholm (Photograph by Istvan Hargittai).

Acknowledgments One of us (IH) is grateful to William R. Ott, NIST, for a gracious invitation to give a seminar at NIST, and for making thoughtful and generous arrangements for the visit. Both authors thank Dr. Ott and the Chief of the Metallurgy Division of NIST, Frank W. Gayle for hospitality extended to us during our visit. We appreciate Dr. Gayle's comments on and helpful suggestions for the manuscript and for letting us reproduce his beautiful quasicrystal images in this Editorial.

References

1. Hargittai I (2011) Struct Chem 22:745–748
2. Hargittai I (2011) Israel J Chem 51:1144–1152
3. Hargittai I (2011) Drive and Curiosity: What fuels the passion for science. Prometheus Books, Amherst, New York, pp 155–172
4. Hargittai B, Hargittai I (2005) Candid Science V: Conversations with famous scientists. Imperial College Press, London, pp 25–27 ("John Conway"); 37–41 ("Roger Penrose"); 56–75 ("Alan Mackay"); 76–93 ("Dan Shechtman")
5. Dan Shechtman being quoted in Hargittai B and Hargittai I (2005) Candid Science V, p 84
6. Frank Biancaniello, private communication by e-mail, November 3, 2011
7. Dan Shechtman being quoted in Hargittai B and Hargittai I (2005) Candid Science V, p 85
8. Dan Shechtman being quoted in Hargittai I, Hargittai M (2000) In our own image: Personal symmetry in discovery. Kluwer Academic/Plenum Publishers, New York, p 163
9. Shechtman D, Blech IA (1985) Metall Trans 16A:1005–1012
10. Dan Shechtman, private communication by e-mail, December 26, 2011
11. John Cahn being quoted in Hargittai I and Hargittai M (2000) In our own image, p 166
12. Shechtman D, Blech I, Gratias D, Cahn JW (1984) Phys Rev Lett 53:1951–1953
13. John Cahn being quoted in Hargittai I and Hargittai M (2000) In our own image, pp 167–168
14. Dan Shechtman being quoted in Hargittai B and Hargittai I (2005) Candid Science V, pp 90–91
15. Lidin S (2011) Presentation speech of the Nobel Prize in Chemistry. The Nobel Prize: Translation of the speeches at the 2011 Nobel Prize award ceremony, The Nobel Foundation, Stockholm, pp 21–25

Atomic Structure of Quasicrystals[a,b]

Marc de Boissieu[c]

Abstract
The 2011 Nobel Prize in chemistry has been awarded to Dan Shechtman for his discovery of quasicrystals. The discovery has indeed been a breakthrough in crystallography and solid state physics and chemistry. After a brief introduction to the subject, we review some of the recent advances in the understanding of the atomic structure of icosahedral quasicrystals and in particular for the binary Cd–Yb type quasicrystal. Thanks to a combined analysis of periodic approximant, high quality synchrotron data, and the superspace approach, a detailed insight in the crystal chemistry of this binary quasicrystal has been achieved.

Keywords
Quasicrystal • Atomic structure • Phason

Introduction

The 2011 Nobel Prize in chemistry has been awarded to Dan Shechtman for his discovery of quasicrystals. Indeed the discovery of quasicrystals [1] has been a breakthrough in crystallography and solid state physics and chemistry. Quasicrystals are long range ordered materials with diffraction patterns displaying sharp Bragg peaks, yet with symmetry, such as the icosahedral one, incompatible with lattice translation. Quasicrystals are nowadays obtained in a large number of systems, in some cases as equilibrium phases, in which case large single crystals, with a high degree of structural quality, can be grown, as exemplified in Fig. 1. Although most quasicrystals have been obtained in intermetallic compounds, few examples have been also obtained in soft-condensed matter. Quasicrystal presents a new paradigm for long range order structures and the discovery has opened a wide field of researches ranging from mathematics to applied science. Quasicrystals belong to the broader class of aperiodic crystals [3], which includes incommensurately modulated structure, incommensurate composite and quasicrystals. The very notion of crystal has been redefined by the International Union of Crystallography in 1992 [4]. It is worth to quote the full definition which is given in reciprocal space. "By 'crystal' we mean any solid having an essentially discrete diffraction diagram, and by 'aperiodic crystal' we mean any crystal in which three dimensional lattice periodicity can be considered to be absent."

That the notion of crystal is defined in reciprocal space has been an important shift with respect to previous definitions. The "essentially discrete" nature of the diffraction pattern, means that most of the scattering in the diffraction pattern is contained in sharp Bragg peaks and that those Bragg peaks can be indexed by a linear combination of a finite number n of independent basis vectors. When n is larger than 3, this is an aperiodic crystal.

Quasicrystals are generally defined as aperiodic crystals whose diffraction pattern and symmetry is incompatible with lattice translation. Besides the icosahedral symmetry, which leads to three dimensional quasiperiodicity, there are symmetries leading to structures which are a periodic packing of quasiperiodic planes. This is the case of decagonal,

[a] *Structural Chemistry* 2012, 23(4):965–976.
[b] This paper is dedicated to the 80th birthday of Prof. Sten Andersson.
[c] SIMaP, Grenoble-INP, CNRS, UJF, 1130 rue de la Piscine, BP75, 38402 Saint Martin d'Hères Cedex, France
e-mail: Marc.de-Boissieu@simap.grenoble-inp.fr

I. Hargittai, B. Hargittai (eds.), *Science of Crystal Structures: Highlights in Crystallography*,
DOI 10.1007/978-3-319-19827-9_16

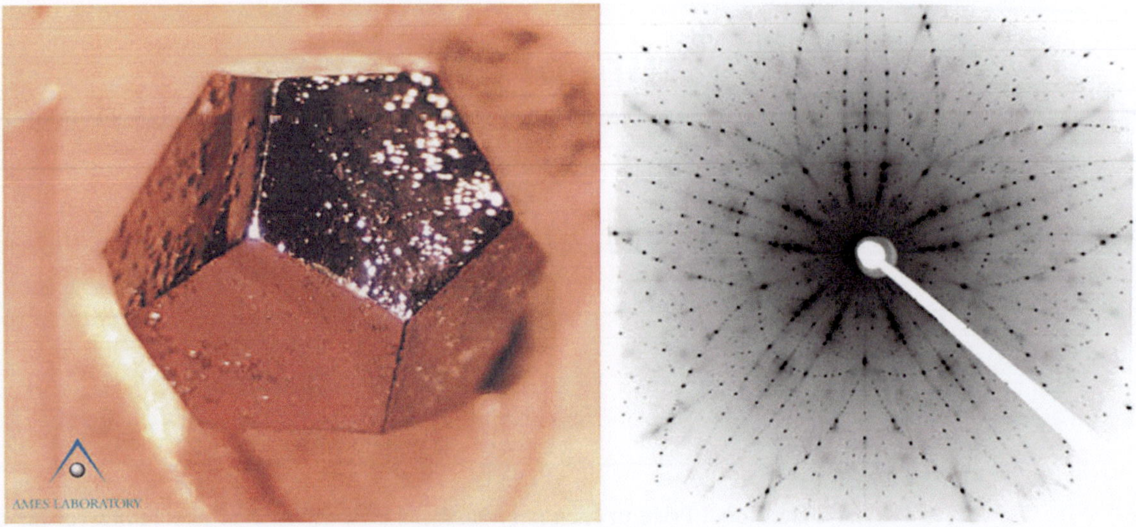

Fig. 1 *Left panel* single crystal obtained by the flux method in the Zn–Mg–Y system (Courtesy of P. Canfield [2]). *Right panel* fivefold Laue diffraction pattern obtained in the icosahedral AlPdMn phase (Courtesy of W. Steurer).

dodecagonal, and octagonal quasicrystals which all have been experimentally observed.

In this paper, we review some of the results obtained recently for elucidating their atomic structure. After a presentation of the different approaches used to describe quasicrystals and their atomic structure, we illustrate the structure determination of quasicrystal with the case of the icosahedral CdYb phase. In particular we will show how local order and crystal chemistry of quasicrystal can be dealt with within the superspace formalism. The last section present diffuse scattering phenomenon and in particular those related to phason fluctuations.

How Can the Atomic Structure of Quasicrystals Be Solved?

Solving the atomic structure of quasicrystal is challenging since there is no periodicity. Two main approaches have been used to describe quasicrystals: decorated quasiperiodic tiling whose prototypal example is the Penrose tiling and the superspace approach. The later one is an extension of what was developed for incommensurately modulated crystals. The same superspace approach can be applied to all aperiodic crystals, although specific difficulties have to be dealt with in the case of quasicrystals.

In both the tiling and the superspace approach, the notion of clusters and local order has been crucial and widely used in the modeling.

Tilings

Quasicrystalline tiling have been discovered by Penrose [5, 6] well before the discovery of Dan Shechtman. The Penrose tiling is constructed with two rhombuses, having angles equal to 72 or 144°, and packed in such a way that there is no periodicity. However, the tiles are arranged in a very orderly way and present locally a fivefold symmetry patterns, as exemplified in Fig. 2. It is interesting to note that what was considered as a "mathematical game" turned out to be the basis of many models for decagonal quasicrystals. This tiling has been generalized to 3D by R. Amman. A.L. Mackay using optical diffraction, showed that the diffraction pattern of the Penrose tiling is composed of sharp Bragg peaks, and thus this could be a new state of ordered matter for which he introduces the term quasi-lattice in 1982 [7]. The name quasicrystal was coined by D. Levine and P. Steinhardt [8] in a

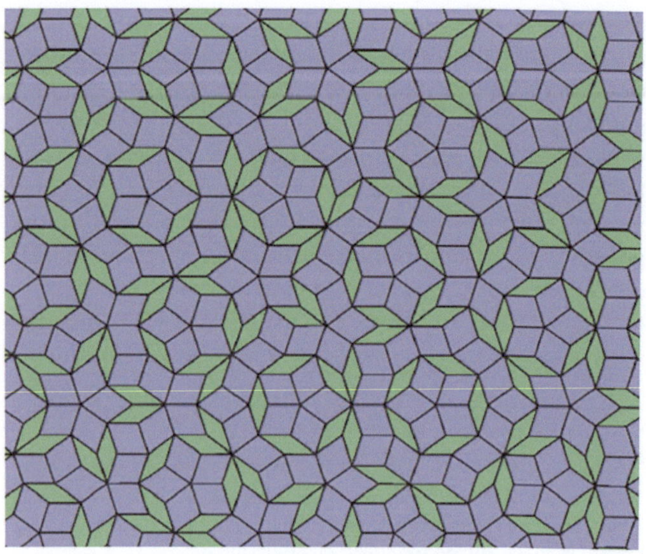

Fig. 2 Penrose tiling made with two tiles. Fivefold local environment are visible all around the picture.

paper published just after the seminal paper by D. Shechtman et al., in which they present the 3D Penrose tiling and its numerical diffraction pattern. Since then many different tilings, with different symmetries, have been proposed. The modeling of the atomic structure of quasicrystal is obtained by decorating by atoms the tiles of the tiling.

There are three main difficulties in this approach: the first one concern the choice of the underlying tiling. In general for a given symmetry, several choices of tiling with different tiles shape are possible. The second problem concerns the size of the tile. Indeed because of the inflation property of quasicrystal, the tile size cannot be determined experimentally within a scale factor. The third one concern diffraction data analysis: it is not possible to "extract" tiling decoration information from the X-ray or electron diffraction pattern alone, the only possibility being a trial and error approach. This is where the use of the superspace crystallography approach brings in essential information.

Superspace Crystallography

The superspace crystallography was invented by de Wolff [9, 10], Janner and Janssen [11–13] to describe the atomic structure of incommensurately modulated structure and latter on for incommensurate composite crystals. Non periodic modulated structures have been known for a long time and are examples of non-periodic long range ordered structures. However, they can be described as a "perturbation" (at least for the simpler one's) of a periodic structure. In a very elegant way, P.M. de Wolff, A. Janner and T. Janssen have shown that an incommensurately modulated structure can be described as a periodic structure in a superspace having dimensions larger than 3. The number of dimensions is equal to $3+n$, where n is the number of independent modulation q-vectors. This approach allows to classify the space groups, and because of the periodicity, the computation of the structure factor, an important step for any structural refinement. This has leaded to the development of softwares which are well established and currently used for the determination of incommensurately modulated and incommensurate composite crystals. Importantly, crystal chemistry can be analyzed within this superspace approach in a very detailed way: interatomic bonds lengths, coordination numbers, and occupation probabilities are quantities directly accessible from the superspace approach, allowing a detailed understanding of the crystal chemistry of the systems.

For quasicrystals a similar approach has been proposed by several groups right after their discovery [14–17]. A simple illustration of this approach, for the case of a one dimensional (1D) quasicrystal, is given in Fig. 3. The Fibonacci chain is a 1D quasiperiodic sequence of long (L) and short (S) segment whose length are in the ration $\tau/1$, where τ is the golden mean equal to $(1+\sqrt{5})/2=1.618\ldots$. It can be obtained by a substitution rule such that S is transformed in L and L in LS. It can also be obtained as a section of a decorated 2D square lattice. The square lattice is decomposed in two 1D subspaces: the physical space, called parallel space, and the perpendicular space. The slope of the parallel space is irrational. The square lattice is decorated by segment lines, lying in the perpendicular direction. The resulting 1D quasiperiodic structure is obtained as a section of the decorated 2D lattice: each time the line crosses the parallel space an atomic position is generated. When the segment line length is equal to the projection of the square, the Fibonacci chain is generated. A similar procedure is obtained in reciprocal space. The Fourier transform of the 2D decorated lattice is shown in Fig. 3d, the 1D diffraction pattern is obtained as a projection of the 2D diffraction pattern on the Q_{par} line. Because the 2D Fourier transform involves the Fourier transform of the shape function of the segment line, the intensity of the Bragg peaks decays rapidly as their perpendicular component increases. As a result the strongest peaks have a small perpendicular component (Fig. 3d). The 1D diffraction pattern is indexed with two integer indices. Of course the number of Bragg peaks is infinite but only a finite number subsists if an intensity threshold is given.

This superspace embedding can be generalized for 3D quasicrystals. In the case of the icosahedral symmetry, a 6D space is required to describe the quasicrystal. This is related to the six 5-fold axes of the icosahedrons. The segment lines are now 3D objects, which are generally described as polyhedra. The structure is fully described by the position and the detailed shape of these polyhedra (called atomic surfaces or existence domain). This is a difficult problem, but there is a direct connection between the shape and the chemical content of the polyhedra, and the resulting local environments generated in the 3D quasiperiodic structure. It is thus possible somehow to connect the "crystal chemistry" principles with the detailed shape of the atomic surfaces [18, 19].

To this end the comparison with periodic approximant has been crucial.

Periodic Approximant, Clusters and Local Order

A very important notion for structural analysis of quasicrystals is the one of periodic approximant. This concept was used immediately after the seminal publication by D. Shechtman to propose the first realistic atomic models [20, 21]. It can be understood in the superspace description of the quasicrystal: if the slope of the cut space is rational instead of irrational, this generates a periodic approximant. This is equivalent to approximating τ by successive rational Fibonacci numbers such as 1/1, 2/1, 3/2 …. In real systems

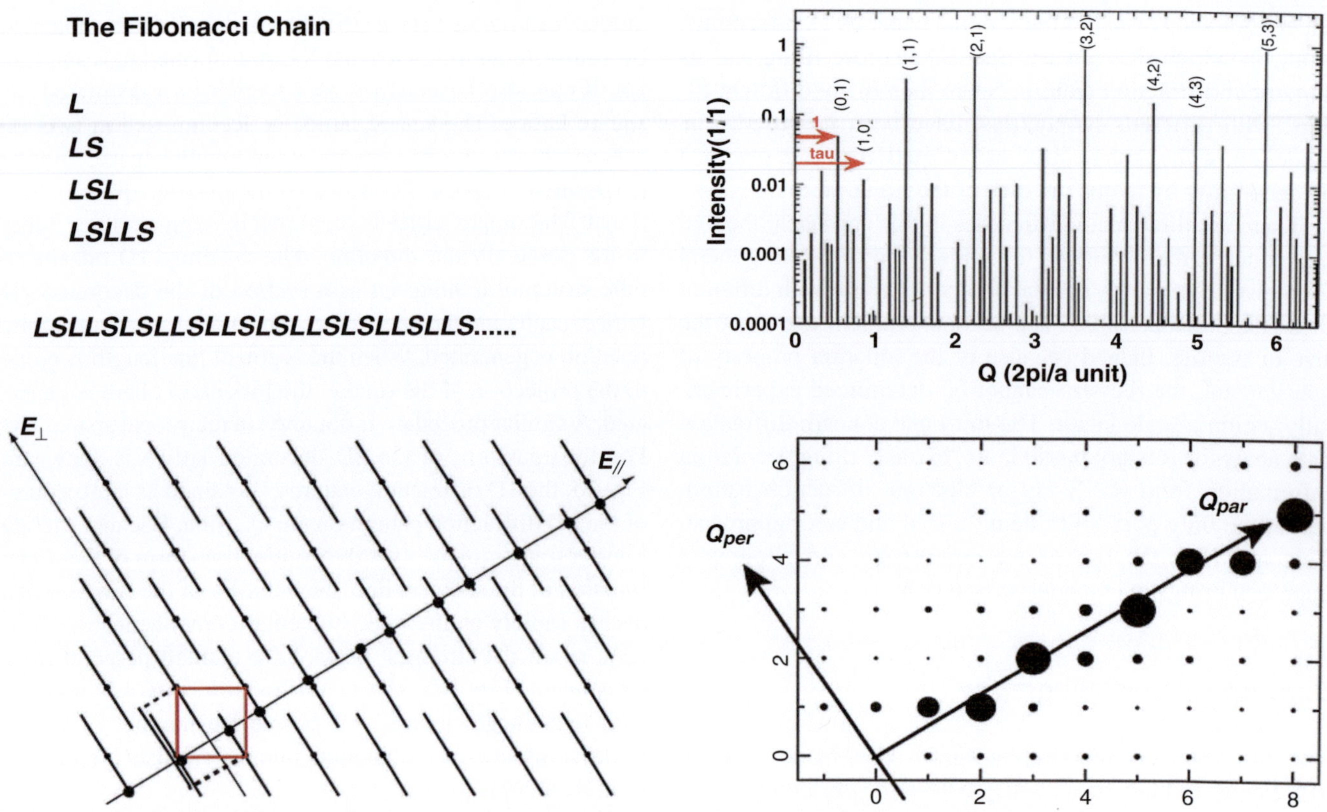

Fig. 3 Schematic presentation of the superspace approach in the case of the 1D Fibonacci chai. *Left panel*, direct space: substitution rule for the Fibonacci construction. 2D embedding of the Fibonacci chain (see text). *Right panel* reciprocal space. The *bottom panel* displays the 2D diffraction pattern, then projected on the 1D (*top panel*) axis. The 1D diffraction pattern can be indexed with two integer indices.

1/1 and 2/1 approximant have been observed. In the case of icosahedral quasicrystal, the approximant is a periodic crystal with cubic symmetry. In both the 1/1 approximant and the parent quasicrystal one can find the same local environments, which is conveniently described in term of atomic clusters.

The notion of clusters to describe structures was used for a long time in the field of intermetallic compounds. Initiated by the work of Frank and Kasper [22, 23], Pauling [24], and Samson [25] this allowed to solve the structure of complex systems with large unit cells. One of the main ingredients in this approach is the tetrahedral close-packing of atoms considered as spheres. For instance in the case where the structure is composed of two atoms with different sizes (a large and a small one as in the Bergman AlZn-Mg phase) large clusters are built up with the Friauf polyhedron as basic unit. In the case of the Bergman AlZn-Mg phase [24], which is a 1/1 approximant to a quasicrystal, this lead for instance to a large "soccer ball" cluster built up with 20 Friauf polyhedra optimizing the dense packing of a large (Mg) and a small (Al/Zn) atom.

This approach has been applied successfully to quasicrystals. In the case of icosahedral quasicrystal, three different families of clusters and thus structures have been identified so far: the Mackay cluster (found in the AlMnSi phases [20, 21, 26]), the Pauling triacontahedron found in the Bergman AlZn-Mg phase and the Tsai type cluster found in the CdYb type 1/1 and 2/1 approximant [27, 28]. Not only the clusters but the way they are connected is also an important ingredient: usually the cluster inter-connect along high symmetry axes. Once clusters and their linkages rules are identified, this is used to model the quasicrystal along the two routes indicated above: either using decoration of tiles or using the superspace approach which lead to more general type of framework.

This cluster approach is thus very important, as it introduces the crystal chemistry in the quasicrystal modeling. This is illustrated for the case of the CdYb approximant, for which an extremely detailed structural analysis is available (Fig. 4) [29]. Whereas all shells have almost an icosahedral symmetry, the first shell is a tetrahedron of Cd atoms which is disordered at room temperature. The next shell is a Cd dodecahedron, followed by a Yb icosahedron, a Cd icosidodecahedron, and a decorated rhombic triacontahedron unit (RTH unit hereafter). This constitutes a large cluster, containing 158 atoms with a radius of about 1.5 nm. This RTH cluster is packed periodically in the so-called 1/1 and 2/1 approximant [28, 30]. In the Cd_6Yb 1/1 approximant the packing is extremely simple: the RTH units are packed on a bcc lattice. They are connected along the twofold axis where they share a face and along the threefold axis where they overlap forming an obtuse rhombohedron (OR) decorated with Cd on vertices and mid-edges. The packing in the larger unit cell $Cd_{5.8}Yb$ 2/1 approximant is a bit more complicated but what is striking is that the linkages between the clusters are exactly the same: sharing a face along the twofold axis and overlapping along the threefold axis. Because the number

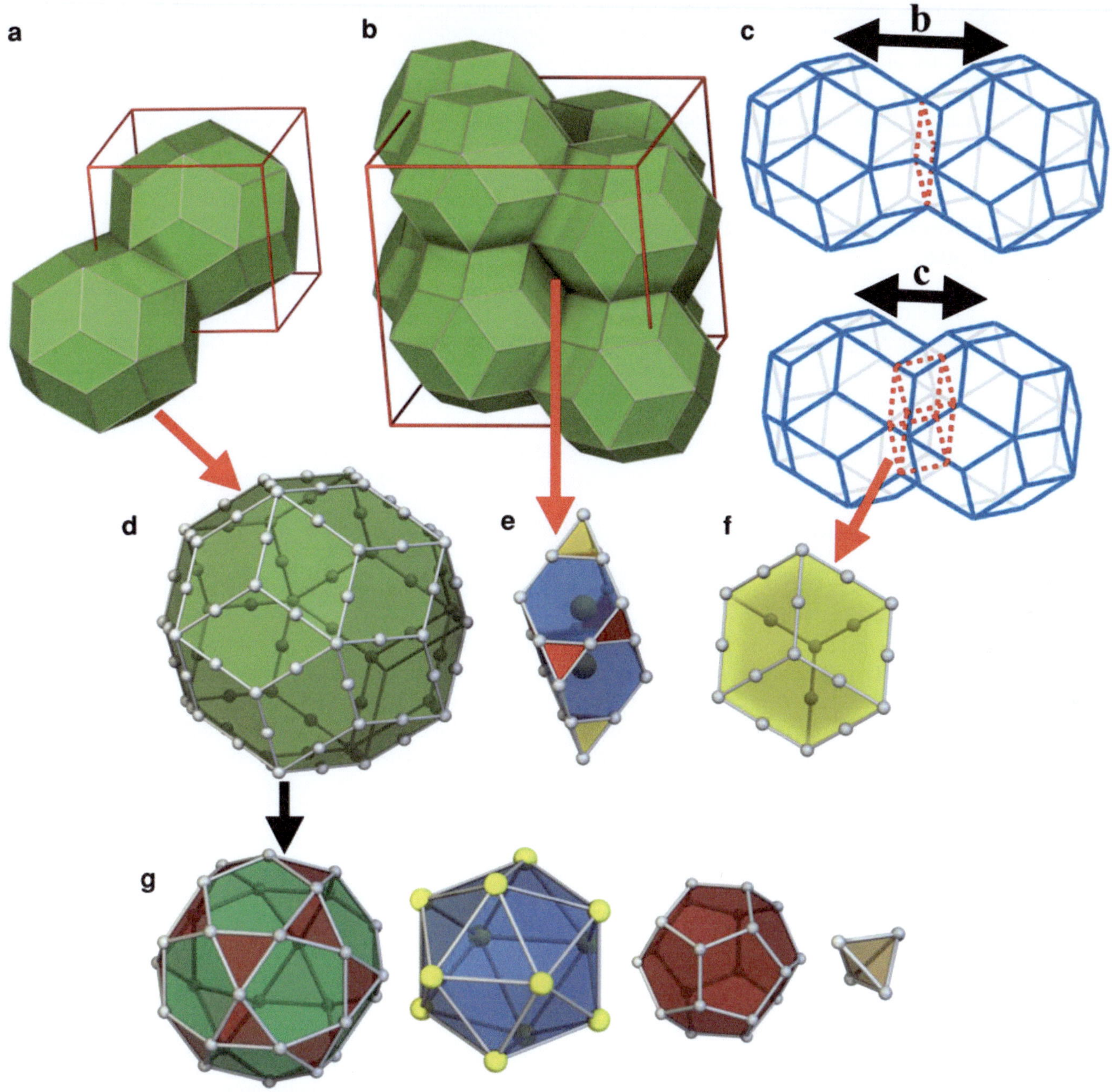

Fig. 4 1 Approximants, building units and linkages. (**a**) The body-centered-cubic packing of RTH units in the 1/1 cubic approximant. (**b**) The packing of RTH units in the 2/1 cubic approximant. (**c**) The two different types of linkages between the RTH units. The ***b***-and ***c***-linkage occurs along the two- and threefold directions, respectively. (**d–f**) The three fundamental building units: RTH (**d**), AR (**e**) and OR (**f**), and their corresponding atomic decorations. All rhombic faces of these three building units are decorated with Cd atoms on the vertices and mid-edges. (**g**), Atomic sub-shells inside the RTH unit (92 atoms, $R=0.78$ nm along its twofold direction). From *left* to *right*: a Cd icosidodecahedron (30 atoms, $R=0.65$ nm), a Yb icosahedron (12 atoms, $R=0.56$ nm), a Cd dodecahedron (20 atoms, $R=0.46$ nm), and the inner Cd tetrahedron (4 atoms, occurring in different orientations). All Cd atoms are shown in *gray* and Yb atoms in *yellow*. The *red arrows* indicate where in the approximant structures the corresponding building units can be found (From Ref. [29]).

of twofold and threefold bonds is different, a supplementary cluster appears to fill in a remaining space: it is constituted of a double Friauf polyhedron, which is a decorated acute rhombohedron (AR) (see Fig. 4). It is important to note that the chemical disorder is rather small in this compound. Indeed, Cd and Yb have a metallic radius of 0.157 and 0.174 nm respectively, and cannot be exchanged easily. There is thus a strong Cd/Yb ordering and all sites are either Cd or Yb. A few sites in the 2/1 approximant are partially occupied Cd sites. This is very much different to all previously known ternary alloys, where generally chemical disorder is present.

Analysing Diffraction Data: The CdYb Icosahedral Case

We present in this section the structural analysis of the CdYb quasicrystal. Combining X-ray synchrotron data, the detailed knowledge of the 1/1 and 2/1 approximant, a sophisticated six-dimensional modeling a very detailed structural refinement could be achieved. It is worth mentioning that the resulting crystal chemistry is well understood.

Superspace Crystallography Analysis and Modeling

Collecting a large number of diffraction data is a key point when analyzing the atomic structure of quasicrystals. Indeed the "details" of the atomic structure is contained in the weakly diffracting data which have a large Q_{perp} component of their diffracting vector. This requires measuring Bragg peak intensity with a large dynamical range, generally with more than 7 orders of magnitude. In the case of the CdYb quasicrystal, about 5000 unique reflections have been measured. This is by far the largest data set available for icosahedral quasicrystal, especially if one takes into account the very high symmetry of the icosahedral point group. Having a data set with both high Q_{par} and high Q_{per} components turned out to be crucial for the structural solution of the CdYb icosahedral quasicrystal [29].

Similarly to periodic crystal, one is faced with the phase problem when analyzing the data. Whereas the early quasicrystal solution made use of the Patterson analysis, phasing methods developed for periodic crystals have been proposed for quasicrystalline and aperiodic materials. Although no rigorous demonstration of their validity has been really proposed, they have been tested on different model systems and on real one, giving pretty good results. All the methods rely on a back and forth Fourier transform of the initially randomly phased structure factor. Using a constraint in the direct space density map such as, the low density elimination, the charge flipping or the minimum charge criterion, a convergence to a solution is obtained. This allows a proper phasing of the strongest reflections and thus a first 6-D density map. In the case of the CdYb case, the obtained density map clearly shows that: (i) the same large triacontahedron unit was present in the icosahedral and (ii) that it is connected in the same way along threefold and twofold direction, but on a quasiperiodic network.

Modeling of the structure has been achieved considering an ideal network of twofold and threefold bonds which was proposed by Henley and is a subset of the 3D Penrose tiling, the so-called 12-fold vertices sphere packing, which favor the two- and threefold linkages [31]. A large TRH unit is placed on every vertex of this network. However some free space is remaining, which is filled by the double Friauf polyhedron unit. The detailed 6D modeling is done in such a way as to allow the position of the atoms constituting the cluster to be "relaxed" from their ideal position. This is a very important step which allows the local configuration to be adjusted.

The resulting model is parameterized with about 250 variables and fitted against the 5,000 observed intensities, leading to a **R** factor equal to 0.09. This is already a good indication that the resulting structure is correct, which is confirmed by the analysis of the local structure.

Analysis of the Resulting Structure

The resulting quasiperiodic structure can be described in different ways. The first one is resulting from its very construction principle that is in term of clusters: 93.8 % of the atoms belong to the RTH unit. This is a very important point, as this lead to a very small number of atoms filling the remaining space. As explained above, the remaining space is filled-in by the prolate (double Friauf polyhedron) and the oblate rhombodedron. Because the network is quasiperiodic, new local environment are showing up as compared to the periodic 1/1 and 2/1 approximants. This is exemplified in Fig. 5 which shows a portion of the resulting quasicrystal. Two main new units have to be considered: (i) a stellate polyhedron, which is made by the union of 20 AR, and whose interior form the so-called Bergman cluster; (ii) a combination of 10 OR and 10 AR named "cog wheel complex" whose structure contains part of the previous stellate polyhedron. Both structures are favoring the arrangement of Friauf polyhedra known to occur in Frank–Kasper phases.

It is also interesting to look at the long range organization of the clusters. Because of the quasiperiodicity, inflation and hierarchical organization can be shown to occur as exemplified in Fig. 6, where only the cluster centers are represented. As can be seen a cluster of cluster can be defined (yellow disk), which in turn, when the scale is "inflated" by the factor

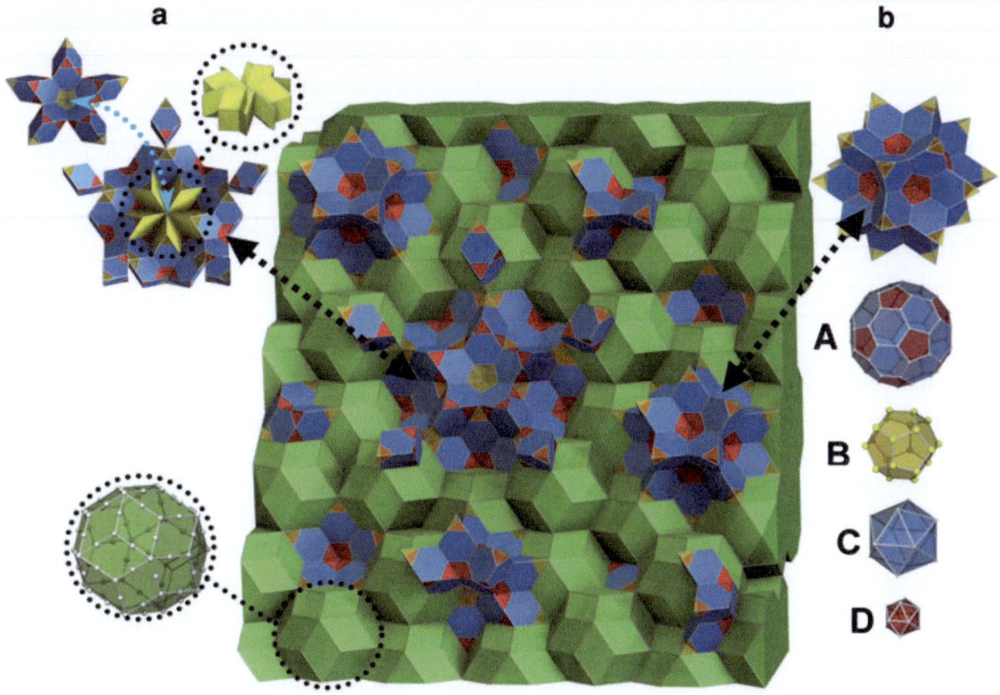

Fig. 5 Ideal structure of the i-YbCd5.7 QC. A slab of i-YbCd5.7 in 3D physical space with its normal vector parallel to a fivefold axis. The building units and atoms are shown using the same *color codes* as in Fig. 1. The "cog wheel complex" is seen near the centered of the main image. This is a large arrangement of ARs that serves as a host for a stellar arrangement of ten ORs, as seen in the *top left inset*, where the contents of the cog wheel complex are shown. The stellate polyhedron composed of 20 ARs is also a recurring arrangement in the structure. A complete stellate polyhedron is seen in the *top right inset*; the first four atomic sub-shells (labeled **A–D**) of this arrangement correspond to a Bergman cluster. The RTH units are shown in *green*, as seen in the *bottom left inset* (From Ref. [29]).

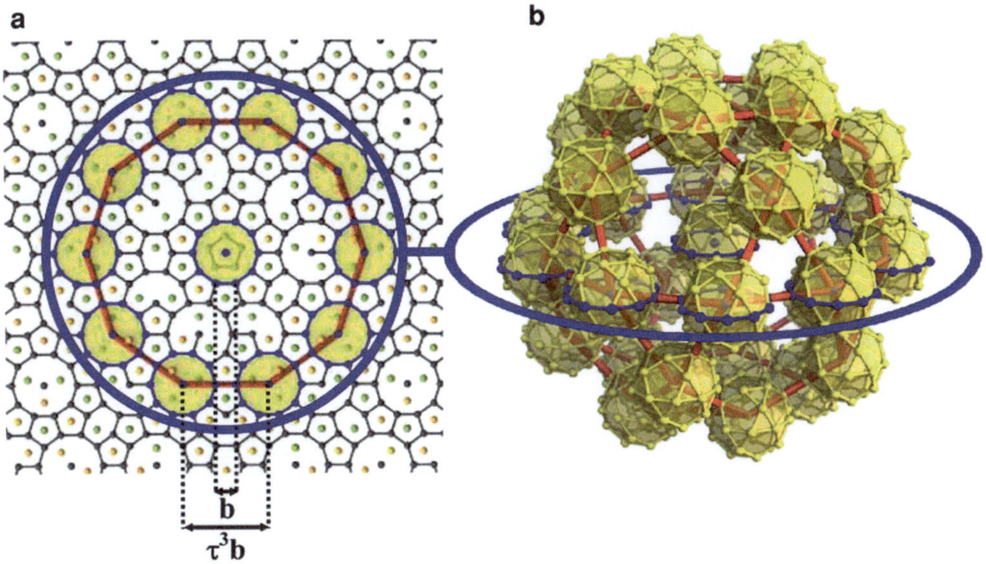

Fig. 6 Inflation properties of the i-YbCd5.7 QC. (**a**) A dense plane of RTH units is seen along the fivefold axis. Only the centers of the RTH units are shown as colored spheres. The dense plane is slightly puckered; the *orange spheres* are above the ideal plane, *green spheres* are below and *dark gray* (and *blue*) spheres are in the plane. The bonds connecting the *dark gray spheres* are twofold *b*-linkages (see Fig. 1c). In the center of the image a *yellow* decagon outlined by *blue* RTH centers and *b*-linkages is seen. This represents a "cluster of RTH units." A large τ^3-inflated decagon cluster composed of "clusters of RTH units" at its vertices is outlined by *thick red* bonds. (**b**) A 3D perspective view of the τ^3-inflated cluster; its radius is about 12 nm. It is seen that the τ^3-inflated cluster and its constituent "clusters of RTH units" form an icosidodecahedron. The *blue circle* and decagons in (**a**) and (**b**) indicate the position of the cut plane. RTH centers seen in *orange*, *green* and *dark gray* have been omitted in ***b*** for clarity (From Ref. [29]).

τ^3, gives a cluster of clusters of clusters. This hierarchical packing can be continued up to infinity and might be at the origin of some of the specific properties of quasicrystals.

The resulting model thus fulfill the same crystal chemistry requirements than the one deduced from the structure analysis of periodic approximants, yet with a quasiperiodic organization. Why is the quasiperiodic order selected is still an open question, but both in the approximant and in the quasicrystal one of the main "driving" force is the close-packing of a large and a small atom. This is very much confirmed by the low temperature structural study of the ZnSc 1/1 approximant whose structure has been determined recently [32]. The ZnSc 1/1 approximant phase is isostructural to the CdYb one [33]. Similarly to the CdYb 1/1 approximant, whereas the inner tetrahedron is disordered at room temperature, an ordering of the tetrahedra occurs below about 160 K [34]. The structure goes from a cubic to a monoclinic cell, with a doubling of the cell along the (110) direction, along which direction the tetrahedral are oriented in an anti-parallel way. The structural solution of this low T phase, which was solved by Ishimasa et al. [32], is extremely interesting as it gives clue on the atom packing. The central tetrahedron induces a strong distortion of the successive icosahedral shells so as to be as close as possible to a close-packing of the atoms, and respecting the "usual" Zn–Zn distances. For instance, the central tetrahedron induces a very strong distortion of the surrounding dodecahedral shell with edge lengths ranging from 6.7 to 7.3 Å (see Fig. 7). However, with this configuration the Zn–Zn lengths are in the range 2.5–2.8 Å which is encountered in simpler Zn–Sc compounds. In the same way, the Sc icosahedron and the Zn icosidodecahedron also display strong distortion from perfect icosahedral symmetry, all resulting from the close-packing of atoms. Even more interesting, the distortion can be also associated with the twofold and threefold linkages. In particular, where the threefold linkages occur, the dodecahedron atom is displaced away from the center and the triangular part of the icosidodecahedron "opens" up as shown in Fig. 7. This gives a clear interpretation of "how" the orientation information is mediated from one tetrahedron to the next one located at a distance of 1.2 nm.

A similar distortion has been found in the refined quasicrystalline structure, although the number of twofold and threefold bonds around a central tetrahedron is different [29].

We have thus a picture which is fully consistent with the know crystal chemistry.

This is also confirmed by atomistic simulations carried out using oscillating pair potentials. Atomistic simulation for the quasicrystal is demanding and can in fact only be carried out on large unit cell approximants. Because of the large number of atoms it is not possible to carry out ab initio calculation. Ab initio calculations have been limited to the case of the 1/1 approximant, with an ordering of the tetrahedra. This simulation showed the importance of Cd–Yb (or Zn–Sc) hybridization in the 1/1 approximant crystal stability [35, 36]. Recently, oscillating pair potentials have been developed for the study of the Zn–Sc and Cd–Yb systems. The pair potentials are obtained by fitting energy and forces on a set of ab initio data and are displayed in Fig. 8 for the Zn–Sc system [37]. As expected the pair potential minimum is at higher distance for Sc–Sc than for Zn–Zn. Although a direct interpretation is not possible, the deeper minimum for

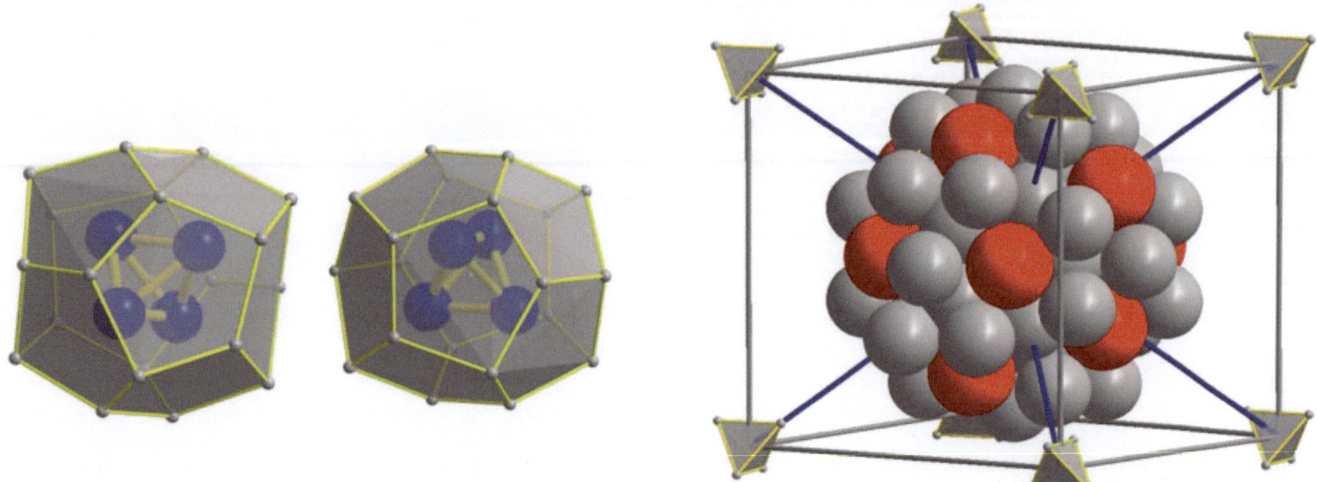

Fig. 7 Cluster structure of the low temperature monoclinic phase in the Zn–Sc 1/1 approximant (From Ref. [32]). *Left* illustration of the dodecahedron distortion seen along two different orientations. The inner tetrahedron atoms are in *blue* for clarity. *Right* external icosidodecahedron shell displayed in space filling, with Zn and Sc radii equal to 1.4 and 1.6 Å, respectively. Gray and *red color* stand for Zn and Sc atoms. The threefold connections to the neighboring tetrahedra are displayed in *blue*. The high temperature cubic cell is shown in *gray*.

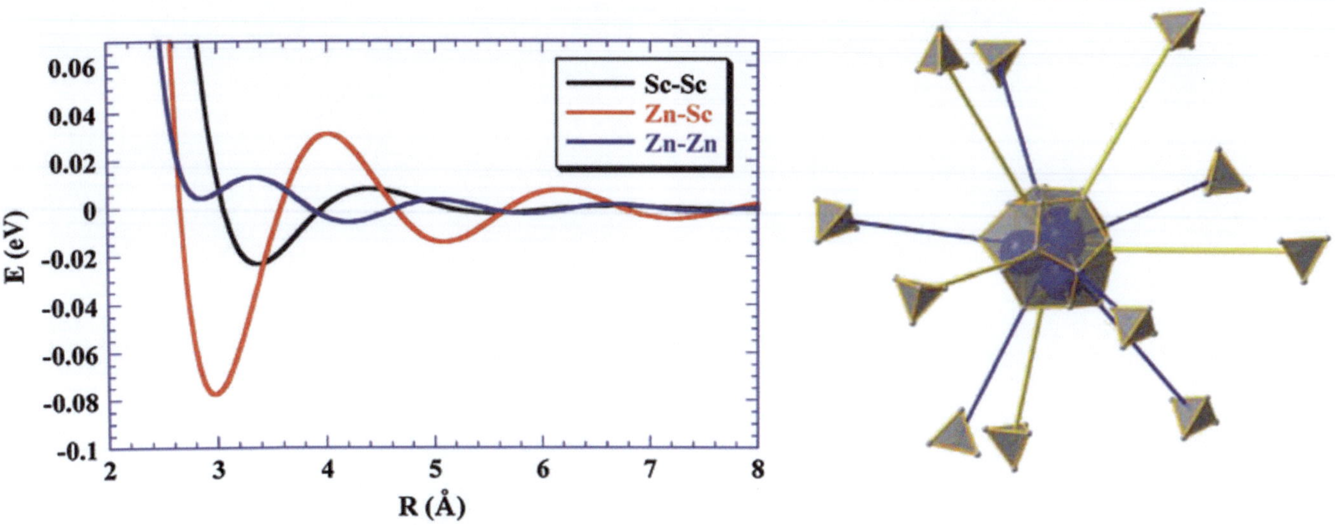

Fig. 8 *Left* oscillating pair potential in the Zn–Sc system. *Blue*, *red* and *black colors* stand for Zn–Zn, Zn–Sc, and Sc–Sc, respectively (From [37]). *Right* local environment obtained in the quasicrystal simulation. The dodecahedron distortion is clearly visible (From Ref. [38]).

Zn–Sc interaction is certainly a signature of the sp-d Zn–Sc hybridization. These pair potentials have been showed to accurately reproduce the lattice dynamics of the ZnMg-Sc quasicrystal and its 1/1 approximant. In particular, a quantitative agreement was obtained when comparing the measured inelastic neutron (or X-ray) scattering data with the simulation [37]. This is a particularly severe validity test for the inelastic neutron scattering cross section is related to the pattern of vibrations of the atoms. In particular, the differences between the quasicrystal (simulated using a 3/2 approximant) and the 1/1 approximant were perfectly reproduced. This implies that both the atomic modeling and the pair potential interactions are realistic. The resulting relaxed structure, displays a similar distortion of the icosahedral cluster as shown on Fig. 8 [38].

Beyond the Average Structure: Diffuse Scattering and Phason Modes

As for periodic crystals the structure determination of quasicrystal lead to an "average" picture related to the one point correlation function. The solution does not take into account the eventual two points correlation related to disorder, which is best measured by diffuse scattering. As for periodic crystal we may have different sources of diffuse scattering: thermal vibration (thermal diffuse scattering TDS hereafter), short range chemical order, dislocations…. There however is a supplementary contribution, which is specific to aperiodic crystals, named phason. It is beyond the scope of this review to give a detailed introduction on phasons modes in aperiodic crystals which can be found in ref [3, 39].

Long wavelength phason modes can be introduced within the hydrodynamic theory of quasicrystals which deals with long wavelength excitations and relies on symmetry breaking analysis [14, 40–42]. In the high dimensional picture of quasicrystal, the hydrodynamic theory relies on the invariance of the free energy F of the system with rigid translation of the cut space along the perpendicular direction. Equivalently it is related to a phase invariance when expressed in term of density waves. Within this theory phason modes are the long wavelength diffusive excitation, which are specific to all aperiodic crystals (incommensurately modulated phases, incommensurate composites and quasicrystal). The hydrodynamic theory is a continuous coarse grained theory, and does not give clues on the atomistic realization of phason modes. For quasicrystal a simple picture can be taken from the 2D image of the Fibonacci chain. If the cut E_{par} is translated along the perp direction, some configurations LS will be changed to SL. This reorganization involves the "jump" of an atom between two almost energetically equivalent positions located at a distance S/τ, and has been named phason jump or atomic jump in the quasicrystal jargon. One should note that this local rearrangement might not be specific to quasicrystals as it can also occur for approximant. However, the long-wavelength phason excitation is really specific to quasicrystals, as an irrational slope of the cut space is required for the free energy F to be invariant.

For the case of icosahedral quasicrystals, the hydrodynamic theory leads to a generalized elasticity theory involving five elastic constants: the two Lame coefficient related to shear and compressive strain (the icosahedral symmetry lead to isotropic tensors), K1 and K2 the phason elastic constants

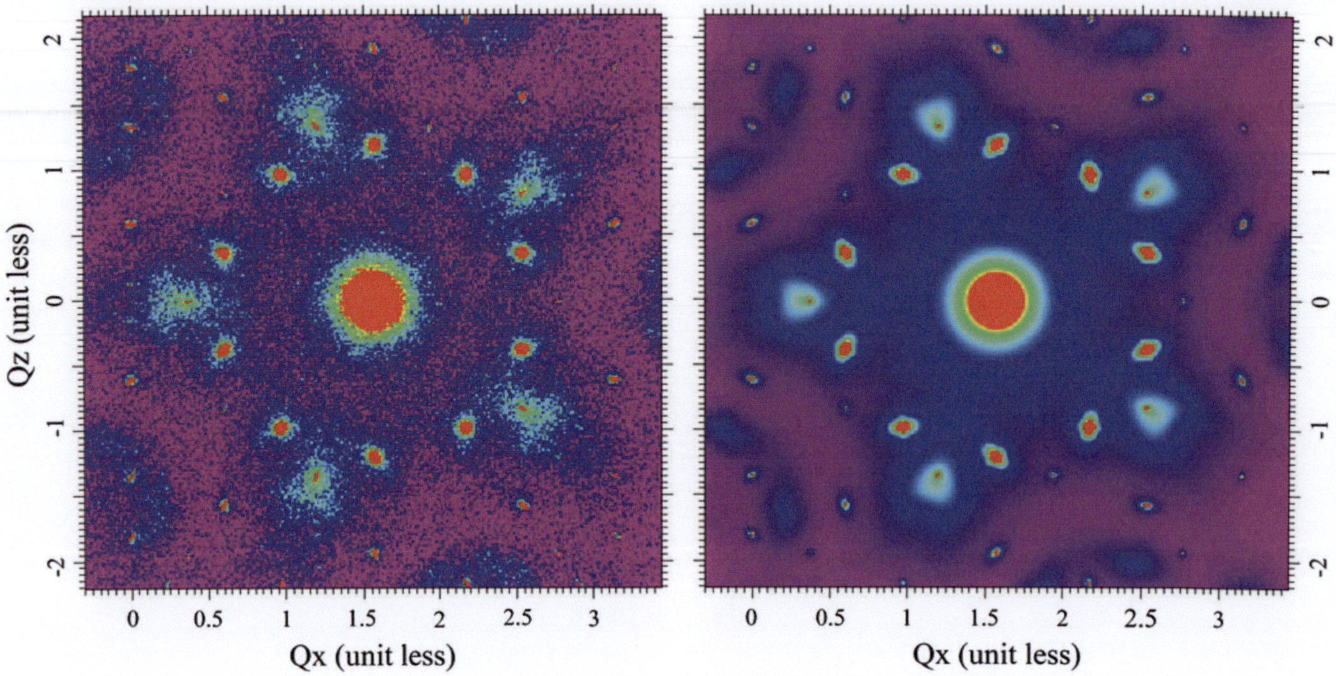

Fig. 9 Diffuse scattering in the i-AlPdMn phase. *Left panel* diffuse scattering measured around the strong fivefold 18/29 reflection, in a plane perpendicular to the fivefold axis. Measurements where made on a rotating anode (Cu target) 4 circles diffractometer using a point detector. *Right panel* simulation of the diffuse scattering taking into account the contribution from out of plane Bragg reflections. Both the shape and the intensity distribution of the diffuse scattering are well reproduced: note for instance the lobes like distribution seen on the outer part of the diffraction pattern (From Ref. [49]).

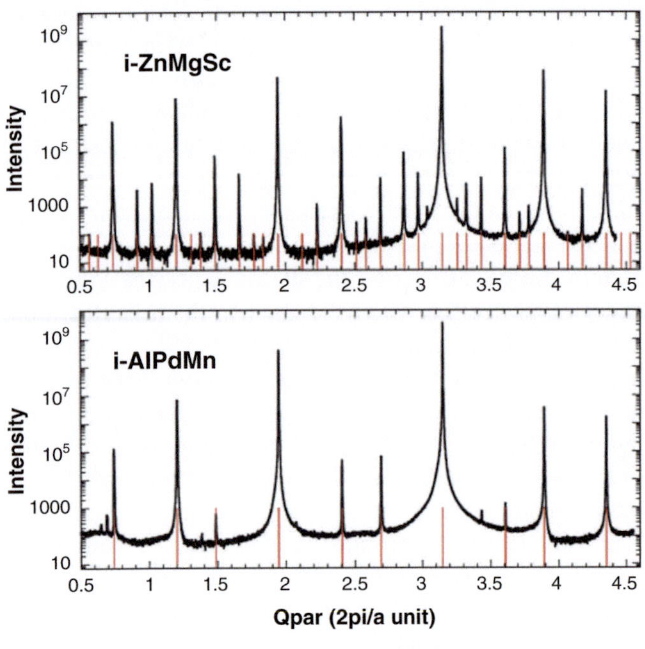

Fig. 10 Comparison of the scattered intensity signal measured in the i-ZnMgSc and i-AlPdMn phases along a twofold axis. The amount of diffuse scattering is lower in i-ZnMgSc and the number of large Q_{per} weak reflections larger. *Bars* stand for the calculated Bragg peak positions considering a maximum Q_{per} value of 7 in i-ZnMgSc and 2.5 in i-AlPdMn (From Ref. [48]).

and K3 a phonon–phason coupling term. In a way similar that phonons give rise to thermal diffuse scattering, phason fluctuations also give rise to diffuse scattering. For icosahedral quasicrystal, the phason diffuse scattering is located close to the Bragg peaks [43–45] and has an intensity which goes as

$$I\left(\mathbf{Q}_{par}+\mathbf{q}\right)=I_{Bragg}.Q_{per}^{2}.F\left(K2/K1,\mathbf{e}_{per},\mathbf{q}\right)/\mathbf{q}^{2} \quad (1)$$

where I_{Bragg} is the Bragg peak intensity, $\mathbf{Q}_{par}$ and $\mathbf{Q}_{per}$ the par and perp components of the reciprocal vector, F a functional form involving the so-called hydrodynamical matrix, and $\mathbf{e}_{per}$ the polarization vectors of the phason modes.

This expression gives a very specific shape to the diffuse scattering which can be recognized from the TDS contribution, when this one cannot be suppressed using neutron energy analysis.

It is remarkable that all icosahedral quasicrystals studied so-far do display some diffuse scattering in their diffraction pattern, diffuse scattering which can always be interpreted as originating from long wavelength phason fluctuations [46–48].

This is exemplified in Fig. 9 which shows the measured and simulated diffuse scattering recorded for the i-AlPdMn quasicrystal [46, 49, 50]. All the observed diffuse scattering can be reproduced using the expression 1, for which a single parameter is needed i.e., the ratio of K2/K1. It was also shown using coherent X-ray scattering that in this icosahedral phase the diffuse scattering is dynamical in nature above 600 °C, although with a very slow dynamics.

Diffuse scattering has also been observed in the Cd–Yb type quasicrystal. In particular in the ZnMg–Sc icosahedral phase, although very weak, a clear supplementary phason contribution is observed in the diffuse scattering when compared to the 1/1 approximant [48]. The diffuse scattering being much smaller than in the i-AlPdMn phase a large number of weak Bragg peaks with a high Q_{per} component is observed as shown in Fig. 10.

Conclusion

We have shown in this review that the structure of quasicrystal is rather well understood, at least for the Cd–Yb type icosahedral phase. A good understanding of the crystal chemistry is achieved thanks to the availability of crystal approximant. The superspace approach and description of quasicrystal allows a very deep insight in this crystal chemistry. The quasicrystal is viewed as a quasiperiodic packing of a large cluster connected along twofold and threefold axes. One must not be misleading by this cluster description, which remains based on the tetrahedral closepacking of a large and a small atom. In that respect, the central tetrahedron which breaks the icosahedral symmetry certainly plays a crucial role. We have also shown that atomic scale simulation is now feasible using oscillating pair potential. This certainly opens up the route for the understanding of the relationship between the structure and physical properties, a question which is still opened. Long wavelength phason modes are fluctuations specific to quasicrystals and more generally to aperiodic crystals. It is remarkable that they are observed through diffuse scattering measurements for all known quasicrystals.

One of the most interesting and challenging questions remains the mechanisms that allow the long-range quasiperiodic order to set in.

References

1. Shechtman D, Blech I, Gratias D, Cahn JW (1984) Metallic phase with long-range orientational order and no translational symmetry. Phys Rev Lett 53:1951–1953
2. Fisher IR, Islam Z, Panchula AF, Cheon KO, Kramer MJ, Canfield PC, Goldman AI (1998) Growth of large grain R–Mg– Zn quasicrystals from the ternary melt (R=Y, Er, Ho, Dy, Tb). Phil Mag B 77:1601
3. Janssen T, Chapuis G, de Boissieu M (2007) Aperiodic Crystals. From modulated phases to quasicrystals, vol 20. IUCr Monographs on Crystallography. Oxford University Press, Oxford
4. Report of the Executive Committee for 1991 (1992). Acta Crystallographica Section A 48 (6):922–946. doi:10.1107/S0108767392008328
5. Gardner M (1977) Mathematical games. Sci Am 236:110–121
6. Penrose R (1979) Pentaplexity. A class of non-periodic tilings of the plane. Math Intelligencer 2:32–37
7. Mackay AL (1982) Crystallography and the Penrose pattern. Physica A114:609–613
8. Levine D, Steinhardt PJ (1984) Quasicrystals: a new class of ordered structures. Phys Rev Lett 53:2477–2480
9. de Wolff PM, Aalst v (1972) The four dimensional space group of γ-Na_2CO_3. Acta Cryst A 28:s111
10. de Wolff PM, Janssen T, Janner A (1981) The superspace groups for incommensurate crystal structures with a one-dimensional modulation. Acta Cryst A 37:625–636
11. Janner A, Janssen T (1979) Superspace groups. Physica A99:47–76
12. Janner A, Janssen T (1980) Symmetry of incommensurate crystal phases. II. Incommensurate basic structures. Acta Cryst A36:408–415
13. Janner A, Janssen T (1980) Symmetry of incommensurate crystal phases. I. Commensurate basic structures. Acta Cryst A36:399–408
14. Kalugin PA, Kitaev AY, Levitov LS (1985) $Al_{0.86}Mn_{0.14}$: a six-dimensional crystal. JETP Lett 41:145–149
15. Duneau M, Katz A (1985) Quasiperiodic patterns. Phys Rev Lett 54:2688–2691
16. Elser V (1985) Indexing problems in quasicrystal diffraction. Phys Rev B 32:4892–4898
17. Gratias D, Cahn JW, Mozer B (1988) Six-dimensional Fourier analysis of the icosahedral Al73Mn21Si6 alloy. Phys Rev B 38:1643–1646
18. Duneau M, Oguey C (1989) Ideal AlMnSi quasicrystal: a structural model with icosahedral clusters. J Phys (France) 50:135–146
19. Yamamoto A, Hiraga K (1988) Structure of an icosahedral Al–Mn quasicrystal. Phys Rev B 37:6207–6214
20. Guyot P, Audier M (1985) A quasicrystal structure model for Al–Mn. Phil Mag B 52:L15–L19
21. Elser V, Henley CL (1985) Crystal and quasicrystal structures in Al–Mn–Si alloys. Phys Rev Lett 55:2883–2886
22. Frank FC, Kasper JS (1958) Complex alloy structures regarded as sphere packings. I. Definitions and basic principles. Acta Cryst 11:184–190
23. Frank FC, Kasper JS (1959) Complex alloy structures regarded as sphere packings. II. Analysis and classification of representative structures. Acta Crystallogr A 12:483–499
24. Bergman G, Waugh JLT, Pauling L (1957) The crystal structure of the metallic phase Mg32(Al, Zn)49. Acta Cryst 10:254
25. Samson S (1958) The crystal structure of the intermetallic compound Mg3Cr2Al18. Acta Cryst. 11:851
26. Mackay AL (1962) A dense non-crystallographic packing of equal spheres. Acta Crystallogr 15:916–918
27. Tsai AP, Guo JQ, Abe E, Takakura H, Sato TJ (2000) A stable binary quasicrtystal. Nature 408:537–538
28. Gomez CP, Lidin S (2003) Comparative structural study of the disordered MCd/sub 6/quasicrystal approximants. Phys Rev B 68:024203\024201–024209
29. Takakura H, Gomez CP, Yamamoto A, de Boissieu M, Tsai AP (2007) Atomic structure of the binary icosahedral Yb–Cd quasicrystal. Nat Mater 6:58–63
30. Gomez CP, Lidin S (2001) Structure of Ca13Cd76: a novel approximant to the MCd5.7 quasicrystals (M.Ca, Yb). Angew Chem Int 40:4037–4039
31. Henley CL (1986) Sphere packings and local environments in Penrose tilings. Phys Rev B 34:797–816
32. Ishimasa T, Kasano Y, Tachibana A, Kashimoto S, Osaka K (2007) Low temperature phase of the Zn–Sc approximant. Phil Mag 87:2887–2897
33. Kaneko Y, Arichika Y, Ishimasa T (2001) Icosahedral quasicrystal in annealed Zn–Mg–Sc alloys. Phil Mag Lett 81(11):777–787

34. Tamura R, Nishimoto K, Takeuchi S, Edagawa K, Isobe M, Ueda Y (2005) Universal low-temperature phase transition in Zn- and Cd-based crystalline approximants. Phys Rev B 71(9):092203
35. Ishii Y, Fujiwara T (2001) Hybridization mechanism for cohesion of Cd-based quasicrystals. Phys Rev Lett 87(20):206408/206401–206404
36. Ishii Y, Fujiwara T (2002) Electronic structures of Cd-based quasicrystals and approximants. J Alloys Compd 342:343–347
37. de Boissieu M, Francoual S, Mihalkovic M, Shibata K, Baron AQR, Sidis Y, Ishimasa T, Wu D, Lograsso T, Regnault LP, Gähler F, Tsutsui S, Hennion B, Bastie P, Sato TJ, Takakura H, Currat R, Tsai AP (2007) Lattice dynamics of the Zn–Mg–Sc icosahedral quasicrystal and its Zn–Sc periodic 1/1 approximant. Nat Mater 6:977–984
38. Mihalkovic M, Francoual S, Shibata K, de Boissieu M, Baron AQR, Sidis Y, Ishimasa T, Wu D, Lograsso T, Regnault LP, Gahler F, Tsutsui S, Hennion B, Bastie P, Sato TJ, Takakura H, Currat R, Tsai AP (2008) Atomic dynamics of i-ScZnMg and its 1/1 approximant phase: experiment and simulation. Phil Mag 88(13):2311–2318
39. de Boissieu M, Currat R, Francoual S (2008) Phason modes in aperiodic crystals. In: Fujiwara T, Ishii Y (eds) Handbook of metal physics: quasicrystals. Advances in metal physics. Elsevier Science, Amsterdam, pp 107–169
40. Bak P (1985) Symmetry, stability and elastic properties of icosahedral incommensurate crystals. Phys Rev B 32(9):5764–5772
41. Bak P (1985) Phenomenological theory of icosahedral incommensurate (quasiperiodic) order. Phys Rev Lett 54:1517–1519
42. Lubensky TC, Ramaswamy S, Torner J (1985) Hydrodynamics of icosahedral quasicrystals. Phys Rev B 32(11):7444–7452
43. Jaric MV, Nelson DR (1988) Diffuse scattering from quasicrystals. Phys Rev B 37:4458–4472
44. Ishii Y (1992) Phason softening and structural transitions in icosahedral quasicrystals. Phys Rev B 45(10):5228–5239
45. Widom M (1991) Elastic stability and diffuse scattering in icosahedral quasicrystals. Phil Mag Lett 64(5):297–305
46. de Boissieu M, Boudard M, Hennion B, Bellissent R, Kycia S, Goldman AI, Janot C, Audier M (1995) Diffuse scattering and phason elasticity in the AlPdMn icosahedral phase. Phys Rev Lett 75(1):89
47. de Boissieu M, Takakura H, Bletry M, Guo JQ, Tsai AP (2002) Phason fluctuations in i-AlPdRe and i-CdYb phases. J Alloys Compd 342:265–270
48. de Boissieu M, Francoual S, Kaneko Y, Ishimasa T (2005) Diffuse scattering and phason fluctuations in the Zn–Mg–Sc icosahedral quasicrystal and its Zn–Sc periodic approximant. Phys Rev Lett 95:105503
49. de Boissieu M, Francoual S (2005) Diffuse scattering and phason modes in the i-AlPdMn quasicrystalline phases. Z Kristallogr 220:1043–1051
50. Létoublon A, de Boissieu M, Boudard M, Mancini L, Gastaldi J, Hennion B, Caudron R, Bellissent R (2001) Phason elastic constants of the icosahedral Al–Pd–Mn phase derived from diffuse scattering measurements. Phil Mag Lett 81:273–283

Some Properties of Three-Periodic Sphere Packings[a,b]

Michael O'Keeffe[c]

Abstract

Some properties of sphere packings are reviewed. Candidates for the least-dense packings of one kind of sphere for different coordination numbers are identified and described. Some topological properties of sphere packing nets are also described.

Keywords

Circle packing • Crystal structure • Nets • Sohere packing • Tilings

I have knocked on the doors of chemistry (Ego namque, pulsatis chymiae foribus)… J. Kepler [1].

Introduction

The problem of the finding the structure of the densest packing of equal spheres has a long history in mathematics, culminating in the definitive study of Hales [2] in 1998 proving that what was long known to scientists as "closest packing" in three dimensions was indeed just that, as originally conjectured by Kepler [1] in 1611. The geometrical arrangement of 3-periodic closest packings (cubic, hexagonal, etc., known collectively as Barlow packings) and the patterns of their holes (interstices) play a well-known central role in descriptive inorganic crystal chemistry [3, 4]. In this context Brunner [5] has remarked that the same structures arise as the solution to an apparently different problem: What is the arrangement of points (atoms) in space that for a given density maximizes the shortest distances between them? Stated this way the problem is a sort of Tammes problem and lends insight into the frequent occurrence of close packing in crystal structures [6].

An enormously valuable compilation of 3-periodic homogeneous sphere packings (those with just one kind of sphere) has been made by Fischer and associates [7–11]. This is almost complete, but lacks full results for monoclinic and orthorhombic symmetries. They number in the thousands.

Sphere packings carry a *net* (a simple periodic connected graph) in which the vertices correspond to centers of the spheres, and pairs of spheres in contact define the edges. One is then lead to ask about topological properties such as number of topologically different kinds of edge (*edge transitivity*) and the sizes of cycles (closed paths) of the net.

In general (but not always!) the nets of interest in chemical structures have embeddings in which all edges are equal and correspond to the shortest intervertex distance. Accordingly these nets are the graphs of packings of equal-size, but not necessarily topologically equivalent, spheres. For example all the 4-coordinated nets of zeolites are of this type.

A sphere packing in n dimensions is said to be locally *stable* (or *locally jammed*) if each sphere is in contact with at least $n+1$ others and the contacts are not all on the same hemisphere (in this context a hemisphere includes the great circle dividing the sphere). The general question of jamming is complex and depends on the nature of boundary conditions, such as a rigid box [12]. I refer only to local stability

[a] *Structural Chemistry* 2012, 23(4):1079–1087.

[b] Dedicated to Sten Andersson as an 80th birthday gift.

[c] Department of Chemistry and Biochemistry, Arizona State University, Tempe, AZ 85287, USA
e-mail: mokeeffe@asu.edu

I. Hargittai, B. Hargittai (eds.), *Science of Crystal Structures: Highlights in Crystallography*,
DOI 10.1007/978-3-319-19827-9_17

(jamming) here. In this study a more relevant consideration is whether a low density sphere packing can be smoothly decreased in volume to approach arbitrarily close to a higher density and higher coordination while remaining a sphere packing. Many examples of this have been given elsewhere [3, 4], I adduce just one simple example here. The 6-coordinated sphere packing with sphere centers at the points of a primitive cubic lattice has symmetry $Pm\bar{3}m$. Reducing the symmetry to $R\bar{3}m$ allows the rhombohedral angle to vary. As the angle is reduced from 90° the density increases with spheres in 6-coordination until at an angle of 60° there are suddenly 12 contacts and the structure is that of cubic closest packing with symmetry $Fm\bar{3}m$.

My interest in such situations arose from reading a paper describing a candidate for "densest possible equivalent periodic sphere packing with 11-coordination". However the structure in question (see below) can be smoothly transformed in a lower symmetry setting to come arbitrarily close to hexagonal closest packing all the time keeping 11 nearest neighbors in contact, and the high symmetry configuration is in fact the lowest (known) density 11-coordinated sphere packing [13].

Some years ago Slack [14] published a list of the "most-dense and least-dense" packings of symmetry-related spheres as a function of coordination number. For the reasons given above, I feel that the most-dense is not a real problem. What one can ask is for the packing with the largest minimum density for a given coordination number. However the least-dense appears more interesting, and here I adduce several improvements to most of Slack's list.

Data for 3-periodic structures (sphere packings, tilings, nets etc.) have been collected in a searchable database known as the reticular chemistry structure resource (RCSR) [15]. Nets are identified by their RCSR symbols composed of bold lowercase letters.

In order to facilitate comparisons with structures that are not sphere packings, the density of a circle or sphere packing is expressed as $\rho = N/V$ where V is the volume of a unit cell and N is the number of non-overlapping unit diameter circles or spheres in that cell. The fraction of space filled by the circles or spheres is $\pi\rho/4$ or $\pi\rho/6$ respectively.

Generally the lowest density configuration of a sphere packing is at the maximum symmetry. Interestingly this is not universally the case as Koch et al. [16] show with a counterexample. This structure (**sme** in the RCSR database) with a uninodal net has minimum density equal to 0.312 in the maximum symmetry $R\bar{3}c$ but two embeddings in $R\bar{3}$ with density 0.7 % lower. Despite this counterexample I have assumed for the structures described below that the minimum density is at maximum symmetry.

A second reason for interest in low density sphere packings originated in work on 4-coordinated nets of low topological density [17]. Then I realized that a certain 4-coordinated sphere packing that was locally jammed had lower density than a structure described by Heesch and Laves [18] of which no less an authority than Hilbert and Cohn-Vossen [19] said "there are reasons to believe that this is the loosest possible packing"; however they did not state those reasons. The same Heesch and Laves packing was also cited in several compilations as the "rarest" sphere packing [17]. If the great Hilbert could be found wrong, perhaps this is an interesting field! Actually I think it reflects the general ignorance among mathematicians of periodic structures, familiar to crystal chemists, other than densest sphere packings—a topic which we have found recent occasion to comment on in connection with a different example [20].

Circle Packings

The densest packing of equal circles is well-known to be the unique structure with each circle in contact with six others with density $\rho = 2/\sqrt{3} = 1.155$, Fig. 1a. For packings with one kind of circle (i.e., with vertex transitive nets) the least-dense packing ($\rho = 12/(12+7\sqrt{3}) = 0.4974$) is that shown in Fig. 1b. One can easily make lower density packing ($\rho = 0.3476$) by replacing each circle by a group of three circles as in generation of Fig. 1c from Fig. 1b. However in this instance, the new pattern is not locally stable and there are two kinds of circle, i.e. with a binodal (vertex 2-transitive) net. Figure 1d shows a locally stable low density ($\rho = 0.4137$) structure in which there are three kinds of circle, all 3-coordinated. This has lower density than the lowest density packing with just one kind of circle.

In disproving an old conjecture, it was shown long ago that in fact with more than one kind of circle one can construct locally stable circle packings of arbitrarily low density but in these there are spheres with 3 and 4 contacts [21, 22]. As far as I know no locally stable circle packings of uniform coordination and arbitrarily low density have yet been discovered. It appears to be a challenging problem. (I am grateful to Ruggero Gabrielli for correspondence on this point).

Low Density Sphere Packings in Three Dimensions

In what follows candidates for the lowest density uninodal sphere packings were found by a search of the RCSR database. They are all, as is to be expected, structures from the Fischer et al. enumeration [8–11]. However I have not seen most of them explicitly recognized before. They form a remarkably diverse set with, I submit, some intrinsic beauty. The structures are identified by their RCSR symbol (lowercase bold letters). Each structure has a Fischer symbol of the form *n*/*m*/*pi*. Here *n* is the coordination number, *m* is the size

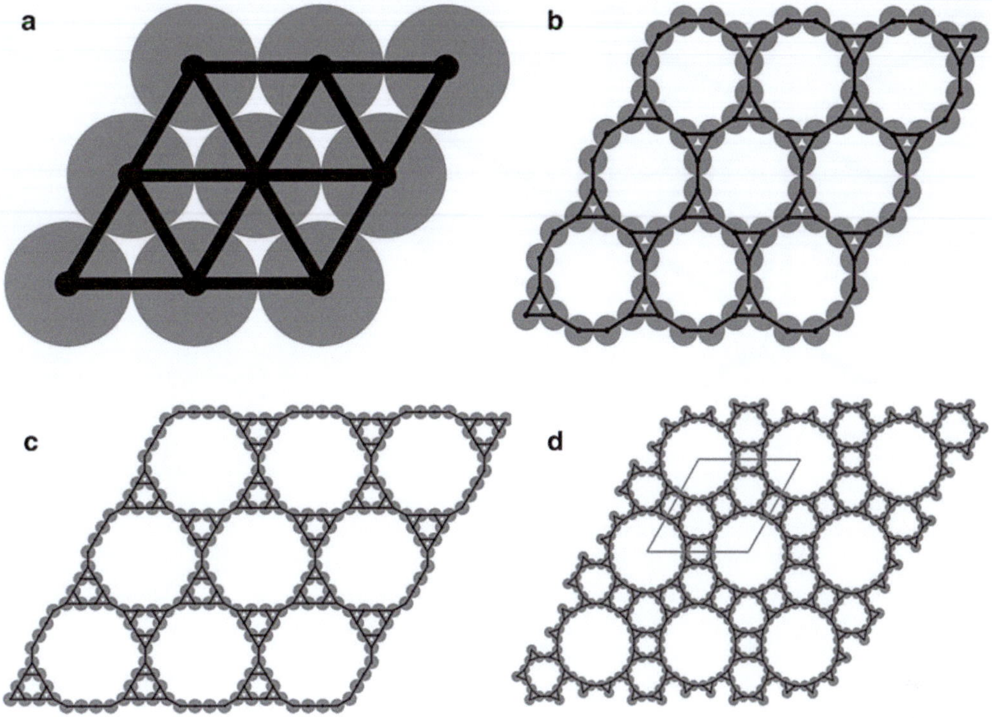

Fig. 1 Circle packings discussed in the text.

of the shortest cycle in the graph, *p* is a letter denoting the crystal system (e.g. *c* for cubic, *h* for hexagonal) and *i* is an arbitrary serial number.

12-Coordination

The only uninodal sphere packings with 12-coordination are of course cubic and hexagonal closest packing with $\rho=\sqrt{2}=1.4142$. As they are the only such packings they are also, trivially, those of lowest density for 12-coordination.

11-Coordination

There are six uninodal 11-coordinated sphere packings and they all have the same minimum density $\rho=8/(3+\sqrt{8})=1.3726$. Two of them are on Slack's list [14]. One (**ela** 11/3/*t*1) is the 11-coordinated structure referred to above [13]. As shown in Fig. 2 in **ela** (symmetry $P4_2/mnm$) there are columns of edge-shared octahedra linked by edges to leave tunnels of square cross section. Concerted rotations of the columns collapse the tunnels and produce a twelfth contact in hexagonal closest packing (symmetry $P6_3/mcm$). The intermediate symmetry is the common subgroup *Pnnm* [13].

The other five 11-coordinated structures are **elb** 11/3/*o*2, **elc** 11/3/*m*1, **eld** 11/3/*h*2, **ele** 11/3/*o*1, **elf** 11/3/*h*1. These are all built up of slabs of two close-packed layers (each sphere in 9-coordination) packed in such a way that each sphere has two extra contacts between slabs [4] as shown in Fig. 3.

Fig. 2 *Left* the 11-coordinated packing **ela**. *Columns* of edge-haring octahedra are shown in *blue*. *Right* the 12-coordinated **hcp** illustrated similarly (Color figure online).

10-Coordination

The unique lowest density 10-coordinated is **tci** (10/3/*h*1) density $\rho=6\sqrt{2}/7=1.2122$. It is made up of parallel $3^4.6$ layers (Fig. 4) each layer is 4-coordinated and has 6/7 the density of a close-packed layer. The layers are stacked in *ABC* fashion co that each sphere has 3 contacts with spheres on neighboring layers. The symmetry is $R\,\overline{3}$. This structure is on Slack's list of least-dense sphere packings [14].

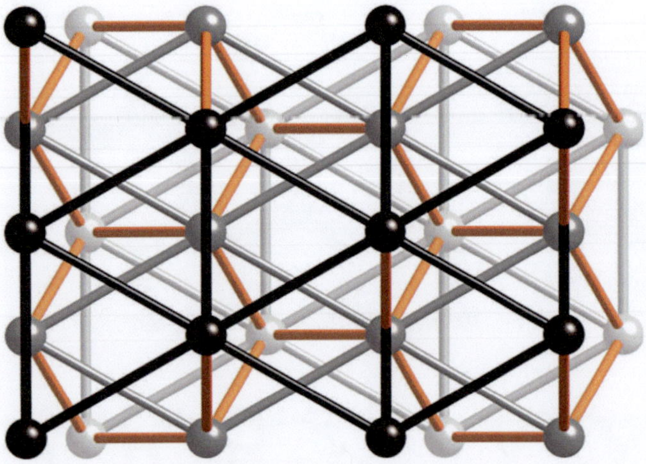

Fig. 3 The three-layer sequence of 3^6 layers found in **elb**, **elc**, **eld**, **ele**, **elf**. The *spheres* in the middle layer have three neighbors in the layer *below* and two in the layer *above*.

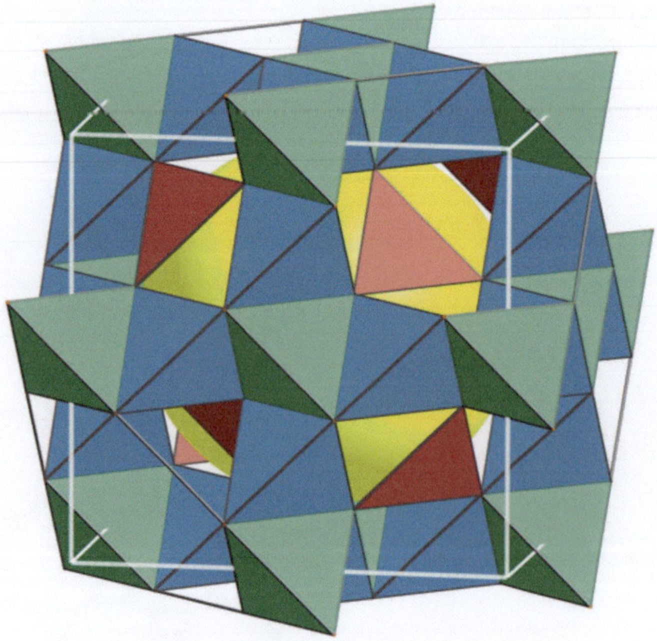

Fig. 5 The 9-coordinated structure **ncc**.

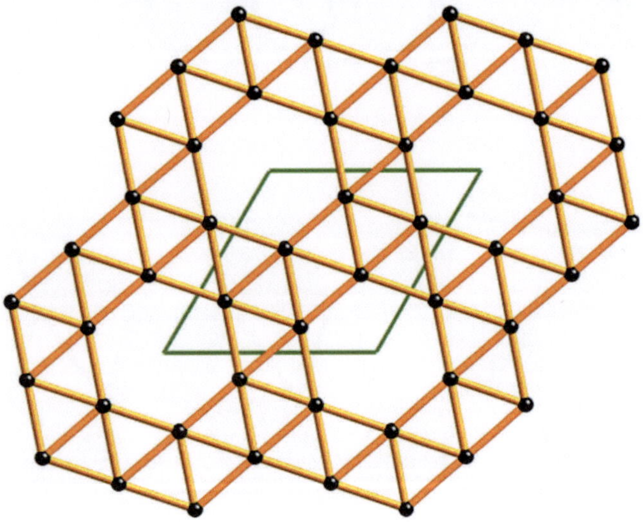

Fig. 4 One layer of the packing in **tci**. *Spheres* have four neighbors in the same layer.

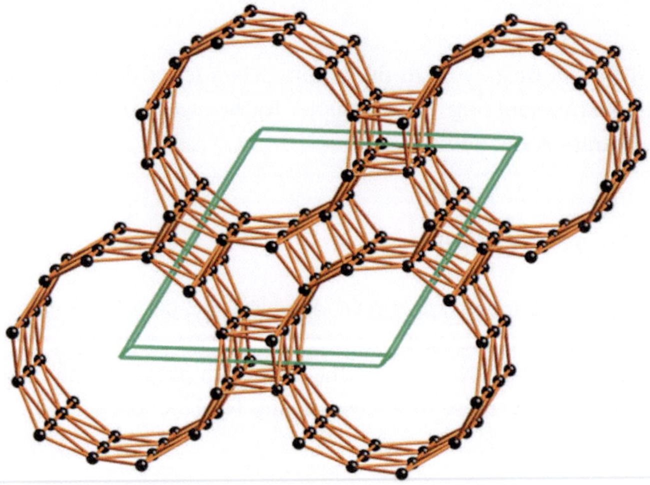

Fig. 6 The 8-coordinated structure **ecq**.

9-Coordination

The unique lowest density is **ncc** (9/3/*c*4), $\rho = 3/2\sqrt{2} = 1.0607$. The cubic structure (symmetry $I\bar{4}3m$) made up of linked regular triangles and tetrahedra is shown is Fig. 5.

8-Coordination

The unique lowest density is **ecq** (8/3/*h*15), $\rho = 16/(9+6\sqrt{3}) = 0.82507$. In projection this structure resembles the plane Archimedean tiling 4.6.12. However, the links are not coplanar, but form zig-zags normal to the layer as shown in Fig. 6. A structure with this density appears on Slack's list [14], presumably it is the same (details are not available in that article).

7-Coordination

The unique lowest density is **svu** (7/3/*c*4), $\rho = 324/(3+\sqrt{5}+2\sqrt{2})3 = 0.6178$. In this structure regular icosahedra are linked by octahedra in a diamond net, Fig. 7. This was already identified by O'Keeffe and Hyde [4] as a candidate for least-dense 7-coordinated sphere packing.

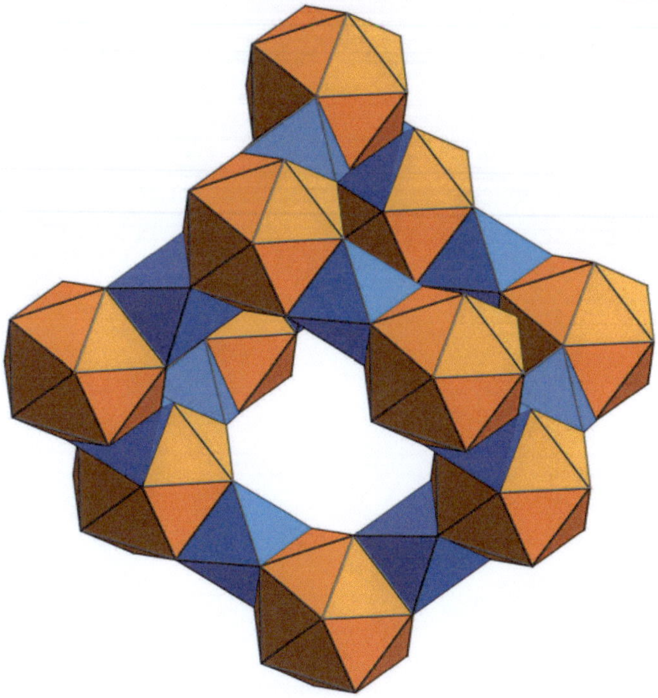

Fig. 7 The 7-coordinated structure **svu**.

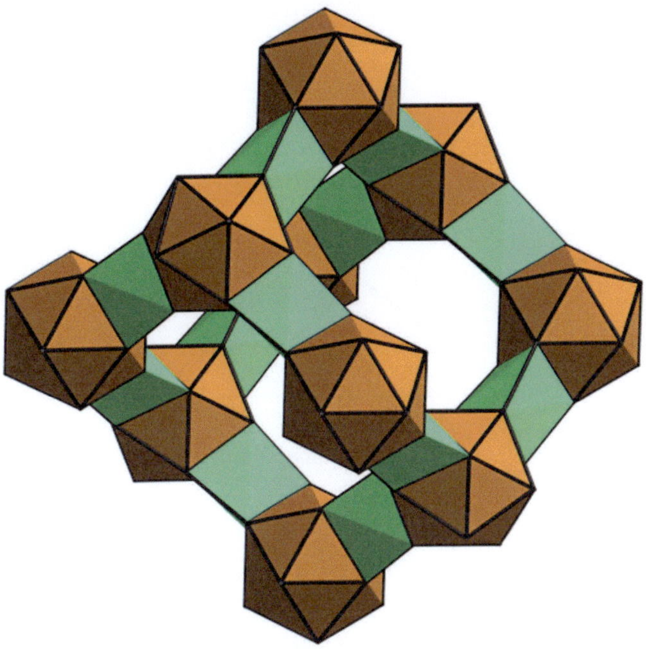

Fig. 8 The 6-coordinated structure **snb**.

6-Coordination

The lowest density is **snb** (6/3/*c*25), $\rho=0.4992$. This is closely related to the 7-coordinated structure in which icosahedra are linked by octahedra in a diamond net. The

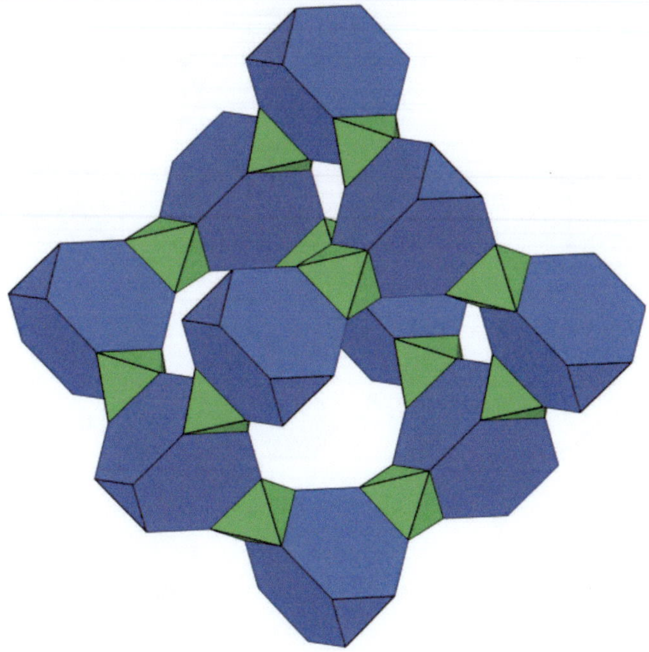

Fig. 9 The 5-coordinated structure **crs-a**.

difference is now the icosahedra are linked by slightly irregular trigonal prisms to produce a chiral structure (symmetry $F4_132$) as shown in Fig. 8.

5-Coordination

The lowest density locally stable structure is **crs-a** (5/3/c10), $\rho=81\sqrt{2}/343=0.3340$. This (Fig. 9) is again reminiscent of the 7-coordinated structure, but now it is truncated tetrahedra that are linked by octahedra into a diamond net. This structure is in Slack's compilation [14].

There is a lower density 5-coordinated structure that is not locally stable. This is **srs-f** (5/3/*c*30), $\rho=3/8\sqrt{2}=0.2651$. In this structure the vertices of the 3-coordinated net **srs** are replaced by octahedra kinked to three others by pairs of edges as shown in Fig. 10.

4-Coordination

The lowest density locally stable structure is **sod-a** (4/3/*c*10), $q=48/(2+3\sqrt{2})^3=0.1973$, Fig. 11. It is derived from the net of sodalite (**sod**) by replacing each vertex by a group of vertices that are arranged as in the vertex figure of the original net, in this case a tetrahedron. This process, called *augmentation*, is analogous to *truncation* for finite polyhedra. The candidate for lowest density canvassed by Hilbert and Cohn-Vossen [19] is the augmented diamond net (**dia-a**) which has density $4/(\sqrt{2}+2/\sqrt{3})^3=0.2359$.

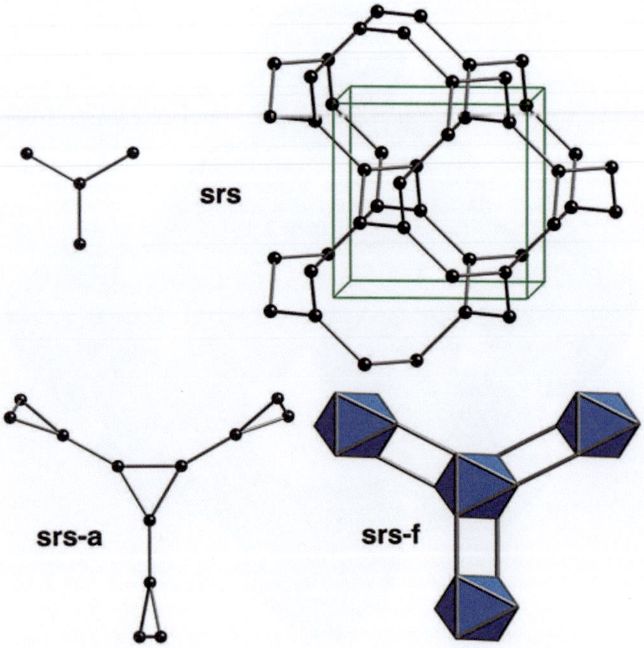

Fig. 10 *Top right* the 3-coordinated **srs** net with, on the *left*, a fragment of the structure (one vertex and its neighbors). *Bottom left* a corresponding fragment of the 3-coordinated **srs-a** net. *Bottom right* a similar fragment of the 5-coordinated **srs-f** net.

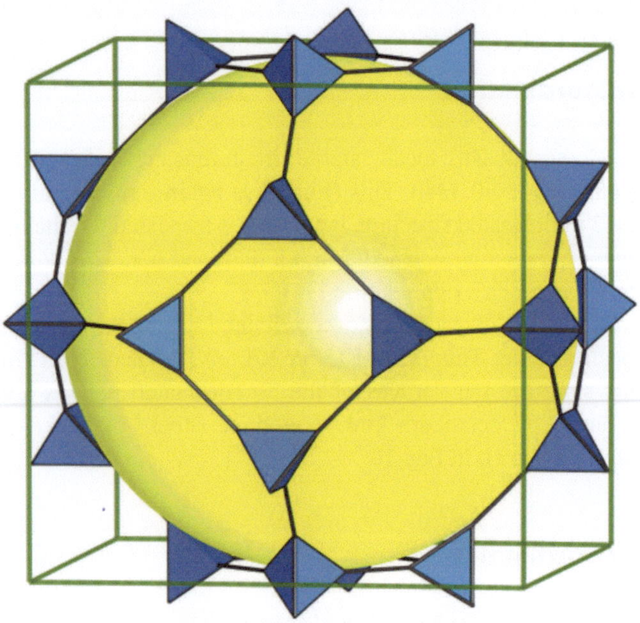

Fig. 11 The 4-coordinated net **sod-a**.

There is also a lower density 4-coordinated sphere packing that is not locally stable. This is **lcv-a** (4/3/c27), $\rho=6/(2+\sqrt{2})^3=0.1507$, Fig. 12. It is the corresponding augmented version of the 4-coordinated net **lcv**.

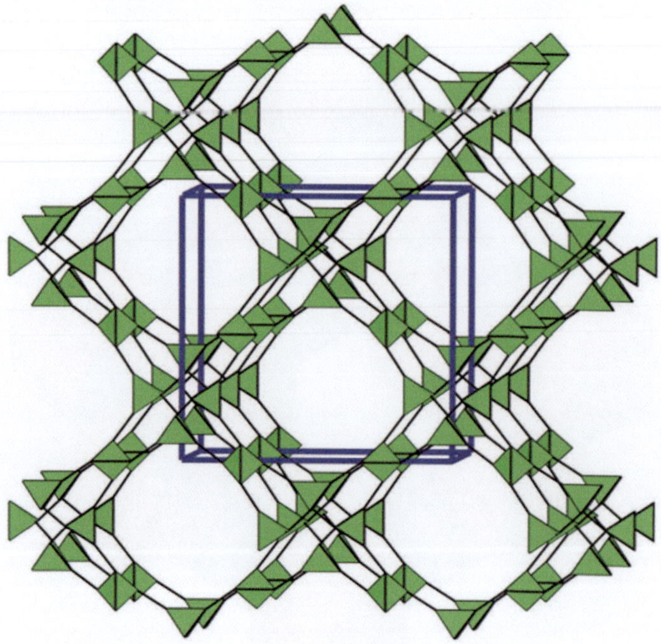

Fig. 12 The 4-coordinated net **lcv-a**.

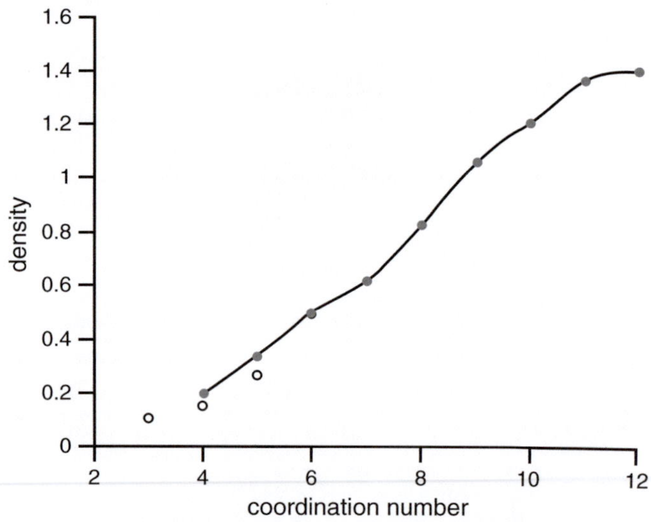

Fig. 13 Density as a function of coordination number for least-dense sphere packings. *Open circles* refer to unstable packings.

3-Coordination

3-coordinated sphere packings are necessarily not locally stable. The lowest density structure is **srs-a** (3/3/c1), $\rho=3/(\sqrt{2}+2\sqrt{2}/\sqrt{3})^3=0.1060$ Fig. 10. This structure has long been known as the best candidate for the least-dense packing of equivalent spheres [17].

Of the 15 structures described here, only 6 are on Slack's [14] list of least-dense sphere packings.

The dependence of minimum density on coordination number is shown in Fig. 13; it may be seen to be approximately linear.

The above results refer specifically to packings with one kind of sphere. With more than one kind of sphere questions of the existence of locally stable packings with arbitrarily low density occur. Trivially the stable circle packings can be converted to stable sphere packings by prismatic stacking of the former. But are there stable packings of arbitrarily low density with, for example, 4-coordination?

Some Properties of Sphere Packing Graphs

Symmetric (Vertex- and Edge-Transitive) Graphs

Delgado-Friedrichs and coauthors [23, 24] reported 20 vertex- and edge-transitive (*symmetric* to graph theorists) nets that had embeddings in which edges lengths were the shortest

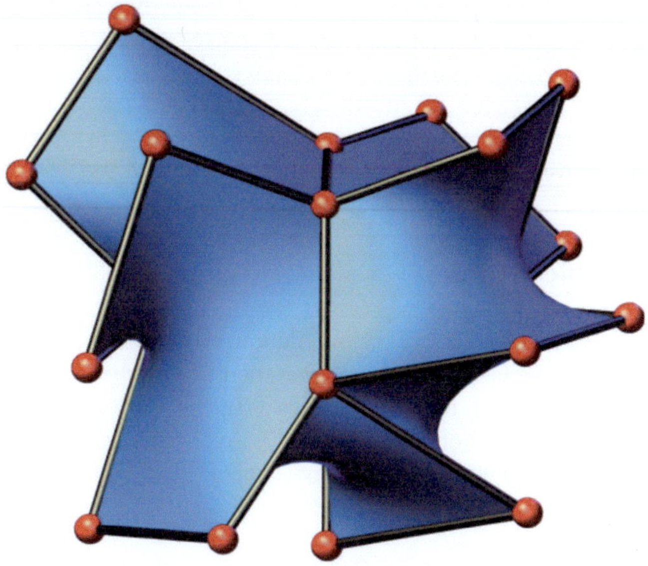

Fig. 14 A space-filling (isohedral) tile [7^8] of a tiling that carries the **itv** net.

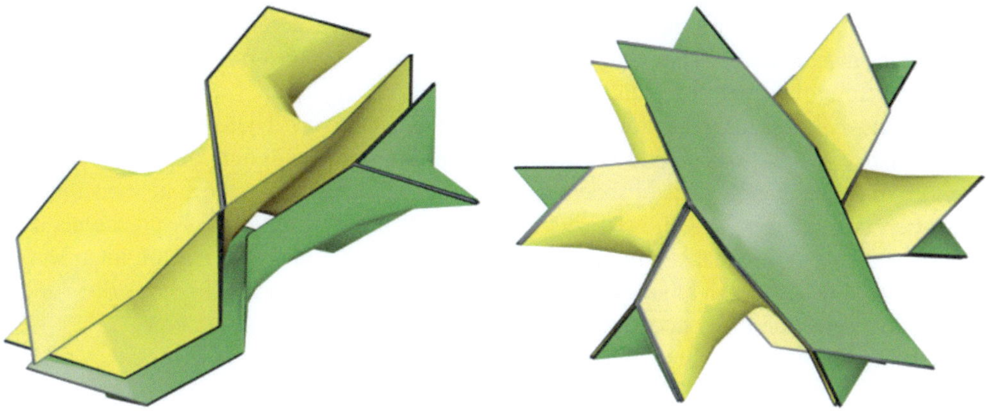

Fig. 15 *Left* the two tiles [4^2,10^2] and [10^4] of the **uoo** net, *right* part of the tiling.

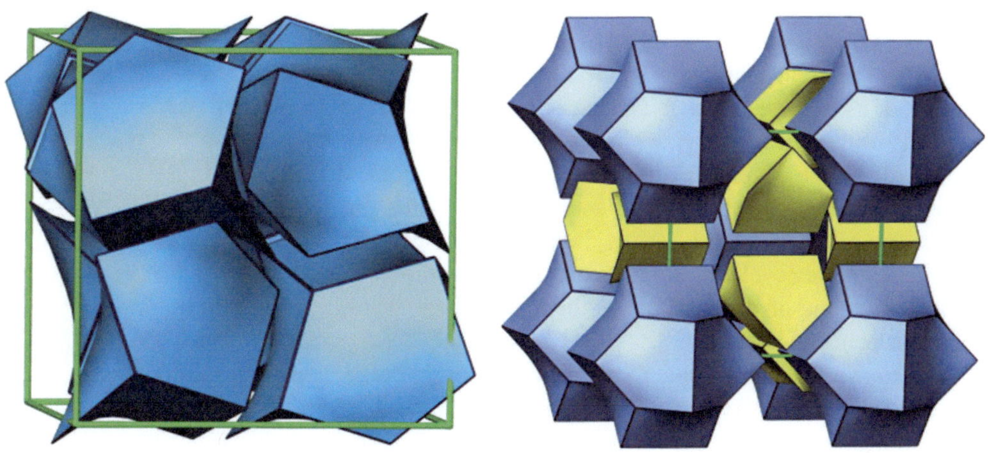

Fig. 16 *Left* the isohedral [5^6] tiling of the **fsn** net. *Right*, the 3[5^4]+[5^{12}] tiling of the **ith-d** net. Both tilings are face transitive.

distance between vertices. For two of these (**bcs** and **hxg**) there are also shortest intervertex distances that do not correspond to edges, so these two are not sphere packings. Thus I believe there are exactly 18 sphere packings with symmetric graphs. These are of particular importance in crystal chemistry [23].

Triclinic Graphs

The Fischer et al. [11] compilation of sphere packings includes the fascinating result that there is exactly one such structure that has triclinic symmetry. Such a result raises the question of what other uninodal intrinsically triclinic graphs are there, and what are their provenances?

Ring Sizes for 4-Coordinated Sphere Packings

4-Coordinated nets are especially important in chemistry. Thus they are the nets of covalent structures such as diamond and lonsdaleite, the nets of zeolites and framework aluminosilicates. They are also the nets of simple tilings which are tilings by simple polyhedra in which three tiles meet at an edge and four at a vertex, and a simple polyhedron is one in which three edges meet at each vertex. Simple tilings are ubiquitous as the structure of foams and cellular materials and clathrate hydrates etc. All these have maximum symmetry embeddings with equal edges that correspond to the shortest intervertex distance; they are then the nets of sphere packings. Questions that often arise concern the allowed ring sizes in such nets. The rings of interest are really *strong* rings which are cycles of the net that are not the sum of smaller cycles.

In this connection the concept of a natural tiling for a net is relevant [23]. A *proper* tiling of a net is one that preserves the full symmetry. For simple high symmetry nets this is sufficient to uniquely define a tiling. For others we add the constraint that no one face of a tile be larger than the rest. This is equivalent to requiring the faces of the tiles be strong rings. For complicated low symmetry structures, such as some zeolite nets, further rules are necessary to arrive at a unique net [25]. The important point is that the faces of tiles can be thought of as *essential* rings in the sense that all other rings are sums of these (they provide a basis set of rings). This is in accord with the general approach to cyclic molecules where one seeks a smallest set of smallest rings. Thus for cubane five of the six 4-rings form a complete basis set in the sense that all other cycles of the structure (including the sixth 4-ring) are sums of some or all of those five rings. Thus cubane is pentacyclooctane.

It is known that for the nets of simple tilings the average ring size (average tile face size) $\langle n \rangle$ is >9/2 and <6 [26]. The lower limit is approached for example by repeated augmentation of the sodalite net. Thus $\langle n \rangle$ for **sod** 5.14, **sod-a** 4.65, **sod-a-a** 4.54, **sod-a-a-a** 4.51. The density is rapidly decreasing, for the last it is only $\rho = 0.04$.

It was thought that nets of 4-coordinated sphere packings in general had at least one ring with six or fewer edges, but a nice counterexample was found [27] in the structure of ice XII which has a net of a binodal sphere packing. This net (**itv**) has an isohedral natural tiling in which all the rings are 7-rings (face symbol $[7^8]$, Fig. 14. In general for 4-coordinated nets, not necessarily those of sphere packings, there is no upper limit to the minimum ring size, RCSR includes nets **ten** and **elv** with shortest rings that are 10-rings and 11-rings respectively.

For uninodal sphere packings the largest average ring size is in **uoo** (4/4/*h*3). There are two tiles (Fig. 15) that occur in ratio 1:1 and their face symbols are $[4^2.10^2]$ and $[10^4]$. The average face size (i.e., average essential ring size) is 17/2 = 8.5.

A curiosity, not original to me (I forget who I got it from), is the observation that there is no known 4-coordinated net in which all the rings are 5-rings. This is clearly impossible for the net of a simple tiling as the only simple polyhedron with 5-ring faces is the pentagonal dodecahedron and this is well-known to tile the 3-sphere rather than Euclidean space, and simple tilings must include tiles with faces that are 6-rings or larger. It is not obvious, however, that the necessity to have at least one essential ring that is a 6-ring or larger should apply to all 4-coordinated nets, but it appears to.

There are however, at least two nets of sphere packings with only one kind of ring that is a 5-ring. These are the nets of binodal sphere packings **fsn** and **ith**. The tilings are shown in Fig. 16. **fsn** is the net of the cubic $(Pa\overline{3})$ form of SiP_2O_7 and **ith-d** is of interest as the net of low density metal–organic frameworks [28]. These nets are not however 4-coordinated.

Acknowledgments My interest in geometry applied to crystals and related periodic structures was developed in a magical summer long ago on Singö with Sten Andersson and Bruce Hyde who have remained mentors, collaborators and friends ever since. My work in Tempe on the geometrical basis of crystal chemistry is supported by the U. S. National Science Foundation, Grant DMR 0804828.

References

1. Kepler J (1611) Strena seu de nive sexangula (translated as The six-cornered snowflake by Bromberg J). Paul Dry Books, Philadelphia (2010)
2. Hales TC (2005) Ann Math 62:1065
3. Hyde BG, Andersson S (1989) Inorganic crystal structures. Wiley, New York
4. O'Keeffe M, Hyde BG (1996) Crystal structures I: patterns and symmetry. Miner Soc Am
5. Brunner GO (1971) Acta Crystallogr A 27:388
6. O'Keeffe M (1977) Acta Crystallogr A 33:924
7. Fischer W (2004) Acta Crystallogr A 60:249

8. Fischer W (2005) Acta Crystallogr A 61:435
9. Sowa H, Koch E (2006) Acta Crystallogr A 62:379
10. Sowa H, Fischer W (2010) Acta Crystallogr A 66:292
11. Fischer W, Koch E (2003) Acta Crystallogr A 59:509
12. Donev A, Torquato S, Stillinger FH, Connelly R (2004) J Appl Phys 95:989
13. O'Keeffe M (1984) Mater Res Bull 19:1433
14. Slack GA (1983) Zeits Krist 165:1
15. O'Keeffe M, Peskov MA, Ramsden SJ, Yaghi OM (2008) Acc Chem Res 41:1782
16. Koch E, Sowa H, Fischer W (2005) Acta Crystallogr A 61:426
17. O'Keeffe M (1991) Zeits Krist 196:21
18. Heesch H, Laves F (1933) Zeits Krist 85:443
19. Hilbert D, Cohn-Vossen S (1932, 2005) Anschauliche Geometrie (Translated as Geometry and the imagination). Springer, Chelsea (1952)
20. Hyde ST, O'Keeffe M, Proserpio DM (2008) Angew Chem Int Ed 47:7996
21. Böröczky K (1964) Ann Univ Sci Budapest Eötvös Sect Math 7:79
22. Kahle M. www.math.ias.edu/~mkahle/sparse.pdf
23. Delgado-Friedrichs O, O'Keeffe M, Yaghi OM (2003) Acta Crystallogr A 59:22
24. Delgado-Friedrichs O, O'Keeffe M, Yaghi OM (2003) Acta Crystallogr A 59:515
25. Blatov VA, Delgado-Friedrichs O, O'Keeffe M, Proserpio DM (2007) Acta Crystallogr A 63:418
26. Leo F, Stong R (1993) Trans Am Math Soc 337:891
27. O'Keeffe M (1998) Nature 392:879
28. Furukawa H, Ko M, Go YN, Aratani N, Choi SB, Choi E, Yazaydin AO, Snurr RQ, O'Keeffe M, Kim J, Yaghi OM (2010) Science 329:424

Cluster Packing from a Higher Dimensional Perspective[a]

Walter Steurer[b] and Sofia Deloudi[b]

Abstract

The way to find the optimum packing of quasicrystal-constituting clusters is discussed based on the projected-cell approach. We illustrate why the quasiperiodic arrangement of partially overlapping clusters with decagonal or icosahedral symmetry is the most efficient one, by relating it to the packing of unit cells in hypercubic lattices.

Keywords

Quasicrystal • Cluster • Packing • Higher-dimensional approach

Introduction

Quite a few stable decagonal and icosahedral quasicrystals (QC) have been identified so far in binary and ternary intermetallic systems (see, e.g., [1, 2] and references therein). Their structures, properties and stability have been intensively studied during more than a quarter century as testified by more than 10,000 publications to date. Nevertheless, it is still not fully understood why and how QC form, particularly in view of the fact that the same kind of clusters can constitute both QC and their periodic counterparts, the approximants. Furthermore, for many not directly involved in QC research, the powerful but arcane description of quasiperiodic structures as projections from hyperspace has still an aura of mystery. In the following, we will try to demystify the higher-dimensional approach by looking at the well-known strip-projection method from a different angle, demonstrating that its physical basis is in the cluster shape. This unusual view should also help sketching the big picture of QC formation.

We want to emphasize that the term "quasi" in connection with "crystal" has nothing to do with lack of order, it rather indicates the particular kind of order. The structural perfection of QC, as reflected in the width of X-ray diffraction peaks, for instance, is comparable to that of other complex intermetallic phases. This means that the global, overall structural correlation length can reach tens or hundreds of micrometers, despite the existence of some intrinsic local structural disorder. The idealized structure of QC is usually described in terms of a quasiperiodic tiling decorated by particular atomic arrangements or, equivalently, as quasiperiodic packings of multishell clusters. Here, the term "clusters" denotes fundamental structural units which may or may not be stabilized by chemical bonding [3, 4].

While the symmetry of the outer cluster shells usually reflects the overall symmetry of the QC, that of the inner shell(s) is frequently lower giving rise to structural disorder. We focus in this study only on the outer cluster shell(s), which determine the way of packing via shared structural subunits. We will explore how clusters with non-crystallographic symmetry can be packed in the topologically best possible way, i.e., without gaps and "glue atoms" filling them. One crucial boundary condition is that the local composition has to be as close as possible to the overall one.

It is well known that some regular and semiregular polyhedra, and even multi-shell clusters, can be described as projections of higher-dimensional polytopes (see e.g., [5, 6]). Consequently, it is obvious to treat the problem of the topologically best cluster packing from a higher dimensional perspective. For instance, every three-dimensional (3D) zonohedron with

[a] *Structural Chemistry* 2012, 23(4):1115–1120.

[b] Laboratory of Crystallography, Department of Materials, ETH Zurich, Wolfgang-Pauli-Strasse 10, 8093 Zurich, Switzerland
e-mail: steurer@mat.ethz.ch

I. Hargittai, B. Hargittai (eds.), *Science of Crystal Structures: Highlights in Crystallography*,
DOI 10.1007/978-3-319-19827-9_18

octahedral or icosahedral symmetry can be described as orthogonal projection of an *n*D hypercube [5]. Thereby, the number of edge directions of the respective zonohedron defines the dimension *n*; this gives, for instance, *n*=6 for the triacontahedron. The assembly of such zonohedra with optimum density (frequency in a given volume) for a given chemical composition is a non-trivial problem that can be tackled much more easily in higher dimensions. Columnar clusters with, in projection, octagonal, decagonal or dodecagonal shape can be related to 2D projections of *n*D (*n*=4, 5, 6) hypercubes.

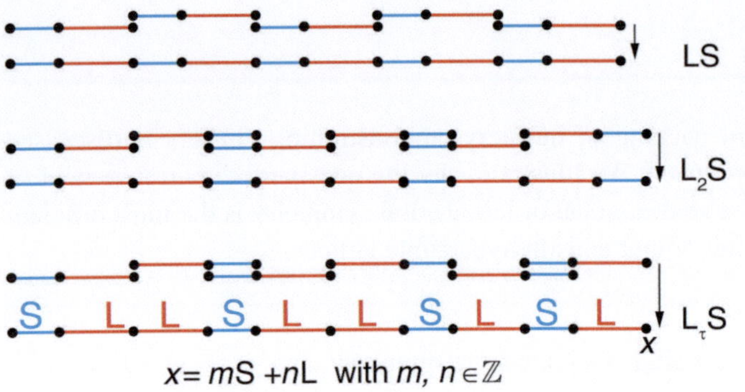

Fig. 1 Sequences with overall composition LS (*top*), L_2S (*middle*) and $L_\tau S$, i.e. the Fibonacci sequence (*bottom*), with the decomposition into covering clusters of the type (LS) in both orientations. The upper two sequences are rational approximants of the Fibonacci sequence.

In order to introduce our basic concept, we will start the discussion with a cluster leading to the simplest quasi periodic structure, the 1D Fibonacci sequence (FS). Subsequently, we will move on to decagonal and triacontahedral clusters whose packing results in the 2D Penrose (PT) and 3D Ammann tiling (AT), respectively.

(LS) Cluster and Fibonacci Sequence

Let us assume that we have an energetically favorable 1D cluster, (LS), consisting of atoms at the vertices of a vertex-sharing assembly of a long interval L and a short interval S, with length ratio $L=\tau S$ and $\tau=2\cos(\pi/5)$. In a structure (sequence, tiling), this cluster can occur in both orientations, (LS) and (SL). While …LL… and …LS… neighbours in a structure should be energetically favorable, …SS… neighbors should be strongly discouraged. Depending on the overall stoichiometry, different structures result as illustrated with the three examples shown in Fig. 1.

On top of Fig. 1, a part of the infinite sequence …(LS)(LS)(LS)… = …LSLSLS… is depicted for an overall composition LS. The clusters, with the same stoichiometry as the whole sequence, are put together just by sharing vertices. All alternative structures with the same stoichiometry that contain …SS… neighbors are energetically less favorable. The most disadvantageous case would be the sequence …(LS)(SL)(LS)(SL)… = …LSSLLS SL… with 50 % SS neighbors. In the middle of Fig. 1, a sequence is drawn with overall composition L_2S, which differs from the cluster stoichiometry LS. The only way to realize this composition is by allowing cluster overlaps of the type (L(S)L) yielding the sequence …(L(S)L)(L(S)L) … = … LSLLSL… Finally, at the bottom of Fig. 1, the sequence with overall composition $L_\tau S$ is shown. It is part of the Fibonacci sequence, a substitutional sequence resulting from iterative substitution operations acting on an alphabet (L, S): L ⇒ LS, S ⇒ L. For approaching locally the overall composition as close as possible, an (LS) cluster assembly with both vertex and S sharing is needed. However, it would be difficult, if not impossible, to piece together the FS based on this information alone without knowledge of the substitution rule. Fortunately, this problem can be easily solved in 2D space.

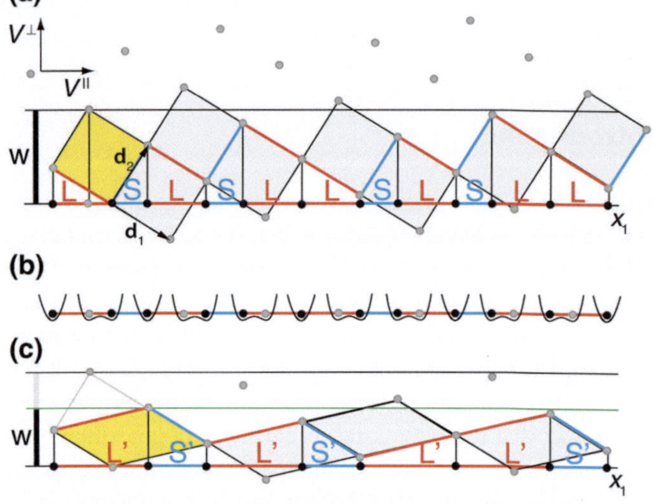

Fig. 2 Fibonacci sequence in the 2D description. (**a**) Projection of a sequence of vertex and edge connected shaded squares gives the FS plus some additional vertices (*grey*) at flip positions in a double-well potential (**b**). The size W of the strip defines the minimum distance between projected lattice points as well as the unit tile sequence; in our case, W is chosen so that the topmost vertex of the *yellow square* is outside the strip. Decreasing the acceptance window by a factor τ^{-1} leads to a scaling of the intervals L and S by a factor τ yielding the sequence L′S′L′S′L′L′S′, with L′=L+S and S′=L (**c**).

For this purpose, we describe the covering cluster (LS) as projection (shadow) of a properly oriented square (shaded in Fig. 2a). Of course, we could use any parallelogram provided it had the same shadow. The topmost and the lowermost vertices of the square project into the interval defined by the left and right vertex and create so-called flip positions. Accepting the interval S as the shortest possible interatomic distance, the two flip positions, with a distance S/τ cannot be occupied at the same time. This is illustrated in Fig. 2b by a double-well potential containing a black dot for an occupied flip position and a grey dot for an unoccupied one. If an atom is excited and jumps to the alternative position in the double well potential, then this corresponds to a so-called simpleton flip (LS) ⇔ (SL).

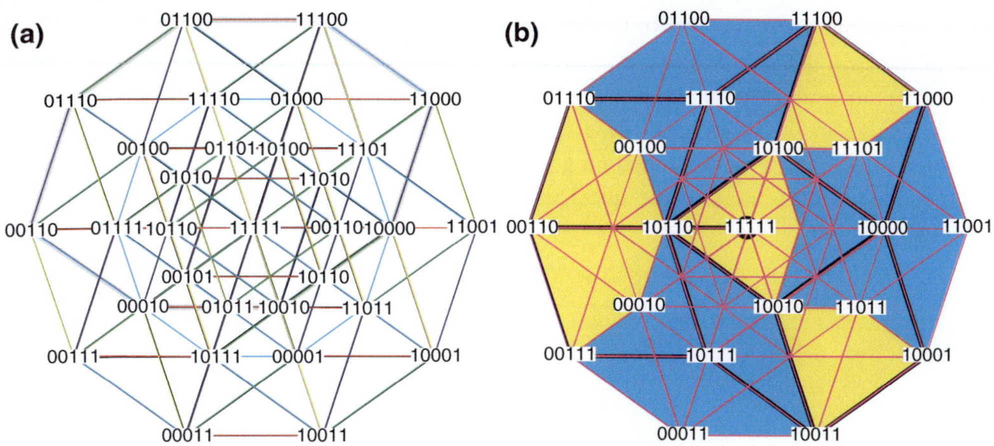

Fig. 3 5D hypercubic unit cell projected along its body diagonal [11111] onto 2D par-space. The vertices are indexed based on the 5-star of basis vectors, with each basis vector in a different color. The thick outlined polygon in (**a**) corresponds to the projection of the 4D face (subcube) with the last index equal to zero. Applying the minimum distance criterion by adjusting the acceptance window, we obtain a vertex distribution related to a decorated rhomb PT, for instance (**b**). The *blue-* and *yellow-colored* regions in (**b**) mark the overlap regions of the Gummelt decagon, a covering cluster of the PT.

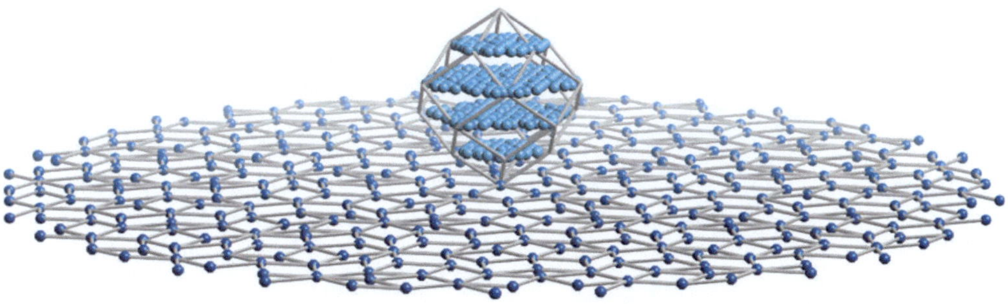

Fig. 4 Visualization of 5D space showing 2D par-space and 3D perp-space projections, i.e. a decorated PT (*dark blue spheres*) and its acceptance window (an *elongated rhombic icosahedron*), connected via the common origin. The nodes of the 5D hyperlattice, with par-space values in the range of the PT shown, form in their perp-space projection the four pentagons inside the acceptance window (*light-blue spheres*).

As illustrated in Fig. 2a, two-cluster sequences of the type (SL) (LS) = SLLS can be obtained by projecting the vertices of compounds of two vertex-sharing squares. The cluster density amounts to $d_{\text{cluster}} = 2/2 = 1$. Overlaps of the type (L(S)L) = LSL result from two squares sharing an S-generating edge. The resulting local cluster density is higher in this case, $d_{\text{cluster}} = 2/3$. The cluster-generating squares can be seen as unit cells of a 2D lattice spanned by the basis vectors $\mathbf{d}_1 = a\,(cos\pi/5, -sin\pi/5)_V$ and $\mathbf{d}_2 = a\,(sin\pi/5, cos2\pi/5)_V$ (D-basis), defined on a Cartesian 1 coordinate system (V-basis); x_1^V is the V-basis coordinate in 2 physical or par(allel)-space $V^{\parallel}$, x_2^V refers to perp(endicular)-space $V^{\perp}$. One obtains the cluster components $L = \mathbf{Pd}_1$; $S = \mathbf{Pd}_2$ with the projection matrix.

$$\mathbf{P} = \begin{pmatrix} 1 & 0 \\ 0 & 0 \end{pmatrix}. \tag{1}$$

The projection of one unit cell onto the perp-space gives the acceptance window W and, therewith, the width of the strip defining the subset of lattice nodes of the tiling. Consequently, its size also determines the minimum distance between the projected lattice nodes. In order to eliminate flip positions, W has to be chosen so that either the topmost (as in our example) or the lowermost vertex of the yellow square comes to lie outside the strip. Then the FS can be obtained as projection of the vertices of an infinite vertex- and edge-connected, respectively, sequence of squares out of a 2D lattice lying within a strip of width W (Fig. 2a).

Generally, the vertices of quasiperiodic tilings correspond to a subset of a Z-module (vector module with integer coefficients) of rank n, depending on the number n of basis vectors needed to index the vertices with integers. Any Z-module can be seen as proper projection of an nD lattice onto the dD par-space. The minimum distance between the tiling vertices defines the width W of the strip that selects the lattice nodes to be projected. The acceptance window W is defined in the (n-d)D perp-space. A change in the strip width W always entails a change in the unit cell shape without changing the unit cell volume. This results in a scaling of the quasiperiodic structure as shown in Fig. 2c.

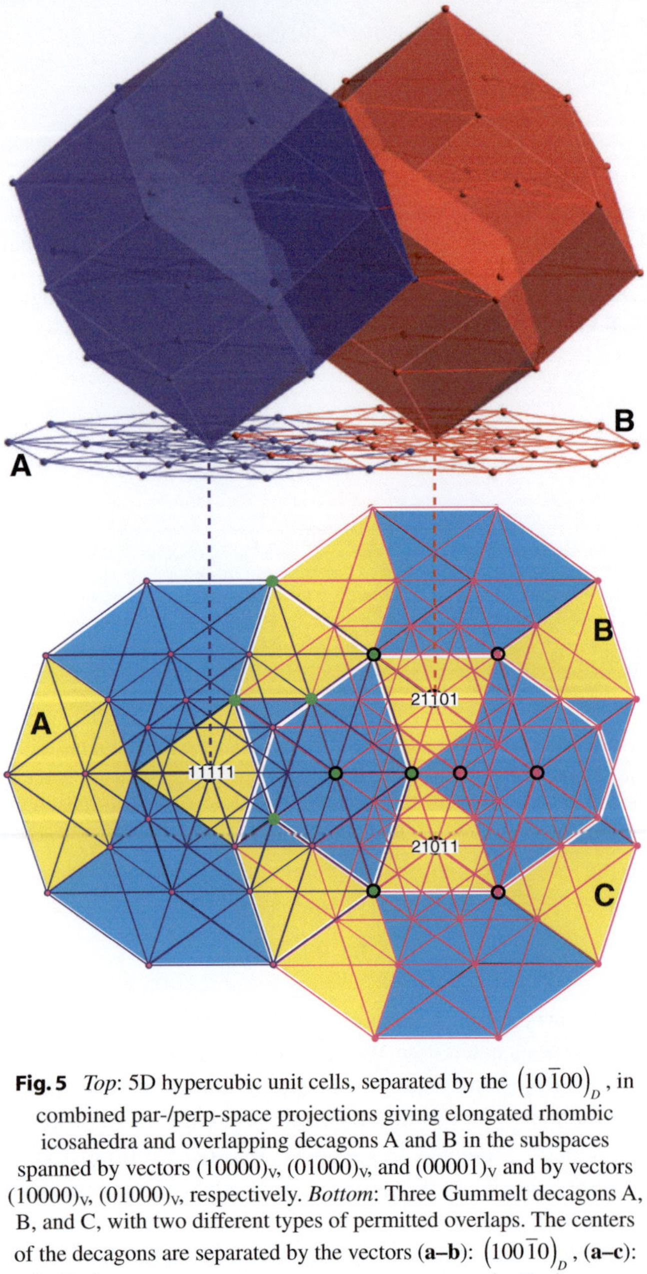

Fig. 5 *Top*: 5D hypercubic unit cells, separated by the $(10\bar{1}00)_D$, in combined par-/perp-space projections giving elongated rhombic icosahedra and overlapping decagons A and B in the subspaces spanned by vectors $(10000)_V$, $(01000)_V$, and $(00001)_V$ and by vectors $(10000)_V$, $(01000)_V$, respectively. *Bottom*: Three Gummelt decagons A, B, and C, with two different types of permitted overlaps. The centers of the decagons are separated by the vectors (**a–b**): $(100\bar{1}0)_D$, (**a–c**): $(10\bar{1}00)_D$ and the τ^{-1} times smaller one for (**b–c**): $(00\bar{1}10)_D$. The corresponding 5D hypercubes share 3D faces containing 8 lattice points each. *Green dots* mark the shared vertices of the hypercubes related to the decagons (**a**) and (**b**), *black circles* those of (**b**) and (**c**).

A strip, meandering with a limited amplitude around the ideal par-space direction, not only leads to local changes in the structure (disorder) but also in the local L/S ratio. However, if the global stoichiometry is kept constant, the disordered structure will remain quasiperiodic on average.

Decagon Cluster and 2D Penrose Tiling

Decagonal QC can be described as packing of partially overlapping decaprismatic columnar clusters. For the study of the different ways to pack these clusters in the quasiperiodic directions, it is sufficient to consider their 2D projections along the tenfold axis, i.e., the periodic direction. The resulting 2D cluster of decagonal shape can be described as projection of a 5D hypercube along one of its fivefold axes. 22 of its 32 vertices project into the interior of the resulting decagon (Fig. 3).

As before, we take this hypercube as unit cell of a 5D hyperlattice with basis vectors $\mathbf{d}_i=(cos2\pi i/5,\ sin2\pi i/5,\ cos4\pi i/5,\ sin4\pi i/5,\ 1/\sqrt{2}\ )_V$, $1\leq i \leq 5$ (D basis); xi^V with $1\leq i\leq 3$ are Cartesian coordinates in parspace, those with $i=4$, 5 in perp-space. The projection of the 5D hypercubic unit cell onto 3D perp-space yields the acceptance window W for the generation of the 2D PT, an elongated rhombic icosahedron (Fig. 4). All hyperlattice points, which fall into this acceptance window if projected onto perp-space, are located inside the strip and generate the PT by par-space projection. The rhombic icosahedron is elongated along its fivefold axis compared to the zonohedron with the same name. Furthermore, its faces with a vertex on the fivefold axis are not congruent to the other ones.

In order to find the most efficient way of packing decagon clusters, we have to analyze the different connectivities of the unit cells in the strip cut out of the 5D hypercubic lattice by the acceptance window W. In 3D, neighboring unit cells can share 2D faces, 1D edges or 0D vertices. In a 5D hyperlattice, they can additionally share 4D or 3D faces with 16 or 8 joint vertices, respectively. The overlap rules of the 2D decagons result from the ways the projected hypercubic unit cells can overlap. In case of the Gummelt decagon [7], the assemblies with allowed overlaps, as defined by the rocket decoration (blue areas in the overlapping decagons shown in Fig. 5), all result from the projection of 5D hypercubic unit cells sharing 3D faces (8 lattice nodes). A higher decagon cluster density would result from the projection of hypercubic unit cells sharing 4D faces (16 lattice nodes), which are shifted against each other by one lattice translation of the type $(10000)_D$. Such a shared area is shown in Fig. 3a. It is not allowed in the ideal PT, but it can be observed as defect in decagonal quasicrystals.

In Fig. 5 (top), a combination is shown of the par- and perp-space projections of two 3D-face-sharing 5D unit cells, separated by the vector $(00\bar{1}10)_D$: the elongated rhombic icosahedra result from the projection onto the subspace spanned by vectors $(10000)_V$, $(01000)_V$, and $(00001)_V$ and the decagons from the projection onto the $(11000)_V$ plane. The vertices that are projected onto equidistant hyperplanes inside the elongated rhombic icosahedra stem from 4D hyperlattice planes with indices $(m_1\, m_2\, m_3\, m_4\, m_5)_D$ with $m_5 = 0, 1, 2, \ldots$

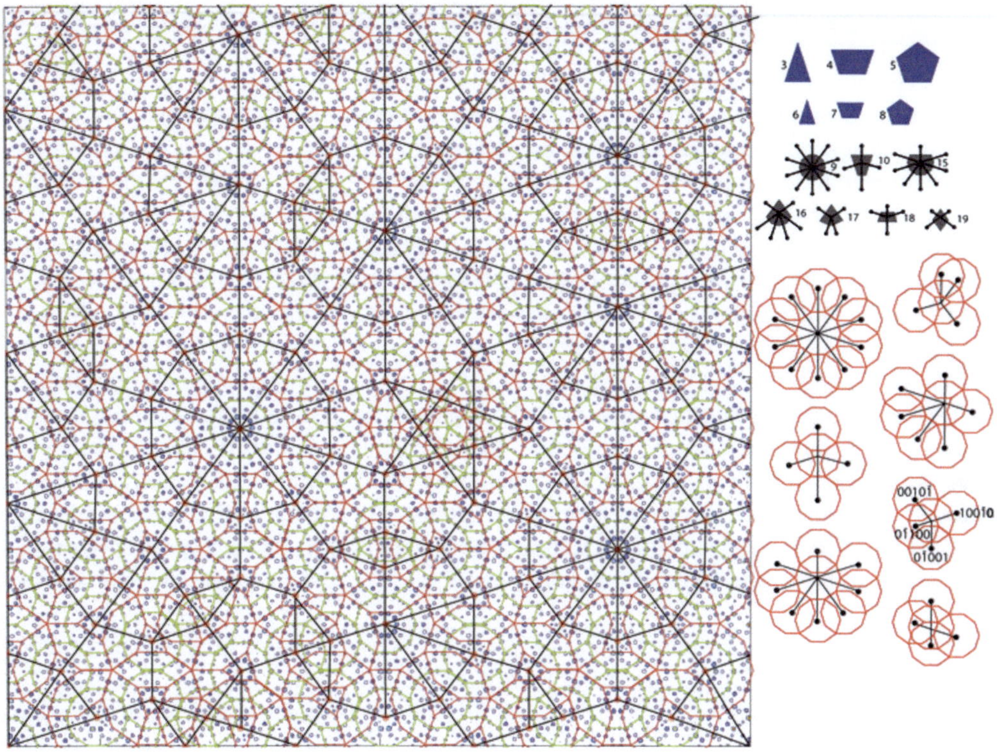

Fig. 6 Projected electron density distribution function of decagonal Al–Ni–Rh (Courtesy of D. Logvinovich, LFK, ETHZ) with ≈ 20 and ≈ 12 Å decagon clusters marked red and green, respectively. The decagon centers are located on the vertices of a DT_5/VT_{13} Masakova tiling [8]. At *upper right*, the Delone and Voronoi tiles are shown (numbering according to [8]), *shaded blue and grey*, respectively, and below, the different vertex configurations and decagon cluster overlaps. For one vertex configuration the lattice translations are given between neighboring 5D unit cells corresponding to the respective overlapping decagons.

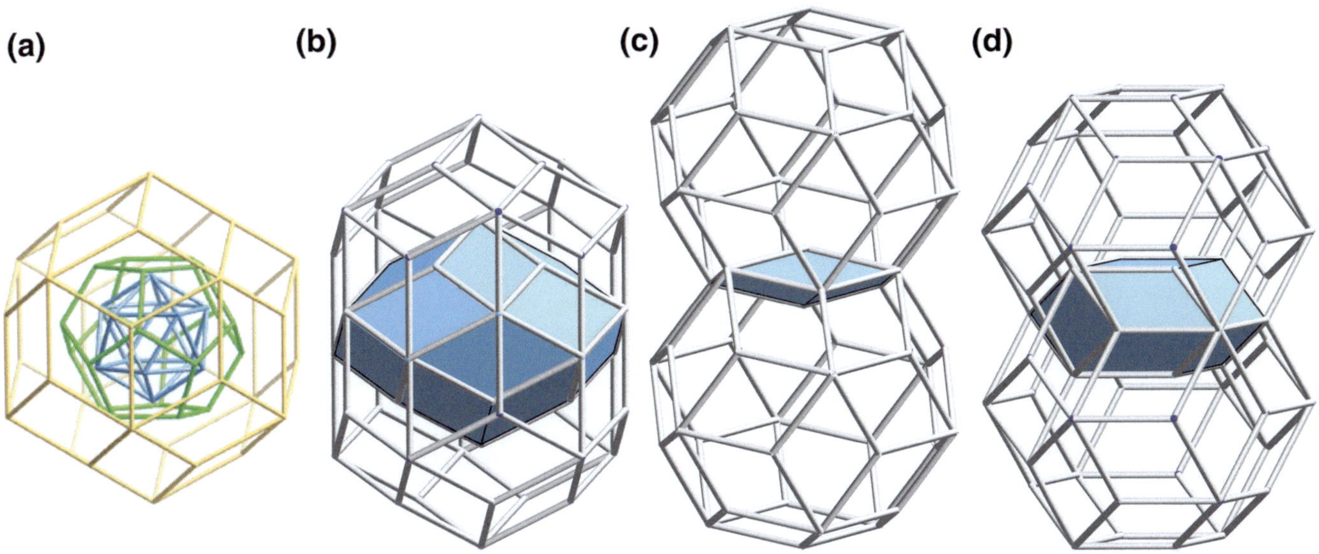

Fig. 7 (**a**) 6D hypercube projected onto 3D par-space. Its 64 vertices project onto the 32 vertices of a triacontahedron (*yellow*), to the 12 vertices of the second shell, a dodecahedron (*green*), and to the 20 vertices of the innermost shell, an icosahedron (*blue*). (**b–d**) Shared regions of triacontahedra overlapping along the fivefold, threefold and twofold axes, respectively, (**b**) rhombic icosahedron, (**c**) oblate rhombohedron, and (**d**) rhombic dodecahedron.

Let us have a look at an example of a real decagonal QC in the system Al–Ni–Rh and see how the cluster concept applies to it. As shown in Fig. 6, the projected electron density distribution function can be covered by copies of a decagonal cluster with ≈ 20 Å diameter (red). The centers of the decagons form the vertices of a DT_5/VT_{13} Masáková tiling [8], created from one decagonal acceptance window in the 5D description. The Rh atoms are located on τ^{-1} times smaller decagons (green) centered at the same positions as the larger ones (red). The shapes of the overlap areas of the large (red) covering decagon clusters are the same as those of the Gummelt decagons, however, their locations on the decagons and combinations are different. The smaller (green) decagons overlap less frequently and only with hexagon overlap areas; mostly they just share edges. The different cases of overlapping decagons can be related to neighboring unit cells of the 5D hypercubic lattice. For one vertex configuration the respective lattice translations are listed in Fig. 6 (bottom right).

The packing of the decagonal clusters can be explained by competing interactions. On one hand, the large decagons have very large overlap regions, on the other hand small decagons just share edges occupied with Rh atoms. The edges of the Voronoi cells can be seen as local reflection lines that relate overlapping decagons by mirror symmetry.

Triacontahedron Cluster and 3D Penrose (Ammann) Tiling

The triacontahedron, an Archimedean solid dual to the icosidodecahedron, can be described as projection of a 6D hypercube along its fivefold axes onto 3D space (Fig. 7a). It is bounded by 12 5D faces, i.e. 5D hypercubes, which give in 3D projection rombic icosahedra. The lower-dimensional hyperfaces project into the respective diminished zonohedra: the 60 4D faces into rhombic dodecahedra and the 160 3D cells into cubes.

The possible overlap regions of two triacontahedra in each case are shown in Fig. 7b–d. The overlap along the fivefold axis results in a shared rhombic icosahedron. It corresponds to the projection of a compound of two 6D hypercubes, separated by a vector of the type $(100000)_D$ and sharing a 5D face with 32 vertices. The overlaps along the three- and twofold directions lead to an oblate rhombohedron and a rhombic dodecahedron, respectively, as shared volumes. The vectors between the 6D hypercubes are of the type $\left(00110\bar{1}\right)_D$ and $\left(0010\bar{1}0\right)_D$, corresponding to shared 3D faces with 8 vertices and 4D faces with 16 vertices, respectively. Consequently, taking the 6D hypercubes as unit cells of a hypercubic lattice and a strip analogously as defined before, then we obtain the densest possible packing of triacontahedral clusters, such as the Bergmann clusters, for instance.

If we remove one zone from the triacontahedron, we obtain a rhombic icosahedron. In projection along the fivefold axis onto 2D, a decagon cluster results as discussed before (see top of Fig. 5). The 6D embedding of decagonal phases allows to reveal structural relationships between decagonal and icosahedral phases in a straightforward way [9].

Conclusions

It is remarkable that the structures of decagonal and icosahedral QC are all closely related to the 2D PT and 3D AT, respectively, although they differ considerably in their chemical composition. However, this is not so surprising if we consider the structural similarities of their fundamental building clusters and, in particular, the way they can pack and overlap for particular stoichiometries. Obviously, a necessary but not sufficient condition for the formation of decagonal and icosahedral QC is the existence of clusters that can be described as proper projections of nD hypercubes, unit cells of nD hyperlattices. The optimum arrangement of the respective clusters in physical space corresponds to nD strips containing vertices of interlinked nD unit cells of the hyperlattice. Since the projection of such a strip gives a quasiperiodic structure, the densest arrangement of these particular clusters results to be quasiperiodic. However, if optimum cluster packing means the best packing of complete clusters then the strip cannot be straight but has to follow the boundaries of the nD unit cells. Of course, the average slope of this zigzag course has to be the same as that of the straight strip. In this case, the true structure could only be derived from electron microscopic images, which reflect the local structure properly. Higher-dimensional structure analysis is already based on the assumption of a straight strip. Therefore, it can give averaged structure information only.

References

1. Steurer W, Deloudi S (2008) Acta Crystallogr A 64:1–11
2. Steurer W, Deloudi S (2009) Crystallography of quasicrystals. Concepts, methods and structures. Springer series in materials science, vol 126. Springer, Heidelberg
3. Steurer W (2006) Philos Mag 86:1105–1113
4. Henley CL, de Boissieu M, Steurer W (2006) Philos Mag 86:1131–1151
5. Coxeter HSM (1973) Regular polytopes. Dover Publications, Inc, New York
6. Berger RF, Lee S, Johnson J, Nebgen B, Sha F, Xu J (2008) Chem Eur J 14:3908–3930
7. Gummelt P (1996) Geom Dedic 62:1–17
8. Masakova Z, Patera J, Zich J (2005) J Phys A 38:1947–1960
9. Mandal RK, Lele S (1991) Philos Mag B 63:513–527

Part III

Structural and Materials Properties

Magnetic Properties of Quasicrystals: Effects of Disorder on Fe Magnetic Moment Formation[a]

R.A. Dunlap[b], R.C. O'Handley[c], M.E. McHenry[d], and V. Srinivas[b]

Abstract

Aluminum-based icosahedral quasicrystals display a wide variety of magnetic behaviors, which result from the presence of chromium, manganese, or iron. The formation of localized Fe magnetic moments in these materials is discussed in the context of the intrinsic disorder present in these structures as well as structural disorder caused by processing conditions.

Introduction

An icosahedral quasi-crystalline phase has been observed in a wide variety of alloy systems. Most common among these are aluminum–transition metal-based materials [1]. Most of these alloys are metastable phases, which require nonequilibrium processing conditions such as rapid solidification [2] or particle beam bombardment [3]. Recently, however, certain aluminum-based [4, 5] alloys have been found to produce stable quasicrystals that can be prepared by equilibrium processing conditions. The magnetic properties of quasicrystals have shown behaviors ranging from diamagnetism [6] to Curie-like [7] and Pauli-like [8] paramagnetism to ferromagnetism [9, 10]. When a magnetic moment forms, it is due to the presence of either chromium [11], manganese [12], or iron [8]. On the basis of a variety of measurements on many different aluminum-based quasicrystals that show local moment formation, it now seems clear that atomic disorder plays an important role in establishing magnetic behavior [13]. Changes in the magnetic properties of a quasi-crystalline alloy in the form of enhanced magnetic moment formation can result from the presence of disorder in several ways. For example, moments may be enhanced by the formation of transition metal–transition metal neighbor pairs [14] or by the formation of physically larger transition metal site "cages" [15]. Here we discuss the importance of disorder in quasicrystals in relationship to the formation of magnetic moments on Fe atoms.

Appearance of Disorder in Quasi-Crystalline Structures

Quasicrystals possess long-range orientational order but no translational order. In this sense they are intermediate between well-ordered crystalline materials and disordered amorphous materials, and experimental evidence is consistent with this picture. In some cases (e.g., specific heat, sound velocity, positron annihilation spectroscopy), a closer relationship with crystalline materials is observed [16–18], while in other cases (e.g., NMR and Mössbauer spectroscopy, neutron scattering) the quasi-crystalline materials appear more amorphous-like [19, 20]. These results are in no way inconsistent, since the importance of disorder on the measurement of a particular physical property is highly sensitive to the relative length scale of the disorder and that probed by the experimental technique.

It is customary to view the quasi-crystal structure as a three-dimensional Penrose tiling (3DPT) [21, 22]. On the basis of atomic arrangements in analogous crystalline

[a] *Structural Chemistry* 1991, 2(5):501–505.

[b] Correspondence should be directed to R.A. Dunlap, Department of Physics, Dalhousie University, Halifax, NS B3H 3J5, Canada

[c] Department of Materials Science and Engineering, Massachusetts Institute of Technology, Cambridge, MA 02139, USA

[d] MST-6, Los Alamos National Laboratory, Los Alamos, NM 87545, USA

I. Hargittai, B. Hargittai (eds.), *Science of Crystal Structures: Highlights in Crystallography*, DOI 10.1007/978-3-319-19827-9_19

compounds, a variety of decorations for the Penrose tiles have been suggested [23, 24]. Recently, however, Janot and co-workers [21, 26] have used neutron diffraction measurements to establish experimentally the atomic positions. These are determined on the basis of a cut and projection to three dimensions (3D) from a six-dimensional cesium chloride-type structure [19]. Since the projected occupancy fraction of the 3D sites is typically less than unity, the structures possess intrinsic atomic disorder. In addition, as well, stoichiometries that deviate from the ideal $Al_{80.9}$ $TM_{19.1}$ (TM = transition metal) yield an additional source of disorder. Thus a consideration of the effect of icosahedral order on the behavior of any physical property of quasicrystals must take into account the presence of the varying degrees of disorder. The formation of magnetic moments on Cr, Mn, or Fe atoms is a property that is sensitive to the degree of disorder in those alloys, and here a systematic study of Fe-moment formation in aluminum-based quasicrystals is presented.

Fe Magnetic Moment Formation in Quasicrystals

Magnetic moments on Fe atoms are not observed in icosahedral Al-Fe alloys and icosahedral alloys of Al-Fe with other early 3D transition metals [15, 27, 28]. On the other hand, icosahedral phase formation has been reported in rapidly quenched Al-Mo-Fe [29], Al-Ta-Fe [30], and Al-Cu-Fe [4], and subsequent magnetic measurements [8, 31] have shown the presence of a localized Fe moment. Al-Cu-Fe can also be prepared as a single-phase quasicrystal by annealing [4], and this allows for the comparison of the properties of alloys of the same composition prepared by different methods. For the magnetic susceptibility of rapidly quenched $Al_{80}Mo_9Fe_{11}$, $Al_{70}Ta_{10}Fe_{20}$, and $Al_{65}Cu_{20}Fe_{15}$, see Figs. 1, 2, and 3, respectively. Samples of Al-Mo-Fe and Al-Ta-Fe were prepared for this work by rapid solidification as described previously [2, 7, 8]. Details of the Al-Cu-Fe sample preparation are given elsewhere [31]. Localized Fe magnetic moments extracted from these measurements on the basis of the low temperature Curie-like behavior are given in Table 1. Fu-kamichi et al. [31] have also found that the local Fe moment in annealed $Al_{65}Cu_{20}Fe_{15}$ was near zero. Consistent results were reported by Stadnik et al. [32].

Figure 4 shows that there is a well-defined linear relationship between Fe content and the size of the local Fe moment in the rapidly quenched alloys. Although these data show a highly linear behavior, they can also be fit quite well to a quadratic dependence which is constrained to pass through the origin as has been suggested for Al-Mn quasicrystals by Hauser et al. [14]. This suggests that the magnitude of the Fe moment is related to the Fe–Fe pairing probability. On the other hand, magnetization measurements of rapidly quenched and annealed Al-Cu-Fe as given in Table 1 show that differences in atomic order (as well as differences in chemical composition) play an important role in establishing the magnetic properties of these alloys.

It has been proposed by Srinivas et al. [33] that ^{57}Fe Mössbauer spectroscopy is an effective method of establishing the degree of local disorder around Fe sites in quasicrystalline alloys. These investigators have used an extension of the shell model of Czjzek [34] as proposed by Eibschutz et al. [20], of the form

$$P(\Delta) \propto \frac{\Delta^{n-1}}{\sigma^n} \exp\left(-\frac{\Delta^2}{2\sigma^2}\right)$$

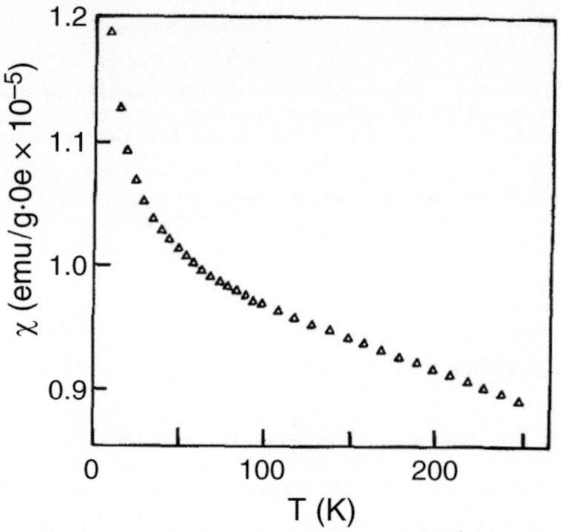

Fig. 1 Magnetic susceptibility of icosahedral $Al_{80}Mo_9Fe_{11}$ measured in an applied field of 10 kOe (Data from Ref. [8]).

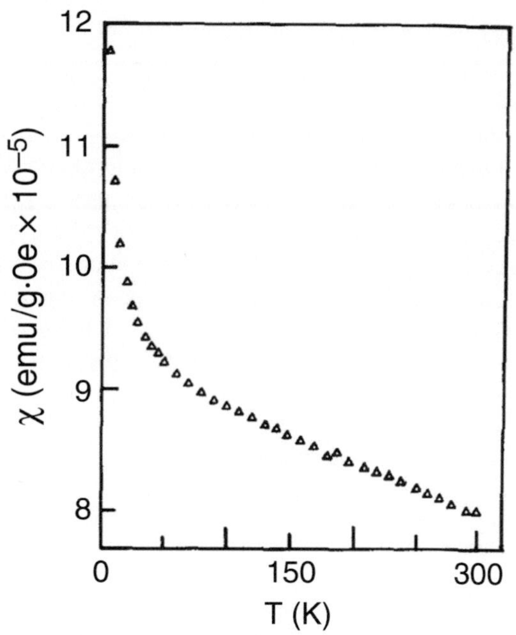

Fig. 2 Magnetic susceptibility of icosahedral $Al_{70}Ta_{10}Fe_{20}$ measured in an applied field of 10 kOe (Data from Ref. [8]).

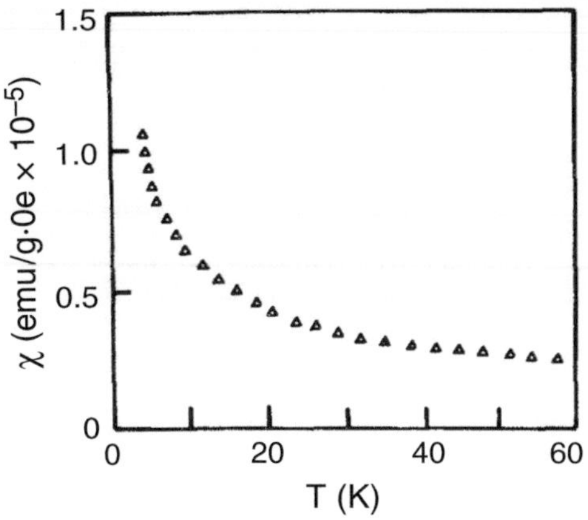

Fig. 3 Magnetic susceptibility of icosahedral $Al_{65}Cu_{30}Fe_{15}$ measured in an applied field of 20 kOe (Data from Ref. [30]).

Table 1 Properties of icosahedral iron-containing alloys

Alloy	Preparation	$\mu(\mu_B/Fe)$	σ (mm/s)	Refs.
$Al_{80}Mo_9Fe_{11}$	rq	0.19	0.351	[8], this work
$Al_{70}Ta_{10}Fe_{20}$	rq	0.48	0.363	[8], this work
$Al_{65}Cu_{20}Fe_{15}$	rq	0.32	–	[4, 30]
	rq	–	0.267	[6, 31]
	Annealed	~0	–	[4, 30]
	Annealed	~0	0.256	[6, 31]

rq rapidly quenched

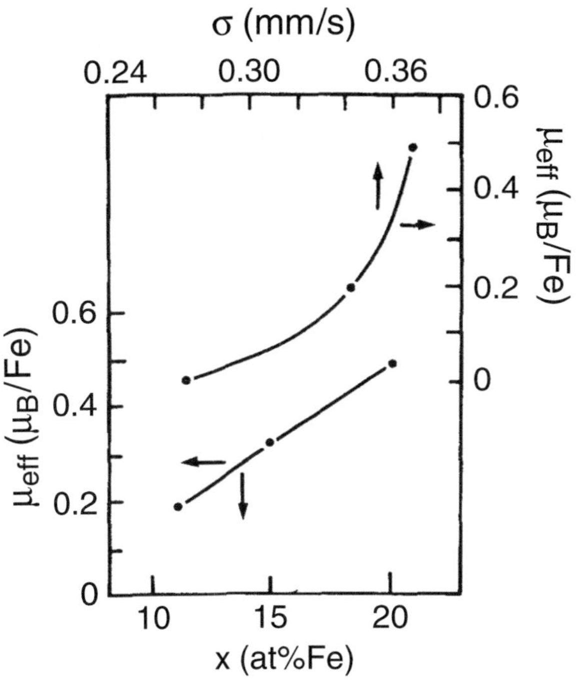

Fig. 4 Measured local Fe moment in Al-TM-Fe quasicrystals (TM = Mo, Ta, Cu) as a function of x (Fe content) for rapidly quenched alloys and as a function of the Mössbauer effect parameter σ (see text) (Data from sources given in Table 1).

where n and σ are fitted parameters. Srinivas et al. [33] have suggested that the degree of disorder in quasicrystals can be inferred from the magnitude of the parameter σ. Measured room temperature ^{57}Fe Mössbauer effect spectra of Al-Mo-Fe and Al-Ta-Fe are shown in Fig. 5. Results of shell model fits to these data are given in Table 1. Mössbauer spectra for rapidly quenched and annealed $Al_{65}Cu_{20}Fe_{15}$ from Ref. [35] are illustrated in Fig. 6.

Subsequent measurements by Stadnik and Stroink [32] have shown results consistent with those of Kataoka et al. [35] for the as-quenched sample. On the other hand, Stadnik and Stroink have found very little distinction between the spectra of rapidly quenched Al-Cu-Fe and that of annealed Al-Cu-Fe. This suggests that the quenched and annealed samples studied by Stadnik and Stroink [32] had very similar microstructures while the two samples studied by Kataoka et al. [35] were quite different from each other. On the basis of electron diffraction studies, He et al. [36] have suggested that annealed $Al_{65}Cu_{20}Fe_{15}$ is, in fact, a decagonal quasicrystal. The structure of the ^{57}Fe Mossbauer effect spectrum for annealed Al-Cu-Fe shown in Fig. 6 is more similar to that reported for decagonal Al-Fe by Gupta et al. [37] than it is to that reported in typical icosahedral quasicrystals. It is, therefore, likely that the details of the microstructure of the Al-Cu-Fe alloys can be variable and are sensitive to the processing conditions. Since the differentiation between icosahedral and decagonal phases on the basis of X-ray and electron diffraction patterns is not always straightforward, caution is necessary in identifying phases in these materials. For samples on which both magnetic moment measurements and a shell model analysis of an Fe

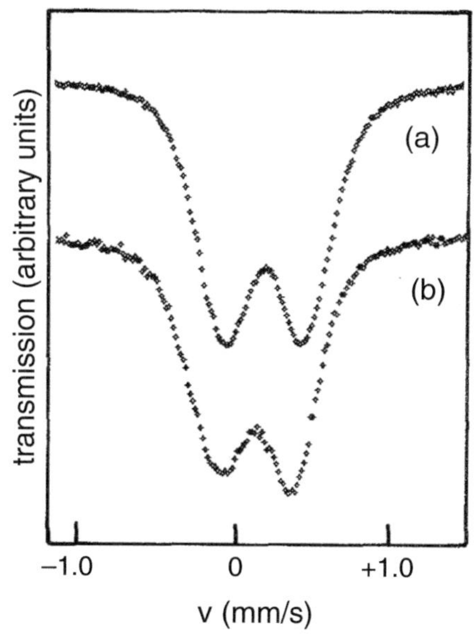

Fig. 5 Room temperature ^{57}Fe Mössbauer effect spectra of icosahedral $Al_{80}Mo_9Fe_{11}$ (*a*) and $Al_{70}Ta_{10}Fe_{20}$ (*b*).

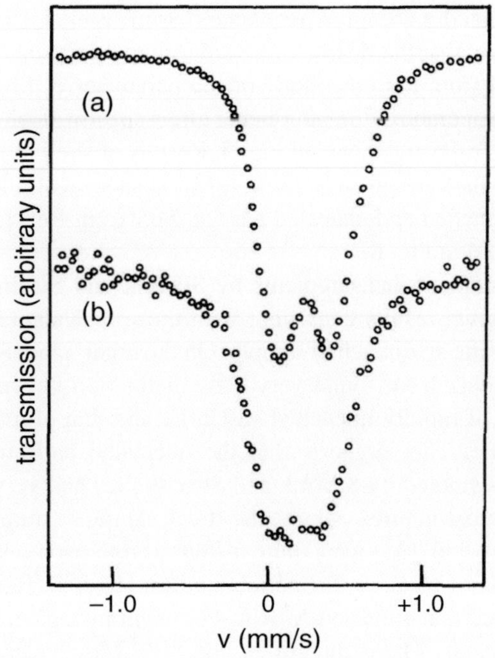

Fig. 6 Room temperature ^{57}Fe Mössbauer effect spectra of icosahedral $Al_{65}Cu_{20}Fe_{15}$. (*a*) Rapidly quenched sample and (*b*) annealed sample (Data from Ref. [35]).

Mössbauer spectrum have been made, values of $\mu_{\rm eff}$ per Fe atom are plotted in Fig. 4 as a function of the parameter σ. This illustrates the correlation between $\mu_{\rm eff}$ and σ. The degree of disorder in any quantitative sense is difficult to determine on the basis of σ However, clear evidence for a relationship between these quantities has been demonstrated [33]. In fact, as Fig. 4 illustrates, $\mu_{\rm eff}$ is a much more sensitive function of σ than it is of *x*. This is not unexpected, since σ is presumably a better indication than *x* of variations in the local Fe environment. A qualitative evaluation of the features shown in spectrum (b) of Fig. 6 is certainly consistent with a high value of σ; moreover, the associated disorder that is implied is manifested by the considerable increase in $\mu_{\rm eff}$ over the annealed sample.

In conclusion, in Al-TM-Fe quasicrystals we observe a correlation between Fe magnetic moment function and both the quantity of Fe present in the alloy and the degree of atomic disorder as evidenced by Mössbauer measurements. The present investigation also demonstrates the wide variations in microstructure (as evidenced by Mössbauer effect moments) that can occur in samples prepared under nominally similar processing conditions and the extent to which this can influence the resulting magnetic properties.

Acknowledgments This work was supported in part by grants from the Natural Sciences and Engineering Research Council of Canada.

References

1. Bancel, P. A.; Heiney, P. A. *Phys. Rev. B*, **1986**, *33*, 7917–7922.
2. Dunlap, R.A.; Dini, K.*J.Phys. F:Metal Phys.*, **1986**, *16*, 11–16.
3. Budai, J.D.; Aziz, M. J. *Phys. Rev.*, **1986**, *B*, 2876–2878.
4. Tsai, A. P.; Inoue, A.; Masumoto, T. *Jpn. J. Appl. Phys.*, **1987**, *26*, LI505–LI507.
5. Tsai, A. P.; Inoue, A.; Masumoto, T. *Jpn. J. Appl. Phys.*, **1988**, *27*, LI587–L1590.
6. Staknik, Z. M.; Stroink, G.; Ma, H.; Williams, G. *Phys. Rev. B*, **1989**, *39*, 9797–9804.
7. McHenry, M. E.; Srinivas, V.; Bahadur, D.; O'Handley, R. C.; Lloyd, D. J.; Dunlap, R. A. *Phys. Rev. B*, **1989**, *39*, 3611–3615.
8. Srinivas, V.; McHenry, M. E.; Dunlap, R. A. *Phys. Rev. B*, **1990**, *40*, 9590–9594.
9. Dunlap, R. A.; McHenry, M. E.; Srinivas, V.; Bahadur, D.; O'Handley, R. C. *Phys. Rev. B*, **1989**, *39*, 4808–4811.
10. Tsai, A. P.; Inoue, A.; Masumoto, T.; Kataoka, N. *Jpn. J. Appl. Phys.*, **1988**, *27*, L2252–L2255.
11. McHenry, M. E.; Dunlap, R. A.; Srinivas, V.; Bahadur, D.; O'Handley, R. C. *Phys. Rev. B*, **1990**, *41*, 6933–6938.
12. McHenry, M. E.; Dunlap, R. A.; Chatterjee, R.; Chow, A.; O'Handley, R. C. *J. Appl. Phys.*, **1988**, *63*, 4255–4257.
13. O'Handley, R. C.; Dunlap, R. A.; McHenry, M. E. *Philos, Mag. B*, **1990**, *61*, 677–690.
14. Hauser, J. J.; Chen, H. S.; Waszczak, J. V. *Phys. Rev. B*, **1986**, *33*, 3577–3580.
15. Eibschutz, M.; Lines, M. E.; Chen, H. S.; Waszczak, J. V.; Papaefthymiou, G.; Frankel, R. B. *Phys. Rev. Lett.*, **1987**, *57*, 2443–2446.
16. Graebner, J. E.; Chen, H. S. *Phys. Rev. Lett.*, **1987**, *58*, 1945–1948.
17. Birge, N. O.; Golding, B.; Haemmerle, W. H.; Chen, H. S.; Parsey, J. M., Jr. *Phys. Rev. B*, **1987**, *36*, 7685–7688.
18. Dunlap, R. A.; Lawther, D. W.; March, R. H. *J. Phys. F: Metal Phys.*, **1987**, *17*, L39–L43.
19. Warren, W. W.; Chen, H. S.; Espinosa, G. P. *Phys. Rev. B*, **1986**, *34*, 4902–4905.
20. Eibschutz, M.; Chen, H. S.; Hauser, J. J. *Phys. Rev. Lett*, **1986**, *56*, 169–172.
21. Janot, C.; DeBoissieu, M.; Dubois, J. M.; Pannetier, J. *J. Phys: Condens. Matter*, **1989**, *1*, 1029–1048.
22. Penrose, R. *J. Inst. Math. Appl.*, **1974**, *10*, 266–271.
23. Henley, C. L.; Elser, V. *Philos. Mag. B*, **1986**, *53*, L59–L66.
24. Elser, V.; Henley, C. L. *Phys. Rev. Lett.*, **1985**, *55*, 2883–2886.
25. Dunlap, R. A.; O'Handley, R. C.; McHenry, M. E.; Chatterjee, R. *Phys. Rev. B*, **1988**, *37*, 8484–8487.
26. Janot, C.; Dubois, J. M. *J. Non-Crystalline Solids*, **1988**, *106*, 193–206.
27. Dunlap, R. A.; Lloyd, D. J.; Christie, I. A.; Stroink, G.; Stadnik, Z. M. *J. Phys. F: Metal Phys.*, **1988**, *18*, 1329–1341.
28. McHenry, M. E. Electronic structure and magnetism in metallic alloys exhibiting local icosahedral order. Ph. D. thesis, Massachusetts Institute of Technology, Cambridge, 1988.
29. Hui, M.; Chen, X. *Solid State Commun.*, **1988**, *68*, 813–816.

30. Tsai, A. P.; Inoue, A.; Masumoto, T. *Jpn. J. Appl. Phys.*, **1988**, *27*, L5–L8.
31. Fukamichi, H.; Goto, T.; Komatsu, H.; Wakabayashi, H.; Tsai, A.; Inoue, A.; Masumoto, T. *J. Phys. Colloq.*, **1988**, *49*, CS–239–CS–240.
32. Stadnik, Z. M.; Stroink, G. *Phys. Rev. B*,**1988**, *38*, 10447–10453.
33. Srinivas, V.; Dunlap, R. A.; Bahadur, D.; Dunlap, E. *Philos. Mag. B*, **1990**, *61*, 177–188.
34. Czjcek, G. *Phys. Rev. B*, **1982**, *25*, 4908–4910.
35. Kataoka, N.; Tsai, A. P.; Inoue, A.; Masumoto, T.; Nakamura, Y. *Jpn. J. Appl. Phys.*, **1988**, *27*, L1125–L1127.
36. He, L. X., Wu, Y. K.; Kuo, K. H. *J. Mater. Sci. Lett.*, **1988**, *7*, 1284–1286.
37. Gupta, A.; Jayaraj, M. E.; Rama Reddy, K. *Phys. Rev. B*, **1989**, *39*, 8757–8759.

Overlooked Opportunities in Stereochemistry. Part III: Alfred Werner's Awareness of Spontaneous Resolutions and of the Meaning of Hemihedral Faces in Optically Active Crystals[a]

Ivan Bernal[b,d] and George B. Kauffman[c]

Keywords
Hemihedrism • Spontaneous resolution • Optically active crystal

In previous articles in this series [1, 2] we traced the work of Alfred Werner (1866–1919) in an effort to answer the following questions: Why did Werner, who possessed a reputation for working with such rapidity, require such a long time (14 years) to attain his goal of resolving an optically active coordination compound in order to confirm his postulated octahedral configuration for cobalt(III) and other transition-metal complexes? Also, why did he not make use of conglomerate crystallization techniques, which had been known since 1848, when Louis Pasteur (1822–1895), in one of the truly classic chemical experiments of all time, demonstrated that crystallization of a racemic solution of sodium ammonium tartrate tetrahydrate, $NaNH_4\,(tart)\cdot 4H_2O\ \left(tart = C_4H_4O_6^{2-}\right)$, produced a mechanical mixture of crystals, each of which was made up exclusively of the (*R, R*) or the (*S, S*) antipode (conglomerate crystallization) [3, 4]?

An examination of Werner's articles on optical activity and resolutions reveals that in 1914 he and one of his students, Jacob Bosshart, carried out an examination [5, 6] of the crystallization behavior of bis(1,2-ethanediamine) oxalatocobalt(III) bromide, $[Co(Ox)(en)_2]\,Br\ \left(Ox = C_4O_4^{2-}\right)$, upon seeding a saturated solution of its racemate with crystals of the pure (+)-$[Co(Ox)(en)_2]Br$.

Remark. The article, received by the journal on May 22, 1914 (5 months after Werner had received the 1913 Nobel Prize for chemistry; see Ref. [5]), was based on the experimental work of Jacob Bosshart, as reported in his dissertation [6]. The few facts available about him are as follows: Bosshart, born July 20, 1890, in Stürzikon bei Embrach (Kanton Zürich), studied chemistry at the Eidgenössische Technische Hochschule (ETH), Zürich (fall 1909–spring 1910), and the Universität Zürich (fall 1910–spring 1914). He spent the summer semester of 1912 in military service. During vacation periods he was employed at a federal agricultural–experimental station and a patent bureau. From the beginning of 1913 to the end of February 1914 he carried out his doctoral research under Werner's direction.

Werner reported having observed a large difference in solubility between the racemate and the optically active salt, and therefore they attempted to isolate the less soluble optically active one by precipitation with ethanol from an aqueous solution of this salt enriched with crystals of the (+)-antipode. The experiment succeeded, and examination of the rotatory power of the mother liquor showed that it was enriched with the (–)-antipode. Concentration of the solution, followed by addition of ethanol, yielded a crop of crystals of the (–)-antipode, along with a mother liquor now enriched with the (+)-antipode, and so on. Werner then applied this technique to salts of compounds whose enantiomers are less soluble than their racemates. For example, he reported that the method was successful with bis(1,2-ethanediamine) dinitrocobalt(III)chloride, $[Co(NO_2)_2(en)_2]Cl$, and with bis(1,2-ethanediamine)oxalatochromium(III) bromide, $[Cr(Ox)(en)_2]Br$.

[a] *Structural Chemistry* 1993, 4(2):131–138.
[b] Department of Chemistry, University of Houston, Houston 77204-5641, TX, USA
[c] Department of Chemistry, California State University, Fresno, CA, USA
[d] Correspondence should be directed to Ivan Bernal, University of Houston, Houston, Texas 77204-5641
e-mail: bernalibg@gmail.com

I. Hargittai, B. Hargittai (eds.), *Science of Crystal Structures: Highlights in Crystallography*,
DOI 10.1007/978-3-319-19827-9_20

Werner's report [5] shows that he very clearly understood the nature of the process involved, and in all fairness to him it should be emphasized that this approach requires a pure optically active seeding agent, which, of course, he did not possess until after his first successful resolution in 1911 [7, 8]. Intellectually, however, he possessed the open mind and mental acuity to grasp clearly the concepts involved in the procedure, and he applied the method within 3 years of his first resolution.

Werner's work was largely forgotten until a score of years later when Duschinsky applied the method to the resolution of histidine monohydrochloride, $C_6H_9N_3O_2 \cdot HCl$ [9].

The phenomenon of conglomerate crystallization has long been thought to be extremely rare, but it is not. (Ivan Bernal has found several thousand crystalline conglomerates which are currently being incorporated as part of the Cambridge Crystallographic Data Centre File.) Hence, it would not be surprising if Werner did not consider the possibility of its occurrence with any of his coordination compounds. This may not have been the case for some years, but in 1914, in his second article on optically active rhodium compounds, Werner reported an investigation of tris(oxalato)rhodates(III), $M_3[Rh(Ox)_3]$ [10]. (This article, based on work carried out experimentally by J. Poupardin, was received by the journal on the same day (May 22, 1914) and therefore probably submitted at the same time as the previously cited article on $[Co(Ox)(en)_2]X$ (X = Cl or Br) [5]. We have been unable to obtain biographical information on Poupardin, who is not listed among Werner's *Doktoranden*.) Werner synthesized the racemic potassium salt, $K_3[Rh(Ox)_3] \cdot 4.5H_2O$, according to Leidié's [11] procedure and resolved the anion, using strychninium nitrate as the resolving agent. He obtained the optically active potassium, sodium, and rubidium salts by metathesis.

Werner describes the spontaneous resolution (*spontane Spaltung*) of the potassium salt, which crystallizes as a conglomerate [[10], pp 1959–1960]. The racemic salt was dissolved in boiling water. The concentrated solution was quickly cooled to 90 °C, and then slowly to room temperature, whereupon small, triclinic crystals exhibiting left- and right-handed hemihedral faces were deposited. These were separated, à la Pasteur, under the microscope and their morphology studied under the optical goniometer. Beautifully drawn pictures are given [[10], p 1960] (reproduced in our Fig. 1), the only time that Werner depicted crystals of optical isomers; all his other measurements were made in solution rather than in the solid state. Solutions obtained from these individual crystals yielded optical rotation values corresponding to whether they were left-handed or right-handed.

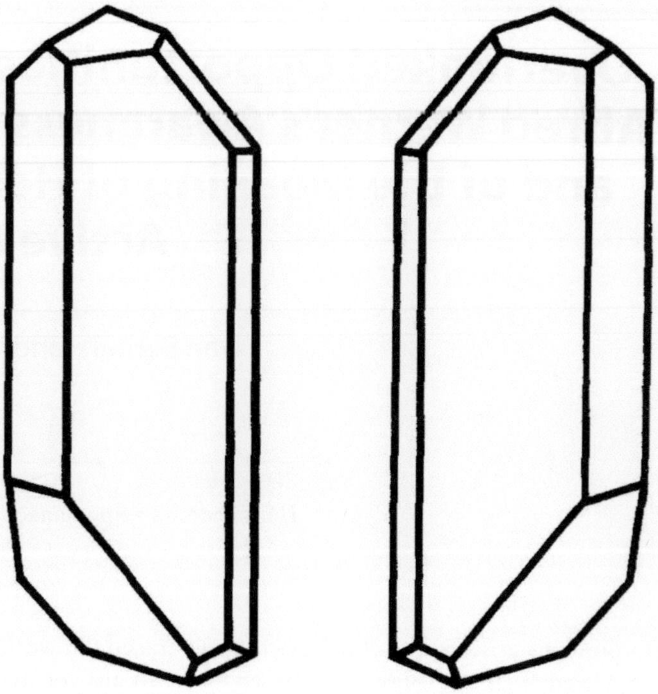

Fig. 1 Hemihedral crystals of $K_3[Rh(C_2O_4)_3]$.

Werner's article does not mention the work of Daniel Delafosse (1796–1878), William Jackson Pope (1870–1939), or any of the other pioneers of optical measurements of hemihedrism and their consequences for optical activity; the only work cited is that of Leidié [11]. Nevertheless, even though Werner's discovery of resolution by conglomerate crystallization was a belated one, occurring after he had resolved coordination compounds by diastereoisomer formation with resolving agents and after he had received the Nobel Prize, we have at least answered the question raised in the previous articles [1, 2] as to whether he was able to accept, understand, and use the concepts involved in this method. The answer is clearly a resounding yes. The tantalizing question of why he did not use this method earlier, however, still remains. A more critical evaluation of these two papers of Werner's (1914) [5, 10] throws light on other facts highly pertinent to this second question.

For reasons that will become apparent later, it is useful to remind the reader that the mode of crystallization of tartrate salts is dependent on (a) the choice of the charge-compensating cations and (b) the temperature at which the salt is crystallized. For example, it was known to Pasteur [4] that none of many salts studied by Frédéric Hervé de la Provostaye (1812–1863) [12] crystallized as a conglomerate. The only one that crystallized as such, $NaNH_4(tart) \cdot 4H_2O$, was reported in 1844 by Eilhard Mitscherlich (1794–1863) in a letter to Jean Baptiste Biot (1774–1862), which the latter published in a memoir to the French Académie des Sciences [13].

Sometime later, it was shown that there is a pronounced thermal dependence of the crystallization pathway followed by $NaNH_4$ (tart) and KNa(tart) salts, as follows [[14], pp 204–205]:

$$\underset{\text{conglomerate}}{NaNH_4(\text{tart})\cdot 4H_2O} \leftarrow 28°C \rightarrow \underset{\text{racemate}}{NaNH_4(\text{tart})\cdot H_2O + 3H_2O}$$

$$\underset{\text{racemate}}{2NaNH_4(\text{tart})\cdot H_2O} \leftarrow 35°C \rightarrow \underset{\text{racemic compounds}}{Na_2(\text{tart})\cdot H_2O + (NH_4)_2(\text{tart})\cdot H_2O + 2H_2O}$$

and

$$\underset{\text{conglomerate}}{KNa(\text{tart})\cdot 4H_2O} \leftarrow -6°C \rightarrow \underset{\text{racemate}}{KNa(\text{tart})\cdot 3H_2O + 3H_2O}$$

$$\underset{\text{racemate}}{2KNa(\text{tart})\cdot 3H_2O} \leftarrow 41°C \rightarrow \underset{\text{racemic compounds}}{Na_2(\text{tart}) + K_2(\text{tart})\cdot 2H_2O + 4H_2O}$$

Moreover, it was clearly known that the thermodynamic condition for a solution of the racemate to crystallize as a conglomerate is that the solubility of the enantiomer must be less than that of the racemate by a prescribed amount [[14], pp 185–186]. Inasmuch as Werner measured and reported the solubilities of the racemate and the enantiomer of $[Co(Ox)(en)_2]Br$, he must have been aware of the reason why such data were important to measure. Thus, by the time that Werner entered this arena, all such facts were clearly established. Moreover, in the thesis of Werner's student [[6], pp 23–26] there is a summary of Emil Erlenmeyer's (1825–1909) work on the conglomerate crystallization of asparagine in which Bosshart cites the solubility difference between crystals of the racemate and enantiomers of asparagine. Thus, it is natural to assume that Bosshart and Werner were aware of the significance of this fact.

Now, let us first analyze in more detail the results of seeding of metal–ammine oxalates and of the conglomerate crystallization of potassium trisoxalatorhodate(III).

In the published article [5], based on Bosshart's thesis [6], Werner dubbed this technique "a new method of resolution for racemic inorganic compounds" (*eine neue Spaltungsmethode für racemische anorganische Verbindungen*), and he cited no earlier work except for several publications of his own. It was probably for this reason that Jacques, Collet, and Wilen [14] refer to Werner's work as a "rediscovery" of the work of Gernez [15], who modified Pasteur's resolution by introducing the technique of selective crystallization induced by seeding. However, in his dissertation [6] Bosshart cites both Gernez's work [[15], p 18] as well as Pasteur's earlier work [3] and Jungfleisch's later confirmation [18] of Gernez's work [[15], p 18]. Bosshart calls the method "resolution by precipitating" (*Spaltung durch Ausfällen*) [[6], p 42].

Remark. Désiré Jean Baptiste Gernez (1834–19I0) [16, 17] was a student, assistant, and intimate friend and collaborator of Pasteur's and helped him with his work on the diseases of wine and silkworms. He worked on the properties of substances in unstable equilibrium, which he called *hors d'equilibre.* These studies involved the crystallization of supersaturated solutions of solid substances, the solidification of supercooled substances, the supersaturation of gaseous solutions, the retardation of boiling, and allotropic or polymeric transformations. He also investigated black phosphorus and the transformation of allotropes of sulfur into one another. An officer of the Legion of Honor for 20 years, he was a lecturer at the École Normale in Paris, a post that he held simultaneously with professorships at other educational institutions. He retired in 1904 and in 1906 was elected to the seat in the Physics Section of the Académie des Sciences left vacant by the death of Pierre Curie (see *Compt. Rend. Acad. Sci.* **1906,** *142*, 1492).

The method that we have been discussing was later called "resolution by entrainment" (*dédoublement par entraînment*) by Amiard [19], and it is also sometimes known as resolution by preferential crystallization [20]. It was first mentioned in a five-sentence excerpt from a letter written in 1866, the year of Werner's birth, by Gernez to Pasteur [15]:

Separation of Levorotatory Tartrates and Dextrorotatory Tartrates by Means of Supersaturated Solutions: by M. Gernez (Extracted from a letter addressed to M. Pasteur).

> I have discovered that a supersaturated solution of levorotatory sodium ammonium double tartrate does not crystallize on contact with a fragment of the same dextrorotatory hemihedral salt; and, *vice-versa,* the supersaturated solution of the dextrorotatory salt does not deposit crystals when seeded with the levorotatory salt.
>
> This fact has led me to study the inactive solution of sodium ammonium double racemate. With the racemic acid that you have kindly placed at my disposal, I have made a supersaturated solution of this salt. Seeded by a particle of dextrorotatory salt, it has deposited only dextrorotatory crystals; a portion of the same liquid, in contact with a levorotatory crystal, has produced a deposit of levorotatory salt. Thus there is a simple means of separating at will from the sodium ammonium double racemate one or another of the two salts which constitute it.

In 1882 Jungfleisch confirmed [18] Gernez's earlier work and pointed out the disadvantages of this method of resolution: only a small yield is obtained if the solution is not strongly supersaturated, but it is difficult to prevent the crystallization of the nonseeded salt from such a strongly supersaturated solution.

Remark. Émile-Clément Jungfleisch (1839–1916) [21–23], a Parisian of Alsatian origin (Werner was born in Mulhouse in Alsace), received his *Docteur ès Sciences* degree in 1868 with a dissertation titled *Recherches sur les dérivés chlorés de la benzine.* He collaborated with Marcellin Berthelot on the law of partition (1872); they showed that the relative quantities of solutes divided into equal volumes of two immiscible solvents are constant, and they defined a "partition coefficient." He also worked with Paul Émile Lecoq de Boisbaudran on the isolation of gallium and indium. He succeeded Berthelot as professor of organic chemistry at the École Supérieure de Pharmacie (1877) and the College de France (1908), both in Paris. He demonstrated that the different forms of tartaric acid are transformed into each other when heated in aqueous solution and that an equilibrium exists between them. He converted ethylene into succinic acid, showed that camphoric acid exists in four isomeric forms, and resolved inactive malic and lactic acids into their enantiomers. Most of his 99 articles involve organic or pharmaceutical chemistry.

Comments on Werner's Seeding Experiments with Bosshart [5, 6]

Piutti prepared *d*-asparagine using Pasteur's procedures. Using this *d*-asparagine and natural *l*-asparagine, he prepared a racemic solution from which he separated the components by fractional crystallization of a seeded solution, the first crop of which was said to be the pure *l*-enantiomer [24]. An interesting variant of the seeding method was discovered by Ostromisslensky [25], who seeded a saturated solution of racemic asparagine with a crystal of nonenantiomorphic glycine (space group $P2_1/c$) [26].

Regarding Ostromisslensky's experiments: (1) Glycine crystallizes in both racemic and optically active crystals. (2) The common racemic (α-form) is obtained by crystallization from water solutions allowed to evaporate at room temperature (α-form, space group $P2_1/c$, was characterized structurally by Albrecht and Corey) [26]. (3) Two optically active forms are known: the β-form (space group $P2_1$, obtained by crystallization from saturated aqueous solution + ethyl alcohol [27]) and the γ-form (space group $P3_1$, obtained by crystallization from solutions which have been acidified with acetic acid or made basic with ammonia [28]).

Although it has been suggested that Ostromisslensky may have used one of the optically active, crystalline forms to seed the asparagine solution, this is unlikely, inasmuch as the crystals of the optically active forms, the β- and γ-polymorphs, are only obtained if the crystals are grown as mentioned above. Moreover, Ostromisslensky's experiments have been repeated and enlarged upon in beautiful detail by Addadi et al. [29], who reported that seeding racemic amino acid solutions with crystals of the racemic, α-form, gives the desired result. The technique here is exposing only one face of the α-glycine crystal to the racemic solution, the isolated faces of the crystals of glycine itself *are chiral surfaces.*

It should further be noted that Ostromisslensky's paper (published in *Berichte* in 1908!) is quite remarkable in other ways regarding Werner's paper [5] based on Bosshart's thesis [6]. Examples are: (a) He correctly assigns the discovery of the seeding method to Gemez and refers to it as the Pasteur–Gernez method. (b) He discusses in detail the work of others in this field, including Kipping and Pope's [30] classic experiments on the conglomerate crystallization of $NaClO_3$, in which these authors proved that it is not necessary for a given, isolated crystallization experiment to produce a 50:50 mixture of crystals of both chiralities. However, the cumulative results of many experiments do give the statistically expected 50:50 mixture. (c) He reviews the physical basis of the method and clearly lists the conditions needed for a crystallization experiment to proceed in a conglomerate crystallization pathway. (d) He discusses Piutti's aforementioned experiments [24]. (e) Ostromisslensky also succeeded in seeding other racemic solutions with extraneous material and in obtaining optically pure crystals. For example, he seeded racemic $NaNH_4$ (tart) solutions with *l*-asparagine and obtained a crop of pure *d*-$NaNH_4$ (tart)$\cdot 4H_2O$. Furthermore, seeds of optically pure NaK(tart), K_2(tart), Na_2 (tart), KH(tart), $(NH_4)_2$(tart), and of *d*-(NH_4)(maleate) gave similar results.

In his thesis [6] Bosshart refers to most or all of the papers that we have mentioned; these preceded Werner and Bosshart's work; moreover, one presumes that Werner had seen these papers, having been active, as pointed out earlier, in the field of chiral resolutions. If not, he surely must have read about them in Bosshart's thesis, which was submitted at the end of February 1914, i.e., ca. 3 or 4 months prior to the submission date of the paper (May 1914).

The point, however, is that Werner does not acknowledge the work of any of these workers in his paper even though it is clear from Bosshart's thesis that he (Bosshart) was aware of them. Moreover, as we have seen, Werner specifically labeled the relevant paragraph of his paper "A New Method of Resolution for Racemic Inorganic Compounds," which it was not.

Comments on Werner and Poupardin's Spontaneous Resolution of $K_3[Rh(C_2O_4)]\cdot H_2O$ as a Conglomerate

First, it is important to remember that since his early years as an organic chemist, Werner had achieved considerable expertise in chirality and optical resolutions, as witnessed by his solution of the problem of the conformations and configurations of hexahydrophthalic acids [31]. Second, there is some confusion concerning Werner's formulation of the potassium salts of the $[Rh(C_2O_4)_3]^{3-}$ anion. Werner and Poupardin claimed to have prepared something labeled as the "racemische rhodium-oxalsaure Kalium" [[10], p 1955] and for which an analysis [[10], p 1956] is given of "$[Rh(C_2O_4)_3]K_3 + 4.5H_2O$." This formulation is correct, and the synthesis and analysis of the product have been repeated by various authors. However, as shown below, Werner and Poupardin described [[10], p 1957] one of the enamiomers as "Kalium-Salze, $[Rh(C_2O_4)_3]K_3 + 1H_2O$. *d*-Salz." This, presumably, is what they claim to have obtained during spontaneous resolution [[10], pp 1959–1960] and described as "triclinic" crystals of sizes up to 1 cm. They described their method of achieving conglomerate crystallization [[10], pp 1959–1960] and gave as evidence of success diagrams purporting to depict enantiomorphic crystals of the antipodes of the compound, crystals which they describe as being *triclinic*. Interestingly, they never quote the work of Zamboni, who 7 years earlier demonstrated the isomorphism of the racemic salts of composition $K_3[Rh(Ox)_3]\cdot 4.5H_2O$ and $K_3[Ir(Ox)_3]\cdot 4.5H_2O$ and had established that these *racemates* were *triclinic* [32]. The correct formulation of the optically active compound, as demonstrated by Herpin's structural analysis [33], is $K_3[Rh(Ox)_3]\cdot 2H_2O$.

Another troubling point is this: Probably every school child in Europe had heard the story about the old, wizened Jean-Baptiste Biot's asking Pasteur to repeat the resolution of $NaNH_4$ (tart)$\cdot 4H_2O$ in front of his very eyes and crying with joy upon seeing the results [34]. Werner, having been born in Alsace, must have known the story; therefore, he must have known that the proof came when Pasteur showed Biot that *solutions made from crystals of the same hemihedrism rotated light, some to the left, their antipodes to the right.* Therefore, why did Werner and Poupardin not record the optical rotation of solutions derived from their crystals when they reported that some were as large as 1 cm—a more than ample supply of chiral material for the preparation of a solution for use in Werner's famous polarimeter (Franz Schmidt & Haensch No. 8142).

Let us therefore now describe the work of Jaeger [35], who prepared racemic $K_3[Rh(Ox)_3]\cdot 4.5H_2O$, resolved it into its antipodes using the strychninium cation, and converted it to the potassium derivative. He measured the wavelength dependence (dispersion) oft he optical rotation of the dextro compound, which he formulated as the $K_3[Rh(Ox)_3]\cdot H_2O$ salt; the solubility of the diasteroisomers of the strychninium salt, as well as the enantiomers and the solubility of the enantiomers and racemate of the potassium salt. Finally, he indexed goniometrically the racemate and the enantiomers of the potassium salt and provided pictures of the morphologies of typical crystals of the three crystalline forms. In what follows, we quote his own words:

> But, as we found during our investigations, that the optically active forms are much more soluble under the same circumstances than the racemic substance is, it seemed highly probable that an error was made here, because, moreover, the facts mentioned cannot be right from the theoretical standpoint. The drawings in WERNER'S paper, intended to give an impression of the crystal forms obtained, rather point to distorted, and accidentally nonsuperposable triclinic crystals of the racemoid being present here, than to the enantiomorphous crystals of true hemihedral symmetry. Moreover, WERNER himself mentions the triclinic symmetry of these crystals, and therefore the validity of Pasteur's principle in his case cannot yet be considered proved by the data given in his paper. Repeatedly we have made attempts in the way indicated by WERNER, to perform a spontaneous fission of the racemic salt; but the solution of it saturated just above 100°C, first being rapidly cooled to 90 °C, and subsequently cooled down to room temperature, never deposited crystals other than the capriciously distorted individuals of the triclinic racemoid. The microscopically small crystals often obtained by very rapid cooling of the hot solution, appeared to be no crystals of the optically active form either; they were rhombic individuals exhibiting prismatic, domatic and basal facets, of a new hydrate of the racemic compound, probably at higher temperatures stable, and containing less water of crystallization. These experiments, if varied in several ways, gave unexceptionally bigger or smaller crystals of the racemic compound. The appreciably greater solubility of the active forms in comparison with that of the racemic salt at all temperatures between 15 °C and 100 °C characterizes the racemoid undoubtedly as the more stable solid phase within this range of temperature, with respect to the mixture of these antipodes. The experiments described by WERNER can therefore never lead to spontaneous fission, and it must be quite impossible, that under these conditions an optically active crystal should increase, when brought at room-temperature into the saturated solution of the racemic salt. This may readily be deduced from BAKHUIS ROZEBOOM'S well-known graphical representations (viz., H. W. Rozeboom, Z. *f phys. Chemie,* **28**, 494 (1899)) [Note the date!!!!] of the solubility relations here prevailing. Indeed, it could be proved on the contrary by often repeated experiments, that a crystal of one of the optically active forms, if brought into a feebly supersaturated or saturated solution of the racemic salt, immediately disintegrates and subsequently disappears completely, and that after some lapse of time, triclinic crystals of the racemic compound are deposited from the solution.

Table 1 Crystallographic Parameters

Racemic compound	Space group	Cell characteristics
$K_3[Ir(Ox)_3]\cdot 4.5H_20$	–	$a=12.5$, $b=6.75$, $c=10.46$ Å
		$\alpha=90.6$, $\beta=102.95$, $\gamma=84.88°$, $z=2$
Chiral compound	Space group	Cell characteristics
$K_3[Ir(Ox)_3]\cdot 2H_2O$	$P3_121$	$a=10.98$, $c=10.38$ Å, $z=3$
$K_3[Rh(Ox)_3]\cdot 2H_2O$	$P3_121$	$a=11.26$, $c=20.06$ Å; $z=6$

Table 2 Crystallographic Parameters

Racemic compound	Space group	Cell characteristics
$K_3[Fe(Ox)_3]\cdot 3H_2O$	$P2_1/c$	$a=7.66$, $b=19.87$, $c=10.27$ Å
		$\beta=105.01°$; $z=4$
$K_3[A1(Ox)_3]\cdot 3H_2O$	$P2_1/c$	$a=7.65$, $b=19.78$, $c=10.32$ Å
		$\beta=106.03°$; $z=4$

Table 3 Crystallographic Parameters

Racemic compound	Space group	Cell characteristics
$K_3[A1(Ox)_3]\cdot 3H_2O$	$P2_1/c$	$a=7.712(1)$, $b=19.518(3)$, $c=10.286(2)$ Å
		$\beta=102.1(3)°$; $z=4$
$K_3[Cr(Ox)_3]\cdot 3H_2O$	$P2_1/c$	$a=7.714(1)$, $b=19.687(4)$, $c=10.361(2)$ Å
		$\beta=108.06(3)°$; $z=4$
$K_3[Mn(Ox)_3]\cdot 3H_2O$	$P2_1/c$	$a=7.72$, $b=19.88$, $c=10.41$ Å
		$\beta=108.22°$; $z=4$

In a more modem description of the racemic and optically active crystals of $K_3[Rh(Ox)_3]$ and related salts of other trivalent metal cations the space group and cell constants of racemic $K_3[Ir(Ox)_3]\cdot 4.5H_2O$ and $K_3[Rh(Ox)_3]\cdot 4.5H_2O$ were determined by Herpin [33], who also reported the crystal structures of optically active (resolved via the use of the strychninium cation salts, as was done by Werner) salts of composition $K_3[Ir(Ox)_3]\cdot 2H_2O$ and $K_3[Rh(Ox)_3]\cdot 2H_2O$ and the past history of these salts. He reported the basic crystallographic parameters of racemic and optically active compounds of $K_3[Rh(Ox)_3]$ (see Table 1).

The chiral crystals were obtained from anions which were resolved by strychninium cations and then converted to the potassium salts; had they been obtained from spontaneous resolutions (conglomerate crystallization) of the unresolved potassium salts, crystals of both enantiomorphs (e.g., $P3_121$ (a left-handed propeller) or $P3_221$ (right-handed propeller)) should have been obtained from such a crystallization procedure. Instead, racemic crystals were obtained, whose cell characteristics are given above. However, the fact is that the author never determined the absolute configuration of either the Ir or Rh compounds by X-ray methods and, thus, had no right to say the crystals chosen for X-ray diffraction studies were either $P3_121$ or $P3_221$. This point must, thus, remain moot. In fact, the first absolute configurational study of a coordination compound ever established by X-ray diffraction, was determined in the same year of 1958 by Saito and associates [36].

Remark. The cell parameters for the Rh racemate are not given; however, Herpin states [[33], p 219] that the "clichés" of the cell (presumably *les clichés de Laue*=Laue

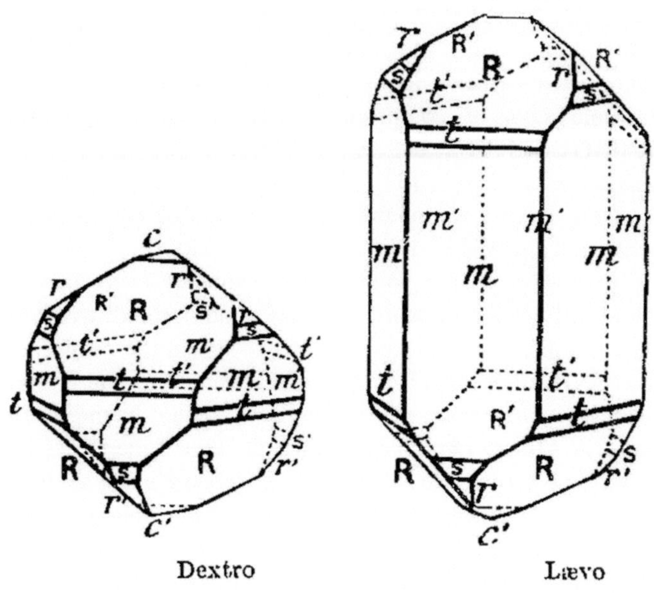

Fig. 2 The dextro (*left*) and levo (*right*) crystals of $K_3[Co(Ox)_3]\cdot 3.5H_2O$.

photographs) are identical to those of theIr analogue and thus isomorphous. Herpin used such a comparison of Laue photographs to demonstrate that the optically active Ir and Rh salts were isomorphous but not isostructural, because for the Ir derivative, $z=3$ molecules/unit cell, while for the Rh salt, $z=6$.

Both structures were solved and refined from three-dimensional data collected on films. Coordinates of the atoms were listed.

In Herpin's second paper [37] on the potassium trisoxalato salts, he gives the space group and cell constants of the Fe and Al salts as well as a complete analysis of the crystal structure of the iron salt, the basic cell parameters for which are shown in Table 2. Herpin also states that $K_3[Cr(Ox)_3]\cdot 3H_2O$, $K_3[Mn(Ox)_3]\cdot 3H_2O$, and $K_3[Ru(Ox)_3]\cdot 3H_2O$ are isomorphous, and presumably isostructural, with the Fe and Al derivatives, a fact that was later confirmed.

Jaeger and Thomas [38], and later Herpin [37], found the Co(III) derivative to crystallize as a racemate of composition $K_3[Co(Ox)_3]\cdot 3.5H_2O$. They all agree that the salt is triclinic as is the racemate. Jaeger and Thomas [38] reported that when the salt is crystallized from water at temperatures above 13.2 °C, it spontaneously resolves, a fact verified by Tutton [39], who checked its composition and reported the goniometric characteristics of the racemic and of the optically active crystals. These are shown in Fig. 2.

In support of Herpin's [37] claim that the A1, Cr, and Ru salts are isomorphous with those of the Fe derivative, whose structure he determined, is the fact that the complete structures of the Al, Cr, and Mn derivatives were determined later. The structures of $K_3\,[Al(Ox)_3]\cdot 3H_2O$ and $K_3\,[Cr(Ox)_3]\cdot 3H_2O$ were determined by Taylor [40], while the crystal structure of the Mn derivative was determined by Lis and Matuszewski [41] (see Table 3).

What these results mean is that (a) with the demonstrated exception of the rather unique $K_3[Co(Ox)_3]\cdot 3.5H_2O$, crystallized under conditions which do not in any way resemble those used by Werner and Poupardin, there is no known salt of the general composition $K_3[M(Ox)_3]\cdot XH_2O$ that crystallizes as a conglomerate. Moreover, the Co salt is photochemically unstable and decomposes upon exposure to room light fairly rapidly. (b) The work of Zamboni [32], Herpin [33, 37], Jaeger [35, 38], Tutton [39], Taylor [40], and Lis [41] is all self-consistent and mutually reinforcing.

These facts derived from their studies [32, 33, 35, 37–41] uniquely add up to the conclusion that Werner and Poupardin made a mistake somewhere; moreover, it was an inexcusable mistake since (1) the published work of Zamboni [32] should have been a warning that their results were highly suspect, and (2) they had single crystals large enough to make solutions to test the chirality of the "resolved" crystals. The origin of the mistake will probably never be ascertained, but, as a mistake, it ranks as a major blunder.

Acknowledgments The authors wish to thank professor L. M. Venanzi, Eidgenössische Technische Hochschule, Zürich; Professors C. H. Eugster and H. R. Ostwald, Universitiit Zurich; and Professor J. P. Adloff, Universite Louis Pasteur, Strassbourg-Cronenbourg; for invaluable information and for helping us locate source materials used in the preparation of this manuscript. Dr. Hans Joachim Steffens kindly translated difficult parts of Bosshart's thesis for us; especially useful was his assistance with the portion dealing with Erlenmeyer's work. Ivan Bernal thanks the Robert A. Welch Foundation for financial support (Grant No. E-594).

References

1. Bernal, I.; Kauffman, G. B. *J. Chem. Educ.* **1987**, *64,* 604.
2. Kauffman, G. B.; Bernal, I. *J. Chem. Educ.* **1989**, *66,* 293.
3. Pasteur, L. *Compt. Rend.* 1848, 26, 535; *Ann. Chim. Phys.* **1850**, *28,* 56.
4. Kauffman, G. B.; Myers, R. D. *J. Chem. Educ.* **1975**, *52*, 777.
5. Werner, A. *Ber.* **1914**, *47*, 2171.
6. Bosshart, J. *Anomale Erscheinungen bei der Spaltung von Racemverbindungen;* Inaugural Dissertation; Conzett: Zürich, 1914.
7. Werner, A. *Ber.* **1911**, *44,* 1887. For a discussion and annotated English translation see Kauffman, G. B.; *Classics in Coordination Chemistry, Part 1: The Selected Papers of Alfred Werner;* Dover: New York, 1968; pp 153–173.
8. Kauffman, G. B.; Lindley, Jr., E. V. *J. Chem. Educ.* **1974**, *51*, 424; *Inorg. Synth.* **1976**, *16,* 93.
9. Duschinsky, R. *Chem. Ind.* (*London*) **1834**, 10; *Festschrift Emil Borel;* F. Reinhardt: Basel, 1936; p 375.
10. Werner, A. *Ber.* **1914**, *47,* 1954.
11. Leidié, E. *Ann. Chim. Phys.* **1889**, *17,* 307.
12. de la Provostaye, F. H. *Ann. Chim. Phys.* **1841**, *3,* 129, 353; **1842**, *4,* 453; **1842**, *5*, 47.
13. Biot, J.-B. *Compt. Rend.* **1844**, *19,* 719.
14. Jacques, J.; Collet, A.; Wilen, S. H., *Enantiomers, Racemates, and Resolutions;* Wiley-Interscience: New York, 1981; pp 223–240.
15. Gernez, D. *Compt. Rend.* **1866**, *63,* 843.
16. Picard, E. *Compt. Rend.* **1910**, *151,* 741.
17. *The Times* (*London*), Nov. 1, 1910, p 11d.
18. Jungfleisch, E. (a) *J. Pharm. Chim.* **1882**, 5, 346; (b) *Bull. Soc. Chim. France* **1884**, *41,* 222.
19. Amiard, G. *Bull. Soc. Chim. France* **1956**, 447.
20. Secor, R. M. *Chem. Rev.* **1963**, *63,* 297.
21. Jordan, C. *Compt. Rend.* **1916**, *162,* 617.
22. Léger, E. *Bull. Soc. Chim. France* **1917**, *21,* 1.
23. Smith, H. M. *Torchbearers of Chemistry;* Academic Press: New York, 1949; p 130.
24. Piutti, A. *Ber.* **1886**, *19,* 1691.
25. Ostromisslensky, I. *Ber.* **1908**, *41,* 3035.
26. Albrecht, G.; Corey, R. B. *J. Am. Chem. Soc.* **1939**, *13,* 35.
27. Iitaka, I. *Acta Crystallogr.* **1960**, *13,* 35.
28. Iitaka, I. *Acta Crystallogr.* **1958**, *11,* 225.
29. For a detailed review see Addadi, L.; Berkovitch-Yellin, Z.; Weissbuch, I.; Lahav, M.; Leiserowitz, L. In Eliel, E. L.; Allinger, N. L.; Wilen, S. H., Eds.; *Topics in Stereochemistry;* Wiley: New York, 1986; Vol. 16, pp. 1–86.
30. Kipping, F. S.; Pope, W. J. *J. Chem. Soc.* **1887**, *71*,611.
31. Werner, A.; Conrad, H. E. *Ber.* **1899**, *32,* 3046.
32. Zamboni, F. Z. *Kristallogr. Min.* **1907**, *16,* 551.
33. Herpin, P. *Compt. Rend.* **1958**, *81,* 201.

34. Pasteur, L. In *Leçons de chimie professées en 1860…*; Société Chimique de Paris: Paris, 1861; pp 1–48; for an English translation see Pasteur, L. *Researches on the Molecular Asymmetry of Natural Organic Products;* Alembic Club Reprint No. 14; E. & S. Livingstone: Edinburgh, Scotland, 1948, pp 20–22.
35. Jaeger, F. M. *Proc. Acad. Sci. Amsterdam* **1917**, *20,* 263.
36. For the structure and absolute configuration of (+)-Λ (*δδδ*) $\{[Co(en)_3]Cl_3\}_2NaCl \cdot xH_2O$; see Saito, Y.; Nakatsu, K.; Shiro, M.; Kuroya, H. *Bull Chem. Soc. Jpn.* **1959**, *8,* 729.
37. Herpin, P. *Compt. Rend.* **1958**, *81,* 145.
38. Jaeger, F. M.; Thomas, W. *Proc. Acad. Sci. Amsterdam* **1919**, *22,* 693.
39. Tutton, A. E. H. *Crystallography and Practical Crystal Measurements,* 2nd ed.; The Indian Printing Works: New Delhi, 1965; reprinted from the original Macmillan, London, 1922, version. See Vol. 2, pp 1282–1283.
40. Taylor, D. *Aust. J. Chem.* **1978**, *31,* 1455.
41. Lis. T.; Matuszewski, J. *Acta Crystallogr. Sect* ***B.*** **1989**, *36,* 1938.

The Geometry of Intermolecular Interactions in Some Crystalline Fluorine-Containing Organic Compounds[a]

Liat Shimoni[a] and Jenny P. Glusker[b,c]

Abstract

The propensity of C—F groups to form C—F···H—C interactions with C—H groups on other molecules has been analyzed. Crystal structures of molecules containing only carbon, hydrogen, and fluorine, but no oxygen, nitrogen, or other hydrogen-bond-forming elements, were chosen for an initial study in which the intermolecular interactions in crystal-structure determinations of polycyclic aromatic hydrocarbons and their analogous fluoro derivatives were analyzed. It is found that C—F···H—C interactions occur, but they are weak, as judged by the intermolecular distances and the angles involved. In a study of crystal structures of molecules containing other elements in addition to carbon, hydrogen, and fluorine, it was found that when an oxygen atom is in a neighboring position on an interacting molecule, a C—O group is more likely than a C—F group to form a linear interaction to the hydrogen atom of a C—H group. Thus, in spite of the high electronegativity of the fluorine atom, a C—F group competes unfavorably with a C—O$^-$, C—OH or C═O group to form a hydrogen bond to an O—H, N—H, or C—H group. It is found, however, particularly for polycyclic aromatic hydrocarbons with substituted CF_3 groups that, in the absence of other functional groups that can form stronger interactions, C—F···H—C interactions may serve to align molecules and give a different crystal packing from that in the pure hydrocarbon (where fluorine is replaced by hydrogen). Thus, C—F···H—X(X = C, N, O) interactions are very weak, much weaker than C═O···H—X interactions, but they cannot be ignored in predictions of modes of molecular packing in complexes and in crystals.

Keywords

C—F···H—C interactions • Fluorine interactions • Polycyclic aromatic hydrocarbons • Fluoro derivatives

Introduction

Fluorine is the smallest and the most electronegative element of the halogen family [1]. It does not, however, behave in the same way as other halogen atoms when in a compound. As a result, it is not possible to predict, with any accuracy, the behavior of fluorine based on the information obtained from analogous compounds containing other halogen atoms.

The manner by which a C—F group interacts with a hydrogen atom in another molecule in the solid state is analyzed in this study. The question asked is to what extent the fluorine in the C—F bond takes part in attractive interactions with other molecules [2]. Some small organic compounds were selected and intermolecular H···F distances and the linearity of the X—H···F group of atoms in these compounds were investigated in order to determine whether there are any basic systematic rules that can be identified (X = C, N, O throughout

[a]*Structural Chemistry* 1994, 5(6):383–397.

[b]The Institute for Cancer Research, Fox Chase Cancer Center, Philadelphia 19111, PA, USA

[c]Correspondence should be directed to Jenny P. Glusker, The Institute of Cancer Research, Fox Chase Cancer Center, Philadelphia, Pennsylvania 19111.
e-mail: Jenny.Glusker@fccc.edu

I. Hargittai, B. Hargittai (eds.), *Science of Crystal Structures: Highlights in Crystallography*,
DOI 10.1007/978-3-319-19827-9_21

this article). The sum of the van der Waals radii of hydrogen and fluorine has been reported [1] as approximately 2.3–2.5 Å. We examined H···F interactions up to 2.8 Å, but extended the range in specific cases to 3.0 Å or more.

Materials and Methods

The Cambridge Structural Database (CSD) [3] was searched for fluorine-containing compounds. Initially, pairs of analogous compounds containing C—H and/or C—F bonds, excluding all compounds containing any other chemical elements, were studied. At a later stage of this analysis, compounds were selected that contained oxygen and/or nitrogen atoms that could compete with the fluorine atom in its intermolecular interactions, such as the formation of hydrogen bonds. A list of the compounds selected from the CSD, and the geometry of the C—F···H—X interactions obtained in this analysis, are given in the supplementary material. The overall scheme of the geometrical analysis is illustrated Scheme 1.

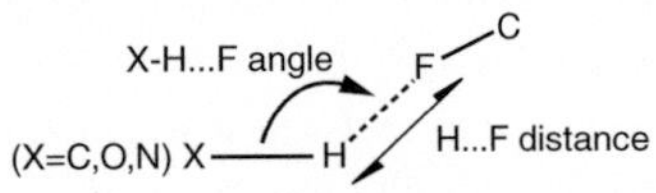

Scheme 1

The CSD input files, containing query information, were stored in instruction files (named *.que, where * is a chosen name). The CSD program QUEST was run to search the database file for structures that matched the input requirements, and to write the coordinates of these crystal structures onto a data file (*.dat). The CSD program GSTAT [3] was then run on this coordinate data file using the instruction file *.geo as input. The effective sum of the van der Waals radii was increased in specific cases from 2.8 to 3.0 Å (implying van der Waals radii (Å): H1.2–1.5, F1.3–1.5). This enabled the determination of the geometry of the intermolecular interactions associated with the selected fluorine-containing fragment. The positions of any hydrogen atoms attached to X = C, N, O and the X—H···Y angle (Y = F, O, N) are normalized to mean neutron distances and X—H···Y angles, based on values listed in a study by Taylor and Kennard [4]. For compounds for which the coordinates of the hydrogen atoms were not available in the database, hydrogen-atom positions were calculated with the computer program CALCAT [5]. Mopac charge calculations were performed with the Ampac/Mopac module in the Insight molecular modeling package (Biosym Technologies, Inc.). The MNDO semiempirical Hamiltonian was used, and charges were fit to the electrostatic potential in a shell beginning at 1.4 times the van der Waals surface of the molecule and extending a thickness of 0.6 Å outward. All illustrations were drawn with the program ICRVIEW [6]. Details of input files and a list of CSD reference codes and bibliographical information on the crystal structures used in this study are given in Supplementary Material Tables 1 and 2.

Results

In this study we investigated the packing in the crystalline state of three specific types of polycyclic hydrocarbons. The first group of molecules contains only carbon, hydrogen, and fluorine atoms (18 crystal structures in all); (Refcodes AFMCPD, BAWPUK, BIBRAF, DFNAPH10, FDPHET, FIXVOX, FIXWOY, FIXWEO, FIXXAL, FLNAPH, FMSTLB, FOHMEU, FOHMIY, MFSTIL10, OCFCHX10, PFBIPH01, PFLBPH01, TFBIPH). The second group contains analogous molecules that lack fluorine but contain only carbon and hydrogen atoms, with hydrogen replacing fluorine in the first group (four crystal structures in all); (Refcodes FEWWEJ02, FIXVIR, FIXWAK, PHENAN08). The third group includes similar molecules that contain carbon, hydrogen, and fluorine together with oxygen and/or nitrogen atoms (287 crystal structures in all; the complete list of Refcodes and bibliography are in the supplementary material).

Compounds Containing C, H, and F Atoms

In the first group, only the hydrogen atoms attached to carbon atoms and the electronegative fluorine atom are available to participate in hydrogen bonds or any other similar types of interactions. Thus, if any interaction occurs it will probably be C—H···F—C. In Fig. 1a the distribution of the H—F bond distances in such compounds, up to 2.8 Å from the fluorine atom, is shown; the mean H···F distance is 2.6(1) Å and the median distance is also 2.6 Å. In Fig. 1b the angular distribution (C—H···F angles) is shown; the mean and the median of this distribution are 136(20) and 132°, respectively.

We divided this first main group (containing only carbon, hydrogen, and fluorine atoms) into three subgroups. One subgroup contained compounds with aromatic six-membered ring systems (seven in all), the second subgroup contained compounds with aliphatic six-membered ring systems (six in all), and the third subgroup contained all remaining compounds and will be referred to as "general" compounds (five in all). The distributions of the H···F distances and of C—H···F angles for each of these subgroups are shown in Fig. 1c–h, and the data are given in the Supplementary Material Table 3. The mean and the median values of these distributions of distances and angles are given in Table 1. Note the somewhat narrower distribution of C—H···F intermolecular angles for aromatic compounds than for other types of compounds. The overall picture appears to be one in which molecules arrange in the crystalline state so that they can accommodate and juxtapose in the best possible manner the slight positive charges on hydrogen atoms and the slight negative charges on fluorine atoms. The result is that the F···H distance is near 2.6 Å and the F···H—C angle in the range 130–145°.

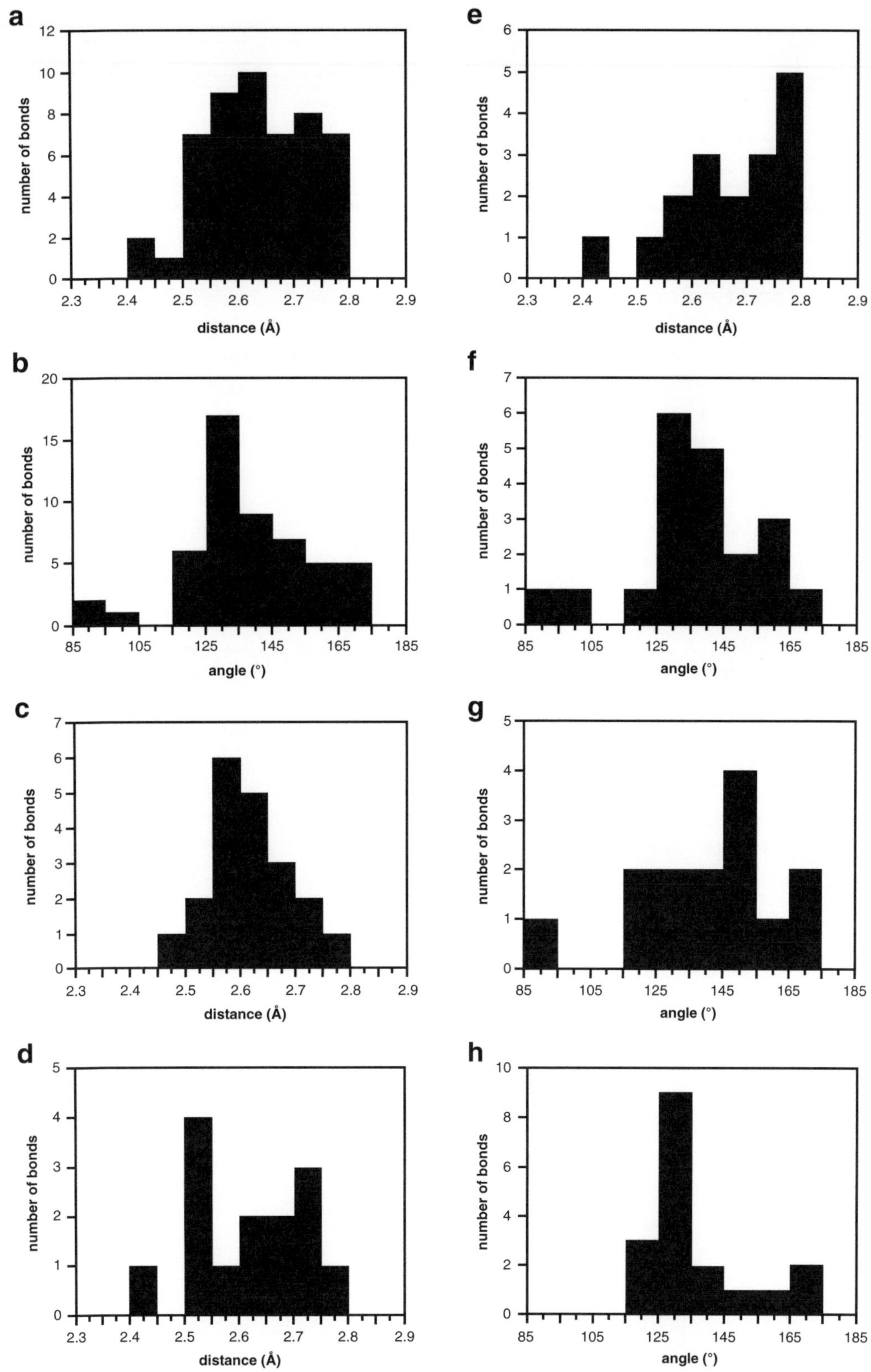

Fig. 1 The distribution of H···F distances (**a**), and (**b**) of C—H···F angles in C—H···F—C interactions for all compounds with C—F as the acceptor group (aromatic and aliphatic and "general"), with C—F as the acceptor group attached to an aromatic ring system (**c**, **f**), with C—F as the acceptor group attached to an aliphatic ring system (**d**, **g**), with C—F as the acceptor group attached to a "general" ring system (**e**, **h**). Distances were measured up to a cutoff value of 2.8 Å.

Table 1 Geometry of C—F···H—C interactions in compounds containing only C—F as a hydrogen acceptor group[a]

	All compounds containing only C—F as hydrogen acceptor group	Aromatic six-membered ring system	Aliphatic six-membered ring system	"General" ring systems
Mean distance F···H(Å)	2.6(1)	2.6(1)	2.6(1)	2.7(1)
Median distance F···H(Å)	2.6	2.6	2.6	2.7
Mean angle F···H—C(°)	136(20)	134(23)	139(21)	136(15)
Median angle F···H—C(°)	132	139	142	132
Number of CSD entries	18	7	6	5

[a]Values in parentheses are estimated standard deviations with respect to the last digit

Comparison of C—H and C—F

We selected four molecules out of the first main group and compared them to four other molecules in which the fluorine atoms were replaced by hydrogen atoms (Refcodes FEWWEJ02, FIXVIR, FIXWAK, PHENAN08). The analysis of the average distances and angles for the eight compounds (2.8 Å from a fluorine atom and 3.0 Å from the analogous hydrogen atom) (Scheme 2) is summarized in Table 2.

In the first pair of molecules, FIXXAL and FEWWEJ02, the H···Y distance in the fluorine containing compound (Y = F) is marginally shorter than that in the hydrocarbon (Y = H). Fluorine is a highly electronegative atom in and therefore might be expected to form a shorter hydrogen bond or interaction compared to the less electronegative hydrogen atom. The similarity of the H···F and H···H distances is therefore surprising, and values of H···F are nowhere near as short as those of O···H, which can be as low as 1.2 Å. The overall trend shown in Table 2 is for a slightly shorter H···Y distance and a significantly larger C—H···Y angle when Y = F rather than when Y = H.

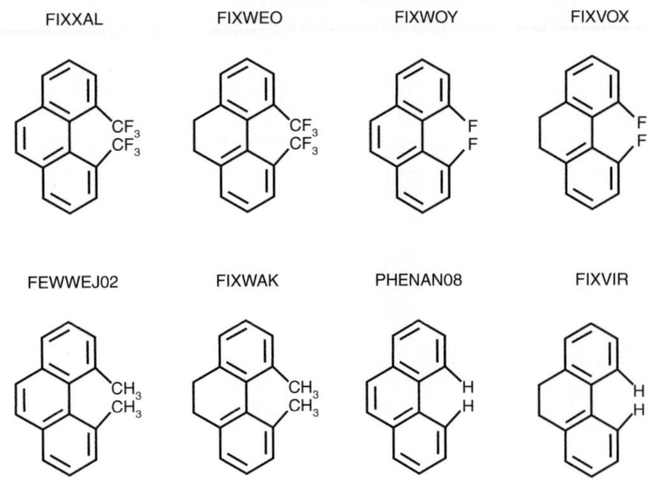

Scheme 2

The overall impression from Table 2 is that the $H(\delta+)\cdots F(\delta-)$ distance is marginally shorter than a $H(\delta+)\cdots H(\delta+)$ distance and that the C—H···F angle tends to a value near 140° (with a range 137–154°), while the C—H···H angle varies more widely (with a range 88–132°). These trends are more noticeable when Y is directly attached to an aromatic ring system than when it is part of a CY_3 group. The charge distribution would be expected to be different for $H(\delta+)\cdots H(\delta+)$ and for $H(\delta+)\cdots F(\delta-)$. The similarity of H···H and H···F distances, although beyond the sum of the van der Waals distances, implies, in view of the larger size of fluorine than hydrogen atoms, that there must be some interaction between hydrogen and fluorine atoms in an H···F interaction.

Graph-Set Analysis of Packing

Graph-set motifs may be assigned to each type of hydrogen bond (referred to in totality as the first level), according to a method proposed by Etter, MacDonald, and Bernstein [7]. Such graph-set motifs for these four fluorine-containing compounds may be designated as summarized in Table 3. The motifs for the four fluorine compounds are shown in Fig. 2a–h (the bond that connects that atoms in a motif is drawn as an open or a thin line to separate it from the nonparticipating atoms for which bonds are drawn as filled double lines).

Table 2 Geometry of C—F(H)···H—C interactions

Refcode[a]	Y=	H···Y distance (Å)	C—H···Y angle (°)	AS[b]	R-factor[c]	Group with Y	Δ distance (Å) F minus H	Δ angle (°) F minus H
FIXXAL	F	2.63	142	1	0.029	CF_3	−0.07	+10
FEWWEJ02	H	2.70	132	2	0.061	CH_3		
FIXWEO	F	2.66	154	2	0.029	CF_3	0.00	+27
FIXWAK	H	2.66	127	1	0.033	CH_3		
FIXWOY	F	2.60	138	1	0.038	F	−0.40	+50
PHENAN08	H	3.00	88	2	0.043	H		
FIXVOX[d]	F	2.72	137	1	0.048	F	−0.09	+35
FIXVIR	H	2.81	102	2	0.048	H		

[a]See Scheme 2
[b]AS is the mean e.s.d. (C—C) in range 1 = 0.001–0.005 Å, 2 = 0.006–0.010 Å. The e.s.d. values for C—F bonds are similar, while those for C—H bonds are about 10 times greater
[c]A crystallographic measure of the relationship of the difference between observed and calculated structure factors to the observed data
[d]The correct space group, confirmed by the author of the paper, is $P2_1/n$, not that quoted

Two compounds with a CF_3 hydrogen-bond donor group (FIXXAL, FIXWEO) have the same motifs (first level) in the aromatic and the aliphatic ring systems; the motif is C(6) C(7)R_2^2(12). In the two compounds with a C—F donor group (FIXWOY, FIXVOX), the aromatic and the aliphatic ring systems both have a C(6) chain motif, but they differ in the second motif, also a chain (C(8) and C(7), respectively). These two compounds, unlike the situation when CF_3 groups are present, do not pack with a ring motif.

A study of the packing diagram (shown in Scheme 3) reveals a unique mode of packing of compounds containing the CF_3 group, that is, FIXXAL and FIXWEO. Both compounds pack in the crystalline state so that all the CF_3 groups are pointing in one direction, as shown in Scheme 3 (the arrow points from the 9, 10 bond to the center of mass of F). As a result, the crystal must be polar, with an appreciable dipole moment. The other six compounds are packed in a way that the fluorine atoms or the analogous hydrogen atoms of different molecules are not aligned in a preferential direction in this way. Since the hydrogen atoms in the position *ortho* to the CF_3 group are not very accessible to any binding interaction, the only accessible hydrogen is the one between the two *meta* positions along the central ring. Therefore by packing the molecule in the manner observed for FIXXAL and FIXWEO, all the fluorine atoms are able to participate in a stabilizing binding interaction.

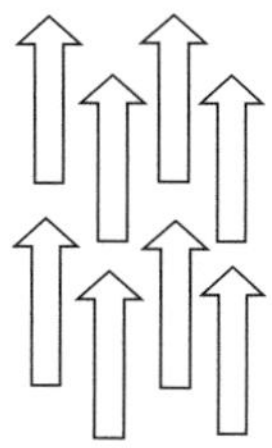

Scheme 3

Table 3 Graph-set analysis of four fluorine-containing compounds

Refcode	Acceptor group	Graph set motifs[a]	Ring system
FIXXAL	CF_3	C(6), C(7), R_2^2(12)	Aromatic
FIXWEO	CF_3	C(6), C(7), R_2^2(12)	Aliphatic
FIXWOY	C—F	C(6), C(8)	Aromatic
FIXVOX	C—F	C(6), C(7)	Aliphatic

[a]C chain, R ring. The total number of atoms involved is indicated in parentheses. Hydrogen-bond acceptors and donors are indicated by superscripts and subscripts, respectively

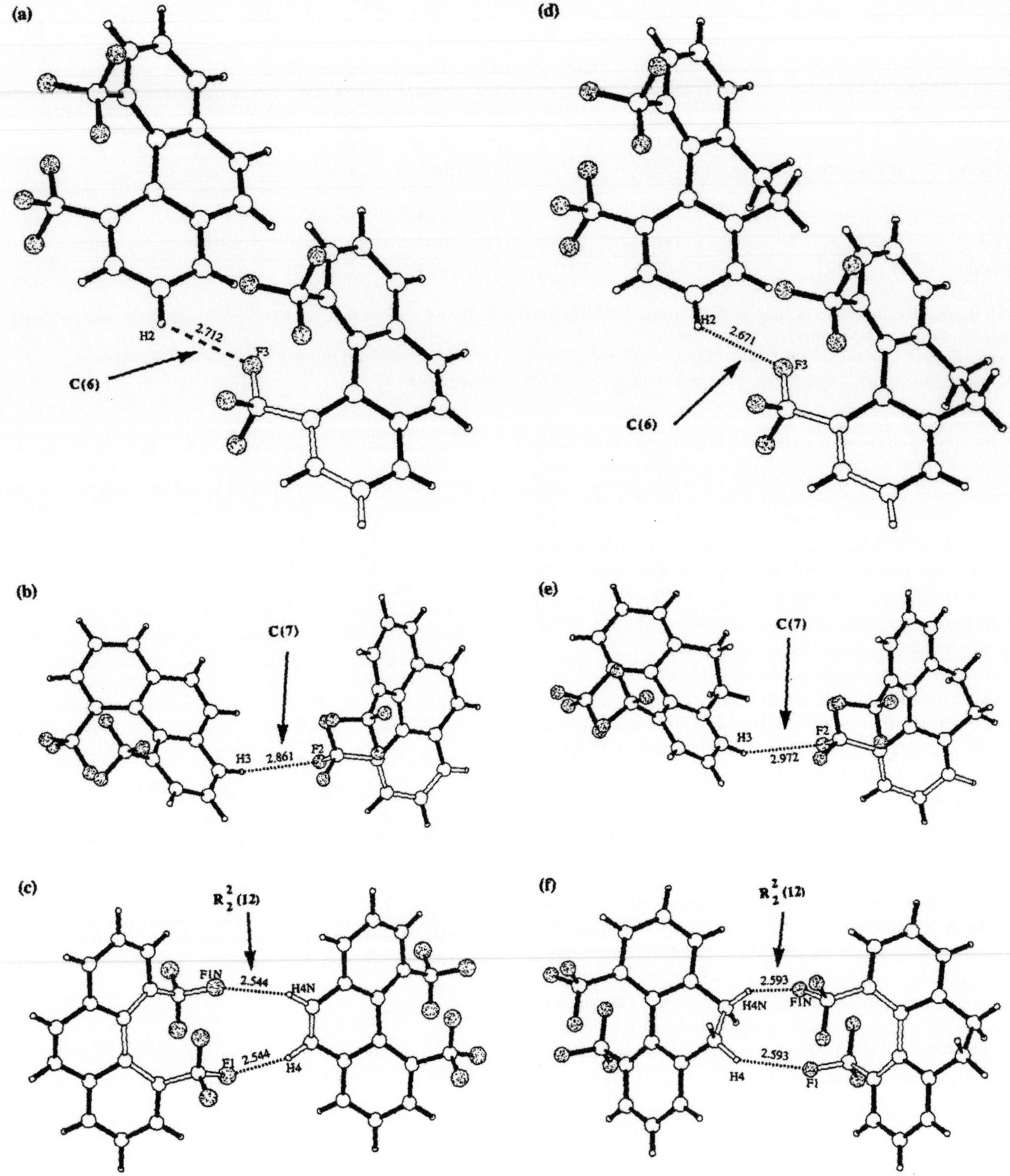

Fig. 2 The binding interactions and the group-set motifs for selected fluorine-containing compounds. (**a**) FIXXAL, chain motif; (**b**) FIXXAL, chain motif; (**c**) FIXXAL, ring motif; (**d**) FIXWEO, chain motif; (**e**) FIXWEO, chain motif; (**f**) FIXWEO, ring motif; (**g**) FOXWOY, two chain motifs; (**h**) FIXVOX, two chain motifs.

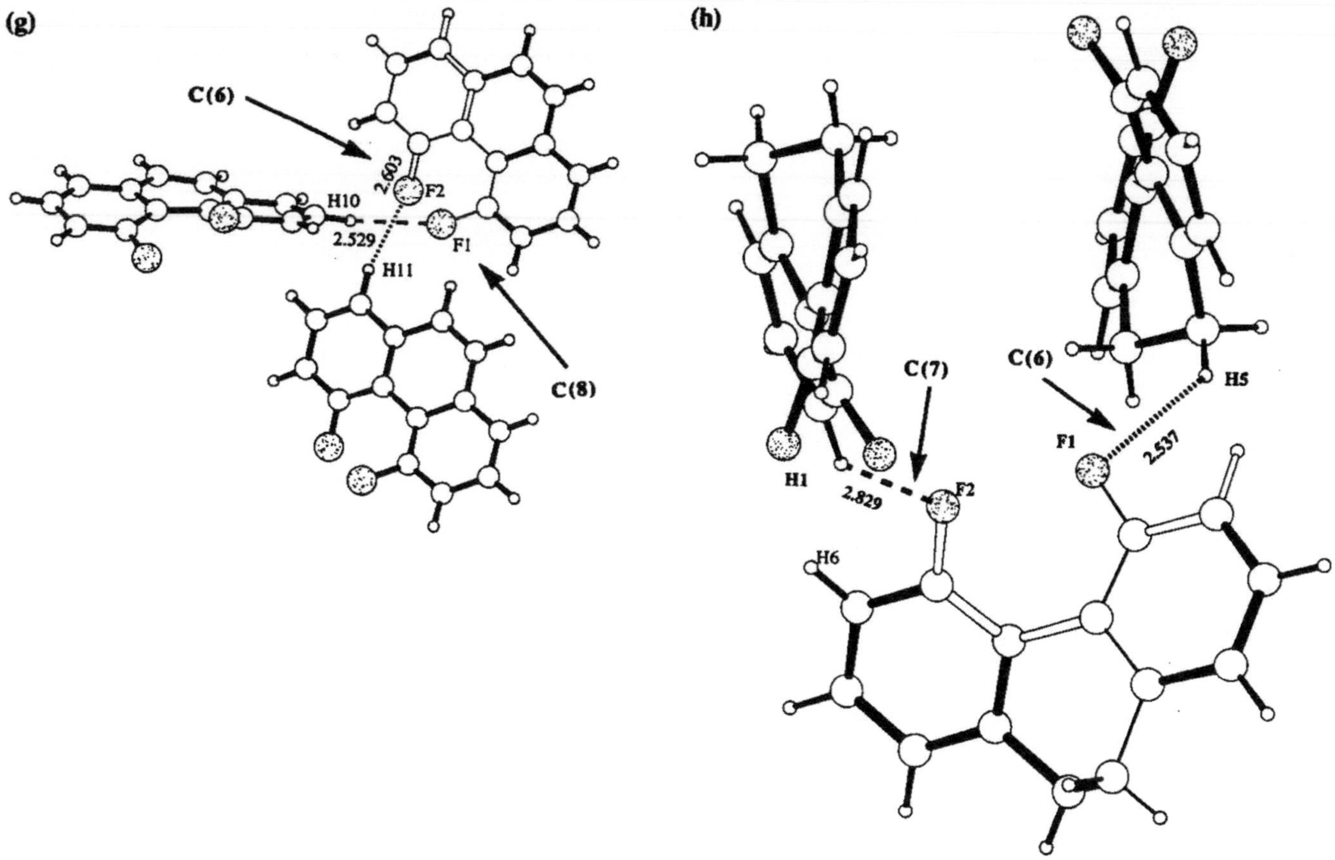

Fig. 2 (continued)

Compounds Containing C, H, F Together with O and/or N Atoms

The C—F group can be compared with the C—OH group as well as with the C—H group. In several examples the biological activity of an enzyme has been inhibited when a hydrogen atom in a substrate is replaced by a fluorine atom [8]. The physical properties of both C—F and C—OH groups are similar with respect to bond lengths, force constants, atomic refractivity, and electronegativity. Since C—F and C—OH groups can form hydrogen bonds in the equivalent directions (when both fluorine and oxygen are hydrogen-bond acceptors), it is possible that the oriented substrate analogue may still bind and, in so doing, exert some significant effects.

We investigated the extent to which the presence of another hydrogen-bond acceptor atom in a molecule affects the geometry of the X—H···F interaction. A group of molecules that contains either oxygen or nitrogen atoms or both, in addition to the fluorine atom, was selected, and an example of such a compound is shown in Fig. 3. Since we were able to find several compounds with the required specific characteristics in the CSD, the list of compounds in the study (up to 3.0 Å from the acceptor group) was divided into two groups, each composed of three subgroups:

1. A fluorine atom attached to
   (a) An aromatic six-membered ring system (103 in all)
   (b) An aliphatic six-membered ring (36 in all)
   (c) A five-membered ring (13 in all)
2. A CF_3 group attached to
   (d) An aromatic six-membered ring system (32 in all)
   (e) An aliphatic six-membered ring (16 in all)
   (f) A five-membered ring (13 in all)

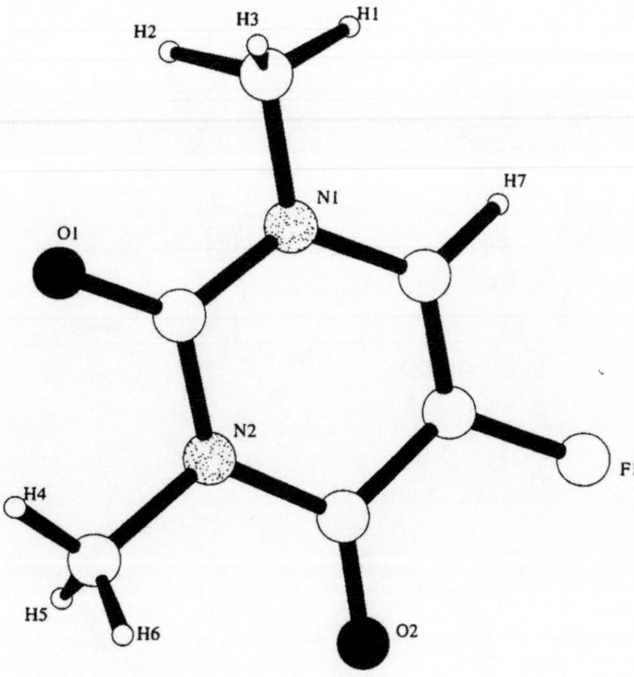

Fig. 3 An example of an organic compound containing oxygen and nitrogen as an acceptor group in addition to the fluorine atom of a C—F bond (Refcode KAMSAS).

The distribution of the H···F distances and of X—H···F angles of two main groups were plotted for compounds with the fluorine atom attached to an aromatic or an aliphatic carbon atom of a five-membered ring in Fig. 4a, b. The second group consists of compounds in which a CF_3 group is attached to an aromatic, or an aliphatic, or five-membered ring in Fig. 4c, d. The distributions of the H···F distance and of X—H···F angles of the six subgroups are given in Figs. 5a–f and 6a–f. The data are given in Supplementary Material Tables 4 and 5. The mean and median values of the distributions of these distances and angles are given in Tables 4 and 5.

The results in Table 4 show that the mean H···F distance tends to remain the same, 2.5–2.6 Å, and the X—H···F angle tends to decrease as the ring system of the compound becomes less aromatic. The results in Table 5 show that when the fluorine atom is a part of a highly electronegative CF_3 group, the mean H···F distance tends to stay the same, 2.5 Å, and so is the X—H···F angle, 137° in all types of ring systems—aromatic, aliphatic, and five-membered.

In order to evaluate the strength of the fluorine atom as an acceptor compared to oxygen and nitrogen atoms on the same molecule, we studied 287 compounds (the complete list of Refcodes and bibliography are in the supplementary material) in which the fluorine atom is involved in a binding interaction with the more acidic hydrogen atom compared to the hydrogen bond to the O or N atom (where amine and hydroxyl hydrogen atoms are considered to be more acidic hydrogen compared to those in aromatic or aliphatic hydrogen groups). Of the 287 compounds included in this study, only seven cases were found in which the C—F group was involved in a binding interaction with the more acidic hydrogen (Refcodes DOLSEC (aliphatic F), FLESTR (aliphatic C—F), GERBUA (aromatic CF3), KINWIN (aliphatic C—F), SOYHIX (aliphatic CF_3), TRFLOC10 (aromatic CF_3), VIKPUA (aliphatic CF_2)); in all the other compounds the oxygen and/or nitrogen atoms were involved in a binding interaction with the more acidic hydrogen in the crystal structure, and the fluorine atom was involved with the same or less acidic type of hydrogen, such as an aliphatic hydrogen atom.

Finally, we studied a "three-center" hydrogen bond system, where one hydrogen-bond acceptor is a fluorine atom and the second acceptor is either an oxygen atom or a nitrogen atom, to see if this arrangement affects the geometry of the H···F interaction. Of the 287 compounds studied, only 11 were found to have this arrangement within distances up to 2.8 Å from the hydrogen atom (Refcodes ADFGLP, BDFACM, BULMAW10, CDFACM, CPFBUR, DEBPSIA, FIYMUV, FMPZTZ10, FURACM, KAMJUX, VIKPUA). None of the compounds had three-center bonds in which both acceptors were fluorine atoms within this distance. Illustrations of a few interactions are shown in Fig. 7a–f. In 8 of these 11 crystal structures the second hydrogen-bond acceptor was an oxygen atom (Refcodes ADFGLP, BDFACM, BULMAW10, CDFACM, CPFBUR, DEBPSIA, FURACM, KAMJUX), and in the other three compounds the second hydrogen-bond acceptor was a nitrogen atom (Refcodes FIYMUV, FMPZTZ10, VIKPUA). The parameters for H···F distance and X—H···F angle are given in Table 6. In this table it is shown that the H···F distance and the X—H···F angle tend to be marginally smaller (2.6 Å, 2.7 Å and 127°, 117° for oxygen and nitrogen as a hydrogen-bond acceptor, respectively) compared to these parameters in the system where a fluorine atom is involved in a two-center interaction system (2.7 Å and 131° respectively). These systems of three-center hydrogen bonds or hydrogen interactions are found to be planar, with minor deviations of the hydrogen atoms from the plane defined by the two acceptors and the donor.

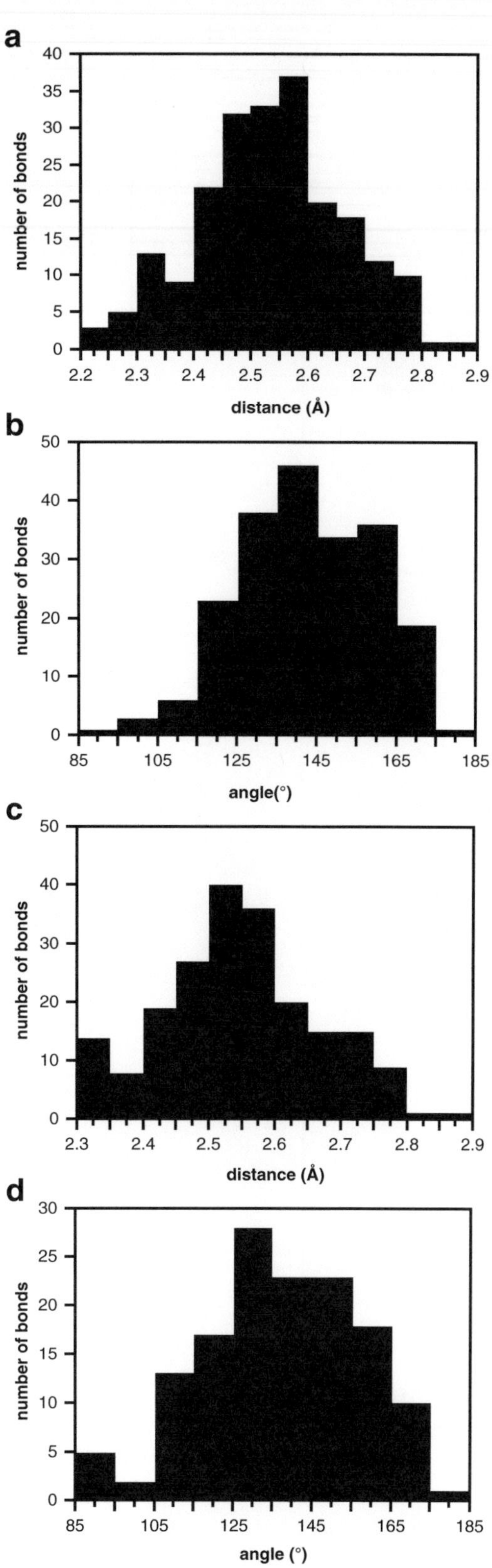

Fig. 4 The distribution of H⋯F distances (**a**) and of X—H⋯F angles (**b**) in X—H⋯F—C interactions with C—F as the acceptor group (in addition to N or O atoms) attached to an aromatic or an aliphatic or a five-membered ring system, and the distribution of H⋯F distances (**c**) and X—H⋯F angles (d) with CF_3 as an acceptor group (in addition to N and/or O atoms) attached to an aromatic or an aliphatic or a five-membered ring system. Distances were measured up to a cutoff value of 3.0 Å.

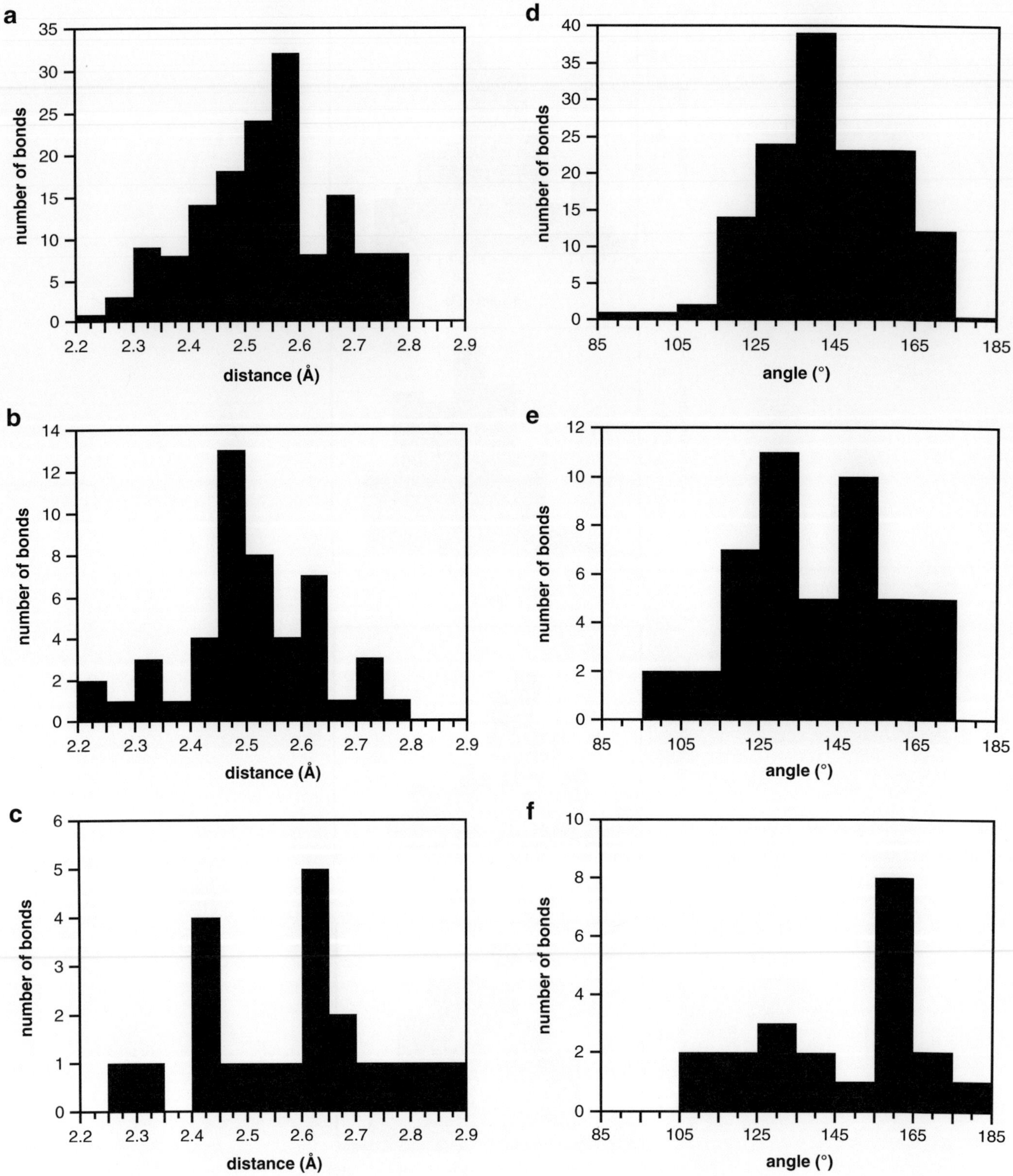

Fig. 5 The distribution of H···F distances and X—H···F angles in X—H···F—C interactions with C—F as the acceptor group (in addition to N and/or O atoms) attached to an aromatic ring system (**a**, **d**), or attached to an aliphatic ring system (**b**, **e**), or attached to a five-membered ring system (**c**, **f**). Distances were measured up to a cutoff value of 3.0 Å.

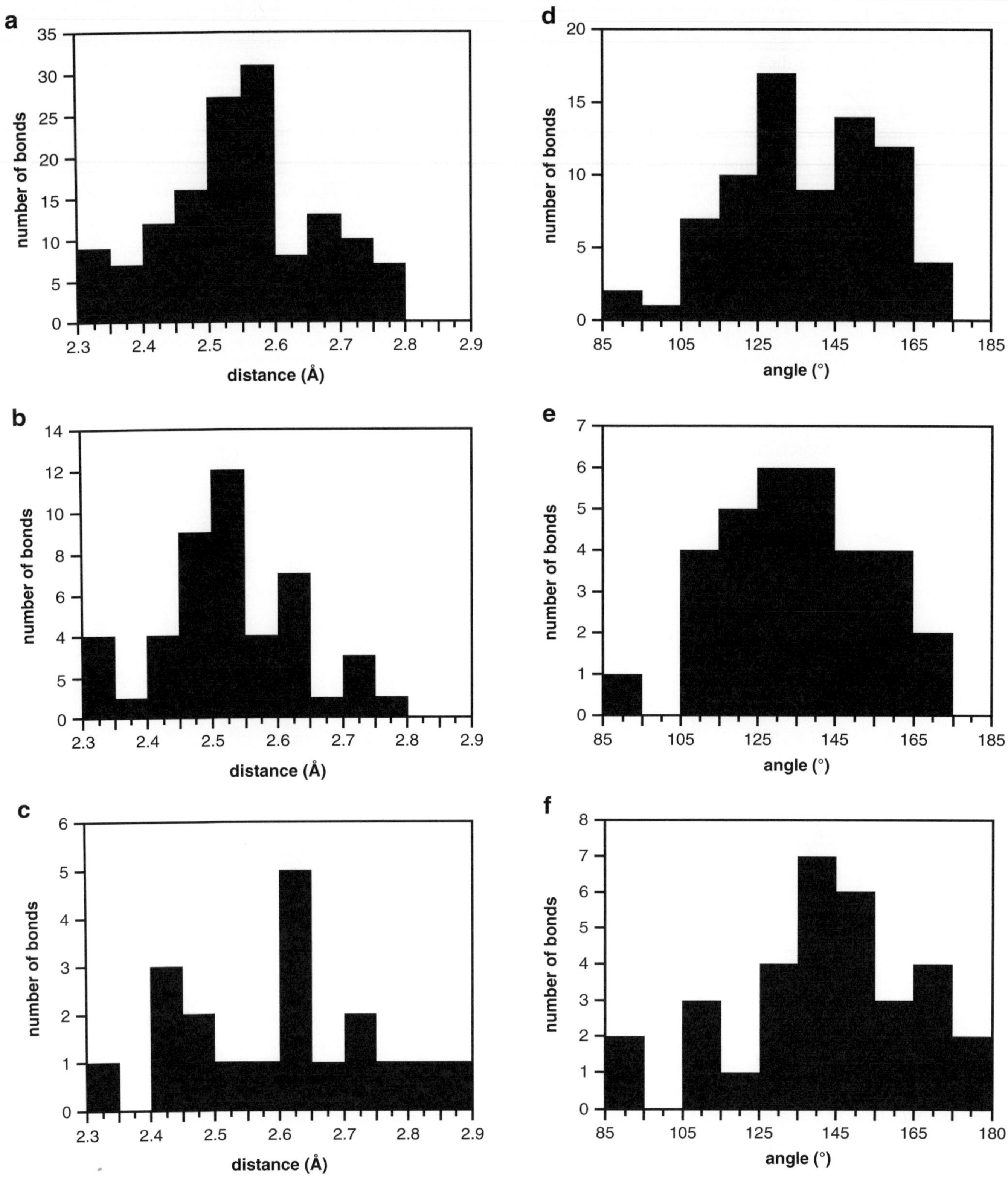

Fig. 6 The distribution of H···F distances and X—H···F angles in X—H···F—C interactions with CF_3 as the acceptor group (in addition to N and/or O atoms) attached to an aromatic ring system (**a**, **d**), or attached to an aliphatic or five-membered ring system (**b**, **e**), or attached to a five-membered ring system (**c**, **f**). Distances were measured up to a cutoff value of 3.0 Å.

In the course of this study we identified a number of compounds that have either an H···F distance or an X—H···F angle, significantly smaller than those for the rest of the compounds (for example, H···F, 2.3 Å or X—H···F, 90°). No correlation was found between short distances and small angles. In attempt to find an explanation to this behavior we studied the partial charge distribution of these compounds and compared the results with those for compounds with "normal" geometry (distances and angles following a normal distribution, i.e., H···F distance, 2.6–2.7 Å and X—H···F, 130–135°). Examples for three compounds from each of the two groups are shown in Figs. 8a–c and 9a–c respectively. We did not find any difference in the charge distribution between the two groups of compounds. In both, the fluorine atom has about the same partial charge (–0.23 to –0.27). But in all the compounds, whether the fluorine atom is the only potential acceptor or whether there is an additional acceptor atom, such as oxygen and/or nitrogen atoms, the partial charge on the fluorine is about the same, and smaller than on the oxygen or nitrogen atoms.

Table 4 Geometry of C—F in H—X interactions in compounds with C—F as the acceptor group in addition to O and/or N atoms

F atom attached to	Mean H···F distance (Å)	Median H···F distance (Å)	Mean X—H···F angle (°)	Median X—H···F angle (°)
All compounds with C—F as the acceptor group in addition to O and/or N atoms	2.5(1)	2.5	141(19)	141
Aromatic six-membered ring	2.5(1)	2.6	141(18)	142
Aliphatic six-membered ring	2.5(1)	2.5	139(19)	138
Five-membered ring[a]	2.6(2)	2.6	146(21)	157

[a]In this group of compounds we extended our study up to 3.0 Å for the H···F distance, since we found valuable information up to that distance

Table 5 Geometry of C—F in H—X interaction in compound with CF_3 as the acceptor group

CF_3 group attached to	Mean H···F distance (Å)	Median H···F distance (Å)	Mean X—H···F angle (°)	Median X—H···F angle (°)
All compounds with CF_3 as the acceptor group	2.5(1)	2.5	137(20)	137
Aromatic six-membered ring[a]	2.5(1)	2.6	137(19)	136
Aliphatic six-membered ring	2.5(1)	2.5	135(20)	135
Five-membered ring	2.6(2)	2.6	138(23)	139

[a]In this group of compounds we extended our study up to 3.0 Å for the H···F distance, since we found valuable information up to that distance

Conclusions

This study provides some insight into the nature of hydrogen bonding. Fluorine is the most electronegative element, but the C—F group is a poor hydrogen-bond former, as was established also from the calculations of the electric charge distribution. On the other hand, C—F···H—O and C—F···H—N interactions (like C—H···O and C—H···N interactions) cannot be ignored in packing considerations. While these interactions are weak, if there is a sufficient number of them they contribute appreciably to the energy of the system [9, 10].

There are two major types of interactions involving the hydrogen-bond interaction, D—H···A; attractive and repulsive interactions [11]. These interactions are

1. Electrostatic interactions that may be either attractive or repulsive
2. Polarization effects
3. Charge transfer
4. Exchange repulsion

Of these the important attractive force relevant to our study, where the acceptor group is a fluorine atom, is the electrostatic interaction, because this type of interaction falls off less

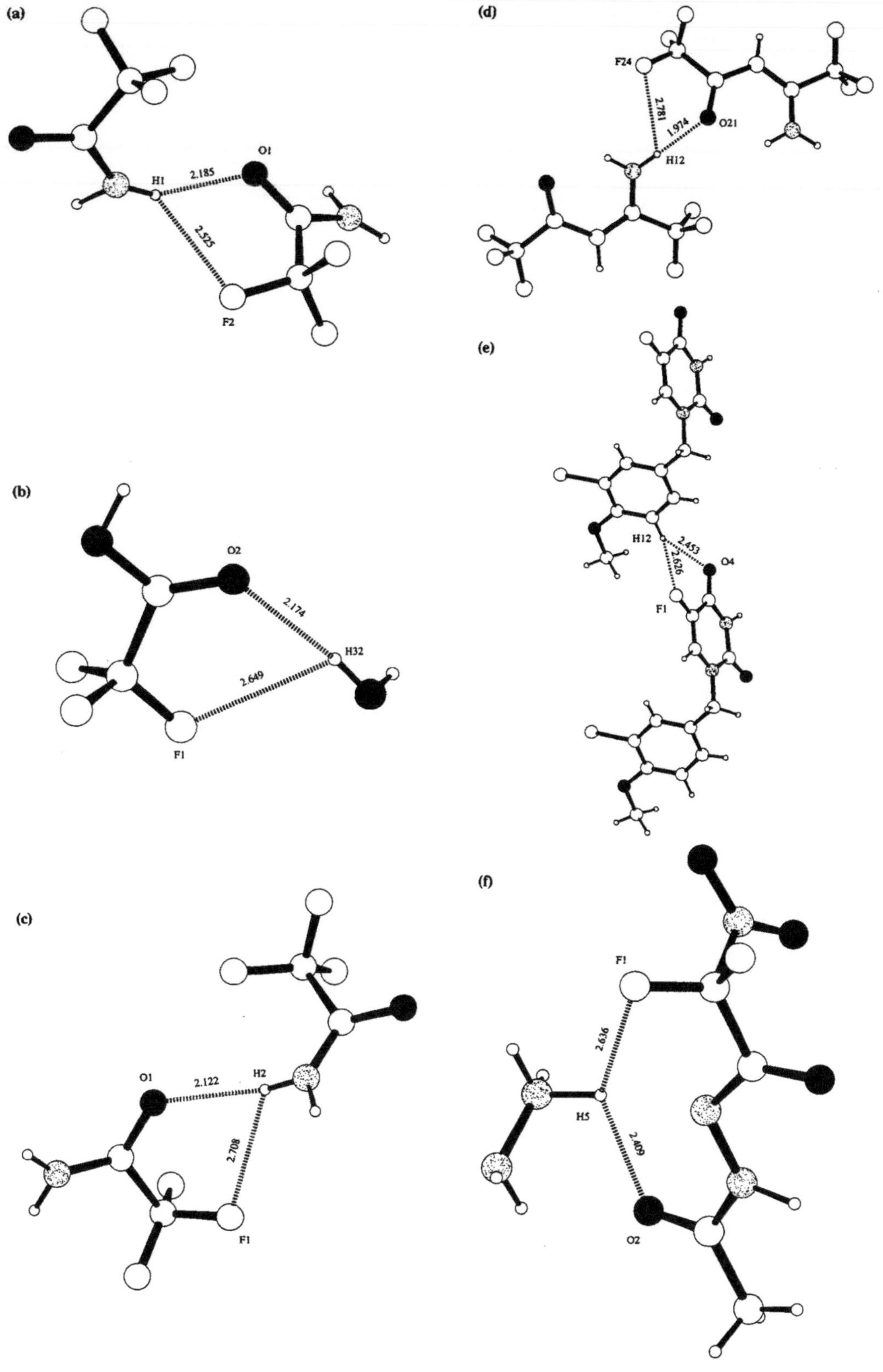

Fig. 7 Examples of selected compounds having three-center bond interactions. (**a**) BDFACM, (**b**) BULMAW10, (**c**) CDFACM, (**d**) DEPSIA, (**e**) FURACM, (**f**) FIYMUV.

Table 6 Geometry of C—F···H—X interaction in compounds that have a three-center hydrogen bond where the second acceptor is an oxygen or nitrogen atom

O atom as the second acceptor in the three-center H bond							N atom as the second acceptor in the three-center H bond						
Refcode	H···F distance (Å)	X—H···F angle (°)	H···O distance (Å)	X—H···O angle (°)	AS[a]	*R*-factor	Refcode	H···F distance (Å)	X—H···F angle (°)	H···N distance (Å)	X—H···N angle (°)	AS[a]	*R*-factor
ADFGLP	2.5	129	2.8	118	1	0.039	FIYMUV	2.6	109	1.9	167	2	0.086
BDFACM	2.5	140	2.2	150	3	0.087	FMPZTZ10	2.6	107	2.5	175	1	0.055
BULMAW10	2.6	110	2.2	167	1	0.051	VIKPUA	2.7	136	2.4	157	2	0.031
CDFACM	2.7	123	2.1	171	2	0.059							
CPFBUR	2.5	138	2.6	149	1	0.042							
DEPSIA	2.8	117	2.0	172	1	0.044							
FURACM	2.6	122	2.4	153	1	0.052							
KAMJOX	2.4	138	2.4	139	2	0.045							
Average	2.6	127	2.3	152				2.7	117	2.3	163		

[a]AS is the mean e.s.d. (C—C) in range 1 = 0.001–0.005 Å, 2 = 0.006–0.010 Å, 3 = 0.011–0.030 Å. The e.s.d. for a C—H bond is about 10 times greater than that for a C—C bond

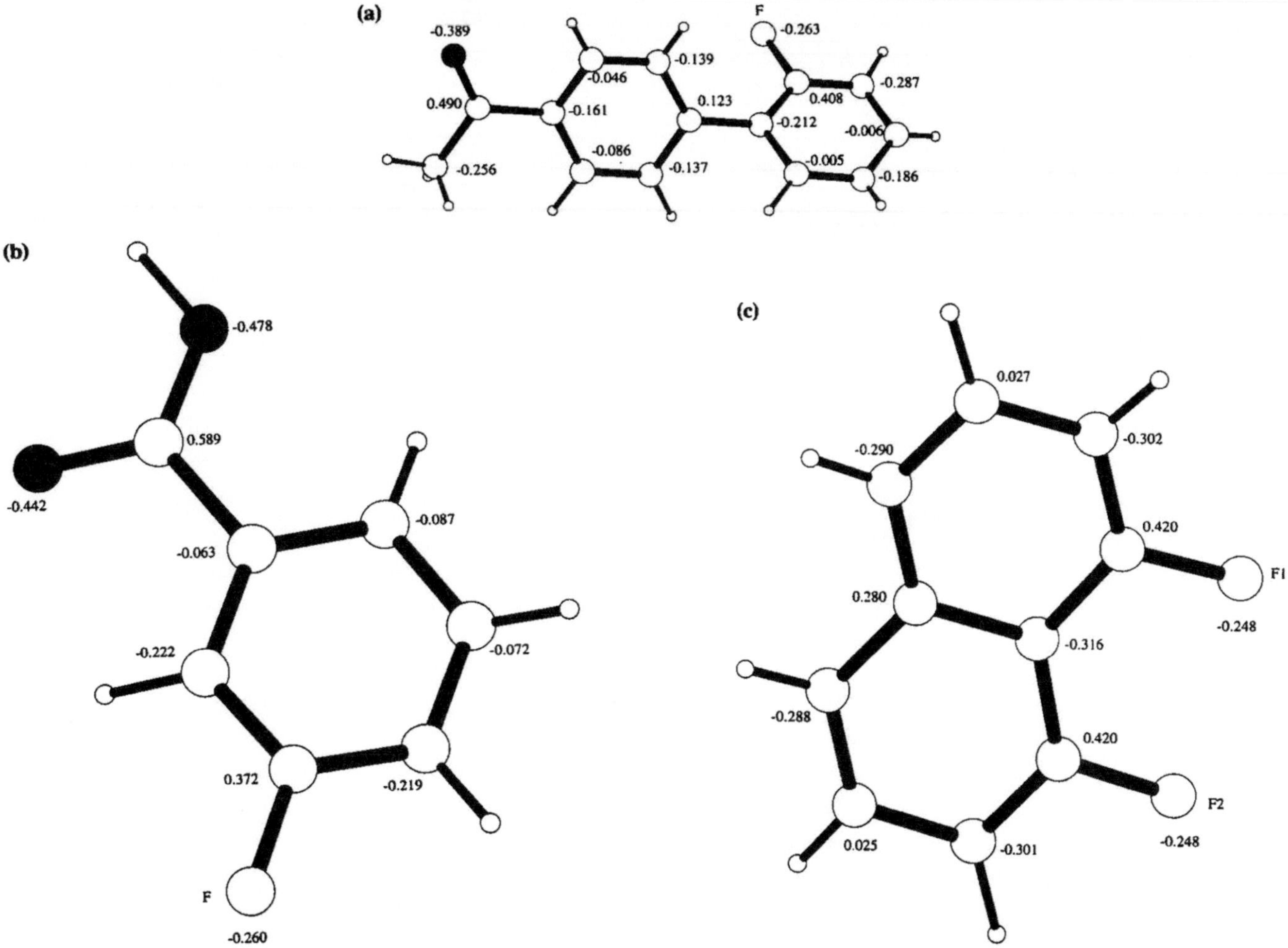

Fig. 8 Example of the electric charge distribution in compounds in which either the H···F distance and/or the X—H···F(X = C,O,N) angle was found to be significantly smaller than most of the compounds studied: (**a**) AFBIPH, (**b**) COVJIG, (**c**) FLNAPH. The filled sphere is an oxygen atom, the speckled sphere is a nitrogen atom, and the fluorine atom is an open sphere with the atom label.

rapidly as a function of distance than do the other attractive interactions. Therefore, at longer distances, as in the case of H···F 2.6 Å, it is the most important type of interaction and the X—H···F binding interaction is mainly an electrostatic one. The exchange repulsion interaction effects mainly the X—H···F angle. The larger repulsion between the two, the more linear the X—H···F will be; that is, the closer it will be to 180°.

In the group of compounds containing only carbon, hydrogen, and fluorine atoms, it was found that H···F distance and the C—H···F angle is about the same in all three different ring systems (aromatic, aliphatic, and "general"). When another acceptor atom, such as oxygen and/or nitrogen, is present in a compound containing a C—F group, the H···F distance tends to be somewhat smaller (about 2.6 Å and 2.5–2.6 Å, respectively) compared to the system containing only fluorine atom as the only acceptor group.

C—F interactions occur whether or not O—H···O, N—H···O interactions are present and these C—F interactions do not appear to be perturbed much by the presence of another hydrogen-bond acceptor. A three-center hydrogen-bond system, where the second acceptor is the oxygen or nitrogen atom in addition to the fluorine atom, was rarely found in the CSD; there were only 11 compounds out of 287 with this type of interaction. This hydrogen-binding interaction is found to be approximately planar with slightly smaller values of the X—H···F angle compared to a two-center bond system where the single acceptor is a fluorine atom.

The importance of C—E···H—C interactions is shown by the crystal structures of FIXXAL and FIXWEO. One side of the molecule contains CF_3 groups, and the other side has many C—H groups. Therefore, the molecules line up head to head and tail to tail. The way these two molecules pack in the crystal indicates that C—F···H—C interactions are acting as stabilizing influences.

This analysis has shown that, while the C—F bond has only a marginal hydrogen-bond acceptor capability, there is

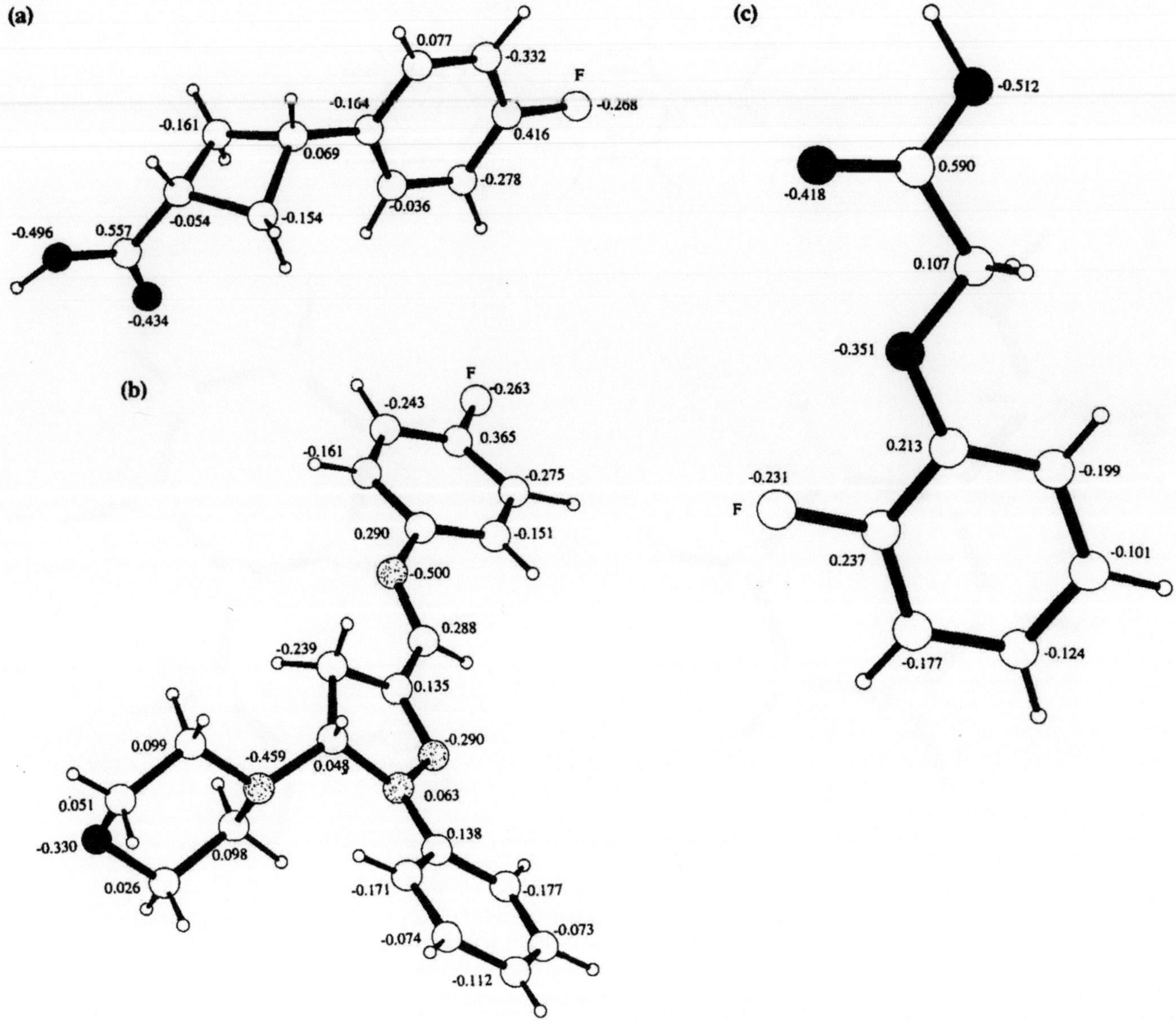

Fig. 9 Example of the electric charge distribution in compounds in which the $H\cdots F$ distance and the $X—H\cdots F(X = C,O,N)$ angle fit in the normal distribution of such values: (**a**) CADMUP, (**b**) DERXUT, (**c**) FADZOZ. The filled sphere is an oxygen atom, the speckled sphere is a nitrogen atom, and the fluorine atom is an open sphere with the atom label.

directionality in its interaction with hydrogen-bond donors, provided better donors are not also present. Such interactions cannot, however, be ignored in predictions of modes of molecular packing in complexes and in crystals.

Acknowledgments We want to thank Ms. Doree Sitkoff (from the Department of Biochemistry and Molecular Biophysics at Columbia University Health Sciences) for calculating the electron charge distribution and for fruitful discussions concerning this paper. This work was supported by NIH grants GM-44360 and CA-06927 and by an appropriation from the Commonwealth of Pennsylvania.

Supplementary Material Available Tables 1, 2, 3, 4, and 5 and the list of CSD reference codes and bibliographical information on crystal structures in this study are available from National Auxiliary Publications Service, c/o Microfische Publications. P.O. Box 3513, Grand Central Station, New York, NY 10163-3513. (NAPS Document No. 05160, consisting of 39 pages.)

References

1. Pauling, L. *J. Am. Chem. Soc.* **1930,** *54,* 3570.
2. Murray-Rust, P.; Stallings, W. C.; Monti, C. T.; Preston, R. K.; Glusker, J.P. *J. Am. Chem. Soc.* **1983,** *105,* 3206.
3. Allen, F. H.; Bellard, S.; Brice, M.D.; Cartwright, B. A.; Doubleday, A.; Higgs, H.; Hummelink, T.; Hummelink-Peters, B. G.; Kennard, O.; Motherwell, W. D. S.; Rodgers, J. R.; Watson, D. G. *Acta Crystallogr., Sect. B* **1979,** *35,* 2331.
4. R. Taylor; Kennard, O. *J. Chem. Soc. Perkin Trans.* **1987,** *2,* 51.
5. Carrell, H. L. CALCAT; Program from The Institute for Cancer Research, Philadelphia, PA, 1977.
6. Erlebacher, J.; Carrell, H. L. ICRVIEW; Program from The Institute for Cancer Research, Philadelphia, PA, 1992.
7. (a) Etter, M. C. *Acc. Chem. Res.* **1990,** *23,* 120; (b) Etter, M. C.; MacDonald, J. M.; Bernstein, J. *Acta Crystallogr., Sect. B* **1990,** *46,* 256.
8. *Carbon-Fluorine Compounds. Chemistry, Biochemistry Biological Activities*; A Ciba Foundation Symposium: Elsevier-Excerpta Medica-North Holland: Amsterdam, London, New York, 1972.
9. (a) Sarma, J. A. R. P.; Desiraju, G. R. *Acc. Chem. Res.* **1980,** *19,* 222; (b) Desiraju, G. R. *Acc. Chem. Res.* **1991,** *29,* 290.
10. Aakeröy, C. B.; Seddon, K. R. *Chem. Soc. Rev.* **1993,** *22,* 397.
11. Umeyama, H.; Morokuma, K. *J. Am. Chem. Soc.* **1977,** *99,* 1316.

Molecular and Crystal Structures of a Class of Push-Pull Quinonoid Compounds with Potential Nonlinear Optical Applications[a]

M. Ravi[b], Shmuel Cohen[c], Israel Agranat[d], and T.P. Radhakrishnan[b,e]

Abstract

Molecular and crystal structures of three compounds with electron donor–conjugation unit–electron acceptor framework are reported where the conjugation unit is formally a quinonoid ring. Semiempirical quantum chemical calculations indicate large hyperpolarizabilities in these push–pull molecules. Since all the crystal structures are found to belong to centrosymmetric space groups, modifications are necessary for quadratic nonlinear optical applications.

Keywords

Push–pull quinonoids • Nonlinear optics • Crystal structures • Hyperpolarizability

Introduction

Molecules with the electron donor–conjugation unit–electron acceptor framework are of interest in the fabrication of a number of molecular electronic devices. The concept of a molecular rectifier has been proposed [1] and tested [2] making use of such molecular entities. Possibility of using such molecules to fabricate molecular devices for solar energy harvesting has been discussed [3]. In recent years there has been an exponential growth in the number of investigations involving the synthesis and systematic study of molecules of this kind, as potential candidates to prepare efficient nonlinear optical (NLO) materials [4]. The fast response to external field stimuli and the large hyperpolarizabilities, β and γ of selected organic molecules and polymers make them prime candidates for NLO applications. For quadratic effects a noncentrosymmetric arrangement of the molecules is essential and this prerequisite has largely limited the choice of systems. However, molecules with large β which form centrosymmetric crystals could be utilized if in appropriate matrices they could be oriented noncentrosymmetrically by poling in strong electric fields or through Langmuir–Blodgett film techniques.

As part of a systematic investigation of a class of molecules having the push–pull framework, we have studied some quinonoid molecules [5] based on the prototypical system, **1**, reported in 1962 by a Du Pont group [6]. The diaminomethylene group "pushes" and the dicyanomethylene group "pulls" the π-electron cloud of the quinonoid ring in these molecules. Semiempirical quantum chemical studies we have carried out showed that these substituted 7,7′-diamino-8,8′-dicyanoquinodimethanes have large hyperpolarizabilities; hence these are of potential interest in the development of novel NLO materials. We consider in this paper the molecules **2**, **3**, and **4** in which the donor amino groups are respectively the six-member saturated ring systems, piperidino, piperazino, and morpholino. These molecules have large ground state dipole moments and crystallize in centrosymmetric crystal lattices. Nevertheless these

[a] *Structural Chemistry* 1996, 7(3):225–232.

[b] School of Chemistry, University of Hyderabad, Hyderabad 500 046, India

[c] Department of Inorganic Chemistry, The Hebrew University of Jerusalem, Jerusalem, Israel

[d] Department of Organic Chemistry, The Hebrew University of Jerusalem, Jerusalem, Israel

[e] Correspondence should be directed to T. P. Radhakrishnan, School of Chemistry, University of Hyderabad, Hyderabad 500 046, India

I. Hargittai, B. Hargittai (eds.), *Science of Crystal Structures: Highlights in Crystallography*,
DOI 10.1007/978-3-319-19827-9_22

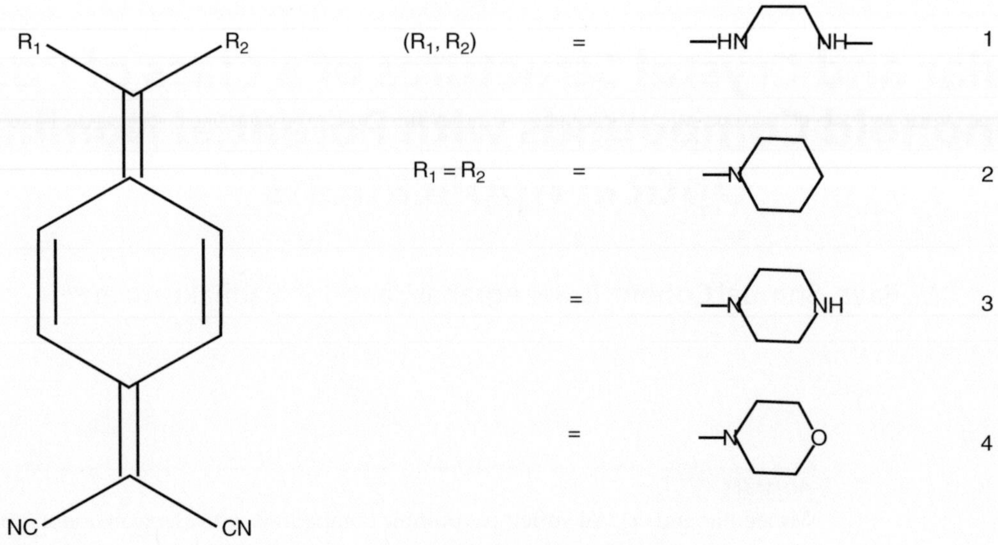

Scheme 1

systems warranted careful study as models based on which modifications to obtain noncentrosymmetric materials could be attempted (Scheme 1).

We present first a brief overview of the theoretical estimate of the β of these systems. The details of the molecular and crystal structure analysis are presented along with the discussion of relevant molecular structure characteristics. The concluding section outlines the future prospects and studies in progress. The experimental section provides the details of synthesis and structure determination.

Results and Discussion

Theoretical Studies of Hyperpolarizability

Semiempirical AM1 method [7] was used for molecular geometry optimizations. The β was calculated using a subroutine, the efficacy of which has been demonstrated by several test calculations [8]. In this procedure, the geometry and the wave functions from the SCF calculations were used in the sum-over-states method employing all single and pair excitations within 12 molecular orbitals (giving rise to a total of 109 configurations). Hyperpolarizabilities were evaluated as the vectorial projection on the strongest dipole axis, based on the common definition

$$\beta_{\text{vec}} = \left(\sum \beta_i^2\right)^{1/2}$$

where

$$\beta_i = \beta_{iii} + \frac{1}{3}\sum_{ji}\left(\beta_{ijj} + \beta_{jji}\right)$$

and β_{iii}, β_{ijj}, etc., are the hyperpolarizability tensor components. The values of β evaluated for excitation energies, $\hbar\omega = 0.0$ eV for the prototypical system, 2-(4-dicyanomethylenecyclohexa-2,5-dienylidene)-imidazolidine, **1**, as well as the molecules **2, 3,** and **4** are provided in Table 1. We note that the only experimental value of β available in this family of compounds ($(-245 \pm 60) \times 10^{-30}$ esu evaluated at $\hbar\omega = 1.17$ eV for **1** [9]) is in excellent agreement with the computed value at that excitation energy (-243×10^{-30} esu). The table provides further, the computed ground state dipole moments and the change in dipole moment associated with the lowest singlet excitation.

The value of β calculated for compounds **1**–**4** are fairly large for the relatively short conjugation lengths they have. For comparison, *p*-nitroaniline has a $\beta(0)$ of -9.1×10^{-30} esu [8]. The large values of β probably arise from the large $\Delta\mu$ associated with the excited states, several of which have large transition dipoles with the ground state [10]. We note that the calculated dihedral angle, ϕ, between the diaminomethylene plane and the quinonoid ring, is found to be generally lower than that found in the crystals. The increased ϕ as well as the more polar environment in the crystal lattice

Table 1 The calculated ground state dipole moment, change in dipole moment on excitation, and hyperpolarizabilities (10^{-30} esu) of 1–4

Molecule	μ_g (D)	$\Delta\mu$ (D)	β_{vec} ($\hbar\omega = 0$ eV)
1	14.8	−14.1	−57.2
2	15.0	−15.8	−73.3
3	12.2	−15.1	−60.5
4	12.5	−15.2	−67.7

are expected to enhance the molecular hyperpolarizabilities. The absorption maxima of **1–4** fall in the range 375–420 nm. It may be noted that the large β values and the relatively high-energy absorptions (near UV) make these molecules attractive candidates for the fabrication of nonlinear optical materials for applications like second-harmonic generation (SHG) in the visible range. However, we note that the large ground state dipole moments (Table 1) increase the probability of obtaining centrosymmetric crystal space groups.

Molecular and Crystal Structures

The space groups, unit-cell details, and features of the X-ray analysis of **2, 3,** and **4** are collected in Table 2. It is seen that all the three compounds belong to centrosymmetric space groups. Though it has been argued [11] that there is no strong correlation between large ground state dipole moments of molecules and their tendency to form centrosymmetric crystal structures, the former property does often encourage a centrosymmetric arrangement of molecules in the crystal lattice. Our systems are typical examples of such trends.

The molecular structure and stereoviews of the unit cells of **2, 3,** and **4** are presented in Figs. 1, 2, and 3 respectively. The significant bond lengths in the three compounds are collected in Table 3 (the atom labelings are as shown in Figs. 1a, 2a, and 3a respectively). In all three molecules, the bonds in the unsaturated six-membered ring that serves as the "conjugation unit" in the push–pull system are found to be nearly benzenoid in character, not quinonoid. The lengths of the exocyclic bonds as well as the N—C bonds at the amino end and C—N bonds at the cyano end also reflect the dominance of the zwitterionic resonance form with strong positive charge on the diaminomethylene unit and negative charge on the dicaynomethylene unit. Our theoretical analysis [10] of the parent system 7,7′-diamino-8,8′-dicyanoquinodimethane with structural changes induced by the influence of varying external fields, indicated a peaking of the β at structures intermediate between the benzenoid and quinonoid extremes but closer to the benzenoid end. This implies that the large hyperpolarizabilities of these strongly benzenoid zwitterionic molecules we have synthesized could still be improved by slightly decreasing the zwitterionic nature by appropriate chemical modifications.

The molecular structures of **2, 3,** and **4** reveal normal bond angles within the benzenoid ring, the exocyclic bonds and the heterocyclic substituents. The unsaturated rings are quite planar in all the structures. Table 3 provides two important dihedral angles ϕ_1 and ϕ_2 in these structures, ϕ_1 is the dihedral angle between the average planes of the dicyanomethylene unit and the benzenoid ring, and σ_2 is the similar angle between the diaminomethylene unit and the benzenoid ring. The ϕ_1 values indicate that the dicyanoquinodimethane moiety is almost coplanar with the benzenoid ring. However, the large values of ϕ_2 (43–49°) indicate that a strong out-of-plane twisting of the diaminomethylene unit occurs in these molecules. The analogy to the well-known [12] twisted intramolecular charge transfer (TICT) systems is not out of place; the important contrast with conventional TICT systems is that the twisting in our systems is a ground state phenomenon. The twisting arises from the overcrowding due to the steric repulsions between the *ortho* H atoms on the benzenoid ring and those on the neighboring carbons of the nitrogens of the heterocyclics. The twisting is further facilitated by the zwitterionic nature of the molecules with the consequent weakening of the C1—C7 bond (Table 3). We believe that this strong molecular twisting could provide isolable chiral molecular structures and attempts to induce noncentrosymmetric lattices exploiting such chiralities are in progress in our laboratory.

We find that the three crystal structures irrespective of the donor amino group are rather similar. Since the steric effect due to the six-member rings are similar and the electronic nature of the amines are practically identical, this is not surprising. The possibility of H-bonding through the piperazine unit in **3** does not appear to have developed any novel features in the crystal packing. However, if one examines the molecular dipole arrangements in the *ac* plane projections (in all the unit cells the longest axis is the *b* axis) in **2, 3,** and **4** the following subtle differences in the packing of dipoles can be observed. In **2**, the dipoles are arranged antiparallel to

Table 2 Crystallographic data for compounds **2, 3,** and **4**.

Compound	2	3	4
Molecular formula	$C_{20}H_{24}N_4$	$C_{18}H_{22}N_6$	$C_{18}H_{20}N_4O_2$
Morphology	Cubes	Plates	Needles
Crystal system	Triclinic	Triclinic	Monoclinic
Space group	$P\bar{1}$	$P\bar{1}$	$P2_1/n$
a, Å	13.540(1)	13.372(1)	12.695(1)
b, Å	16.239(1)	15.614(1)	15.355(1)
c, Å	8.771(1)	8.756(1)	8.756(1)
α, deg.	101.78(1)	99.31(1)	90.0
β, deg.	104.79(1)	105.63(1)	106.44(1)
γ, deg.	95.61(1)	97.06(1)	90.0
V, Å^3	1802.4(5)	1710.5(6)	1656.0(4)
z	4	4	4
ρ(calcd.), g cm^{-3}	1.18	1.25	1.30
μ(CuK$_\alpha$), cm^{-1}	5.24	5.91	6.71
No. of unique reflections	5260	5073	2404
No. of reflections with $I \geq 3\sigma_I$	4635	4591	2171
R	0.034	0.034	0.032
R_w	0.062	0.061	0.056

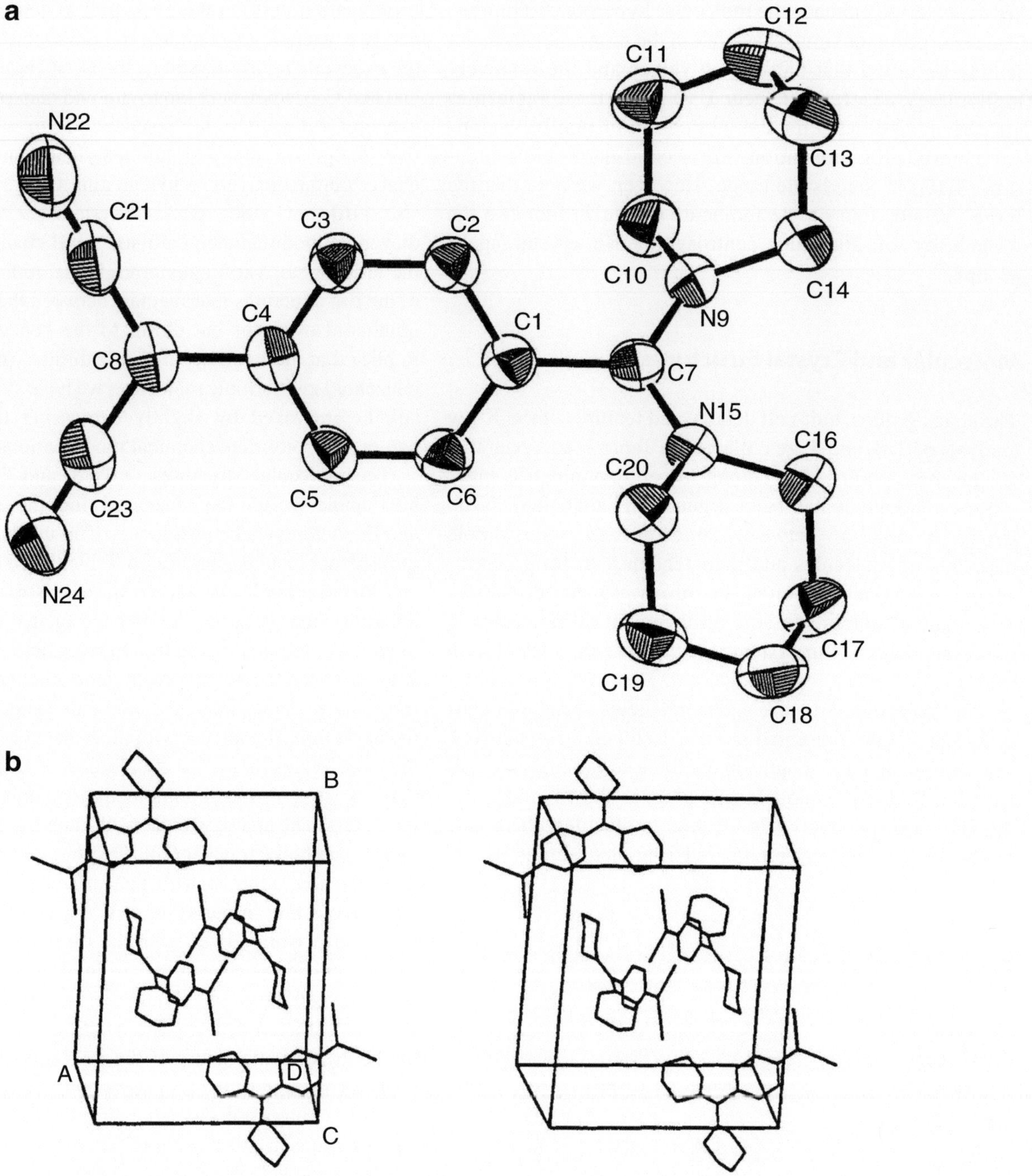

Fig. 1 (**a**) Molecular structure and (**b**) stereoview of the unit cell along the *c*-axis of **2**.

form two distinct dimer units which form a centric array roughly perpendicular to the *b* axis. In **4,** parallel headto-tail arrangement of the dipoles from a noncentric array in the *ac* plane. However, the adjacent plane has opposite orientation leading to an overall centrosymmetric structure. In **3,** the arrangement is more complex; the molecules form zigzag head-to-tail chains approximately in the [101] direction, with adjacent chains out of phase with each other. The effectively noncentric array together with the neighboring arrays of opposite orientation again leads to centrosymmetric lattice. If by inclusion of solvent molecules or cocrystallizing with weak dipolar molecules, the interaction between the noncentric arrays in **3** or **4** could be diminished, there is a possibility of fabricating noncentrosymmetric lattices.

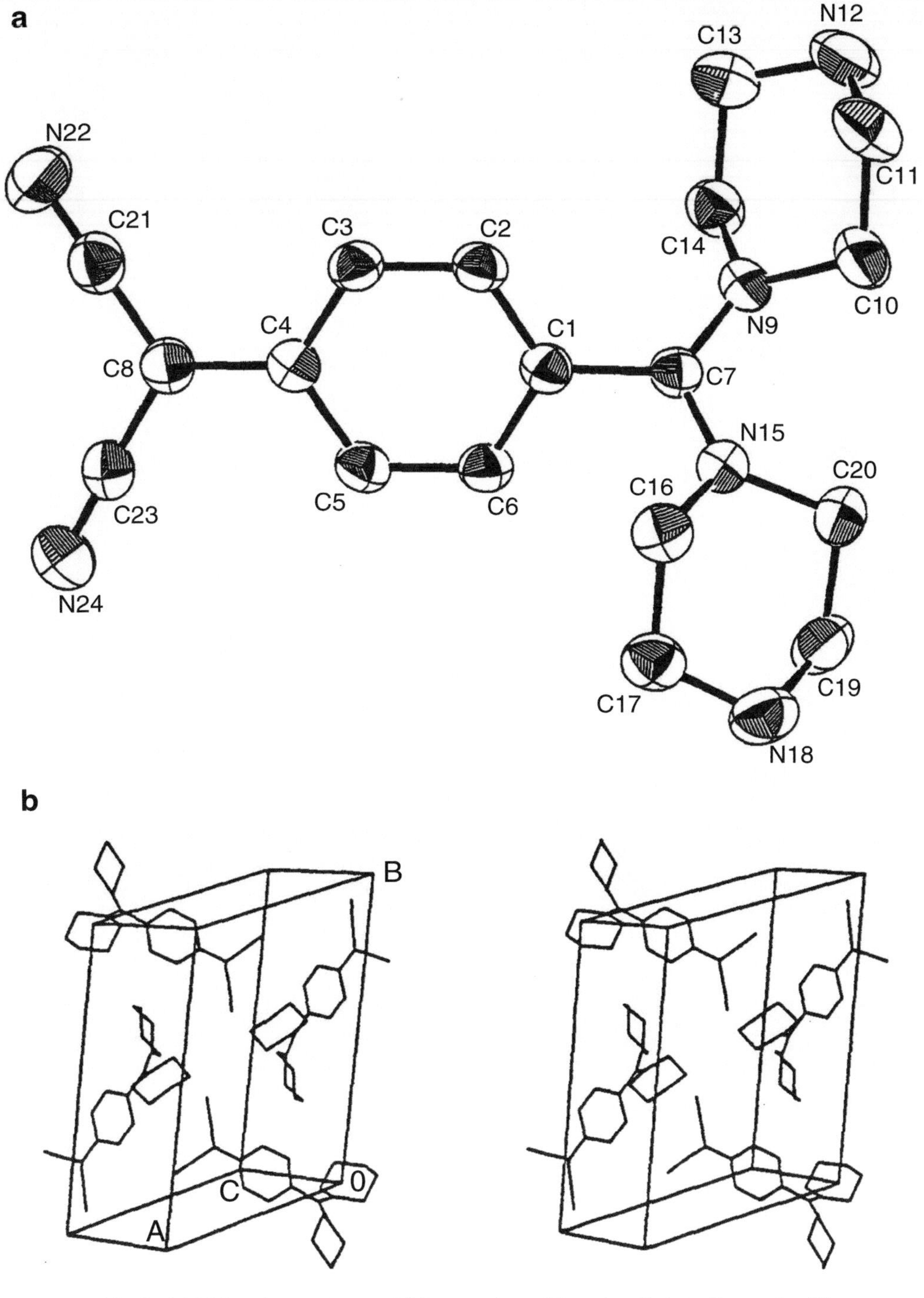

Fig. 2 (**a**) Molecular structure and (**b**) stereoview of the unit cell along the *a*-axis of **3**.

Conclusion

Diaminodicyanoquinodimethane systems provide a novel class of push–pull quinonoid molecules with suitable molecular properties to develop molecular NLO materials. The three molecules **2–4** presented here are typical examples of these easily synthesized compounds. These molecules crystallize quite well; the crystal structure analysis we have reported in this paper, however, indicated centrosymmetric space groups for all the molecules. It is concluded that though these systems have potentially useful nonlinear optical properties at the molecular level, the high ground state dipole moments are a hurdle to noncentric crystal lattice formation. The tendency toward a centrosymmetric arrangement of the

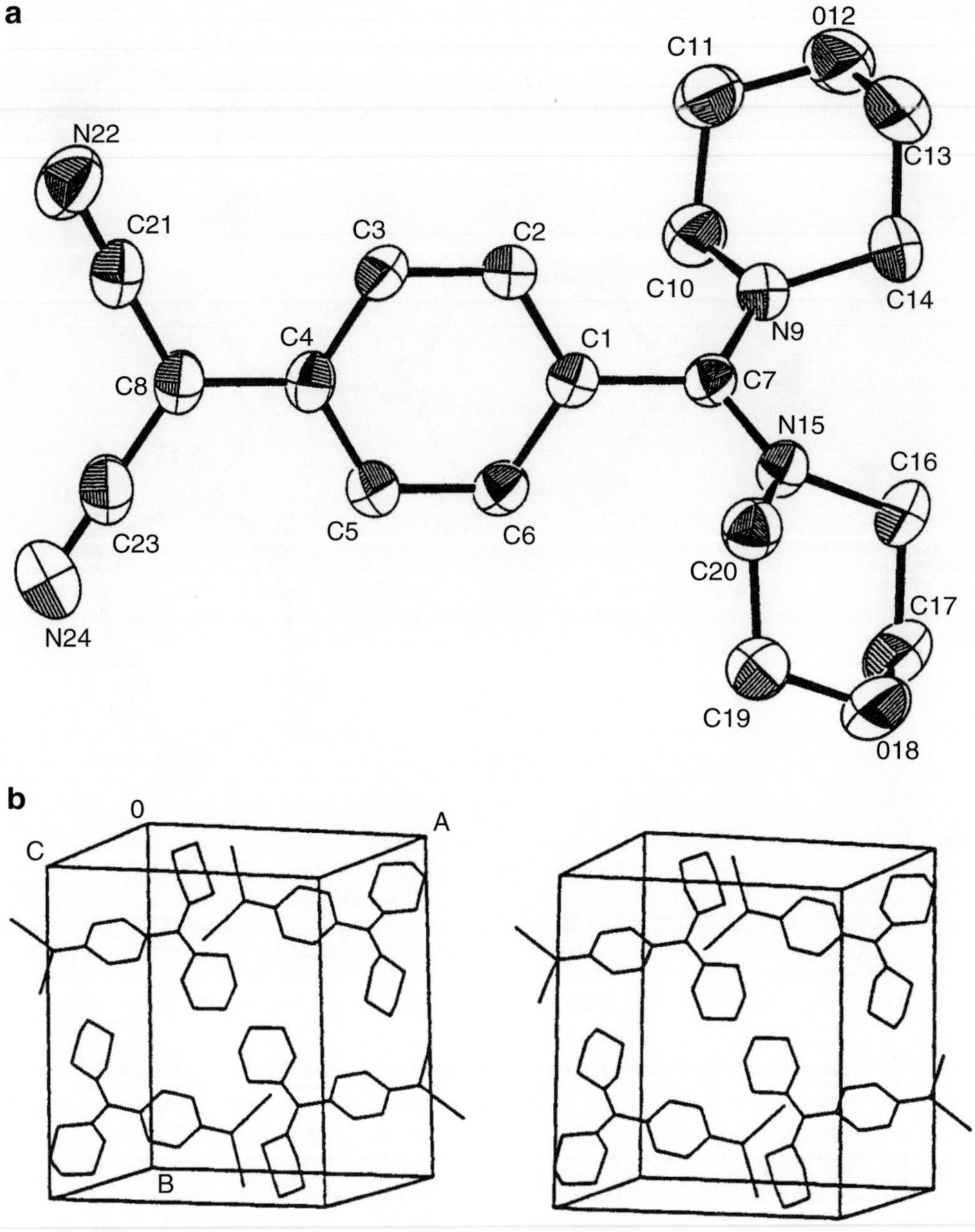

Fig. 3 (**a**) Molecular structure and (**b**) stereoview of the unit cell along the *c*-axis of **4**.

molecules has to be overcome by suitable crystal engineering strategies or poling techniques, so that useful quadratic NLO materials can be fabricated from this class of molecules. Some initial success in observing solid-state second-harmonic generation in crystalline materials based on these molecules utilizing chiral substituents such as (*S*)-2-methoxymethylpyrrolidine has already been achieved [5, 13]. The molecular chirality necessitates for symmetry reasons a noncentric crystal lattice and hence quadratic nonlinear effects. Our current efforts are directed at improving the SHG efficiency.

Experimental

Synthesis

Compounds **2–4** were prepared following the procedure reported by Hertler *et al.* [6]. The preparation procedure for these compounds is exemplified by the case of **3**. Since the synthesis and characterization of **2** has been reported in Ref. [14], we provide the characterization details only for **3** and **4**.

To a warm solution of 100 mg of 7,7′,8,8′-tetracyanoquinodimethane (TCNQ) in 10 ml of tetrahydrofuran, 170 mg

Table 3 Important bond lengths (see Figs. 1, 2, and 3 for labeling) in Å and dihedral angles (see text for definition) in degrees in 2, 3, and 4

Compound	2	3	4
C7–N9	1.333	1.336	1.333
C7–N15	1.335	1.335	1.337
C1–C7	1.467	1.465	1.466
C1–C2	1.394	1.392	1.396
C1–C6	1.395	1.396	1.397
C2–C3	1.370	1.373	1.371
C5–C6	1.369	1.372	1.372
C3–C4	1.406	1.403	1.405
C4–C5	1.405	1.404	1.407
C4–C8	1.438	1.436	1.440
C8–C21	1.399	1.404	1.407
C8–C23	1.405	1.407	1.404
ϕ_1	5.03	6.62	1.37
ϕ_2	46.76	43.01	48.56

of piperazine dissolved in 5 ml was added. A dark green color developed immediately, then disappeared slowly and a yellow crystalline compound began to precipitate. After standing for 5 h the solution was cooled to about 10 °C and the precipitate was filtered out. The filter cake was washed first with cold tetrahydrofuran and then with ether to give 110 mg (70 % yield) of the crystalline compound **3**.

7,7′-Dipiperazino-8,8′-Dicyanoquinodimethane (3)

Recrystallization from acetonitrile gave yellow platelike crystals which did not melt up to 350 °C. The infrared spectrum of the compound (KBr wafer) shows absorptions at 2180, 2140 cm^{-1}, (conjugated nitrile), and 3150 cm^{-1} (N-H absorption). The UV–VIS absorption spectrum in acetonitrile solution shows a band at 415 nm. Elemental analysis (calculated for C, H, N): C, 66.31 % (67.08 %); H, 6.82 % (6.83 %); N, 25.39 % (26.08 %).

7,7′-Dimorpholino-8,8′-Dicyanoquinodimethane (4)

Recrystallization from acetonitrile gave needlelike crystals which did not melt up to 350°. The infrared spectrum (KBr wafer) shows absorptions at 2170 and 2130 cm^{-1} (conjugated nitrile). The UV–VIS absorption spectrum in acetonitrile solution shows a band at 420 nm. Elemental analysis (calculated for C, H, N): C, 66.65 % (66.66 %); H, 6.28 % (6.17 %); N, 16.57 % (17.28 %).

Crystal Structure Determination

The X-ray intensity data were collected on an EN-RAF-NONIUS CAD4 computer-controlled diffractometer. CuKα (λ = 1.54178 Å) radiation with a graphite crystal monochromator in the incident beam was used. The standard CAD4 centering, indexing, and data collection programs were used. The unit-cell dimensions (Table 2) were obtained by a least-squares fit of 24 centered reflections in the range of $21 \leq \theta \leq 30°$. Further details are provided in the supplementary material.

Acknowledgment We thank Professor J. Chandrasekhar for providing the subroutine for the hyperpolarizability calculations and MR thanks the Council of Scientific and Industrial Research and TPR the Department of Science and Technology, New Delhi for financial support.

Supplementary Material Available
A listing of crystallographic data (21 pages) is available from the authors upon request.

References

1. Aviram, A.; Ratner, M.A. *Chem. Phys. Lett.* **1974,** *29,* 277.
2. Martin, A. S.; Sambles, J. R. *Phys. Rev. Lett.* **1993,** *70,* 218.
3. Mirkin, C. A.; Ratner, M.A. *Ann. Rev. Phys. Chem.* **1992,** *43,* 719.
4. (a) Chemla, D. S.; Zyss, J., Eds.; *Nonlinear Optical Properties of Organic Molecules and Crystals;* Academic Press: New York, 1987. (b) Articles in *Chemical Reviews* **1994,** *94* (1).
5. Ravi, M.; Rao, D. N.; Cohen, S.; Agranat, I.; Radhakrishnan, T. P. *Curr. Sci. (India)* **1995,** *68,* 1119.
6. Hertler, L. R.; Hartzler, H. D.; Acker, D. S.; Benson, R. E. *J. Am. Chem. Soc.* **1962,** *84,* 3387.
7. Dewar, M. J.; Zoebisch, E. G.; Healy, E. F.; Stewart, J. J. P. *J. Am. Chem. Soc.* **1985,** *107,* 3902.
8. Clark, T.; Chandrasekhar, J. *Isr. J. Chem.* **1994,** *B33,* 435.
9. Lalama, S. J.; Singer, K. D.; Garito, A. F.; Desai, K. N. *Appl. Phys. Lett.* **1981,** *39,* 940.
10. Ravi, M.; Radhakrishnan, T. P. *J. Phys. Chem.* (in press).
11. Whitesell, J. K.; Davis, R. E.; Saunders, L. L.; Wilson, R. J.; Feagins, J.P. *J. Am. Chem. Soc.* **1991,** *113,* 3267.
12. Rettig, W. *Angew. Chem. Int. Edn. Engl.* **1986,** 25, 971.
13. Ravi, M.; Narayana Rao, D.; Cohen, S.; Agranat, I.; Radhakrishnan, T. P. (submitted for publication).
14. Gomper, R.; Wagner, H.; Kutter, E. *Chem. Ber.* **1968,** *101,* 4144.

X-ray Crystallographic Studies on the Noncovalent Syntheses of Supermolecules[a]

Matthew C.T. Fyfe[b], J. Fraser Stoddart[b,d], and David J. Williams[c]

Abstract

An approach to the supramolecular syntheses of discrete multicomponent aggregates of noncovalently bound molecules, i.e., supermolecules, is described. This approach involved the careful analysis of X-ray crystal structures so as to permit a gradual increase in superstructural complexity. Many elaborate supermolecules were synthesized noncovalently from dialkylammonium-containing cations and crown ethers, following the initial observation that the dibenzylammonium ion threads through dibenzo[24]crown-8 to generate a singly stranded, singly encircled [2]pseudorotaxane, principally as a result of $[N^+—H\cdots O]$ and $[C—H\cdots O]$ hydrogen bond formation. The scope of the fundamental recognition motif obtained from this initial observation was then broadened, through the use of thread-like ions with multiple dialkylammonium centers and/or larger crown ethers, so that multiply stranded and/or multiply encircled pseudorotaxanes could be prepared. Cations bearing both dialkylammonium and crown ether recognition sites were also used for the noncovalent synthesis of a discrete daisy chain supermacrocycle and the basic recognition motif was combined with other motifs for the production of a wide range of novel superarchitectures. As a greater understanding of the noncovalent interactions governing the self-assembly of the complex superarchitectures was acquired, new protocols for the noncovalent syntheses of doubly docked pseudorotaxanes and interwoven supramolecular bundles, including a supramolecular analogue of the photosynthetic special pair, were developed. The discovery that anions can play a prominent role in the solid-state self-assembly of some of the supermolecules was a valuable spinoff of the research.

Keywords

Noncovalent interactions • Pseudorotaxanes • Self-assembly • Solid-state structures • Supramolecular chemistry

Introduction

In the past few years, the noncovalent bond [1] has become an increasingly important constituent of the synthetic chemist's repertoire, partly because the chemistry of the covalent bond has been almost stretched to its limits [2]. As a result, the field of supramolecular synthesis [3], i.e., the preparation of supramolecular [1] entities using noncovalent bonds, has become a flourishing area of contemporary chemical research to the extent that

[a] *Structural Chemistry* 1999, 10(3):243–259.

[b] Department of Chemistry and Biochemistry, University of California, Los Angeles, CA 90095, USA
e-mail: stoddart@chem.ucla.edu

[c] Department of Chemistry, Imperial College, South Kensington, London SW7 2AY, UK

[d] Correspondence should be directed to Professor J. F. Stoddart, Department of Chemistry and Biochemistry, University of California, Los Angeles, California 90095; Telephone: (310) 206-7078; Fax: (310) 206-1843; e-mail: stoddart@chem.ucla.edu

I. Hargittai, B. Hargittai (eds.), *Science of Crystal Structures: Highlights in Crystallography*, DOI 10.1007/978-3-319-19827-9_23

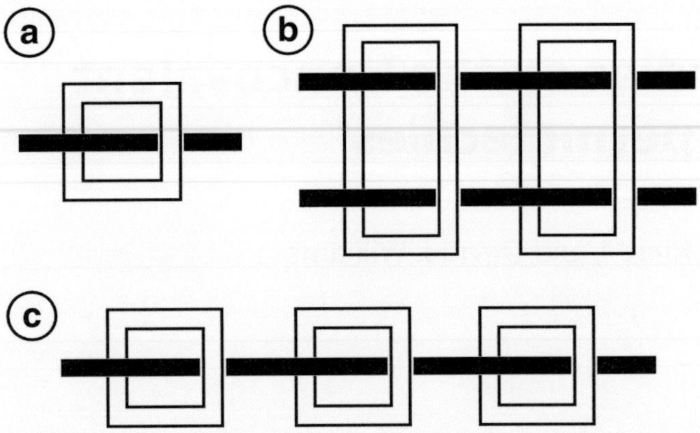

Fig. 1 Diagrams depicting the superstructures of several multicomponent pseudorotaxanes. Pseudorotaxanes are inclusion complexes that have been defined [11] as the noninterlocked counterparts of rotaxanes [12]. They are formed when one or more thread-like molecules/ions interpenetrate the cavities of one or more macrocyclic species. As distinct from their mechanically interlocked congeners, i.e., the rotaxanes, these supramolecular complexes are free to dissociate into their separate components as they do not possess bulky stopper groups. Their analogy in the macroscopic world would consist of one or more beads being threaded onto one or more strings (that have no knots at their ends). Pseudorotaxanes are comprised of *n* components, as indicated by the prefix [*n*]. Thus, the diagrams show (**a**) [2]-, (**b**) [4]-, and (**c**) [4]pseudorotaxanes.

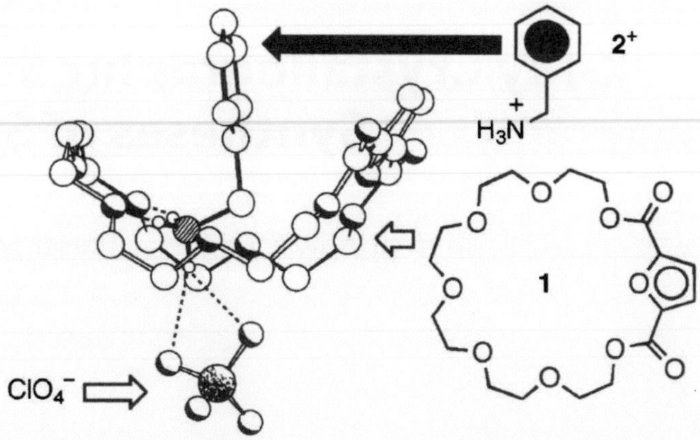

Fig. 2 The crystal structure of the ion-pair pseudorotaxane [**1**·**2**][ClO_4] generated as a result of the bifurcated $\left[N^+—H\cdots O\right]$ hydrogen bonds between one of the NH_3^+ hydrogen atoms and two of the oxygen atoms of the ClO_4^- counterion.

synthetic chemists are now able to utilize noncovalent bonding interactions for the self-assembly [4] of a diverse series of intermolecularly bound entities, ranging from the discrete, oligomeric supermolecules [5] to the infinite, polymeric supramolecular arrays [6]. X-ray crystallography has played [7] a pivotal role in this aspect of synthetic supramolecular chemistry [3]. By carefully analyzing the noncovalent interactions that hold certain superstructures together in the solid state, the synthetic supramolecular chemist has been able to prepare ever more complex and intricate supramolecular architectures noncovalently, with a view to the ultimate creation of devices [8] and materials [9], the likes of which have been observed previously only in the natural world. In this review, we discuss how X-ray crystallography has aided our noncovalent syntheses of discrete supramolecular entities, i.e., supermolecules, using recognition motifs based upon the interaction between crown ethers and secondary dialkylammonium ions [10]. Initially, these recognition motifs were employed solely to synthesize simple pseudorotaxanes (Fig. 1) noncovalently. Subsequently, we have been able to increase superstructural complexity gradually, through the application of X-ray crystallography as a tool that augments our basic understanding of the recognition motifs.

Discovery of a New Recognition Motif

The motivation for our initial investigations [13] on the noncovalent synthesis of pseudorotaxanes from crown ethers and dialkylammonium ions was obtained from the crystal structure [14] of the 1:1 complex formed (Fig. 2) between the [24]crown-8 derivative **1** and benzylammonium perchlorate (**2**·ClO_4). In this complex, only two of the NH_3^+ hydrogen atoms are involved in hydrogen bonds with the oxygen atoms of the crown ether, the other hydrogen atom interacting with the ClO_4^- counterion, via hydrogen bonding *through* the macrocycle's cavity, to generate, what has been termed [13c–d], an ionpair pseudorotaxane. This crystal structure led us to conclude that we could effect the noncovalent syntheses of crown ether–dialkylammonium ion-based pseudorotaxanes by replacing this superstructure's $—H\cdots ClO_4^-$ unit with an alkyl substituent. Indeed, co-crystallization [13a, c–d] of the crown ether dibenzo[24]crown-8 (DB24C8) with dibenzylammonium hexafluorophosphate (**3**·PF_6) furnishes (Fig. 3) two crystallographically independent [2]pseudorotaxanes [DB24C8·**3**][PF_6], where, in both superstructures, the cation is threaded through the macrocycle

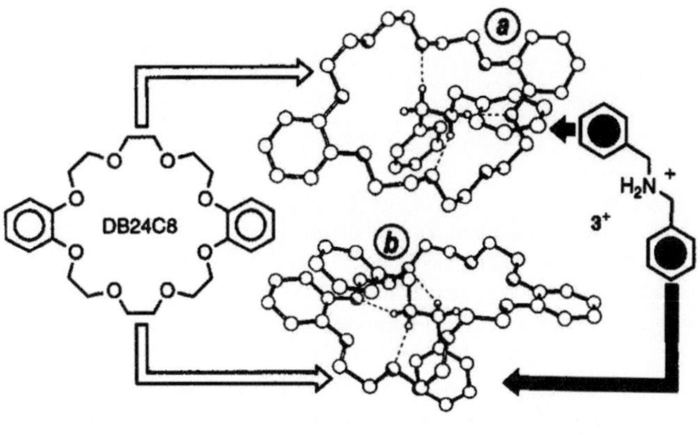

Fig. 3 The two crystallographically independent [2]pseudorotaxanes (*a* and *b*) that self-assemble from DB24C8 and 3^+. In addition to the hydrogen bonds that stabilize both complexes, pseudorotaxane (*b*) is stabilized by a π-π stacking interaction between both components.

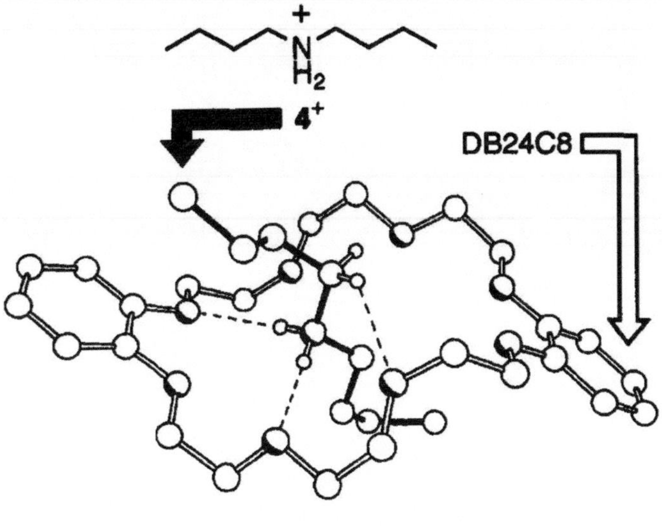

Fig. 4 The crystal structure of [2]pseudorotaxane [DB24C8·**4**]$^+$, which is stabilized by means of $[N^+—H\cdots O]$ and $[C—H\cdots O]$ hydrogen bonds.

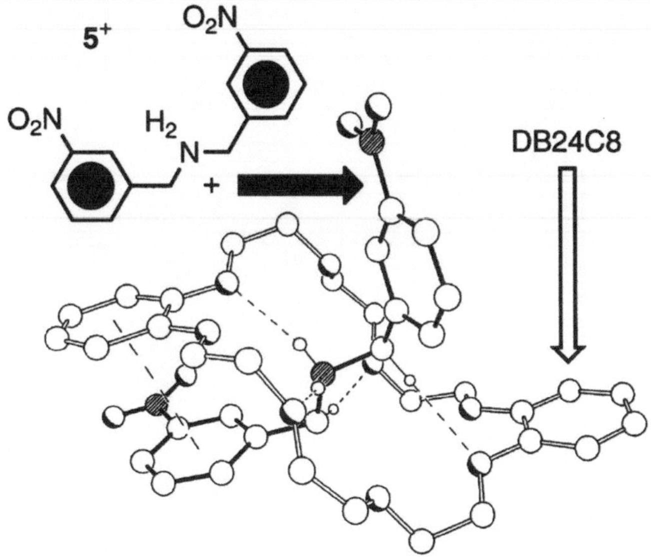

Fig. 5 The crystal structure of the [2]pseudorotaxane formed from DB24C8 and the bis(3-nitrobenzyl)ammonium ion (**5**$^+$). This superstructure is stabilized by two $[N^+—H\cdots O]$ and two $[C—H\cdots O]$ hydrogen bonds, together with a π-π stacking interaction between one of the π-electron-rich catechol rings and one of the π-electron-deficient nitrobenzyl groups.

primarily by means of $[N^+—H\cdots O]$ and $[C—H\cdots O]$ hydrogen bonds from the hydrogen atoms of the cation's NH_2^+ and CH_2 units, respectively, to the polyether oxygen atoms of DB24C8. Thereafter, several similar [2]pseudorotaxanes were synthesized noncovalently from DB24C8 and other dialkylammonium ions, including the di-*n*-butylammonium ion (**4**$^+$) ([13a, c–d], Fig. 4), two dicycloalkylammonium ions [15], and a wide range of substituted dibenzylammonium ions ([16], Fig. 5). Moreover, all of these pseudorotaxanes form in solution, as evidenced by ^{1}H NMR spectroscopic studies. FAB/LSI mass spectrometric analyses have also permitted pseudorotaxane characterization in the gas phase. The observation of pseudorotaxane formation in solution has led to several novel syntheses [17] of rotaxanes and molecular switches.

Extending the Recognition Motif

Recently, we have situated both the DB24C8 and secondary dialkylammonium recognition sites on the same covalent skeleton in order to effect the noncovalent synthesis of daisy chain superarchitectures [18]. The cation **6**$^+$ (Fig. 6) was designed in such a way that its complementary recognition sites would interact with each other in an intermolecular [19], rather than an intramolecular [20], sense, i.e., instead of biting its own tail, the self-complementary monomer **6**$^+$ would be predisposed to self-assemble into a supramolecular daisy chain. However, one question still remained: Would **6**$^+$ oligomerize noncovalently to generate discrete daisy chain supermacrocycles, or would it polymerize, via noncovalent bonds, to produce a macromolecular daisy chain [21]? The answer came from X-ray crystallography, which demonstrated, along with LSI mass spectrometry, that **6**$^+$ dimerizes noncovalently to create a two-component supramolecular macrocycle that is equivalent to the smallest possible cyclic daisy chain. The X-ray crystallographic analysis of **6**$\cdot O_2CCF_3$ reveals (Fig. 7) the formation of this C_2-symmetric daisy chain, in which each component's dialkylammonium-bearing sidearms are threaded concurrently through the complementary DB24C8 recognition sites of their counterparts. The $[(\mathbf{6})_2]^{2+}$ supermolecules are stabilized, once again, by $[N^+—H\cdots O]$ hydrogen bonds, as well as by π-π stacking interactions between the DB24C8 catechol rings that support the dialkylammonium-containing sidearms.

Could the primitive recognition motif exhibited by the [DB24C8·**3**]$^+$ complex be developed further so that multiply encircled pseudorotaxanes, consisting of one oligo(dialkylammonium) chain threaded through several DB24C8 macrorings, could be synthesized noncovalently? These multicomponent complexes would be the precursors of polyrotaxanes [22], each comprised of numerous crown ethers threaded onto a poly(dialkylammonium) strand. The answer to the question posed at the beginning of this paragraph is yes [13b–d], as [3]pseudorotaxanes, composed of a bis(dialkylammonium) strand threaded through two crown ether macrorings, were generated when DB24C8, or its asymmetric congener asym-DB24C8, was co-crystallized

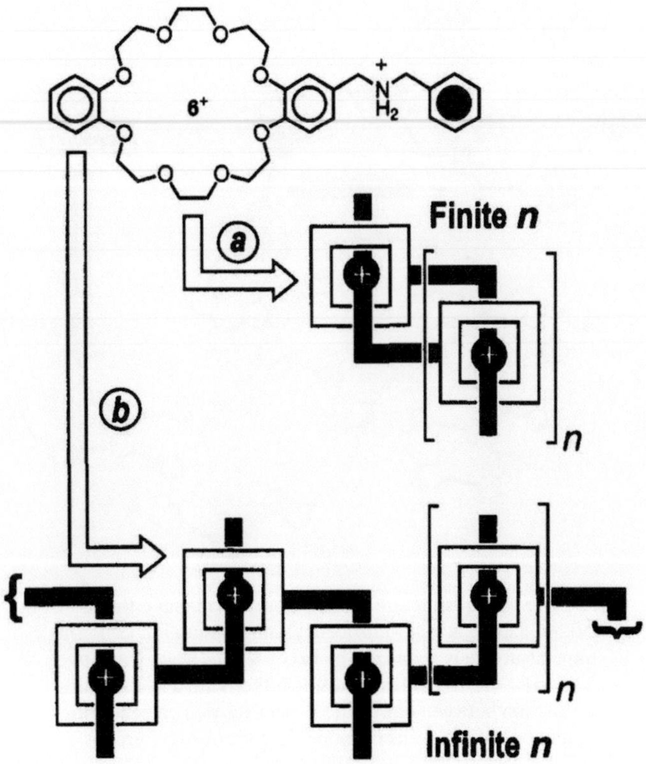

Fig. 6 Two options are open to the self-complementary cation $\mathbf{6}^+$ upon self-assembly. (**a**) A finite number of cations can aggregate, in a cyclic fashion, to generate oligomeric daisy-chain-like supermacrocycles, or (**b**) an infinite number of $\mathbf{6}^+$ monomers can polymerize noncovalently to create a macromolecular daisy chain.

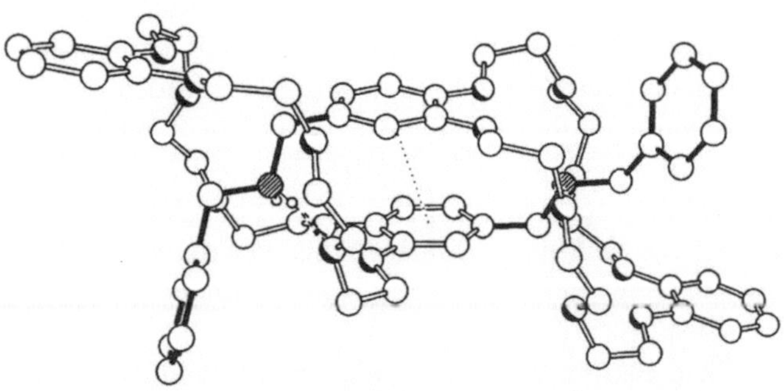

Fig. 7 View of the two-component supermolecule $[(\mathbf{6})_2]^{2+}$ in the solid state. Interestingly, only the C_2-symmetric supermolecule shown and its enantiomer were discovered [18] in the crystal structure, the diastereoisomeric C_i-symmetric (or *meso*) complex not being observed. In other words, the noncovalent dimerization of the cation $\mathbf{6}^+$ proceeds diastereoselectively, at least in the solid state, to give a racemic mixture of supramolecular stereoisomers with C_2 symmetry. The $[(\mathbf{6})_2]^{2+}$ superstructure has been classed as a [*c*2]daisy chain, where the *c* and 2 designate, respectively, that the daisy chain is cyclic and that it is comprised of two components. Using this nomenclature, the macromolecular daisy chain, illustrated in Fig. 6b, is an [a∞]daisy chain, because it is acyclic and is composed of an infinite number of components.

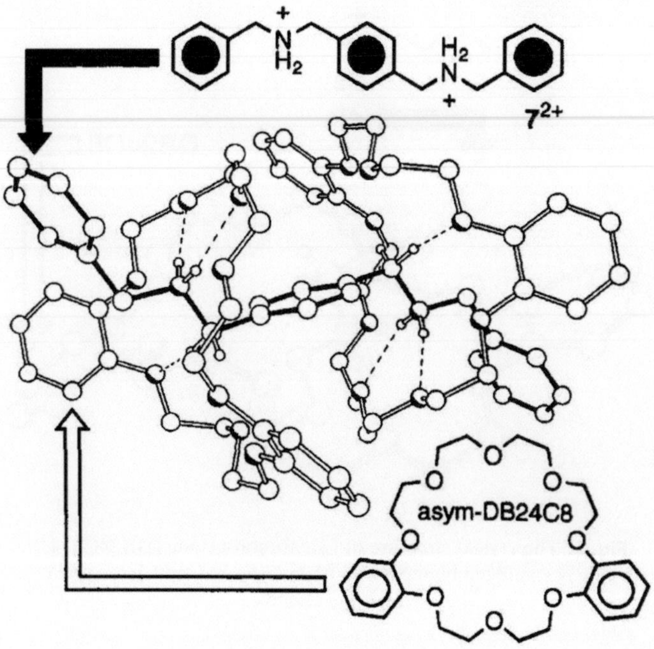

Fig. 8 View of the crystal structure of the [3]pseudorotaxane $[(\text{asym-DB24C8})_2\cdot\mathbf{7}]^{2+}$.

with the salt $\mathbf{7}\cdot 2PF_6$. Figure 8 shows how one $\mathbf{7}^{2+}$ dication pierces the cavities of two asym-DB24C8 macrocycles to generate a [3]pseudorotaxane that is stabilized, like the $[\text{DB24C8}\cdot\mathbf{3}]^+$ complex, by both $[N^+—H\cdots O]$ and $[C—H\cdots O]$ hydrogen bonds. By the same token, the [4] pseudorotaxane $[(\text{DB24C8})_3\cdot\mathbf{8}]^{3+}$ was observed (Fig. 9), in the solid state, when DB24C8 was co-crystallized [23] with the trication $\mathbf{8}\cdot 3PF_6$. Stabilization of this triply encircled pseudorotaxane supermolecule is achieved by a combination of $[N^+—H\cdots O]$ and $[C—H\cdots O]$ hydrogen bonds, supplemented by π-π stacking between the trication's *p*-xylylene rings and one of the catechol rings from each of the DB24C8 macrocycles.

Having established that the recognition motif displayed by the $[\text{DB24C8}\cdot\mathbf{3}]^+$ complex can be extended in one direction, viz., by increasing the number of dialkylammonium centers within the pseudorotaxanes' threadlike components, there was another issue that needed to be addressed, i.e., could we expand the recognition motif in the other direction? Put another way, would several dialkylammonium strands thread their way through the cavity of larger crown ethers to generate multiply stranded pseudorotaxanes? The answer to this question was provided initially [13b–d] by bis-p-phenylene[34]crown-10 (BPP34C10), a crown ether that had been utilized by us previously [12a] for the synthesis of a wide range of catenanes, rotaxanes, and pseudorotaxanes. Gratifyingly, crystals of the complex produced from

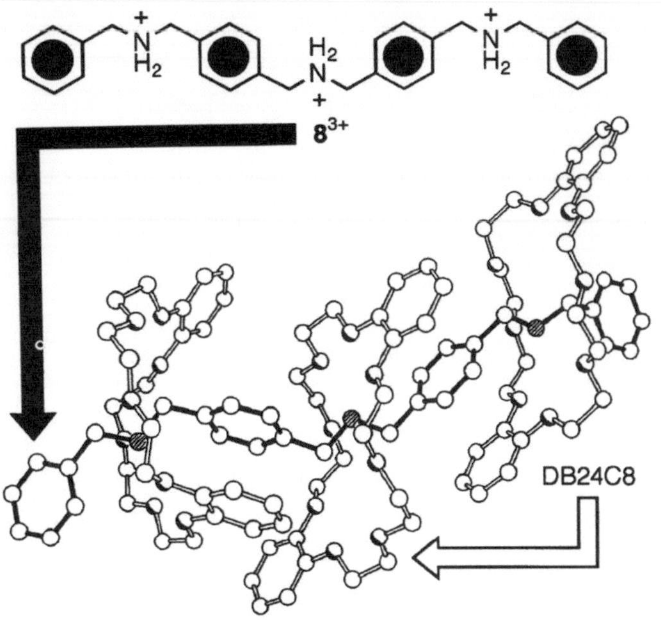

Fig. 9 View of the solid-state structure of the [4]pseudorotaxane generated when the trication $\mathbf{8}^{3+}$ interpenetrates the cavities of three DB24C8 macrorings.

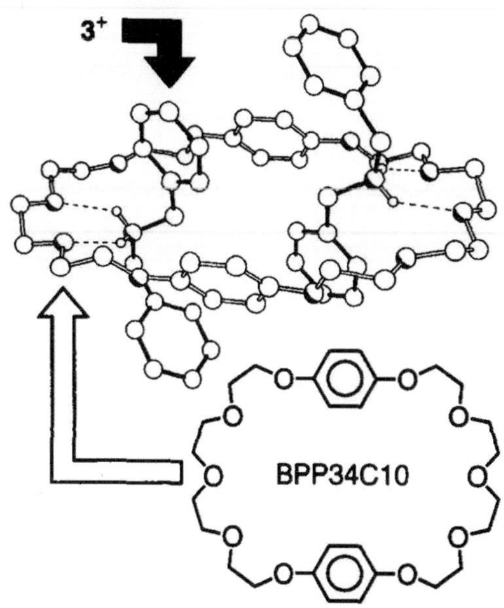

Fig. 10 The solid-state structure of the complex $[\text{BPP34C10}\cdot(\mathbf{3})_2]^{2+}$ in which *two* $\mathbf{3}^+$ cations are threaded through the BPP34C10 macroring by virtue of $[N^+—H\cdots O]$ hydrogen bonds.

BPP34C10 and **3**·PF6 did, indeed, show (Fig. 10) the formation of a doubly stranded [3]pseudorotaxane $[\text{BPP34C10}\cdot(\mathbf{3})_2]^{2+}$ consisting of pairs of $\mathbf{3}^+$ cations threaded symmetrically through the BPP34C10 macroring. The complex is stabilized primarily by means of $[N^+—H\cdots O]$ hydrogen bonds from the NH_2^+ hydrogen atoms to pairs of oxygen atoms located in both of the crown ether's disassociated polyether loops, with secondary stabilization occurring via edge-to-face interactions between one of each cation's phenyl rings and the hydroquinone rings of the crown ether. Notably, however, in contrast with the aforementioned $[\text{DB24C8}\cdot\mathbf{3}]^+$ complex, here no $[C—H\cdots O]$ hydrogen bonds are present to aid in holding the $[\text{BPP34C10}\cdot(\mathbf{3})_2]^+$ superstructure together.

A short while ago, the paradigm, stating that two dialkylammonium strands doubly thread through the cavity of a crown ether bearing two isolated polyether arcs, was developed further [24] in our laboratories. The crown ether 1,5-dinaphtho[38]crown-10 (1/5DN38C10) is similar to BPP34C10 in that it is endowed with two separated polyether loops, but differs in that these polyether loops are separated by 1,5-dioxynaphthalene residues, as opposed to hydroquinone rings. It cocrystallizes with the cation $\mathbf{9}^+$ to generate (Fig. 11) the centrosymmetric doubly stranded [3] pseudorotaxane $[\text{1/5DN38C10}\cdot(\mathbf{9})_2]^{2+}$ that is stabilized, like the related complex $[\text{BPP34C10}\cdot(\mathbf{3})_2]^{2+}$, by virtue of $[N^+—H\cdots O]$ hydrogen bonds and aromatic edge-to-face interactions. Nonetheless, not all crown ethers with two disassociated $OCH_2(CH_2OCH_2)_3CH_2O$ polyether arcs form doubly stranded [3]pseudorotaxanes with dialkylammonium ions. If these arcs are appended to *o*-phenylene units, as in dibenzo[30]crown-10 (DB30C10), the crown ether's cavity becomes so small that it can accommodate only one dialkylammonium ion, resulting in the formation of singly stranded pseudorotaxanes. This fact is illustrated by the crystal structure [25] of complex $[(\text{DB30C10})_2\cdot\mathbf{10}]^{2+}$—prepared from DB30C10 and $\mathbf{10}\cdot 2PF_6$ via the slippage [26] approach—in which (Fig. 12) each of the dication's two dialkylammonium centers is encircled by only one DB30C10 macroring. The C_i-symmetric [3]pseudorotaxane $[(\text{DB30C10})_2\cdot\mathbf{10}]^{2+}$ is stabilized by $[N^+—H\cdots O]$ and $[C—H\cdots O]$ hydrogen bonds, supplemented by both face-to-face and edge-to-face π-π interactions.

What happens when a mixture of DB24C8 and BPP34C10 self-assembles with a thread bearing both dialkylammonium and 4,4′-bipyridinium recognition sites? This question is particularly interesting since BPP34C10 can bind either two dialkylammonium centers or one 4,4′-bipyridinium unit [12a, [27] within its cavity. In a bid to answer this question, we have found [28] that a mixture of DB24C8 and BPP34C10 self-assembles self-selectively with the tetracationic thread $\mathbf{11}^{4+}$ to produce (Fig. 13) the [4]pseudorotaxane supermolecule $[(\text{DB24C8})_2\cdot\text{BPP34C10}\cdot\mathbf{11}]^{4+}$. In this discrete superstructure, the DB24C8 macrorings encircle the dialkylammonium

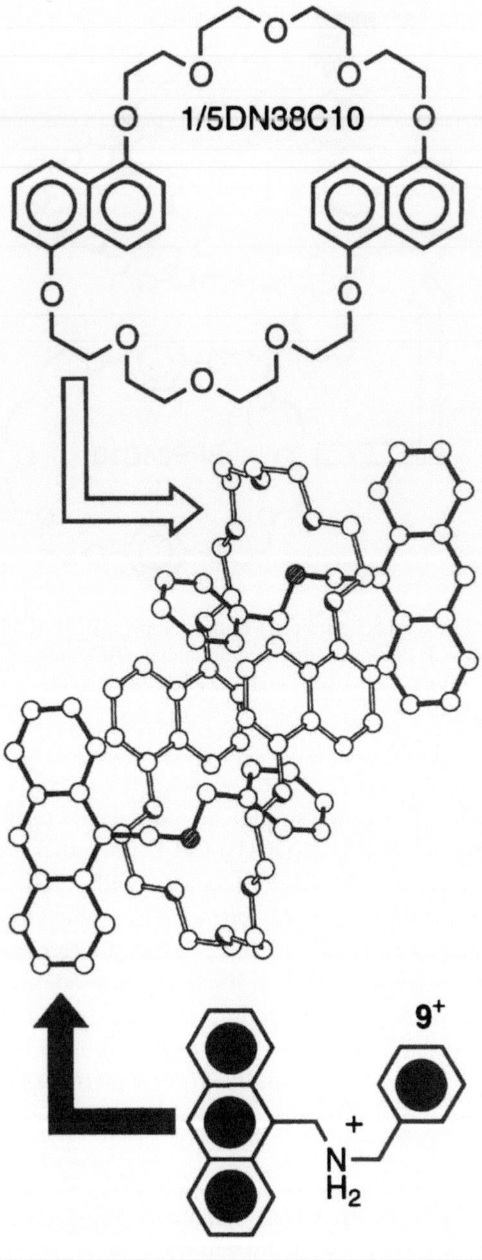

Fig. 11 The 1: 2 doubly stranded pseudorotaxane complex formed by l/5DN38C10 and 9^{+} in the solid state. The intrapseudorotaxane $[N^{+}\cdots N^{+}]$ separation is 9.1 Å in the $[l/5DN38C10\cdot(\mathbf{9})_2]^{2+}$ complex, an increase of 0.7 Å compared to the corresponding distance in [3] pseudorotaxane $[BPP34C10\cdot(\mathbf{3})_2]^{2+}$. This elongation of the distance between the two NH_2^{+} centers is presumably a consequence of the larger size of the l/5DN38C10 host with respect to its congener BPP34C10.

centers by virtue of the usual $[N^{+}—H\cdots O]$ and $[C—H\cdots O]$ hydrogen bonds, while the 4,4′-bipyridinium unit complexes with BPP34C10 through π-π stacking and $[C—H\cdots O]$ hydrogen bonding interactions, in a manner reminiscent of the 1:1 complex formed [27] between

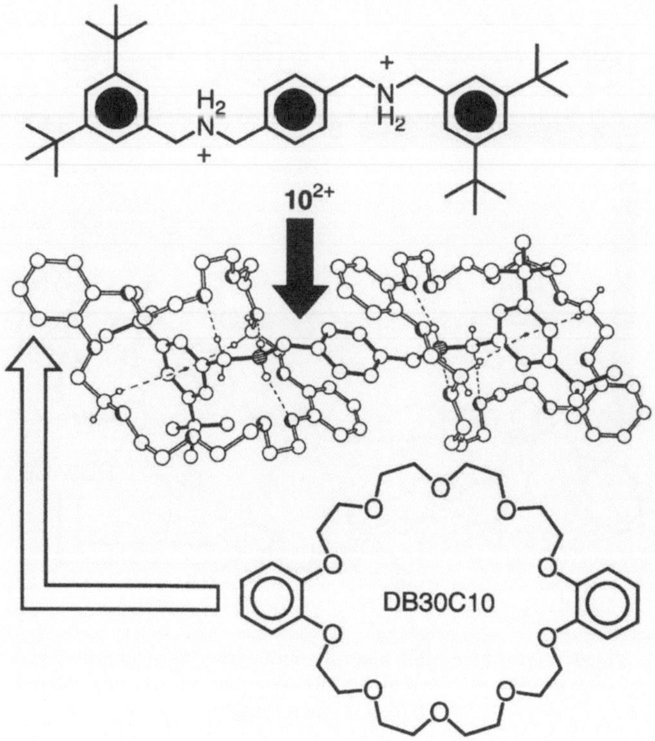

Fig. 12 Doubly stranded pseudorotaxanes are not produced when the size of the spacer unit between two $OCH_2(CH_2OCH_2)_3CH_2O$ polyether arcs is reduced from *p*-phenylene (as in BPP34C10) to *o*-phenylene (as in DB30C10). This fact is highlighted by the crystal structure of the pseudorotaxane $[(DB30C10)_2\cdot\mathbf{10}]^{2+}$.

BPP34C10 and the paraquat (1,1′-dimethyl-4,4′-bipyridinium) dication. The preparation of the complex $[(DB24C8)_2\cdot BPP34C10\cdot\mathbf{11}]^{4+}$ demonstrates adroitly how an intricate superarchitecture can be generated through the operation of two different recognition motifs simultaneously and selectively. It is believed [29] that such concurrent usage of numerous recognition motifs, in a self-selective manner, will become a routine tactic of synthetic supramolecular chemists in future noncovalent syntheses, for the reason that superstructural complexity becomes amplified as the number of intermolecular bonding tools used increases.

Never Forget the Counterions!

During our studies, we discovered an interesting feature associated with some of the complexes formed between crown ethers and dialkylammonium-containing cations, that is, they interact with their associated anions. Since anion binding and recognition play a pivotal role in several essential chemical and biological processes, it is hardly surprising that the design of anion-binding receptors is

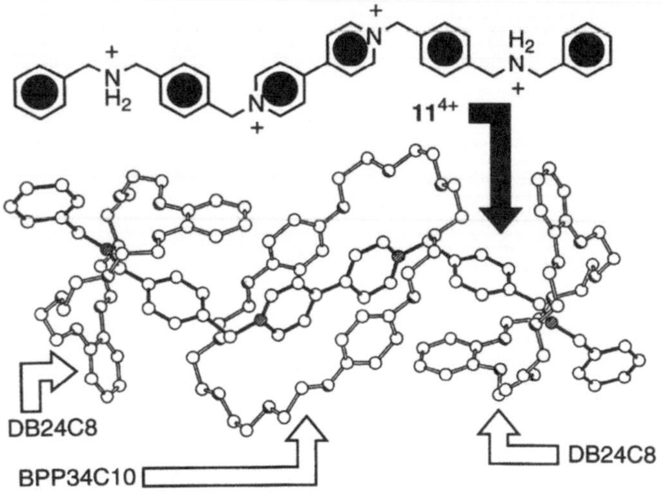

Fig. 13 Operating several distinct recognition motifs judiciously, at the same time, allows the preparation of elaborate superstructures, as demonstrated by the crystal structure of the [4]pseudorotaxane generated when DB24C8 and BPP34CIO interact, in an independent and self-selective manner, with the respective dialkylammonium and 4,4′-bipyridinium portions of the tetracationic thread $\mathbf{11}^{4+}$.

currently [30] arousing considerable attention within the supramolecular community. We were enticed into this field by several serendipitous discoveries relating to anion-assisted self-assembly [31]. All of our work pertains to PF_6^- anions, species that usually adopt random orientations in crystal lattices because of their "spherical" nature. Indeed, these anions are disordered in the crystal structures illustrated in Figs. 3, 4, and 5 and 8, 9, 10, 11, 12, and 13, structures in which there is no interaction with the cationic species. This disorder is relinquished, however, when the PF_6^- anions interact with cationic superarchitectures in the solid state.

It all started during our investigations [23, 32] on the supramolecular synthesis of multiply stranded pseudorotaxanes (Fig. 14) from higher homologues of BPP34C10, viz., tris-*p*-phenylene[51]crown-15 (TPP51-C15) and tetrakis-p-phenylene[68]crown-20 (TPP68-C20), and the cation $\mathbf{3}^+$. We believed that TPP51C15 and TPP68C20 were natural extensions of BPP34C10, whose distinct arrays of three and four hydroquinone-spaced polyether loops should be able to bind to three and four $\mathbf{3}^+$ cations, respectively. Indeed, the X-ray analysis of crystals generated when TPP51C15 was co-crystallized with **3**·PF6 reveals (Fig. 15) the formation of a 1:3 complex, i.e., the triply stranded [4]pseudorotaxane $[\text{TPP51C15}\cdot(\mathbf{3})_3]^{3+}$, wherein three $\mathbf{3}^+$ ions are bound independently to each of the crown ether's separate polyether loops by means of $[N^+—H\cdots O]$ and $[C—H\cdots O]$ hydrogen bonds. A closer investigation of this superarchitecture reveals that one of its affiliated PF_6^- counterions is bound within a cleft, created by the pseudorotaxane's saddle-like co-conformation, by the intermediacy of a series of $[C—H\cdots F]$ hydrogen bonds to the hydrogen atoms of the crown ether's hydroquinone rings and the cation's benzylic CH_2 groups.

The above–mentioned observation led us to believe that we could encapsulate a PF_6^- ion completely within the [5] pseudorotaxane superstructure $[\text{TPP68C20}\cdot(\mathbf{3})_4]^{4+}$. We reasoned that the increased empty space in the pseudorotaxane's core, brought about by the increased dimensions of the crown ether TPP68C20, should allow the complete envelopment of a PF_6^- ion by means of hydrogen bonding and anion-dipole interactions. As expected, the X-ray analysis of crystals obtained from a 1:4 solution of TPP68C20 and $\mathbf{3}\cdot PF_6$ reveals (Fig. 16a) the formation of two, practically identical, crystallographically independent, quadruply stranded [5]pseudorotaxanes $[\text{TPP68C20}\cdot(\mathbf{3})_4]$4+ that are stabilized, yet again, by $[N^+—H\cdots O]$ and $[C—H\cdots O]$ hydrogen bonds. Even more interestingly, however, there are single ordered PF_6^- ions in the cores of both [5]pseudorotaxanes. In each superstructure, the anion is encapsulated completely by the four hydroquinone rings and four $CH_2NH_2^+CH_2$ regions of the pseudorotaxanes, stabilization arising from a series of $[C—H\cdots F]$ hydrogen bonds. Anion envelopment may be viewed as resulting from supramolecular preorganization of the anion recognition sites, i.e., where the macrocycle preorganizes [33] the $CH_2NH_2^+CH_2$ units (anion recognition sites) for PF_6^- binding. The fact that the macrocycle also interacts with the anion means that the [5]pseudorotaxane $[\text{TPP68C20}\cdot(\mathbf{3})_4]^{4+}$ provides (Fig. 16b) elegant concentric spheres of cationic and neutral ligands for complete anion encapsulation.

Anions were also found to play a role in the selfassembly of BPP34C10 with the dication $\mathbf{7}^{2+}$. Originally, we envisaged (Fig. 17) two possible supramolecular outcomes when these two components aggregate with each other, viz., the doubly docked [2]pseudorotaxane $[\text{BPP34C10}\cdot\mathbf{7}]^{2+}$ or the doubly stranded, doubly encircled [4]pseudorotaxane $[(\text{BPP34C10})_2\cdot(\mathbf{7})_2]^{4+}$. In actual fact, we found [13b–d] that the latter four-component supermolecule was generated, through the intermediacy of $[N^+—H\cdots O]$ and $[C—H\cdots O]$ hydrogen bonds, when BPP34C10 and $\mathbf{7}\cdot 2PF_6$ were crystallized together. However, a closer analysis [34] of the crystal structure reveals (Fig. 18) that PF_6^- counterions are located in the cavities formed between 2_1-screw-related [4]pseudorotaxanes, creating an interwoven one-dimensional supramolecular array through $[C—H\cdots O]$ hydrogen bonding.

Is it the PF_6^- ions that direct the solid-state self-assembly of the discrete supermolecule $[(\text{BPP34C10})_2\cdot(\mathbf{7})_2]^{4+}$? We are not entirely sure, to be honest, but this question played a part

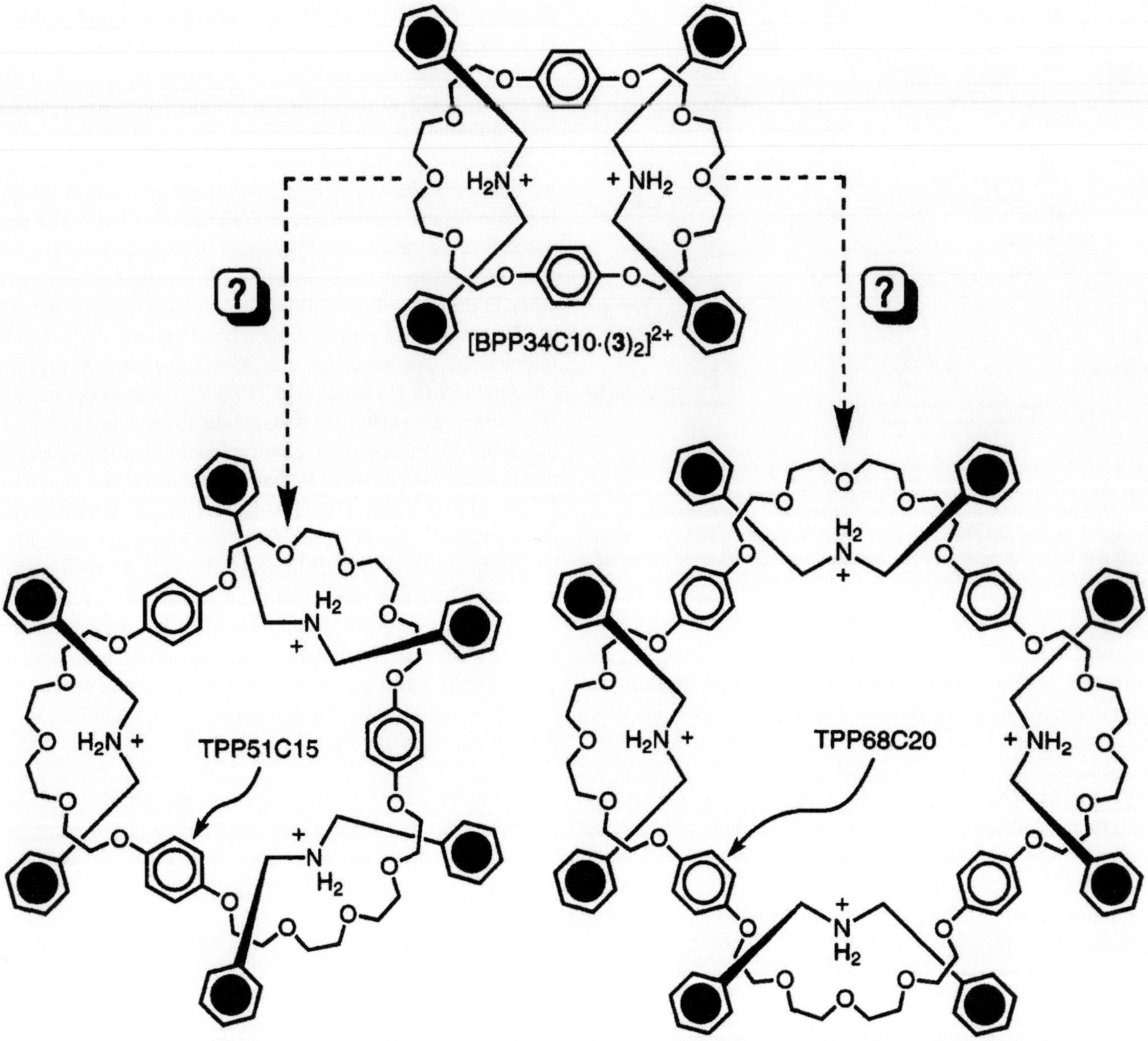

Fig. 14 Complexation of BPP34C10 with 3^+ yields a doubly stranded [3]pseudorotaxane. The question we asked subsequently was: Do larger crown ethers, such as TPP51C15 and TPP68C20, allow the synchronous threading of three and four 3^+ ions through their respective cavities? The answer is yes.

in our thinking when we embarked on a program designed to synthesize doubly docked [2]pseudorotaxanes noncovalently. In this program [34], we utilized bis(dialkylammonium) dications bearing two 3,5-di-*tert*-butylbenzyl termini. These bulky groups were introduced to prevent two separate dications from fitting inside one BPP34C10 macroring and to prohibit the occurrence of solid-state, anion-assisted self-assembly. Indeed, the doubly docked [2]pseudorotaxanes [BPP34C10·**10**]$^{2+}$ and [BPP34C10·**12**]$^{2+}$ were observed (Fig. 19), respectively, in the solid state, when BPP34C10 was crystallized with either of the aryl-spaced bis(dialkylammonium) salts **10**·$2PF_6$ or **12**·$2PF_6$. In both of these C_i-symmetric supermolecules, stabilization is achieved through the agency of the usual hydrogen bonding interactions, together with π-π stacking interactions. Notably, in both examples, no PF_6^- ions associate with the pseudorotaxane supermolecules. Nevertheless, the hydrogen bonding interactions are not optimal in either of the superstructures

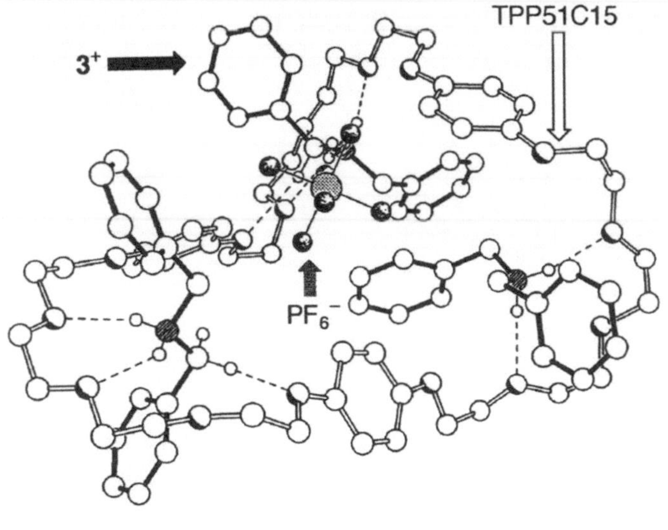

Fig. 15 The solid-state structure of [4]pseudorotaxane [TPP51C15·(**3**)$_3$]$^{3+}$ shows that it partially envelops a PF_6^- counterion through [C—H···F] hydrogen bonding.

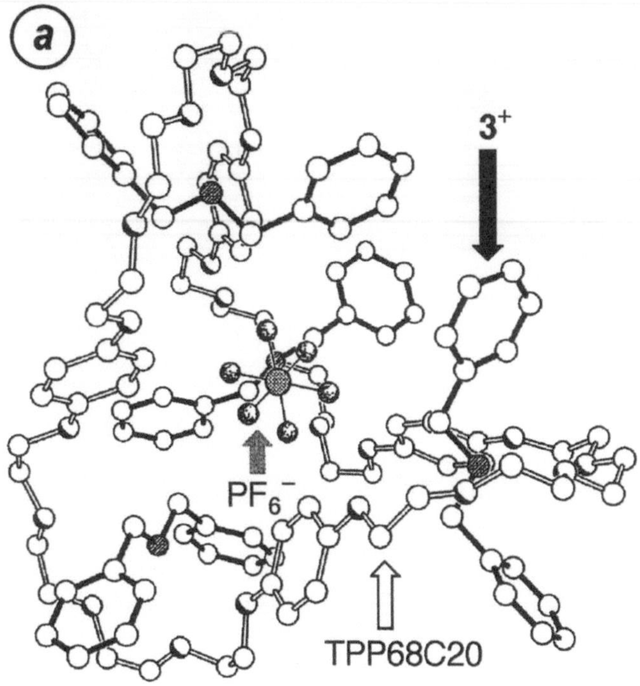

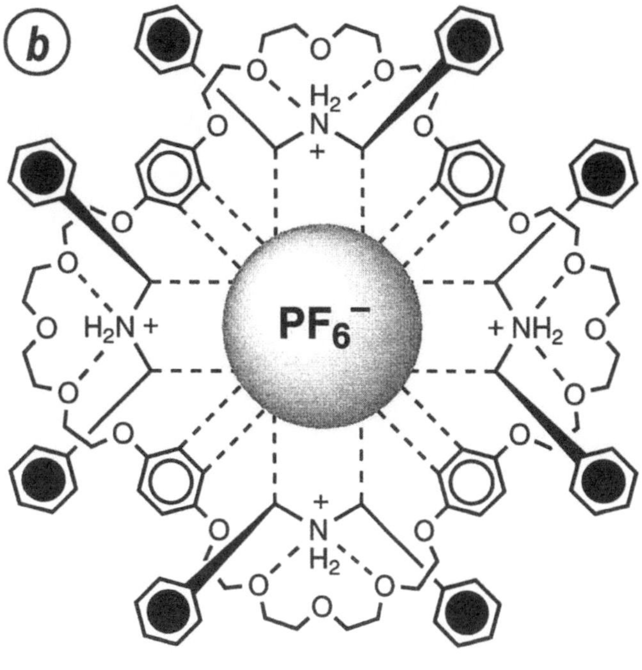

Fig. 16 The two crystallographically independent [5]pseudorotaxanes [TPP68C20·(**3**)4]$^{4+}$ totally encapsulate a PF_6^- ion within their cores. (**a**) Ball-and-stick view of one of the crystallographically independent $[TPP68C20\cdot(\mathbf{3})_4]^{4+}\cdot PF_6^-$ ·complexes. (**b**) Schematic representation of the anion encapsulation by the pseudorotaxane.

[BPP34C10·**10**]$^{2+}$ and [BPP34C10·**12**]$^{2+}$: The former is stabilized partly by hydrogen bonds involving the less basic hydroquinone oxygen atoms, while the latter achieves stabilization through the use of the weaker [C—H···O] hydrogen bonds.

In order to maximize the stabilization imparted to these structures by hydrogen bonding interactions, we then examined the possibility of spacing the dialkylammonium centers with more flexible alkyl units. The X-ray analysis of the 1:1 complex formed between BPP34C10 and the pentyl-spaced dication **13**$^{2+}$ reveals (Fig. 20) the failure to form a pseudorotaxane at all. In this instance, a hot dog-like supermolecule [BPP34C10·**13**]$^{2+}$, possessing a face-to-face [35] co-conformation, is produced as a result of [N^+—H···O] and [C—H···π] hydrogen bonds. This species is formed presumably because of the smaller spacer unit between the NH_2^+ centers, which means that both the BPP34C10 macrocycle and dication **13**$^{2+}$ have to adopt unusually distorted geometries in order that the supermolecule can achieve stabilization via two pairs of [N^+—H···O] hydrogen bonds. Also of note in this case is the fact that the folding of the dication creates, in conjunction with pairs of hydroquinone hydrogen atoms, a hydrophobic cavity, within which one of the PF_6^- counterions is bound through a combination of electrostatic interactions and [C—H···F] hydrogen bonds.

Next, we decided to increase the length of the spacer unit: Studies were instigated on the complexation phenomena associated with BPP34C10 and the hexamethylene-spaced dication **14**$^{2+}$. These investigations revealed that the two components self-assemble in the solid state to form (Fig. 21) a doubly docked [2]pseudorotaxane with the desired hydrogen bonding pattern. The centrosymmetric [2]pseudorotaxane [BPP34C10·**14**]$^{2+}$ is generated when each NH_2^+ center hydrogen bonds to the second and fourth oxygen

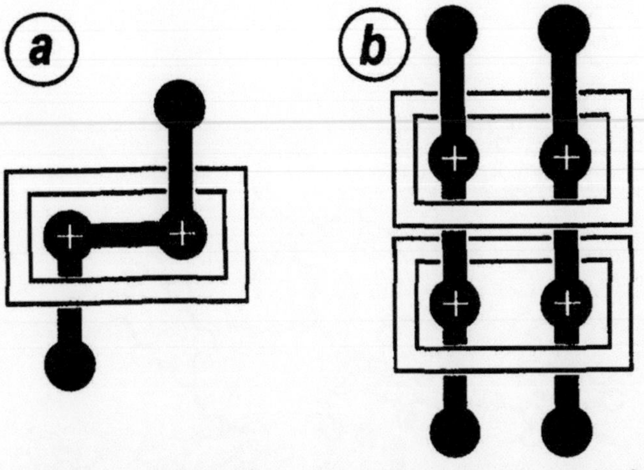

[BPP34C10·7]$^{2+}$ [(BPP34C10)$_2$·(7)$_2$]$^{4+}$

Fig. 17 Diagrams illustrating two of the ways in which BPP34C10 can self-assemble with the dication **7**$^{2+}$. (**a**) A doubly docked [2] pseudorotaxane. (**b**) A doubly stranded, doubly encircled [4] pseudorotaxane.

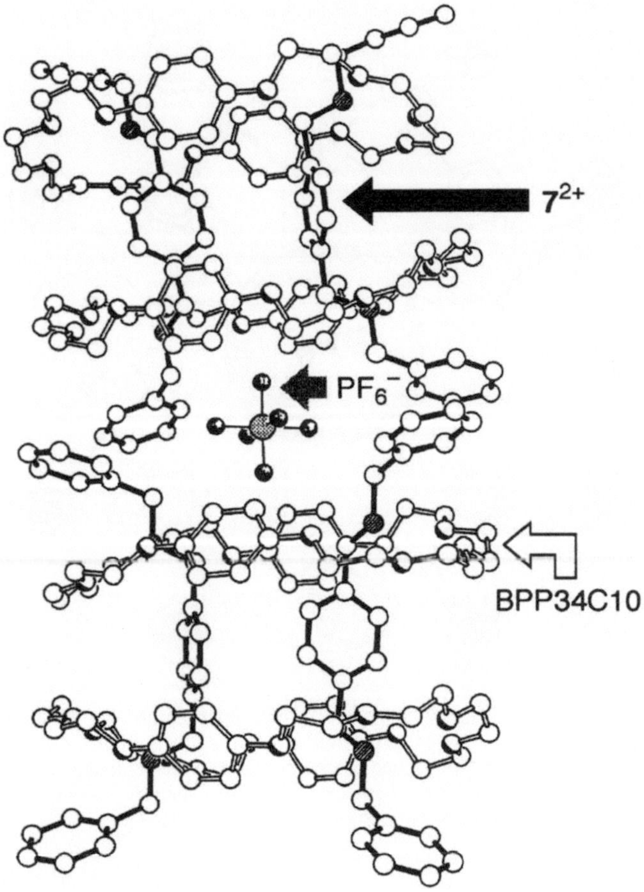

Fig. 18 View of the solid-state structure acquired from co-crystals of BPP34C10 and **7**$^{2+}$. A one-dimensional supramolecular array is created because distinct [(BPP34C10)$_2$·(**7**)$_2$]$^{4+}$ supermolecules interconnect through the intermediacy of hydrogen-bonded PF$_6^-$ ions.

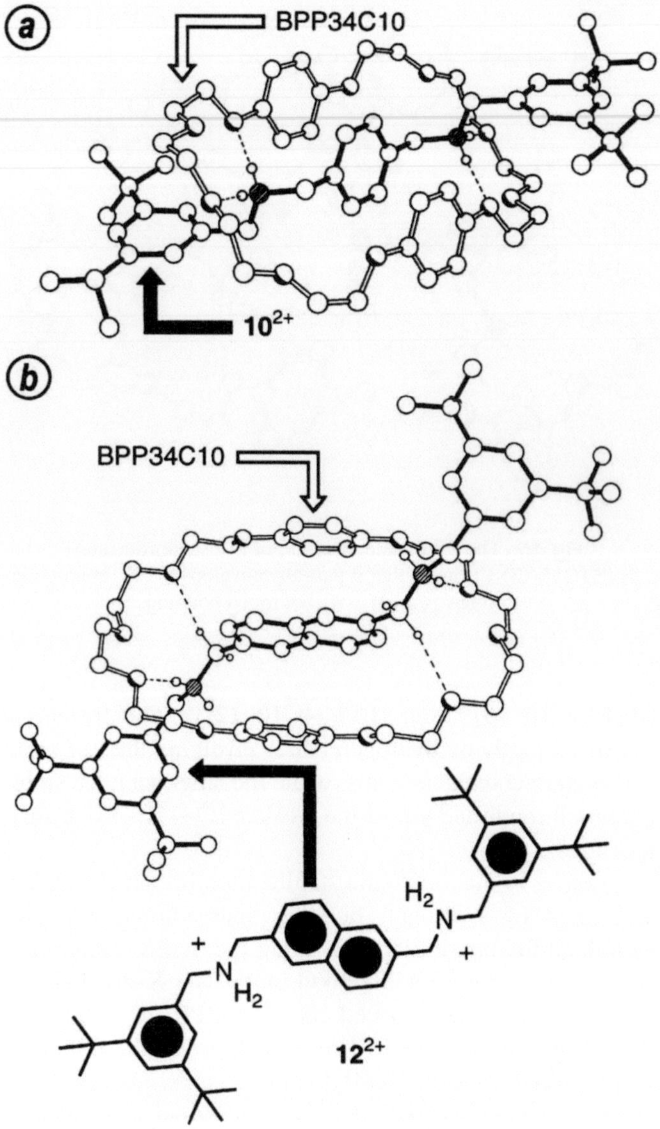

Fig. 19 The crystal structures of two doubly docked [2]pseudorotaxanes that are stabilized by both hydrogen bonding and π-π stacking interactions. (**a**) [BPP34C10·**10**]$^{2+}$ (**b**) [BPP34C10·**12**]$^{2+}$.

atoms of each polyether loop. Supplementary stabilization is conferred upon this two-component supermolecule via [C—H···π] interactions between the dication's central CH_2 groups and the hydroquinone rings of the crown ether. Yet again, PF_6^- ions were not perceived to influence this self-assembly process.

We have also become acquainted with other examples of anion-assisted self-assembly [31, 32] in our studies [36] on crystal engineering [7b] through the noncovalent association of carboxyl-substituted crown ether–dialkylammonium ion-based pseudorotaxanes. In particular, we have discovered that PF_6^- ions can have (Fig. 22) a strong effect on the crystal

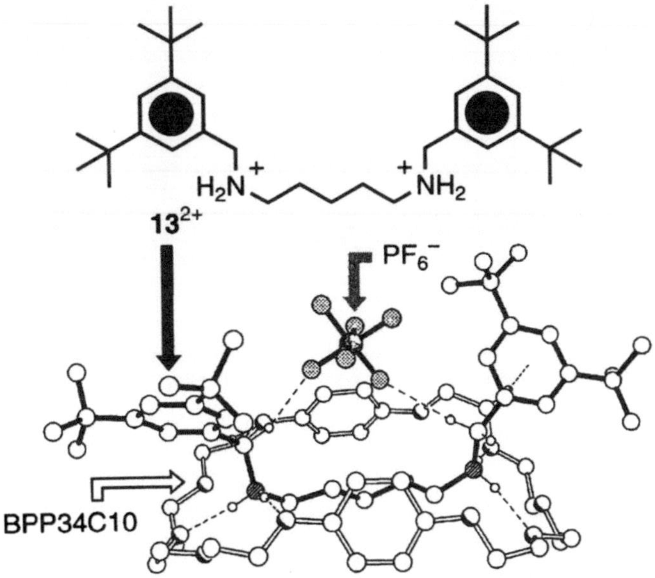

Fig. 20 BPP34C10 and $\mathbf{13}^{2+}$ assume distorted conformations in the solid-state structure of the hot dog-like 1:1 complex [BPP34C10 · **13**]$^{2+}$. These distortions produce a hydrophobic pocket that binds a PF_6^- counterion.

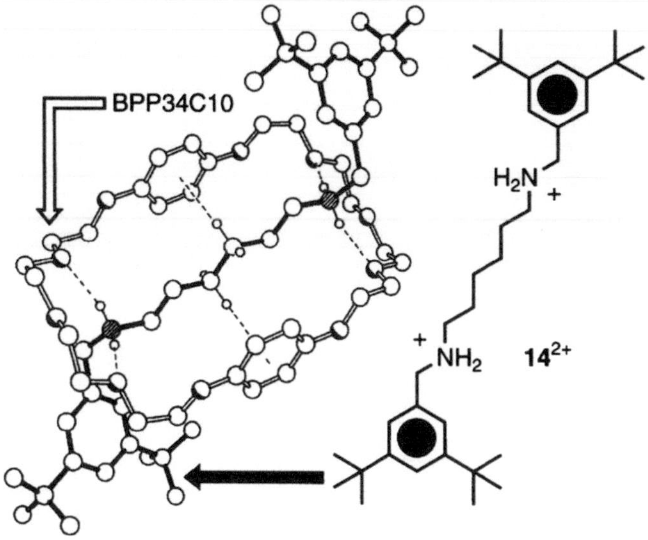

Fig. 21 The crystal structure of the complex [BPP34C10 · **14**]$^{2+}$—a "flawless" doubly docked [2]pseudorotaxane.

structures of the doubly stranded [3]pseudorotaxanes generated from BPP34C10 and secondary dialkylammonium salts possessing one aromatic ring substituted with carboxyl groups. In these structures, BPP34C10 acts as a girdle that holds two dialkylammonium ions together in space, while the directions in which the ensuing [3]pseudorotaxanes' carboxyl-bearing aromatic rings point depend on the PF_6^- counterions. We found that the carboxyl-substituted aromatic rings are oriented in the same direction if the PF_6^- anions associate with the [3]pseudorotaxanes, but that they are guided in opposing directions if no anion-[3]pseudorotaxane interaction is present. For instance, the X-ray analyses of the [3]pseudorotaxanes [BPP34C10 · (**15**)$_2$]$^{2+}$ and [BPP34C10 · (**16**)$_2$]$^{2+}$ reveal (Fig. 23) that both cations are threaded co-directionally through the macrocycle's center as a consequence of the PF_6^- ions that are located, by virtue of $[C—H\cdots F]$ hydrogen bonds, in the clefts formed between the cations' unsubstituted phenyl rings. This important secondary interaction leads to the noncovalent dimerization of pairs of C_i-symmetrically related [3]pseudorotaxanes to generate doubly encircled eight-component supermolecules that are linked through the strong $[O—H\cdots O]$ hydrogen bonds of the carboxyl dimer [37]. In both complexes [BPP34C10 · (**15**)$_2$]$^{2+}$ and [BPP34C10 · (**16**)$_2$]$^{2+}$, the BPP34C10 macrorings and PF_6^- counterions combine to effect supramolecular preorganization of the carboxyl groups for dimerization.

On the other hand, PF_6^- ions were observed not to associate with the [3] pseudorotaxane [BPP34C10 · (**17**)$_2$]$^{2+}$. Here, two cations $\mathbf{17}^+$ thread centrosymmetrically through the BPP34C10 macrocycle with respect to each other, i.e., in an antiparallel orientation, so that their isophthalate units are oriented in opposing directions. By employing carboxyl dimers, these isophthalate units then form the tapes that are present [38] in the crystal structure of isophthalic acid itself to generate (Fig. 24) an interwoven supramolecular cross-linked polymer, in which polymeric isophthalic acid tapes are interlinked through the [3]pseudorotaxane unit. In this instance, the disordered PF_6^- counterions are located in the interstices between the pseudopolyrotaxane sheets.

Interwoven Supramolecular Bundles

The noncovalent syntheses of the pseudorotaxane supermolecules [BPP34C10 · (**3**)$_2$]$^{2+}$ and [(BPP34C10)$_2$ · (**7**)$_2$]$^{4+}$ form the foundations of a new paradigm for supramolecular synthesis. This paradigm states that n molecules of BPP34C10, which possess a total of $2n$ dialkylammonium binding sites, are capable of tying two oligo(dialkylammonium) cations, each bearing n dialkylammonium centers, together to generate discrete multicomponent superarchitectures. We conjectured that, by applying this principle to systems involving branched oligo(dialkylammonium) ions endowed with n cationic "spokes," we could self-assemble multicomponent

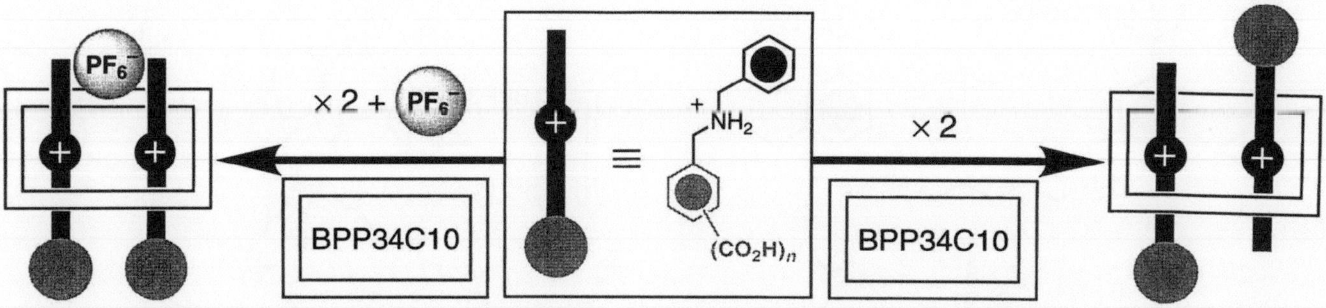

Fig. 22 The BPP34C10 macrocycle supports two dialkylammonium ions within its interior to generate [3]pseudorotaxanes. PF_6^- ions can interact with these pseudorotaxanes to orient recognition sites (in the examples discussed here, these are carboxyl groups) in space for further noncovalent association.

interwoven supramolecular bundle-like architectures, i.e., clusters [32, 39]. In these supermolecules, each of the *n* spokes from one oligo(dialkylammonium) ion would be inserted through half of the cavities of *n* separate BPP34C10 molecules, such that the *n* vacant receptor sites remaining in the crown ethers' cavities would be free to accept another *n* spokes from a second cation. In other words, *n* BPP34C10 macrocycles would clamp two branched oligo (dialkylammonium) ions together, employing, for the most part, $\left[N^+—H\cdots O\right]$ hydrogen bonds.

Our initial studies concentrated [40] on the trifurcated trisammonium trication $\mathbf{18}^{3+}$, whose covalent skeleton consists

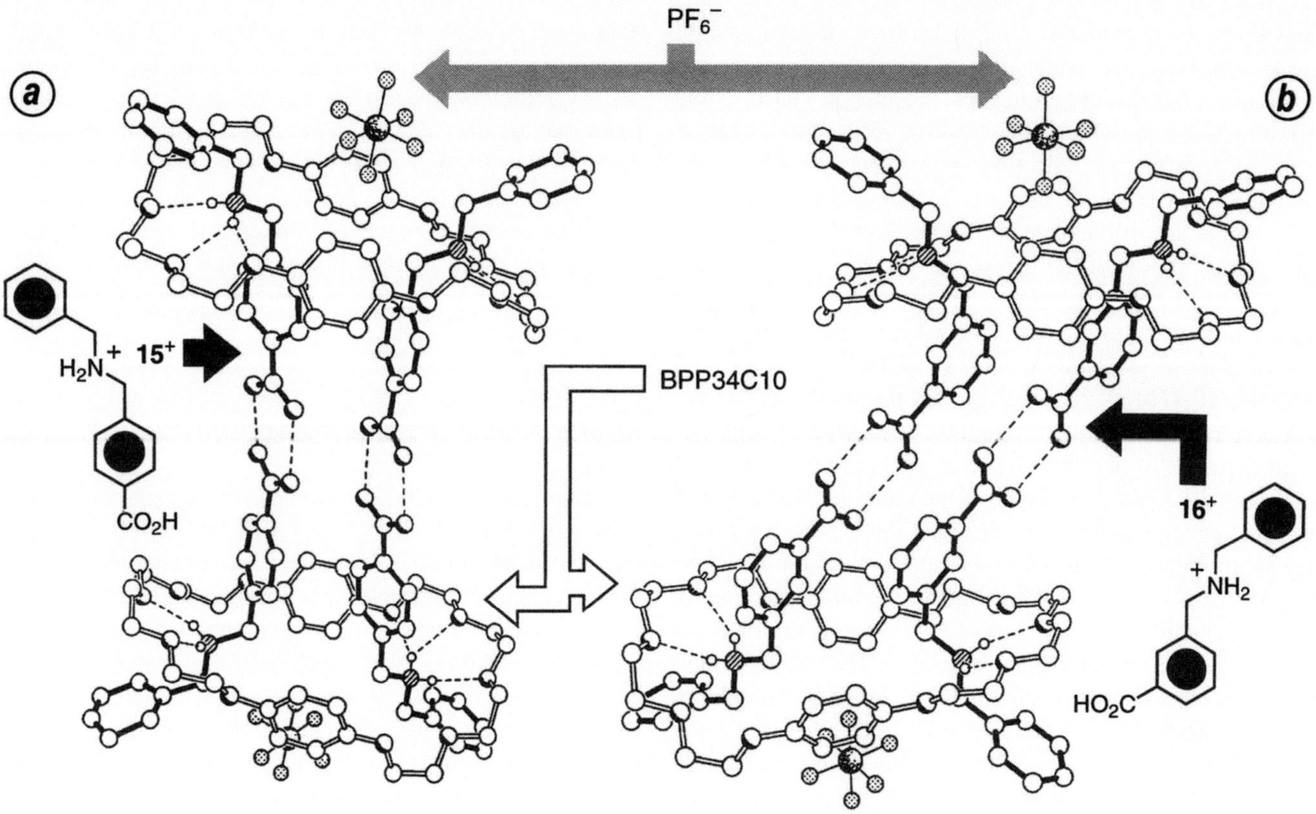

Fig. 23 The crystal structures of two [6]pseudorotaxanes whose self-assembly is assisted by PF_6^- ions. (**a**) $\left\{\left[\text{BPP34C10}\cdot(\mathbf{15})_2\right]^{2+}\right\}_2\cdot 2PF_6^-$. (**b**) $\left\{\left[\text{BPP34C10}\cdot(\mathbf{16})_2\right]^{2+}\right\}_2\cdot 2PF_6^-$.

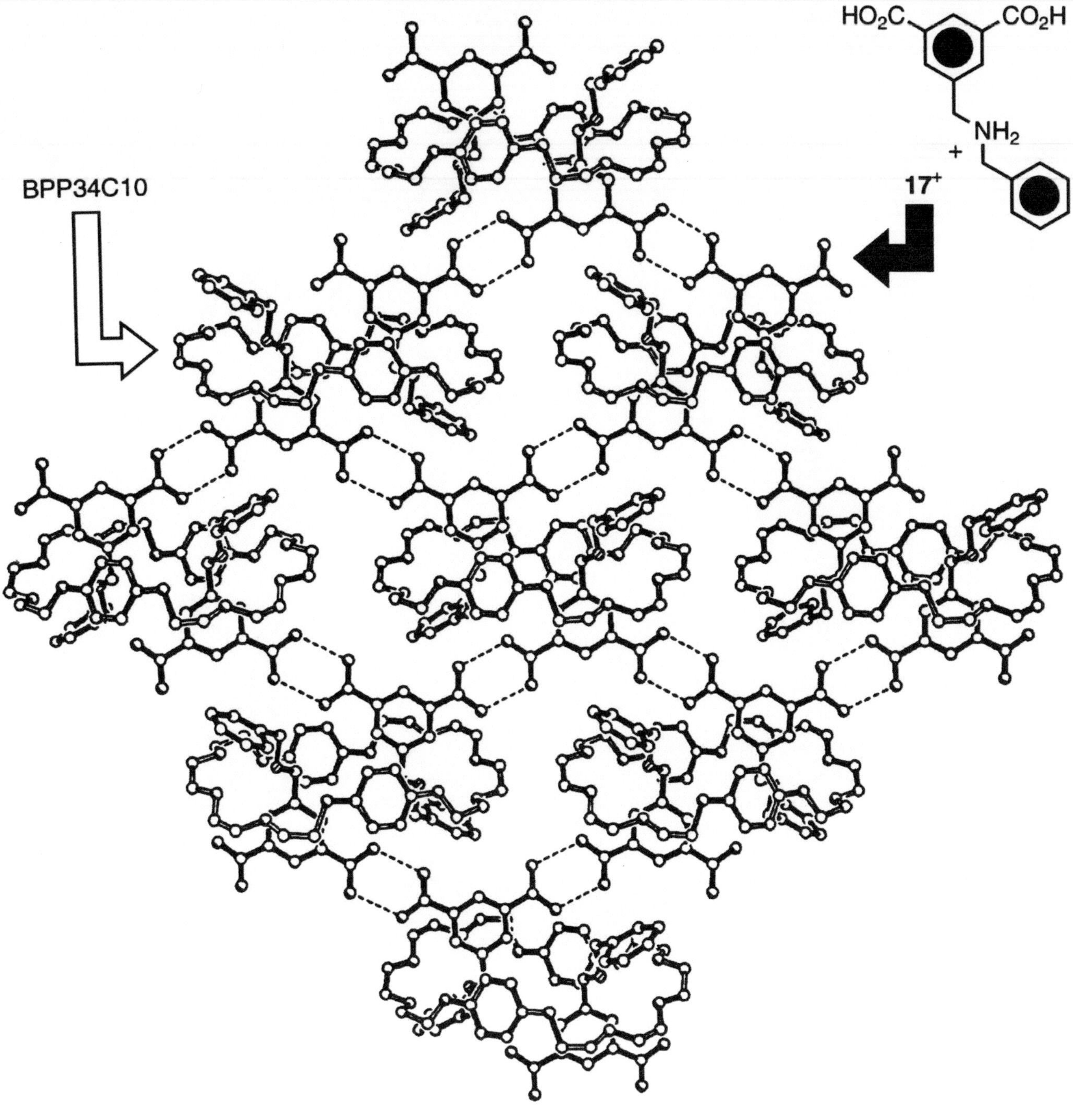

Fig. 24 Portrayal of the interwoven pseudopolyrotaxane sheets $\{[\text{BPP34C10}\cdot(\mathbf{17})_2]^{2+}\}_n$ that form in crystals of BPP34C10 and $\mathbf{17}\cdot PF_6$. This example emphasizes the fact that operating several different recognition motifs—in this case, the carboxyl dimer and double threading motifs—both simultaneously and selectively allows the noncovalent synthesis of diverse and complicated superarchitectures.

of three secondary dialkylammonium branches emanating out from a central polyaromatic core. We hypothesized that two units of this trication would self-assemble with three BPP34C10 molecules to form (Fig. 25) a five-component interwoven superbundle. Gratifyingly, the X-ray analysis of single crystals, obtained from a 3:2 solution of BPP34C10 and $\mathbf{18}\cdot 3PF_6$, revealed (Fig. 26) the creation of a highly symmetrical bundle supermolecule, wherein three BPP34C10 macrocycles tie two $\mathbf{18}^{3+}$ trications together primarily by means of $[N^+—H\cdots O]$ hydrogen bonds. The supermolecule has D_3 symmetry, with the principal axis passing (Fig. 26b) through the centers of the two

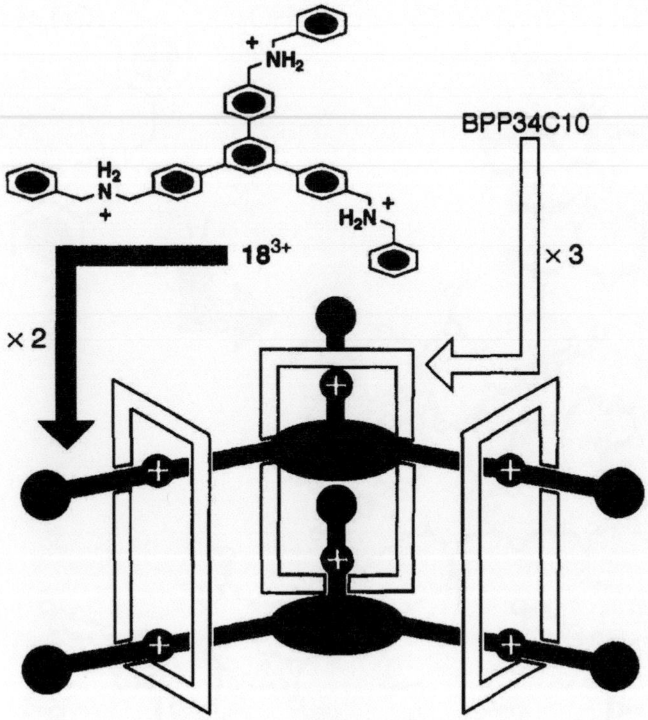

Fig. 25 Schematic diagram illustrating the anticipated [3+2] self-assembly of the interwoven superbundle $[(BPP34C10)_3\cdot(\mathbf{18})_2]^{6+}$, wherein two trications $\mathbf{18}^{3+}$ are fastened together by three BPP34C10 macrocycles.

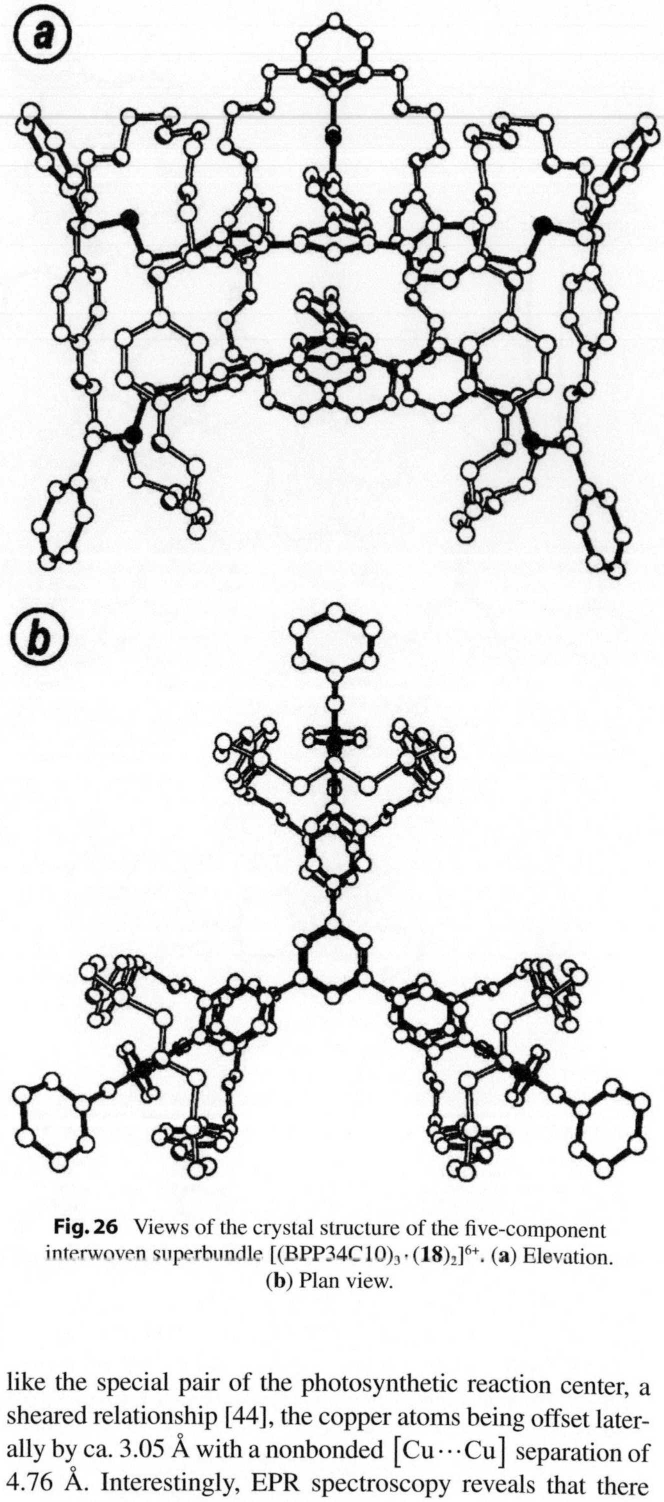

Fig. 26 Views of the crystal structure of the five-component interwoven superbundle $[(BPP34C10)_3\cdot(\mathbf{18})_2]^{6+}$. (**a**) Elevation. (**b**) Plan view.

1,3,5-triarylbenzene rings, which experience a mean interplanar separation of ca. 4.0 Å.

Next, we decided [41] to synthesize noncovalently a supramolecular analogue of the photosynthetic special pair [42, 43] using the protocol that we had developed beforehand. We anticipated that two units of the tetrafurcated tetrakisammonium tetracation $\mathbf{19}^{4+}$, possessing four dialkylammonium spokes fanning out from a central porphyrin core, could be tied together by four BPP34C10 molecules to produce (Fig. 27) a six-component interwoven aggregate bearing two cofacially disposed porphyrinoid nuclei, much like the special pair found in the photosynthetic reaction center [42]. The X-ray analysis of the 4: 2 complex formed from BPP34C10 and $\mathbf{19}\cdot 4PF_6$ revealed (Fig. 28) a six-component supermolecule that had self-assembled in accordance with our design criteria. Pairs of C_i-related $\mathbf{19}^{4+}$ tetracations have their dibenzylammonium sidearms threaded concurrently through the cavities of four cyclically disposed BPP34C10 macrocycles in characteristic fashion. In addition to conventional π-π stacking between the porphyrin nuclei (mean interplanar separation = 3.65 Å), the $[(BPP34C10)_4\cdot(\mathbf{19})_2]^{8+}$ supermolecule is stabilized via $[N^+\!-\!H\cdots O]$ and $[C\!-\!H\cdots O]$ hydrogen bonding between the tetracation's NH_2^+ and CH_2 groups, respectively, and the oxygen atoms of BPP34C10's polyether linkages. The pairs of stacked porphyrin hubs adopt (Fig. 28b), like the special pair of the photosynthetic reaction center, a sheared relationship [44], the copper atoms being offset laterally by ca. 3.05 Å with a nonbonded $[Cu\cdots Cu]$ separation of 4.76 Å. Interestingly, EPR spectroscopy reveals that there are no intermolecular interactions between distinct $\mathbf{19}^{4+}$ porphyrin nuclei in BPP34C10-free solutions, probably as a consequence of charge repulsion. However, these same nuclei do aggregate, under otherwise indistinguishable conditions, in the presence of two molar equivalents of crown ether, indicating that the $[(BPP34C10)_4\cdot(\mathbf{19})_2]^{8+}$ superbundle is formed in solution as well as in the solid state.

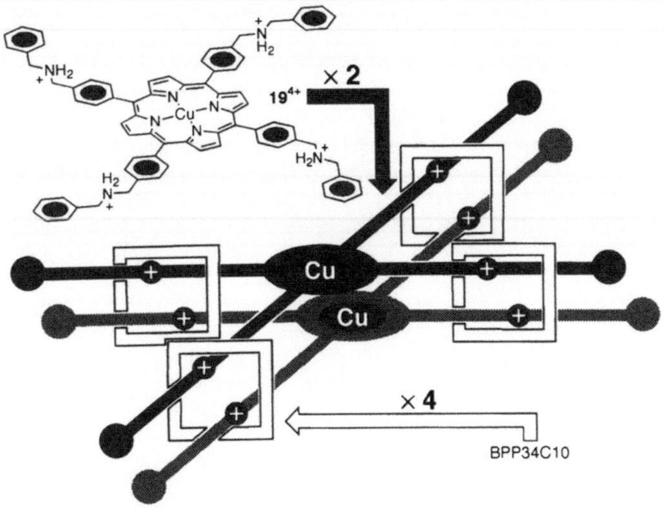

Fig. 27 The noncovalent synthesis of a supramolecular analogue of the special pair of the photosynthetic reaction center. Four BPP34C10 molecules tie two tetracations $\mathbf{19}^{4+}$ together to create a supermolecule bearing a pair of cofacially disposed porphyrinoid nuclei.

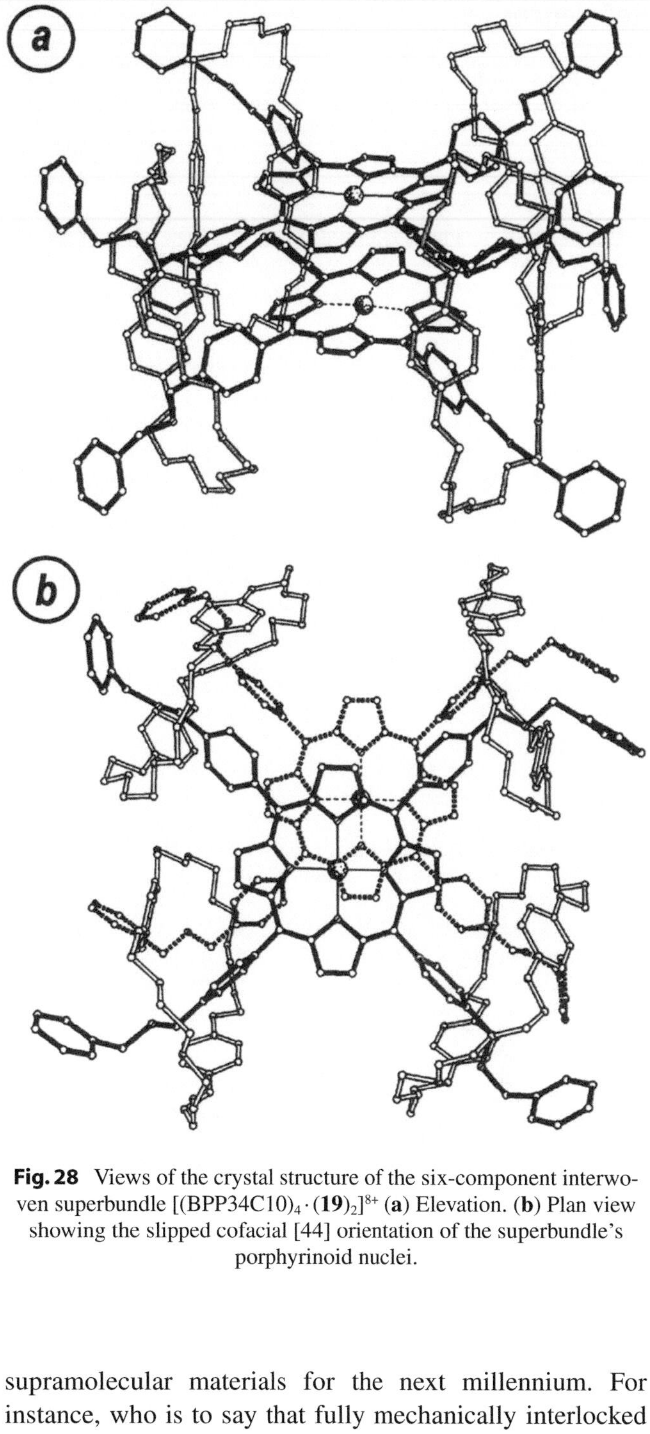

Fig. 28 Views of the crystal structure of the six-component interwoven superbundle $[(\mathrm{BPP34C10})_4 \cdot (\mathbf{19})_2]^{8+}$ (**a**) Elevation. (**b**) Plan view showing the slipped cofacial [44] orientation of the superbundle's porphyrinoid nuclei.

Summary and Outlook to the Future

We have demonstrated how a set of elaborate recognition motifs for the noncovalent synthesis of discrete supermolecules can be developed systematically from an elementary recognition motif, which, in this case, involves the threading of a dialkylammonium strand through a crown ether to generate a pseudorotaxane. The recognition motifs were progressively made more complex by exploiting the intimate relationship between synthetic supramolecular chemistry and X-ray crystallography. Thus, complicated supermolecules, such as daisy chains, interwoven superbundles, and multiply stranded pseudorotaxane anion binders, were synthesized noncovalently following the X-ray crystallographic analysis of the "original" pseudorotaxane $[\mathrm{DB24C8} \cdot \mathbf{3}]^{+}$.

In the future, we intend to continue to develop and refine the recognition motifs based upon crown ethers and dialkylammonium ions so that we can prepare even more intricate molecular (using supramolecular assistance to synthesis) and supramolecular (utilizing supramolecular synthesis) entities that could be incorporated into "intelligent" materials with predetermined functions. By way of illustration, we envisage manipulating the girdle-like properties of the ditopic crown ether BPP34C10 to tie two oligoammonium cavitands [45] together to create supramolecular capsules [4c] that could perhaps be utilized for, inter alia, recognition, transport, and catalysis at the nanoscale level. Additionally, we intend to marry the chemistry that we have profiled in this review with polymer and dendrimer chemistry so as to provide an entry into a whole new range of supramolecular materials for the next millennium. For instance, who is to say that fully mechanically interlocked macromolecular daisy chains [21] or dendrimers [46] endowed with rotaxane-based molecular switches attached to their surfaces will not have remarkable, promethean properties? Likewise, the solid-state noncovalent synthesis of the supermolecule $[(\mathrm{DB24C8})_3 \cdot \mathbf{8}]^{3+}$ has indicated that the preparation of polyrotaxanes is within our grasp. However, a great deal of fundamental research lies ahead of us, both in the solid state and in solution, before these and other goals can be fulfilled.

References

1. (a) Mascal, M. *Contemp. Org. Synth.* **1994,** *I,* 31–46. (b) Lehn, J.-M. *Supramolecular Chemistry;* VCH: Weinheim, 1995.
2. Seebach, D. *Angew. Chem., Int. Ed. Engl.* **1990,** *29,* 1320–1367.
3. Fyfe, M. C. T.; Stoddart, J. F. *Ace. Chem. Res.* **1997,** *30,* 393–401.
4. (a) Lawrence, D. S.; Jiang, T.; Levett, M. *Chem. Rev.* **1995,** *95,* 2229–2260. (b) Philp, D.; Stoddart, J. F. *Angew. Chem., Int. Ed. Engl.* **1996,** *35,* 1154–1196. (c) Conn, M. M.; Rebek, J., Jr. *Chem. Rev.* **1997,** *97,* 1647–1668. (d) Linton, B.; Hamilton, A. D. *Chem. Rev.* **1997,** *97,* 1669–1680. (e) Gillard, R. E.; Raymo, F. M.; Stoddart, J. F. *Chem. Eur. J.* **1997,** *3,* 1933–1940.
5. (a) MacGillivray, L.; Atwood, J. L. *Nature* **1997,** *389,* 469–472. (b) Hasenkopf, B.; Lehn, J.-M.; Boumedine, N.; Dupont-Gervais, A.; Van Dorsselaer, A.; Kniesel, B.; Fenske, D. *J. Am. Chem. Soc.* **1997,** *119,* 10956–10962. (c) Caulder, D. L.; Raymond, K. N. *Angew. Chem., Int. Ed. Engl.* **1997,** *36,* 1440–1442. (d) Timmerman, P.; Vreekamp, R. H.; Hulst, R.; Verboom, W.; Reinhoudt, D. N.; Rissanen, K.; Udachin, K. A.; Ripmeester, J. *Chem. Eur. J.* **1997,** *3,* 1823–1832. (e) Stang, P. J. *Chem. Eur. J.* **1998,** *4,* 19–27. (f) Biradha, K.; Zaworotko, M. J. *J. Am. Chem. Soc.* **1998,** *120,* 6431–6432.
6. (a) Whang, D.; Heo, J.; Kim, C.-A.; Kim, K. *Chem. Commun.* **1997,** 2361–2362. (b) Sijbesma, R. P.; Beijer, F. H.; Brunsveld, L.; Folmer, B. J. B.; Hirschberg, J. H. K. K.; Lange, R. F. M.; Meijer, E. W. *Science* **1997,** *278,* 1601–1604. (c) Russell, K. C.; Lehn, J.-M.; Kyritsakas, N.; DeCian, A.; Fischer, J. *New J. Chem.* **1998,** 123–128. (d) Fyfe, M. C. T.; Stoddart, J. F.; White, A. J. P.; Williams, D. J. *New J. Chem.* **1998,** 155–157. (e) Castellano, R. K.; Rebek, J., Jr. *J. Am. Chem. Soc.* **1998,** *120,* 3657–3663. (f) Batten, S. R.; Robson, R. *Angew. Chem., Int. Ed.* 1998, *37,* 1460–1494.
7. (a) Amabilino, D. B.; Stoddart, J. F.; Williams, D. J. *Chem. Mater.* **1994,** *6,* 1159–1167. (b) Desiraju, G. R. *Chem. Commun.* **1997,** 1475–1482.
8. (a) de Silva, A. P.; Gunaratne, H. Q.; Gunnlaugsson, T.; Huxley, A. J. M.; McCoy, C. P.; Rademacher, J. T.; Rice, T. E. *Chem. Rev.* **1997,** *97,* 1515–1566. (b) Balzani, V.; Gómez-López, M.; Stoddart, J. F. *Acc. Chem. Res.* **1998,** *31,* 405–414. (c) Sauvage, J.-P. *Acc. Chem. Res.* **1998,** *31,* 611–619.
9. Lehn, J.-M. *Makromol. Chem., Macromol. Symp.* **1993,** *69,* 1–17.
10. Fyfe, M. C. T.; Stoddart, J. F. *Adv. Supramol. Chem.* **1999,** *5,* 1–53.
11. Ashton, P. R.; Philp, D.; Spencer, N.; Stoddart, J. F. *J. Chem. Soc., Chem. Commun.* **1991,** 1677–1679.
12. (a) Amabilino, D. B.; Stoddart, J. F. *Chem. Rev.* **1995,** *95,* 2725–2828. (b) Gibson, H. W. In *Large Ring Molecules*; Semlyen, J. A., ed.; Wiley: Chichester, 1996; pp. 191–262. (c) Jäger, R.; Vögtle, F. *Angew. Chem., Int. Ed. Engl.* **1997,** *36,* 930–944.
13. (a) Ashton, P. R.; Campbell, P. J.; Chrystal, E. J. T.; Glink, P. T.; Menzer, S.; Philp, D.; Spencer, N.; Stoddart, J. F.; Tasker, P. A.; Williams, D. J. *Angew. Chem., Int. Ed. Engl.* **1995,** *34,* 1865–1869. (b) Ashton, P. R.; Chrystal, E. J. T.; Glink, P. T.; Menzer, S.; Schiavo, C.; Stoddart, J. F.; Tasker, P. A.; Williams, D. J. *Angew. Chem., Int. Ed. Engl.* **1995,** *34,* 1869–1871. (c) Ashton, P. R.; Chrystal, E. J. T.; Glink, P. T.; Menzer, S.; Schiavo, C.; Spencer, N.; Stoddart, J. F.; Tasker, P. A.; White, A. J. P.; Williams, D. J. *Chem. Eur. J.* **1996,** *2,* 709–728. (d) Glink, P. T.; Schiavo, C.; Stoddart, J. F.; Williams, D. J. *Chem. Commun.* **1996,** 1483–1490.
14. Dalley, N. K.; Bradshaw, J. S.; Larson, S. B.; Simonsen, S. H. *Acta Cryst.* **1982,** *B38,* 1859–1862.
15. Ashton, P. R.; Baxter, I.; Fyfe, M. C. T.; Raymo, F. M.; Spencer, N.; Stoddart, J. F.; White, A. J. P.; Williams, D. J. *J. Am. Chem. Soc.* **1998,** *120,* 2297–2307.
16. Ashton, P. R.; Fyfe, M. C. T.; Hickingbottom, S. K.; Stoddart, J. F.; White, A. J. P.; Williams, D. J. *J. Chem. Soc., Perkin Trans. 2* **1998,** 2117–2128.
17. (a) Kolchinski, A. G.; Busch, D. H.; Alcock, N. W. *J. Chem. Soc., Chem. Commun.* **1995,** 1289–1291. (b) Ashton, P.R.; Glink, P. T.; Stoddart, J. F.; Tasker, P. A.; White, A. J. P.; Williams, D. J. *Chem. Eur. J.* **1996,** 2, 729–736. (c) Ashton, P.R.; Glink, P. T.; Stoddart, J. F.; Menzer, S.; Tasker, P. A.; White, A. J. P.; Williams, D. J. *Tetrahedron Lett.* **1996,** *37,* 6217–6220. (d) Martínez-Díaz, M.-V.; Spencer, N.; Stoddart, J. F. *Angew. Chem., Int. Ed. Engl.* **1997,** *36,* 1904–1907. (e) Kolchinski, A. G.; Alcock, N. W.; Roesner, R. A.; Busch, D. H. *Chem. Commun.* **1998,** 1437–1438. (f) Ashton, P. R.; Ballardini, R.; Balzani, V.; Baxter, I.; Credi, A.; Fyfe, M. C. T.; Gandolfi, M. T.; Gómez-López, M.; Martínez-Díaz, M.-V.; Piersanti, A.; Spencer, N.; Stoddart, J. F.; Venturi, M.; White, A. J.P.; Williams, D. J. *J. Am. Chem. Soc.* **1998,** *120,* 11932–11942.
18. Ashton, P. R.; Baxter, I.; Cantrill, S. J.; Fyfe, M. C. T.; Glink, P. T.; Stoddart, J. F.; White, A. J. P.; Williams, D. J. *Angew. Chem., Int. Ed.* **1998,** *37,* 1294–1297.
19. (a) Sessler, J. L.; Andrievsky, A.; Gale, P. A.; Lynch, V. *Angew. Chem., Int. Ed.* **1996,** *35,* 2782–2785. (b) Zanotti-Gerosa, A.; Solari, E.; Giannini, L.; Chiesa-Villa, A.; Rizzoli, C. *Chem. Commun.* **1996,** 119–120. (c) Dimitrius, M.; Terzius, A.; Coleman, A. W.; de Rango, C. *Carbohydr. Res.* **1996,** *282,* 125–135. (d) Asakawa, M.; Ashton, P. R.; Brown, C. L.; Fyfe, M. C. T.; Menzer, S.; Pasini, D.; Scheuer, C.; Spencer, N.; Stoddart, J. F.; White, A. J.P.; Williams, D. J. *Chem. Eur. J.* **1997,** *3,* 1136–1150.
20. Ashton, P. R.; Ballardini, R.; Balzani, V.; Boyd, S. E.; Credi, A.; Gandolfi, M. T.; Gómez-López, M.; Iqbal, S.; Philp, D.; Preece, J. A.; Prodi, L.; Ricketts, H. G.; Stoddart, J. F.; Tolley, M. S.; Venturi, M.; White, A. J.P.; Williams, D. J. *Chem. Eur. J.* **1997,** *3,* 152–170.
21. Amabilino, D. B.; Parsons, I. W.; Stoddart, J. F. *Trends Polym. Sci.* **1994,** 2, 146–152.
22. (a) Gong, C. G.; Gibson, H. W. *Curr. Opin. Solid State Mater. Sci.* **1997,** 2, 647–652. (b) Gibson, H. W.; Gong, C. G.; Liu, S.; Nagvekar, D. *Macromol. Symp.* **1998,** *128,* 89–98.
23. Ashton, P.R.; Fyfe, M. C. T.; Glink, P. T.; Menzer, S.; Stoddart, J. F.; White, A. J. P.; Williams, D. J. *J. Am. Chem. Soc.* **1997,** *119,* 12514–12524.
24. Ashton, P. R.; Ballardini, R.; Balzani, V.; Gómez-López, M.; Lawrence, S. E.; Martínez-Díaz, M.-V.; Montalti, M.; Piersanti, A.; Prodi, L.; Stoddart, J. F.; Williams, D. J. *J. Am. Chem. Soc.* **1997,** *119,* 10641–10651.
25. Ashton, P. R.; Fyfe, M. C. T.; Schiavo, C.; Stoddart, J. F.; White, A. J. P.; Williams, D. J. *Tetrahedron Lett.* **1998,** *39,* 5455–5458.
26. Raymo, F. M.; Stoddart, J. F. *Pure Appl. Chem.* **1997,** *69,* 1987–1997.
27. Allwood, B. L.; Spencer, N.; Shahriari-Zavareh, H.; Stoddart, J. F.; Williams, D. J. *J. Chem. Soc., Chem. Commun.* **1987,** 1064–1066.
28. (a) Ashton, P.R.; Glink, P. T.; Martínez-Díaz, M.-V.; Stoddart, J. F.; White, A. J. P.; Williams, D. J. *Angew. Chem., Int. Ed. Engl.* **1996,** *35,* 1930–1933. (b) Ashton, P.R.; Ballardini, R.; Balzani, V.; Fyfe, M. C. T.; Gandolfi, M. T.; Martínez-Díaz, M.-V.; Morosini, M.; Schiavo, C.; Shibata, K.; Stoddart, J. F.; White, A. J. P.; Williams, D. J. *Chem. Eur. J.* **1998,** *4,* 2332–2341.
29. Fyfe, M. C. T.; Stoddart, J. F. *Coord. Chem. Rev.* **1999,** *183,* 139–155.
30. (a) Beer, P. D. *Chem. Commun.* **1996,** 689–696. (b) Atwood, J. L.; Holman, K. T.; Steed, J. W. *Chem. Commun.* **1996,** 1401–1407. (c) Schmidtchen, F. P.; Berger, M. *Chem. Rev.* **1997,** *97,* 1609–1646. (d) Antonisse, M.G. M.; Reinhoudt, D. N. *Chem. Commun.* **1998,** 443–448.
31. For leading references on examples of anion-assisted self-assembly, see: (a) Vilar, R.; Mingos, D. M. P.; White, A. J. P.; Williams, D. J. *Angew. Chem., Int. Ed.* **1998,** *37,* 1258–1261. (b) Fender, N. S.; Kahwa, I. A.; White, A. J. P.; Williams, D. J. *J. Chem. Soc., Dalton Trans.* **1998,** 1729–1730.

32. Fyfe, M. C. T.; Glink, P. T.; Menzer, S.; Stoddart, J. F.; White, A. J. P.; Williams, D. J. *Angew. Chem., Int. Ed. Engl.* **1997,** *36,* 2068–2070.
33. (a) Cram, D. J. *Angew. Chem., Int. Ed. Engl.* **1986,** *25,* 1039–1057. (b) Cram, D. J. *Angew. Chem., Int. Ed. Engl.* **1988,** *27,* 1009–1020.
34. Ashton, P.R.; Fyfe, M. C. T.; Martínez-Díaz, M.-V.; Menzer, S.; Schiavo, C.; Stoddart, J. F.; White, A. J.P.; Williams, D. J. *Chem. Eur. J.* **1998,** *4,* 1523–1534.
35. Goldberg, I. In *Inclusion Compounds,* vol. 2; Atwood, J. L., Davies, J. E. D., MacNicol, D. D., eds.; Academic: London, 1984; pp. 261–335.
36. (a) Ashton, P.R.; Collins, A. N.; Fyfe, M. C. T.; Menzer, S.; Stoddart, J. F.; Williams, D. J. *Angew. Chem., Int. Ed. Engl.* **1997,** *36,* 735–739. (b) Ashton, P.R.; Fyfe, M. C. T.; Hickingbottom, S. K.; Menzer, S.; Stoddart, J. F.; White, A. J. P.; Williams, D. J. *Chem. Eur. J.* **1998,** *4,* 577–589.
37. Kolotuchin, S. V.; Fenlon, E. E.; Wilson, S. R.; Loweth, C. J.; Zimmerman, S. C. *Angew. Chem., Int. Ed. Engl.* **1995,** *34,* 2654–2657 and references cited therein.
38. Alcala, R.; Martinez-Carrera, S. *Acta Cryst.* **1972,** *B28,* 1671–1677.
39. Beissel, T.; Powers, R. E.; Raymond, K. N. *Angew. Chem., Int. Ed. Engl.* **1996,** *35,* 1084–1086.
40. Ashton, P.R.; Collins, A. N.; Fyfe, M. C. T.; Glink, P. T.; Menzer, S.; Stoddart, J. F.; Williams, D. J. *Angew. Chem., Int. Ed. Engl.* **1997,** *36,* 59–62.
41. Feiters, M. C.; Fyfe, M. C. T.; Martínez-Díaz, M.-V:; Menzer, S.; Nolte, R. J. M.; Stoddart, J. F.; van Kan, P. J. M.; Williams, D. J. *J. Am. Chem. Soc.* **1997,** *119,* 8119–8120.
42. Deisenhofer, J.; Michel, H. *Science* **1989,** *245,* 1463–1473.
43. For some examples of recent synthetic analogues, see: (a) Kobuke, Y.; Miyaji, H. *J. Am. Chem. Soc.* **1994,** *116,* 4111–4112. (b) Senge, M. O.; Vicente, M.G. H.; Gerzevske, K. R.; Forsyth, T. P.; Smith, K. M. *Inorg. Chem.* **1994,** *33,* 5625–5638. (c) Girolami, G. S.; Hein, C. L.; Suslick, K. S. *Angew. Chem., Int. Ed. Engl.* **1996,** *35,* 1223–1225. (d) Stibrany, R. T.; Vasudevan, J.; Knapp, S.; Potenza, J. A.; Emge, T.; Schugar, H. J. *J. Am. Chem. Soc.* **1996,** *118,* 3980–3981.
44. For an explanation of why stacked porphyrin pairs adopt this sheared, or slipped cofacial, arrangement, see: Hunter, C. A. In *From Simplicity to Complexity in Chemistry—and Beyond;* Müller, A., Dress, A., Vögtle, F., eds.; Vieweg: Braunschweig/Wiesbaden, 1996; Part 1, pp. 113–126.
45. Cram, D. J.; Cram, J. M. *Container Molecules and Their Guests*; The Royal Society of Chemistry: Cambridge, 1994; pp. 85–106.
46. Zeng, F.; Zimmerman, S.C. *Chem. Rev.* **1997,** *97,* 1681–1712.

Asymmetry in the Multiprotein Systems of Molecular Biology[a,b]

Tom L. Blundell[c,d], Victor Bolanos-Garcia[c], Dimitri Y. Chirgadze[c], Nicholas J. Harmer[c], Thomas Lo[c], Luca Pellegrini[c], and B. Lynn Sibanda[c]

Abstract

Signaling in living systems needs to achieve high specificity, to be reversible, and to achieve high signal to noise. Signaling mediated by multiprotein systems has evolved that avoids the requirement for high-affinity binary complexes that would be difficult to reverse and which, in the overcrowded cell, would lead to excessive noise in the system. Symmetrical structures are only occasionally formed. When they are, it is principally to colocate components, for example, the tyrosyl kinases of growth factors, where dimers form. Symmetry is, however, often broken, presumably to create more sensitivity and specificity in the signaling system by assembling other components, into higher-order multiprotein systems. The binding of a single heparin to two 1:1 FGF:FGFR complexes is an example, as is the binding of a single ligase to the Xrcc4 dimer, perhaps so creating a further DNA-binding site.

Keywords

Multiprotein complexes • Symmetry • Signal transduction • Receptor tyrosyl kinases • DNA repair systems

Introduction

"Symmetry signifies rest and binding; asymmetry motion and loosening." These words, written of the visual arts, may equally well be used of living organisms. D'Arcy Thompson (1952) in *On Growth and Form* [1] was well aware that they applied to symmetry and structure at the level of whole organisms and the cell. However, they are also relevant to the fast-growing field of structural molecular biology.

As is well known, point-group symmetries involving rotational axes occur in soluble protein oligomers, for example, insulin hexamers in Fig. 1, in multienzymes complexes, and in spherical viruses (see [2] for review). Mirror planes and centers of inversion cannot occur, because proteins are enantiomers of one hand. Rotational symmetries are also found combined with one-dimensional (1-D) translations as line groups to provide the ropes and girders of biological systems, including the cell cytoskeleton and many fibrous structures. They are combined with two-dimensional (2-D) translations as plane groups to give membranes, but only rarely with three-dimensional (3-D) translations as space groups to give crystals. An example is the storage form of insulin in the α-granules of the islets of Langerhans, which comprise microcrystals of zinc insulin hexamers (Fig. 1c); here the symmetrical zinc ions assemble a stable and symmetrical hexamer complex. The rotational point symmetries confer stability on the macromolecules in an often dynamic, cellular, or extracellular environment; they provide a mechanism for cooperativity between subunits; and they colocalize important components to allow substrate channeling. Combined with translational symmetries, they provide the structure and stability of the cell.

[a] *Structural Chemistry* 2002, 13(3/4):405–412.

[b] This contribution is part of a collection entitled Generalized Crystallography and dedicated to the 75th Anniversary of Professor Alan L. Mackay, FRS.

[c] Department of Biochemistry, University of Cambridge, 80 Tennis Court Road, Cambridge CB2 1GA, UK

[d] To whom all correspondence should be sent; email: tom@cryst.bioc.cam.ac.uk

I. Hargittai, B. Hargittai (eds.), *Science of Crystal Structures: Highlights in Crystallography*, DOI 10.1007/978-3-319-19827-9_24

Abandoning Symmetry in Living Systems

Living systems, however, require more than stability; they must be dynamic. We must ask whether Nature abandons symmetry or alternatively distorts symmetrical structures so that they reach a point of disequilibrium with resulting motion. The aspartic proteinases provide an interesting insight into this question. These enzymes appear to have evolved from a protein composed of two identical copies of a smaller molecule [3]. It is now evident that such enzymes still exist in retroviruses, for example, as HIV proteinase [4, 5]. Elsewhere, evolution appears to have fused two identical proteins and accepted mutations in each half while keeping the basic architecture (or tertiary structure) of the ancestral protein. The net result, compared in Fig. 2a and b, is a pseudosymmetrical (twofold axis) molecule in which the two topologically equivalent catalytic groups (the two "hands" of the molecule) are now in distinctly different environments; one can push, while the other can pull much more effectively. A much better enzyme results and this is found in mammals as pepsin to digest food, as cathepsin D to process proteins within cells, and as renin to control blood pressure.

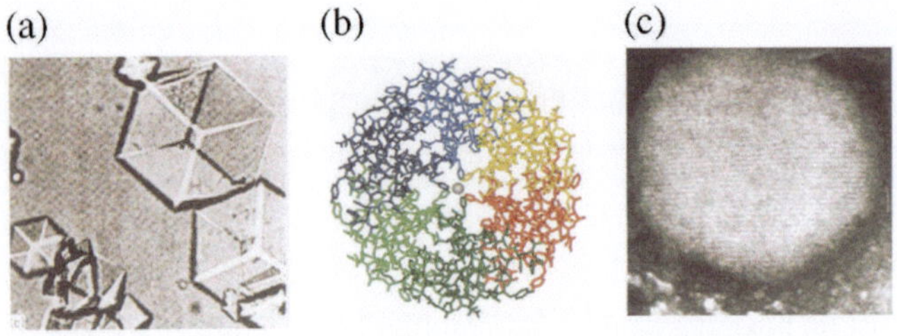

Fig. 1 Crystals (**a**) of 2-zinc insulin hexamers (**b**) resemble the insulin-storage granules (**c**) of the Islets of Langerhans [13].

The Symmetry and Asymmetry of Signaling and Regulatory Systems

More recently the power of molecular and structural biology has turned to another class of macromolecules—those used in cell signaling. These are usually active as multiprotein complexes that assemble and disassemble during the response of cells to external stimuli, both friendly and hostile. They include the cell surface receptors and the multiprotein systems that transduce signals in response to growth factors and hormones, as well as the macromolecular systems that regulate gene expression and repair damaged DNA. These systems are the focus of this review.

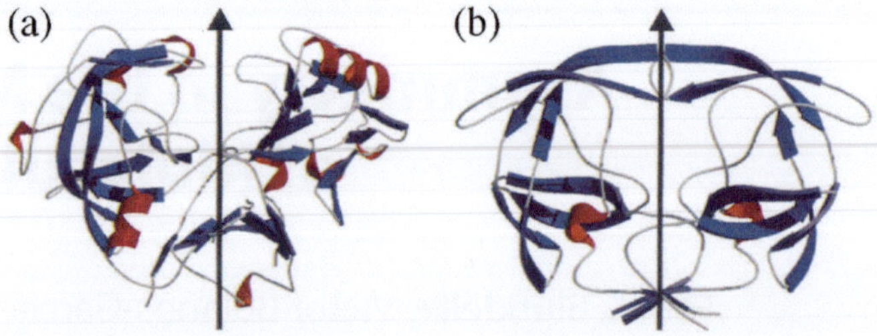

Fig. 2 The structure of the monomeric aspartic proteinase renin (**a**) has pseudo twofold symmetry and is related to the dimeric HIV proteinase (**b**) that has perfect twofold symmetry relating its two subunits [4, 5]. The *arrows* indicate the positions of the twofold symmetry axes.

Signaling Through Growth Factors and Receptor Tyrosyl Kinases

Many growth factors, for example, nerve growth factor (NGF), hepatocyte growth factor (HGF), insulinlike growth factor (IGF), and fibroblast growth factor (FGF), signal through cell surface, receptor tyrosyl kinases. These receptors comprise an extracellular growth factor-binding region, a single transmembrane helix, an intracellular tyrosyl kinase, and a regulatory/multisubstrate docking region. The model for activation most widely accepted is that growth factor binding induces dimerization, or possibly rearrangement of a preformed dimer, so that the intracellular kinases are able to autophosphorylate and/or transphosphorylate each other at multiple tyrosine residues and activate downstream signaling pathways. Whereas the tyrosyl kinases are clearly divergently evolved, the extracellular binding domains and their cognate growth factors have more than one evolutionary origin and thus belong to several unrelated protein families. Furthermore, the mechanism of dimerization appears to vary; stable dimers, dimers induced at the receptor, dimers cross-linked by heparan sulfate, and monomers can all be the active species.

NGF Receptor and the Heterohexameric 7S NGF Storage Complex

Nerve growth factor (NGF) is a neurotropic factor that promotes the differentiation and survival of certain populations of neurons in the central and peripheral nervous systems [6]. The 2.3 Å crystal structure of β-NGF [7] (Fig. 3a) showed that the biologically active dimer [8] has cystine-knot subunits related by a twofold axis. A large, buried, and hydrophobic interface leads to the formation of the extremely stable dimer ($K_d = 10^{-13}$ M).

In mouse submaxillary glands, β-NGF is stored in a protected form known as 7S NGF. This is an $\alpha_2\ \beta_2\ \gamma_2$ complex in which the β-NGF dimer is associated with two α-NGF and two γ-NGF subunits. The 3.1 Å crystal structure of 7S NGF [9] (Fig. 3b) shows that the complete 7S NGF complex adopts the twofold symmetry of the central β-NGF dimer. NGF and its homologs bind high-affinity protein receptors on the surface of target cells. These comprise a family of receptor tyrosine kinases, each specific for one NGF family member. Wiesmann and colleagues [10] crystallized NGF in complex with the membrane proximal extracellular domain of the NGF-specific receptor, TrkA. The 3-D structure (Fig. 3c) demonstrates how the TrkA can be dimerized by NGF dimers, to adopt a complex, also with twofold symmetry. Thus, the symmetrical and stable β-NGF dimer is a very efficient organizer of the symmetry of multiprotein complexes in which it is involved. It brings together receptor dimers in a way that facilitates transphosphorylation of receptor tyrosyl kinases on the other side of the membrane and activates the downstream signaling pathway.

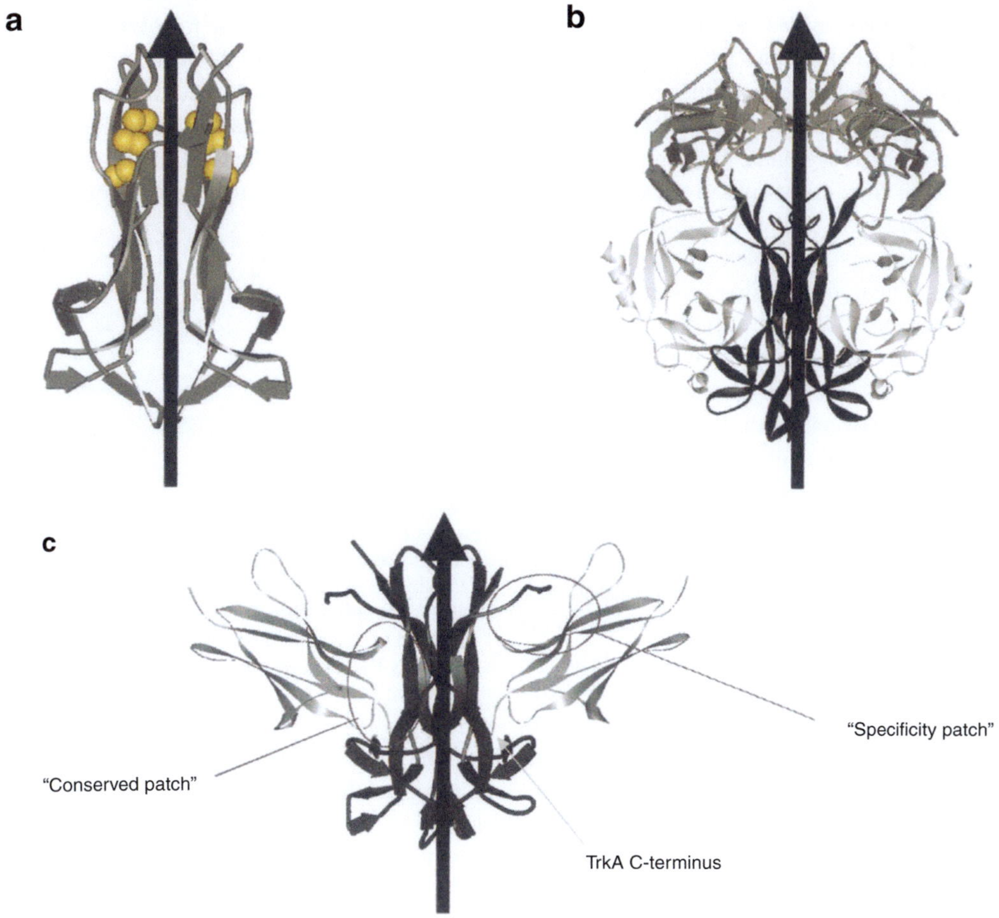

Fig. 3 The symmetrical β-NGF dimer (*black ribbons*) (**a**, [7]) organizes the α-NGF (*ribbons in light grey*), and the γ-NGF (*dark grey*) subunits in the hetero-hexameric 7S-NGF storage complex (**b**, [9]) and the membrane proximal domains (*grey*) in the TrkA receptor complex (**c**; [10]) to have twofold symmetry (*vertical arrows*). The two main interaction regions are shown highlighted. One region is shown on each side for clarity: each TrkA monomer has both regions. The C-terminus of TrkA is highlighted to show the relationship to the membrane.

Insulin, Insulinlike Growth Factor-1 (IGF-1), and Their Receptors

The determination of the crystal structure of insulin [11, 12] provided the first insight into the architecture of a polypeptide hormone/growth factor, but the structure of the complex with the receptor has remained elusive. Knowledge of the sequences, structure, chemical modification, receptor binding, and biological activity [13] indicated that the monomer, rather than the dimer, interacts with the receptor.

The insulin receptor and the homologous type 1 IGF receptor [14, 15] are α_2 β_2-heterotetrameric glycoproteins. The subunit is composed of two disulfide-linked extracellular monomers, each containing a single transmembrane alpha helix, while the subunit consists of a juxtamembrane region, an intracellular tyrosine kinase domain, and a C-terminal regulatory, multisubstrate docking region [16]. Comparative sequence analysis and modeling [17–19] and X-ray analysis [20] suggest that the extracellular regions of the receptors comprise two homologous globular domains (L1 and L2) flanking a cystine-rich domain, followed by three fibronectin III repeats. The L1 and L2 domains of IGFR have β-helical structures and are connected to the cystine-rich domain by flexible linkers, indicating that the relative positions of the L1 and L2 domains in the crystal structure may not be retained in complex with the ligand. Studies with chimeric insulin/IGF-I receptors [21] and alanine mutagenesis [22] suggest that the C-terminus of cystine-rich domain, the LI domain and a C-terminal peptide (residues 692–702) form the ligand-binding site in each subunit. This may rearrange the covalent receptor dimer, with a consequent change of spatial relationship and transphosphorylation of the intracellular receptor tyrosyl kinase enzymes. Thus, activation may occur without any dimer contacts between the two insulin ligands of the dimeric receptor, but with retention of twofold symmetry of the receptor complex.

The structure of the catalytic tyrosine kinase domain of the insulin receptor has been defined at high resolution by X-ray analysis [23]. The overall topology of this kinase domain consists of two lobes, an N-terminal and a C-terminal, with a single connection between them. Both lobes are held apart by steric interactions of a glycine-rich, nucleotide-binding loop and several residues adjacent to the activation loop (A-loop). Since the structures of both inactive and the active forms have been solved, it is now clear that a dramatic conformational change takes place upon autophosphorylation of several tyrosine residues present in the A-loop. In the inactive form, the A-loop traverses the cleft between the N- and C-terminal lobes in such a way that both, protein substrate an ATP-binding sites, are occupied. In the active form, a molecular rearrangement of the A-loop occurs (i.e., the A-loop is displaced approx. 30 Å with respect to the unphosphorylated form) as a result of phosphorylation of tyrosine residues at positions 1158, 1162, and 1163.

Hepatocyte Growth Factor/Scatter Factor, HGF/SF, and Its Met Receptor

Hepatocyte growth factor/scatter factor, HGF/SF, is a cytokine essential for the development of the liver and other cells in the mammalian embryo. Its domain organization resembles that of plasminogen, including an N-terminal domain (N), four kringle domains (K), and a catalytically inactive serine proteinase domain [24–26].

A crystal structure at 2.3 Å resolution of the NK1 fragment, a naturally occurring partial agonist, has been defined [27]. NK1 is a splice variant of the polypeptide growth factor HGF/SF that consists of the N-terminal and first kringle domain, and acts as a partial agonist, requiring heparan sulfate or soluble heparin for activity. In the crystals, two NK1 molecules interact to give a twofold symmetrical dimer through an extensive interface (2010 Å^2), although they are monomeric in solution (Fig. 4a). A surface region, defined by mutagenesis experiments, that appears to interact with the receptor, has been defined [28].

The crystal structures of NK1-heparin complexes show that heparin does not bridge subunits (Fig. 3b), but appears to act electrostatically [28]. The heparin-binding site located in the N-domain (Fig. 3c) is important for dimerization in solution and biological activity of NK1. Thus, HGF/SF NK1 molecules are monomeric in solution, but form dimers when heparin fragments, and therefore presumably heparin sulfate, bind. Interestingly, the same dimers are formed by NK1 on crystallization in several different crystal forms. The dimers are presumably stabilized at the receptor by heparan sulfate or heparin, so leading to dimerization of the receptor.

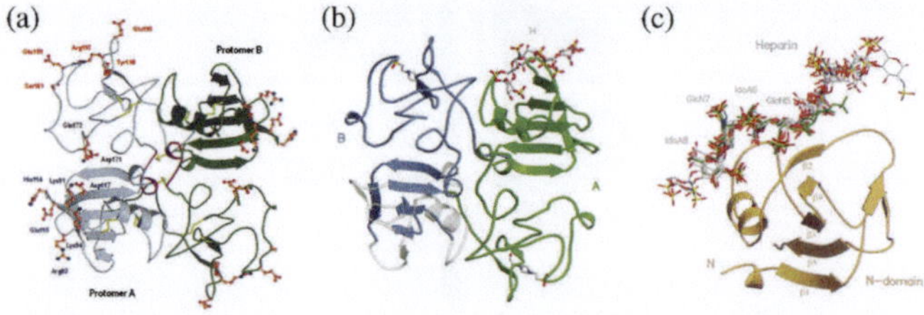

Fig. 4 The 3-D structure of HGF/SF NK1 fragment defined by X-ray analysis: (**a**) dimer of NK1 indicating a twofold axis, with the residues implicated in receptor binding indicated on one protomer [27], (**b**) crystal structure of complex of one heparin fragments with NK1 dimer [28], and (**c**) the superposition of heparin fragments defined independently by X-ray analysis onto one N-domain of HGF/SF [28].

Met is the receptor for HGF/SF and signaling via ligand binding results in the activation of a wide range of biological activities, including cellular motility, angiogenesis, cell growth, and invasion [29]. It has been shown *in vitro* that the multifunctional docking site of Met binds to only one signaling molecule and when several different effector molecules are present at the same time, a hierarchical binding order is observed [30]. Such hierarchical order might be crucial for the Met-mediated multiple signal transduction.

Fibroblast Growth Factor 1 (FGF1) in Complex with Its Receptor (FGFR2) Ectodomain and Heparin

Fibroblast growth factors (FGF) are a family of polypeptides with a pleiotropic range of physiological and pathological activities. Heparan sulfate, or its analog heparin, is required to allow activation of the receptor tyrosine kinase (FGFR) by FGFs [30, 31], and it appears that association of FGF, FGFR, and heparan sulfate in a specific complex on the cell surface is essential for signal transduction. Although a wealth of information has been generated over the past decade by mutational studies, biophysical experiments, and cell growth assays, these data have not provided a reliable model for the FGF–FGFR–heparin complex (see, e.g. [32]). Structural studies have described binary complexes of FGF–heparin [33, 34] and FGF–FGFR [35, 36]. While these structures help confirm the previous data from nonstructural methods, they do not reveal 3-D information about the mode of ligand-induced FGFR oligomerization in the presence of heparin. Schlessinger and colleagues [37] have shown that hexasaccharides can bind to this 2:2 complex, but longer saccharides cannot be seen in this crystal form. We have defined the crystal structure of the FGFR2 ectodomain in a dimeric form, induced by simultaneous binding to FGF1 and a heparin decasaccharide [38]. The complex is assembled around a central heparin molecule linking two FGF1 ligands into a dimer that bridges between two receptor chains (Fig. 5). Although the receptor and ligand moieties are related by an approximate twofold axis, heparin binding is remarkably asymmetric, involving extensive contacts with both FGF1 molecules but only one receptor chain. Thus perfect symmetry appears to be abandoned in the process of activation of the FGF receptor, in contrast to NGF where symmetry is retained. In this case, the presence of a further component, heparan sulfate or heparin, in the process of receptor activation, probably decreases the chance of opportunistic interactions, so increasing signal-to-noise.

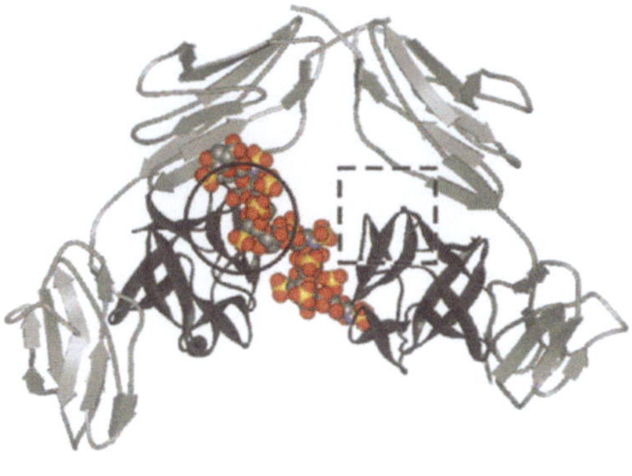

Fig. 5 FGF1, FGFR2 and heparin decamer in complex [38]. FGF1 is displayed as a ribbon drawing in *black*, FGFR2 as schematic in *grey*, and heparin as all atom space-filling model. The C-terminus of FGFR2 (membrane proximal) is at the *bottom* of the picture. Note that heparin binds to one of the two FGFR2 molecules (*ringed*), while in the other protomer, the binding site is vacant (*dashed box*).

DNA Repair Signaling

DNA damage in eukaryotic organisms leads to two major pathways of DNA repair in eukaryotic organisms: homologous recombination and nonhomologous end joining (NHEJ). These pathways have many of the features of signaling from outside the cell; they involve assembly of multiprotein complexes, and more often than not, signal transduction through protein kinases.

Nonhomologous DNA End Joining: The Crystal Structure of an Xrcc4-DNA Ligase IV Complex

A complex of two proteins, Xrcc4 and DNA Ligase IV, plays a central role in DNA nonhomologous end joining (NHEJ) [39, 40], a cellular function required for double-strand break repair and V(D)J recombination [41, 42]. Cells lacking either Xrcc4 or DNA ligase IV exhibit hypersensitivity to ionizing radiation and are defective in V(D)J recombination [43]. In mice, Xrcc4 or DNA ligase IV deficiency causes genomic instability and embryonic lethality [44–47]. The crystal structure of human Xrcc4 bound to a polypeptide [48], corresponding to the DNA ligase IV sequence linking its two BRCA1 carboxyl-terminal (BRCT) domains, shows that a single ligase chain binds to the coiled–coil region of an Xrcc4 dimer (Fig. 6). The helical tails of Xrcc4 undergo a substantial conformational change relative to the uncomplexed protein [48]. The coiled coil appears to "unwind" at the point of ligase binding due to the introduction of a stutter in the supercoil, leading to a flat interaction surface. A buried network of charged hydrogen bonds surrounded by extensive hydrophobic contacts explain the observed tightness of the interaction. The strong conservation of residues at the interface between the two proteins provides evidence that the observed mode of interaction has been maintained in NHEJ throughout evolution.

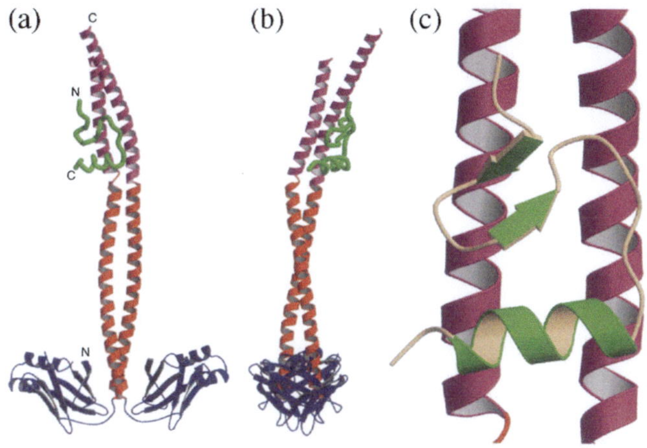

Fig. 6 The structure of an Xrcc4 fragment complexed to the linker region between the BRCT repeats of DNA ligase IV, shown in *green* [48]. The approximate twofold axis of the Xrcc4 dimer, lying roughly vertical in this figure, is distorted by the binding of one repeat to a flattened region of the coiled-coil. (**a**) and (**b**) show the complex from orthogonal views, while (**c**) shows a close-up of the ligase linker binding taken from a view perpendicular to the plane of the flattened helices.

The presence of the stutter in the coiled–coil region leads to a change of hand of the supercoiling, but does not in itself necessarily lead to a lack of twofold symmetry. However, the binding of one ligase linker to the Xrcc4 coiled–coil region appears to lead to a breakdown of the local twofold symmetry, so that a second linker could not bind in the same way. This possibly leads to the formation of a binding site for DNA on the reverse side of the coiled coil, but this idea needs to be tested.

Homologous Recombination: Interactions of Tumor Suppressor Protein BRCA2 with RAD51

BRCA2 is a protein of over 3000 amino acids with essential roles in embryonic proliferation and tumor suppression in adult cells [49]. It binds RAD51, which performs homologous DNA reconstitution and has an evolutionary conserved core resembling bacterial RecA protein. Mutations in the BRCA1 and BRCA2 gene are responsible for the majority of familial cases of breast cancer [50].

Free RAD51 is oligomeric and forms linear fibrous structures in the presence of DNA [51], with approximately six RAD51 monomers per filament turn [52]. Strand exchange occurs after the incorporation of a single-stranded DNA and a homologous double-stranded DNA molecule into the central filamentous axis. Its interaction with DNA is not sequence-specific and thus allows strand exchange and repair of any damaged DNA sequence.

BRCA2 interacts with RAD51 through one or more of eight short BRC repeats, coded within exon 11 [53]. In so doing, it prevents the oligomerization of RAD51 and/or its interaction with DNA. One RAD51 appears to bind one BRC repeat, forming a complex of 1:1 stoichiometry. BRCA2 thus sequesters Rad51, and colocalizes with BRCA1 in a large multiprotein complex at sites of DNA replication, marked by the proliferating cell nuclear antigen, PCNA [54].

Whereas symmetry of the Rad51 filament may be necessary for function (nonspecific), asymmetric binding between Rad51 and BRC repeats is necessary for signaling and targeting at sites of DNA damage (specific). The reversible competition between the two different interactions is key. Breaking symmetry here plays an important role in specificity.

Conclusions

Signaling in living systems needs to achieve high specificity, to be reversible and to achieve high signal-to-noise. Signaling mediated by multiprotein systems has evolved that avoids the requirement for high-affinity binary complexes that would be difficult to reverse and which, in the overcrowded cell, would lead to excessive noise in the system. The robust development and survival of all multicellular organisms is a testament to the success of this process, and structural biology is uniquely placed to provide insight into how this is achieved.

Our brief review indicates that signaling systems do not have the highly symmetrical point symmetries of multienzyme complexes and viruses. Such harmonious and symmetrical motifs are reminiscent of the simple symmetrical designs on early Chinese bronze mirrors that reflected the search for permanent, symmetrical relationships expressed in Confucian philosophy and taken as necessary for the stability of Chinese feudal society. Christianity, once established, also exploited symmetry in its visual arts and architecture to establish a mood of restfulness and permanence; the rose window of the Royal Door of Chartres Cathedral (built in 1150–1160) as well as the Octagon of Ely Cathedral in East Anglia (1322) are two of the finest examples. Muslims have also searched for more complex symmetrical, but nontranslational symmetries, also found in natural systems, as Alan Mackay was first to suggest [55].

Indeed signaling systems are not at rest, but must be responsive to change. Nature appears to abandon symmetry, distorting symmetrical structures so that they reach a point of disequilibrium with resulting motion. There is an analogy with Michelangelo in "Creation," where God and Adam, each in human form, offer a pseudosymmetry; their hands touch in a configuration related by a mirror plane—the right hand of God reflected as the left hand of Adam. Michelangelo disturbed the symmetry gently but firmly to make it clear that life flows strongly into Adam. There is also a parallel with Mao Tse Tung's dialectical reinterpretation of Confucian relationships where the thesis and antithesis are no longer balance (as in yin and yang) and a new synthesis takes place.

Thus signaling systems usually assemble as multiprotein complexes that involve many different protein and other subunits. Symmetrical structures are only occasionally formed. When they are, it is principally to colocate components, for example, the tyrosyl kinases of growth factors, where dimers form. Symmetry is, however, often broken, presumably to create more sensitivity and specificity in the signaling system by assembling other components, into higher-order multiprotein systems. The binding of a single heparin to two 1:1 FGF:FGFR complexes is an example, as is the binding of a single ligase to the Xrcc4 dimer, perhaps so creating a further DNA-binding site.

Nature usually does not select the most restful or harmonious motif in its search for motion in life. Signaling systems that regulate the dynamic processes of cell physiology are good examples; they rarely have very high symmetry. Cell and molecular biologists should consider Chinese bronzes, cathedral windows, and Islamic decorative art by all means; but they should also consider Michelangelo's "Creation," if they wish to understand life and motion at the molecular level.

References

1. D'Arcy Thompson, *On Growth and Form*; Cambridge Univ. Press: Cambridge, 1952.
2. Blundell, T. L.; Srinivasan, N. *Proc. Natl. Acad. Sci. USA* **1996,** *93*, 233.
3. Tang, J.; James, M. N. G.; Hsu, I. N.; Jenkins, J. A.; Blundell, T. L. *Nature* (*London*) **1978,** *271*, 618.
4. Lapatto, R.; Blundell, T. L.; Hemmings, A.; Overington, J.; Wilderspin, A.; Wood, S.; Merson, J. R.; Whittle, P. J.; Danley, D. E.; Geoghegan, K. F.; Hawrylik, S. J.; Lee, S. E.; Scheld, K. G.; Hobart, P. M.; Blundell, T. L. *Nature* (*London*) **1989,** *342*, 299.
5. Miller, M.; Jaskolski, M.; Mohana Rao, J. K.; Leis, J.; Wlodawer, A. *Nature* (*London*) **1989,** *337*, 576.
6. Sofroniew, M. V.; Howe, C. L.; Mobley, W. C. *Annu. Rev. Neurosci.* **2001,** *24*, 1217.
7. McDonald, N. Q.; Lapatto, R.; Murray-Rust, J.; Gunning, J.; Wlodawer, A.; Blundell, T. L. *Nature* (*London*) **1991,** *354*, 411.
8. Angeletti, R. H.; Bradshaw, R. A. *Proc. Natl. Acad. Sci. USA* **1971,** *68*, 2417.
9. Bax, B.; Blundell, T. L.; Murray-Rust, J.; McDonald, N. Q. Structure **1997,** *5*, 1275.
10. Wiesmann, C.; Ultsch, M. H.; Bass, S. H.; de Vos, A. M. *Nature* (*London*) **1999,** *401*, 184.
11. Adams, M. J.; Blundell, T. L.; Dodson, G. G.; Dodson, E.; Vijayan, M.; Hodgkin, D. C. *Nature* (*London*) **1969,** *224*, 491.
12. Blundell, T. L.; Cutfield, J. F.; Cutfield, S. M.; Dodson, G. G.; Dodson, E.; Mercola, D.; Vijayan, M.; Hodgkin, D. C. *Nature* (*London*) **1971,** *231*, 506.
13. Blundell, T. L.; Dodson, G. G.; Mercola, D.; Hodgkin, D. C. *Advan. Protein Chem.* **1972,** *26*, 279.
14. Ullrich, A.; Bell, J. R.; Chen, E. Y.; Herrera, R.; Petruzzelli, L. M.; Dull, T. J.; Gray, A.; Coussens, L.; Liao, Y. C.; Tsubokawa, M.; Mason, A.; Seeburg, P.; Grunfield, C.; Rosen, O.; Ramachandran, J. *Nature* (*London*) **1985**, *313*, 756.
15. Ebina, Y.; Ellis, L.; Jarnagin, K.; Edery, M.; Graf, L.; Clauser, E.; Ou, J. H.; Ma-siarz, F.; Kan, Y. W.; Goldfine, I. D.; Roth, R. A.; Rutter, W. J. *Cell* **1985,** *40*, 747.
16. Hubbard, S. R. *EMBO J.* **1997,** *16*, 5572.
17. Bajaj, M.; Waterfield, M. D.; Schlessinger, J.; Taylor, W. R.; Blundell, T. L. *Biochim. Biophys. Acta* **1987,** *916*, 220.
18. Marino-Buslje, C.; Martin-Martinez, M.; Mizuguchi, K.; Siddle, K.; Blundell, T. L. *Biochem. Soc. Trans.* **1999,** *27*, 715.
19. Ward, C. W. *Growth Factors* **1999,** *16*, 315.
20. Garrett, T. P.; McKern, N. M.; Lou, M.; Frenkel, M. J.; Bentley, J. D.; Lovrecz, G. O.; Elleman, T. C.; Cosgrove, L. J.; Ward, C. W. *Nature* (*London*) **1998,** *394*, 395.
21. Schumacher, R.; Soos, M. A.; Schlessinger, J.; Brandenburg, D.; Siddle, K.; Ullrich, A. *J. Biol. Chem.* **1993,** *268*, 1087.
22. Mynarcik, D. C.; Williams, P. F.; Schaffer, L.; Yu, G. Q.; Whittaker, J. *J. Biol. Chem.* **1997,** *272*, 18650.
23. Hubbard, S. R.; Wei, L.; Ellis, L.; Hendrickson, W. A. *Nature* (*London*) **1994,** *372*: 746.
24. Gherardi, E.; Gray, J.; Stoker, M.; Perryman, M.; Furlong, R. *Proc. Natl. Acad. Sci. USA* **1989,** *86*, 5844.
25. Miyazawa, K.; Tsubouchi, H.; Naka, D.; Takahashi, K.; Okigaki, M.; Arakaki, N.; Nakayama, H.; Hirono, S.; Sakiyama, O.; Takahashi, K.; Gohda, E.; Daikuhara, Y.; Kitamura, N. *Biochem. Biophys. Res. Commun.* **1989,** *163*, 967.
26. Nakamura, T.; Nishizawa, T.; Hagiya, M; Seki, T.; Shimonishi, M.; Sugimura, A.; Tashiro, K.; Shimizu, S. *Nature* (*London*) **1989,** *342*, 440.
27. Chirgadze, D. Y.; Hepple J. P.; Zhou, H.; Byrd, R. A.; Blundell, T. L.; Gherardi, E. *Natur. Struct. Biol.* **1999,** *6*, 72.
28. Lietha, D.; Chirgadze, D. Y.; Mulloy, B.; Blundell, T. L.; Gherardi, E. *EMBO J.* **2001,** *20*, 5543.
29. Sudhalter, J.; Folkman, J.; Svahn, C. M.; Bergendal, K.; D'Amore, P. A. *J. Biol. Chem.* **1989,** *264*, 6892.
30. Stefan, M.; Koch, A.; Mancini, A.; Mohr, A.; Weidner, M.; Niemann, H.; Tamura, T. *J. Biol. Chem.* **2001,** *276*, 3017.
31. Ornitz, D. M.; Yayon, A.; Flanagan, J. G.; Svahn, C. M.; Levi, E.; Peder, P. *Mol. Cell Biol.* **1992,** *12*, 240.
32. Powers, C. J.; McLeskey, S. W.; Wellstein, A. *Endo.-Rel. Cancer* **2000,** *7*, 165.
33. Faham, S.; Hileman, R. E.; Fromm, J. R.; Linhardt, R. J.; Rees, D. C. *Science* **1996,** *271*, 1116.
34. DiGabriele, A. D.; Lax, I.; Chen, D. I.; Svahn, C. M.; Jaye, M.; Schlessinger, J.; Hendrickson, W. A. *Nature* (*London*) **1998,** *393*, 812.
35. Plotnikov, A. N.; Schlessinger, J.; Hubbard, S. R.; Mohammadi, M. *Cell* **1999,** *98*, 641.
36. Stauber, D. J.; DiGabriele, A. D.; Hendrickson, W. A. *Proc. Natl. Acad. Sci. USA* **2000,** *97*, 49.
37. Schlessinger, J.; Plotnikov, A. N.; Ibrahimi, O. A.; Eliseenkova, A. V.; Yeh, B. K.; Yayon, A.; Linhardt, R. J.; Mohammadi, M. *Mol. Cell.* **2000,** *6*, 743.
38. Pellegrini, L.; Burke, D. F.; von Delft, F.; Mulloy, B.; Blundell, T. L. *Nature* (*London*) **2000,** *407*, 1029.
39. Critchlow, S. E.; Bowater, R. P.; Jackson, S. P. *Curr. Biol.* **1997,** *7*, 588.
40. Grawunder, U.; Wilm, M.; Wu, X.; Kulesza, P.; Wilson, T. E.; Mann, M.; Lieber, M. R. *Nature* (*London*) **1997,** *388*, 492.
41. Critchlow, S. E.; Jackson, S. P. *Trends Biochem. Sci.* **1998,** *23*, 394.
42. Jeggo, P. A. *Advan. Genet.* **1998,** *38*, 185.
43. Li, Z. Otevrel, T.; Gao, Y.; Cheng, H. L.; Seed, B.; Stamato, T. D.; Taccioli, G. E.; Alt, F. W. *Cell* **1995,** *83*, 1079.
44. Gao, Y. Sun, Y.; Frank, K. M.; Dikkes, P.; Fujiwara, Y.; Seidl, K. J.; Sekiguchi, J. M.; Rathbun, G. A.; Swat, W.; Wang, J.; Bronson, R. T.; Malynn, B. A.; Bryans, M.; Zhu, C.; Chaudhuri, J.; Davidson, L.; Ferrini, R.; Stamato, T.; Orkin, S. H.; Greenberg, M. E.; Alt, F. W. *Cell* **1998,** *95*, 891.
45. Frank, K. M.; Sharpless, N. E.; Gao, Y.; Sekiguchi, J. M.; Ferguson, D. O.; Zhu, C.; Manis, J. P.; Horner, J.; DePinho, R. A.; Alt, F. W. *Mol. Cell* **2000,** *5*, 993.
46. Gao, Y. Ferguson, D. O.; Xie, W.; Manis, J. P.; Sekiguchi, J.; Frank, K. M.; Chaudhuri, J.; Horner, J.; DePinho, R. A.; Alt, F. W. *Nature* (*London*) **2000,** *404*, 897.
47. Barnes, D. E.; Stamp, G.; Rosewell, I.; Denzel, A.; Lindhal, T. *Curr. Biol.* **1998,** *8*, 1395.
48. Sibanda, B. L.; Critchlow, S.; Begun, J.; Pei, X. Y.; Jackson, S. P.; Blundell, T. L.; Pellegrini, L. *Natur. Struct. Biol.* **2001,** *8*, 1015.
49. Welcsh, P. L.; Owens, K. N.; King, M. C. *Trends Genet.* **2000,** *16*, 69.
50. Ingvarsson, S. *Anticancer Res.* **1999,** *19*, 2853.
51. Benson, F. E.; Stasiak, A.; West, S. C. *EMBO J.* **1994,** *13*, 5764.
52. Yu, X., Jacobs, S. A.; West, S. C.; Ogawa, T.; Egelman, E. H. *Proc. Natl. Acad. Sci. USA* **2001,** *98*, 8419.
53. Wong, A. K.; Pero, R.; Ormonde, P. A.; Tavtigian, S. V.; Bartel, P. L. *J. Biol. Chem.* **1997,** *272*, 31941.
54. Scully, R., Chen, J.; Ochs, R. L.; Keegan, K.; Hoekstra, M.; Feunteun, J.; Livingston, D. M. *Cell* **1997,** *90*, 425.
55. Mackay, A. L. *Phys. Bull.* **1976,** p. 495.

Molecular van der Waals Symmetry Affecting Bulk Properties of Condensed Phases: Melting and Boiling Points[a,b]

Yuri L. Slovokhotov[c], Andrei S. Batsanov[d], and Judith A.K. Howard[d]

Abstract

Qualitative analysis of the trends in melting temperatures T_m and boiling temperatures T_b of organic compounds and related substances supports a concept of van der Waals symmetry suggested earlier by the authors. Data for substituted methanes, ethanes, ethylenes, benzenes and cyclohexanes, reveal linear T_b versus M^α patterns but usually non-uniform T_m versus M^α patterns (where M is the molecular mass, α varies from 0.5 to 1). In the same set of isomers, T_m and T_b often change in different succession. To explain the observed trends, a local van der Waals field U that acts on a molecule in a solid or liquid phase, was represented by a sum of an averaged long-range field U_0 from all molecules, and a short-range "contact" term W determined by a molecular shape. Variations of T_m and T_b among closely related substances correspond to different strength of perturbation of $U_0^{(C)}$ (in a crystal) and $U_0^{(L)}$ (in a liquid) by W. Higher T_b of isomers with medium-symmetry molecules (like 1,2-disubstituted benzenes and cyclohexanes) reflect the better fitting of their molecules to a local molecular environment in a liquid (that contains vacancies). Highly symmetrical quasispherical molecules (e.g., C_2Cl_6 or C_6F_{12}) produce a stabilised solid and destabilised liquid state, hence their molecular crystals easily sublime rather than melt at ambient pressure.

Keywords

Boiling point • Melting point • Molecular condensed phases • Symmetry

"A molecule has a body"
Alexander I. Kitaigorodskii

Introduction

Bulk properties of molecular substances obviously depend on the structure (atomic connectivity) and related parameters (dipole moment, polarisability, etc.) of the molecules. However, at the moment both the mathematical form of this dependence and the underlying physical effects are understood imperfectly and require further investigations. Recently, a number of empirical correlations between chemical/biological activity of substances and their molecular structure have been elucidated using molecular descriptors [1], but theoretical substantiation of this method is still lacking.

Among physical constants accurately determined for large number of organic and inorganic substances are melting points (mp, or T_m) and boiling points (bp, or T_b). Some intriguing regularities, e.g., alteration of mp in even and odd n-alkanes, non-uniform variations of mp among isomers, easy sublimation of compounds with high-symmetric quasispherical molecules, etc., have been empirically known, but not theoretically explained, since mid-twentieth century [2].

[a] *Structural Chemistry* 2007, 18(4):477–491.

[b] Dedicated in memory of Professor Petr M. Zorky

[c] Chemistry Department, Moscow State University, Moscow 119992, Russia
e-mail: slov@phys.chem.msu.ru

[d] Department of Chemistry, University of Durham, South Road, Durham DH1 3LE, UK

I. Hargittai, B. Hargittai (eds.), *Science of Crystal Structures: Highlights in Crystallography*, DOI 10.1007/978-3-319-19827-9_25

In the absence of strong charge separation and specific intermolecular interactions (like H-bonds), bulk properties of molecular condensed phases, including mp and bp, are governed by van der Waals intermolecular forces [3–5]. Most theoretical studies of the structure-property relationships in this class of substances are based on computer simulations and packing analysis (see [5–7] and references therein). A good state-of-art example is a comprehensive theoretical treatment of molecular packing in a set of real and hypothetical crystalline $C_{12}H_{12}$ isomers [8] or in isomers of dichlorobenzene [9].

However, ad hoc packing analysis in organic crystals usually do not give quantitative prediction of their physical constants, because of extreme complexity of the system. At the same time, reported physical data for a huge number of new substances created by synthetic chemistry, have never been analysed in a systematic way. This can be a very promising field of research where new structure-property correlations may be found and, ultimately, explained on physical grounds.

Recently we have found that mp of organic substances reflect the approximate symmetry of their molecular van der Waals surfaces [10], the key factor being a geometry-sensitive short range ("contact") term of the local van der Waals field that acts on a molecule in a crystal. This additive contact term may be regarded as a perturbation in the average local field created by all molecules. We have suggested that the magnitude of the perturbation depends linearly on the deviation of the molecular van der Waals surface from the idealised *van der Waals group* $\widetilde{\mathbf{G}}(\mathbf{U}_0)$ which describes the symmetry of the averaged long-range molecular field.

In the present work, we undertook a systematic survey of T_m and T_b data for substituted methanes, ethanes, ethylenes, benzenes, and cyclohexanes, and applied the hypothesis of van der Waals symmetry to the analysis of the observed trends. The experimental T_m and T_b values show a qualitative agreement with this hypothesis, thus pointing to the probable origins of the van der Waals symmetry in crystals and liquids.

Experimental

Mp and bp of organic and inorganic (organometallic) compounds were extracted from the Beilstein [11] and Gmelin [12] databases, respectively, using the CrossFire software [13]. The relevant crystal structures have been retrieved from the Cambridge Structure Database (CSD) and analysed using ConQuest 1.8, Mercury 1.4.1 and Vista programmes [14]. Graphics were prepared using the SigmaPlot software [15].

Results

As an initial step, we have presented T_m and T_b for several series of organic substances as functions of the molecular mass, M. This revealed some remarkable trends, illustrated in the figures below. In the first place, among structurally similar molecules, the corresponding temperatures often show a linear dependence against M^α, where $1/2 < \alpha < 1$.

Methane Analogues MX_4 and $MX_{4-n}Y_n$

Bp of methane analogues EX_4 (where X=H, alkyl or halogen) correlate linearly with $M^{1/2}$ (Fig. 1). For each series of compounds, the regression line is roughly parallel to that of monoatomic noble gases, but shifted to higher temperatures. There is no drastic difference between molecules with zero and nonzero electric moments, as well as molecules containing X substituents of similar van der Waals dimensions but different masses (e.g., Me and Br) and therefore having different moments of inertia; the corresponding points cluster around the same line. However, bp values of mixed-substituted CMe_nX_{4-n} (X=halogen) with non-zero dipole moment μ and quadrupole moment Q (black triangles in Fig. 1) are elevated by ca. 20–30 K from that of *symm*-EX_4 derivatives. On the other hand, in EEt_4 compounds, bulkier ethyl substituents cause larger increase in bp (by 40–100 K), even though the electric moments of these molecules are much lower than for halides, due to less polar E–C bonds.

The compounds $CR_{4-n}F_n$ with two or more fluorine atoms in a methane-like molecule have much lower bp (Fig. 2a). The depression of bp is also manifested for $EMe_{4-n}Cl_n$

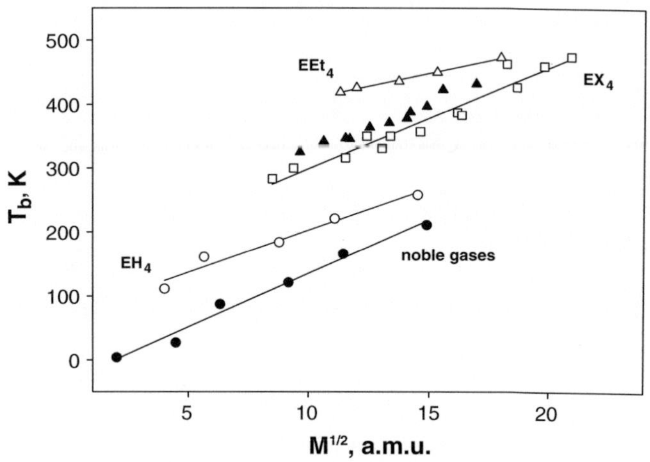

Fig. 1 Boiling points *Tb* versus the square root of molecular mass $M^{1/2}$ in substituted methane analogues EX_4 (E=C–Pb, X=Me, Cl, Br). Noble gases are shown for reference. *Black triangles*: mixed-substituted derivatives $CMe_{4-n}X_n$ with nonzero electric molecular moments.

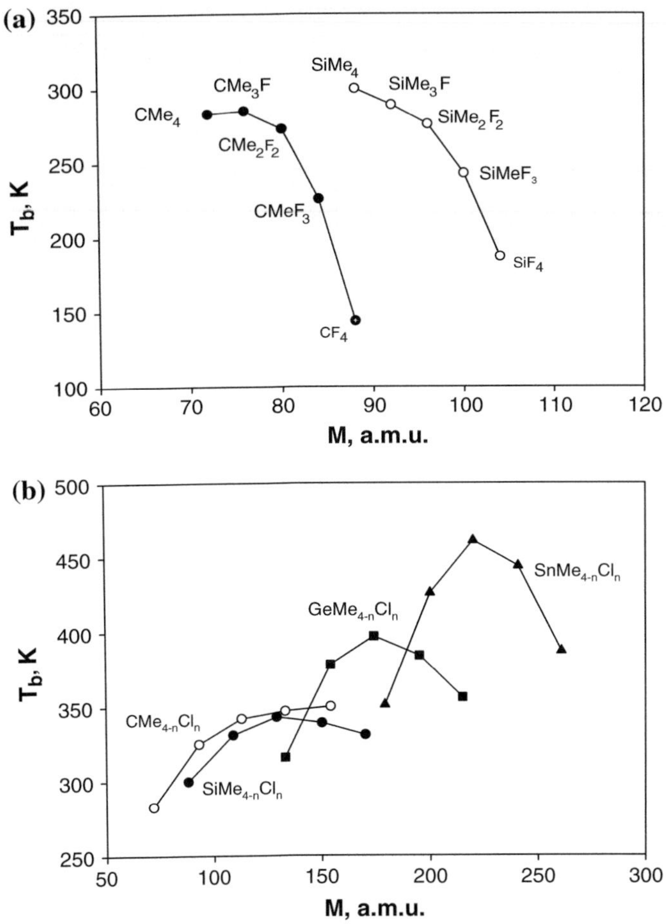

Fig. 2 Electrostatic depression of boiling points in substances composed of tetrahedral molecules with negatively polarised halogen atoms: (**a**) fluorinated derivatives $EMe_{4-n}F_n$ (E=C, Si), (**b**) chlorinated derivatives $EMe_{4-n}Cl_n$ (*open circles*: E=C, *filled circles*: E=Si, *squares*: E=Ge, *triangles*: E=Sn).

derivatives with n=3 or 4, compared to the less-chlorinated analogues with n = 1 or 2 (Fig. 2b). However, one or two polar E-Cl bonds in a molecule bring about an increase of bp, especially for E=Ge and Sn. These two factors cause additional stabilisation of molecular liquids composed of unsymmetrically substituted polar $EMe_{4-n}Cl_n$ molecules, as in the case of E=C (black triangles in Fig. 1). The distinctly nonlinear $T_b(M)$ plot for these compounds cannot be linearised by variation of the exponent in M^{α}. At the same time, bp of tetrafluorides EF_4 increase in the succession CF_4 (144 K) < SiF_4 (187 K) < GeF_4 (237 K), whereas SnF_4 (subliming at 978 K) and PbF_4 (melting at 873 K) are no longer molecular but polymeric solids with hexacoordinated metal atoms [16, 17]. The observed trends of T_b in methane analogues are therefore influenced not only by the molecular van der Waals shape, but also by other factors like electrostatic forces and secondary intermolecular bonding.

Mp of tetrahydrides EH_4 (E=Si to Pb) also increase linearly with $M^{1/2}$ (Fig. 3). This increase is slower than for noble gases; indeed, silane SiH_4 melts 2 K *lower* than methane

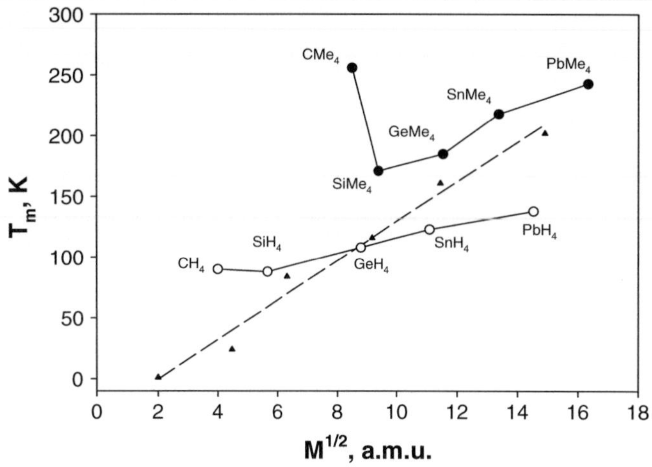

Fig. 3 Melting points T_m in EX_4 derivatives with tetrahedral molecules, as a function of molecular mass. *Open circles*: EH_4, *filled circles*: EMe_4. *Triangles* and *dashed regression line*: noble gases.

CH_4. Larger EX_4 molecules display a strongly non-uniform dependence of T_m upon M. In particular, tetramethylsilane $SiMe_4$ (T_m=171 K) melts 85 K lower than neopentane CMe_4 (256 K), but T_m further increase with the increase of M for the heavier EMe_4 analogues. A similar drop of mp from E=C to E=Si for tetrahaloid EX_4 derivatives is shown in Fig. 4. On the other hand, tetraethyl compounds EEt_4 display a uniform *decrease* of T_m from CEt_4 (240 K) to $PbEt_4$ (143 K).

A remarkable feature in Fig. 4 is a steep increase of T_m from CF_4 to GeF_4 which corresponds the trend in their bp (see above). However, the increase of the number of fluorine substituents in asymmetrically substituted methane analogues progressively destabilises both solid and liquid phases reducing T_m and T_b (*cf.* Fig. 2a and $CMe_{4-n}F_n$ series in Fig. 4).

In the series of chloro-substituted methanes $CH_{4-n}Cl_n$, a monotonic increase of T_b (roughly proportional to $M^{1/2}$) is

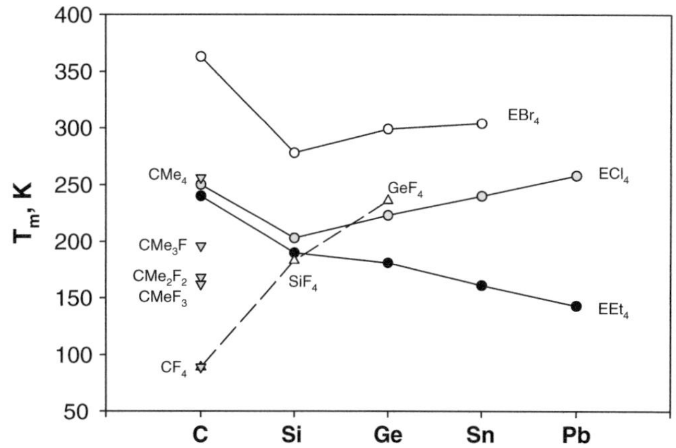

Fig. 4 Melting points of EX_4 derivatives in the C subgroup (scheme). *Open circles*: X=Br, *grey circles*: X=Cl, *black circles*: X=Et. *Turned grey triangles*: CMe_nF_{4-n} series.

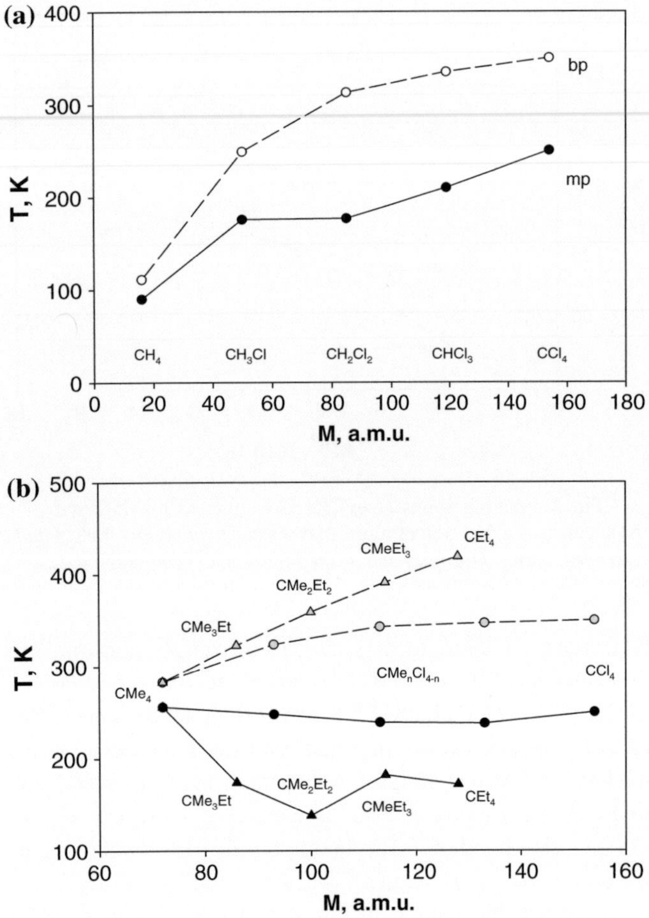

Fig. 5 Plots of mp (*solid lines*) and bp (*dotted lines*) versus mass (**a**) in $CH_{4-n}Cl_n$ set, (**b**) in $CMe_{4-n}Cl_n$ (*circles*) and $CMe_{4-n}Et_n$ sets (*triangles*).

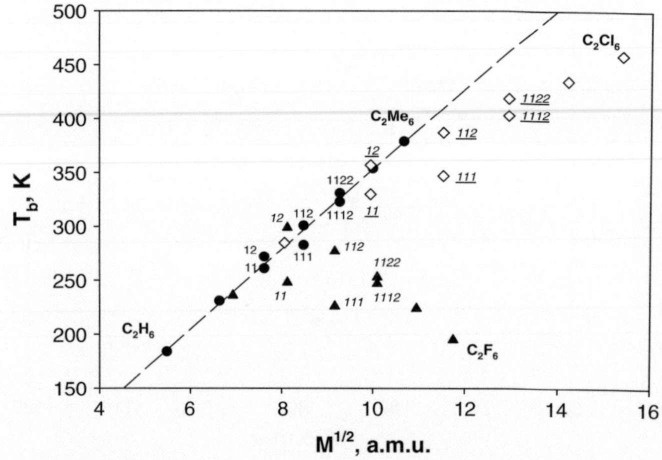

Fig. 6 Boiling point T_b as a function of molecular mass M in substituted ethanes $C_2H_{6-n}X_n$, X = Me (*filled circles, plain numbers*), $C_2H_{6-n}F_n$ (*filled triangles, Italic numbers*) and Cl (*diamonds, underlined Italic numbers*). Isomers are denoted by positions of the substituents at C_2 core (see Scheme 1). *Dashes*: a regression line for $C_2H_{6-n}Me_n$ series.

accompanied by an erratic behaviour of T_m (Fig. 5a). In $CMe_{4-n}Cl_n$ series where the substituents have similar steric dimensions [18], bp and especially mp vary in narrow ranges (283–350 K and 239–256 K, respectively, Fig. 5b). Replacement of one or more of the substituents in *symm-tetrahedral* molecules by heavier and bulkier ones (e.g., Me by Et), often depresses T_m (Fig. 5b). It is noteworthy that T_b and T_m are rather close to each other for CMe_4 (256 and 283 K) but diverge strongly for $SiMe_4$ (171 and 300 K), $GeMe_4$ (185 and 316 K) and other *symm*-tetrasubstituted derivatives with a larger central atom.

Substituted Ethanes $C_2H_{6-n}X_n$ and Ethylenes $C_2H_{4-n}X_n$

Plots of T_b versus $M^{1/2}$ for substituted ethanes are presented in Fig. 6. The $C_2H_{6-n}Me_n$ series shows a good linear dependence of T_b versus. $M^{1/2}$, with a much steeper increase than in the methane series. The plots for Cl- and Br-substituted ethanes deviate from linearity whereas bp of fluoroethanes *increase* from C_2H_6 to $C_2H_4F_2$ and then *decrease* from $C_2H_4F_2$ to C_2F_6. In pairs of di-, tri- and to tetrasubstituted geometrical isomers $C_2H_{6-n}X_n$ (see Scheme 1 for notations), bp differ by up to 51 K (for X = F) in the successions 11- < 12- ($C_2H_4X_2$), 111- < 112- ($C_2H_3X_3$) and 1112- < 1122- ($C_2H_2X_4$; henceforth numbers denote positions of substituents).

Data on silicon analogues Si_2X_6 are more scarce (Table 1). It is noteworthy that whereas ethane $C_2Me_{6-n}Cl_n$ derivatives are subliming solids with $T_m \approx T_b$ at 760 torr, for their $Si_2Me_{6-n}Cl_n$ analogues T_b exceeds T_m by 100–150 K, paralleling the trend mentioned above for the EMe_4 series.

The T_b versus $M^{1/2}$ patterns of substituted ethylenes $C_2Me_{4-n}X_n$ look similar to the patterns of substituted ethanes (Fig. 7). Their main features are almost linear correlation (though with smaller number of points) for X = Me, Cl, Br, a slower growth of bp for Cl- and Br- substituted species as compared to Me-substituted ones and the decrease of bp in fluoroethylenes from $C_2H_2F_2$ to C_2F_4. Differences of bp in three $C_2H_2X_2$ isomers in "*gem-* < *trans-* < *cis-*" succession are larger in halogenated species (up to 58 K in $C_2H_2F_2$).

Plots of T_m as functions of M^α for substituted ethanes (Fig. 8a) and ethylenes (Fig. 8b) differ substantially from the corresponding $T_b(M^\alpha)$ patterns; they are not easy to linearise. Among Me_n-substituted ethanes and ethylenes, the lowest T_m is observed for monosubstituted species, i.e., propane $CH_3CH_2CH_3$ and propene $CH_2{=}CHCH_3$. Among isomers mp vary stronger (up to 72 K in $C_2H_2Cl_2$ and 143 K in $C_2H_3Me_3$) and in a succession different from that of bp, e.g., for trisubstituted ethanes (112- < 111-) and disubstituted ethylenes (*gem-* < *cis-* < *trans-*). Note that $T_b - T_m$ difference for these compounds mostly lies in the range of 100–150 K but becomes very small for highly symmetrical C_2X_6 molecules

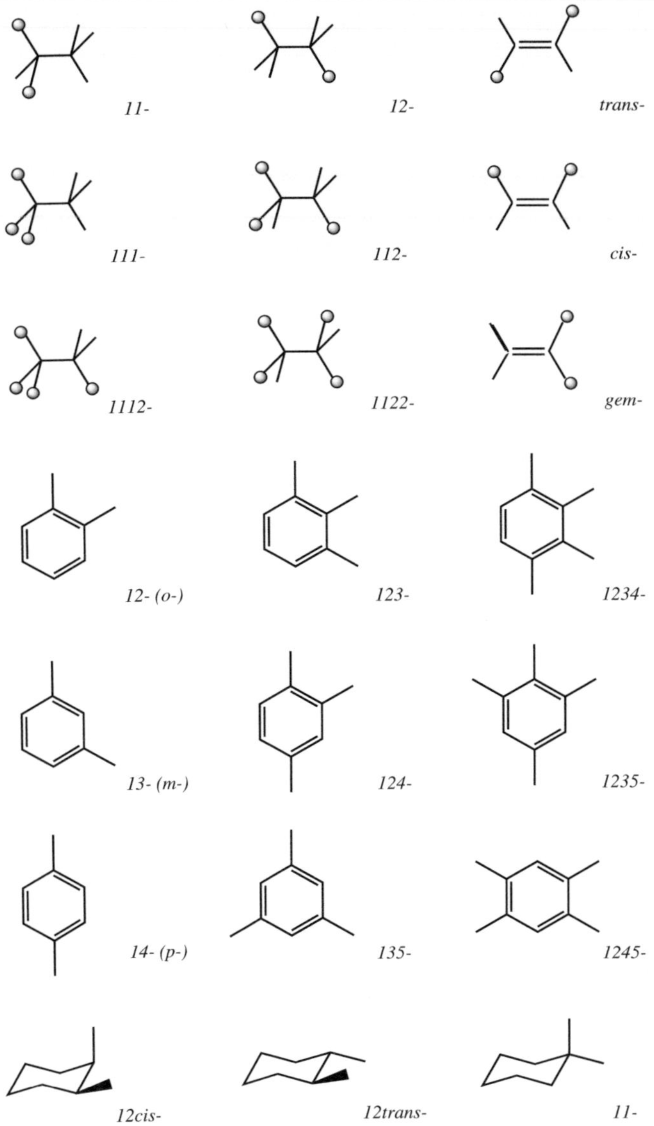

Scheme 1 Notations of isomers.

with bulky X substituents (5 K for C_2Me_6 and <1 K for C_2Cl_6 that sublimes at 459 K under 760 torr, see Table 1). Polyfluorinated derivatives, similarly to their T_b trend, display a strong reduction of T_m, although data are unavailable for some compounds of these series.

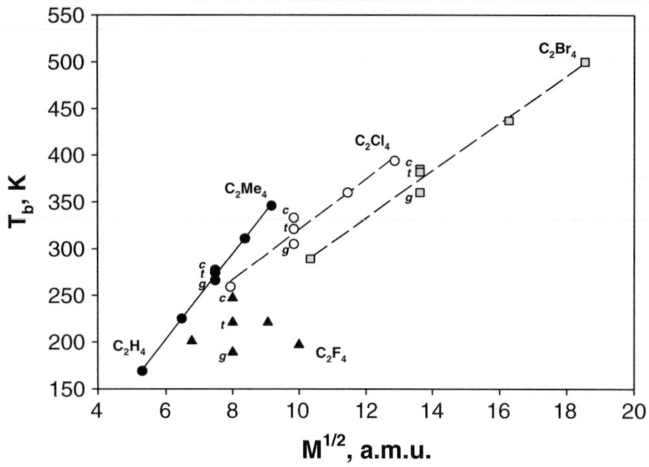

Fig. 7 Boiling points T_b as a function of $M^{1/2}$ in substituted ethylenes $C_2H_{4-n}X_n$, X = Me (*black circles*), F (*black triangles*), Cl (*open circles*), Br (*grey squares*). $C_2H_2X_2$: *c* = *cis*-, *t* = *trans*-, *g* = *gem*-isomer.

Substituted Benzenes $C_6H_{6-n}X_n$

T_m and T_b of some substituted benzenes $C_6H_{6-n}X_n$ are listed in Table 2. Bp linearly increase with M ($\alpha = 1$) for X = Me, Cl, Br, but deviate from linearity for extended X substituents (Fig. 9). Similar to ethanes and ethylenes, T_b for Cl- and Br-substituted species increase slower than for Me-substituted analogues.

Three sets of geometrical isomers in the $C_6H_{6-n}X_n$ series display a small variation of T_b (usually within 10 K), generally in the following successions:

$$\begin{aligned} &C_6H_4X_2: 13- < 14- < 12- \\ &C_6H_3X_3: 135- < 124- < 123- \\ &C_6H_2X_4: 1235- < 1245- < 1234- \end{aligned} \quad (1a)$$

Note that melting points in the same sets of isomers can differ by up to 100–150 K and change in different successions [2, 12]:

$$\begin{aligned} &C_6H_4X_2: 13- < 12- < 14- \\ &C_6H_3X_3: 124- < 123- < 135- \\ &C_6H_2X_4: 1235- < 1234- < 1245 \end{aligned} \quad (1b)$$

Table 1 Melting (T_m) and boiling points (T_b), in K, for high-symmetric ethane-like derivatives $E_2Me_{6-n}Cl_n$ (where E = C, Si)

Compound (isomers)		E_2Me_6	E_2Me_5Cl	$E_2Me_4Cl_2$		$E_2Me_3Cl_3$		$E_2Me_2Cl_4$		E_2MeCl_5	E_2Cl_6
				11-	12-	111-	112-	1112-	1122-		
$C_2Me_{6-n}Cl_n$	T_m	374	405	424	433	–[a]	456	449	445	452	459
	T_b	379	405	425	–	–	449	440	–	–	459
$Si_2Me_{6-n}Cl_n$	T_m	286	–	–	285	297	–	288	279	–	274
	T_b	386	407	417	421	422	421	–	426	–	418

[a]Data not available

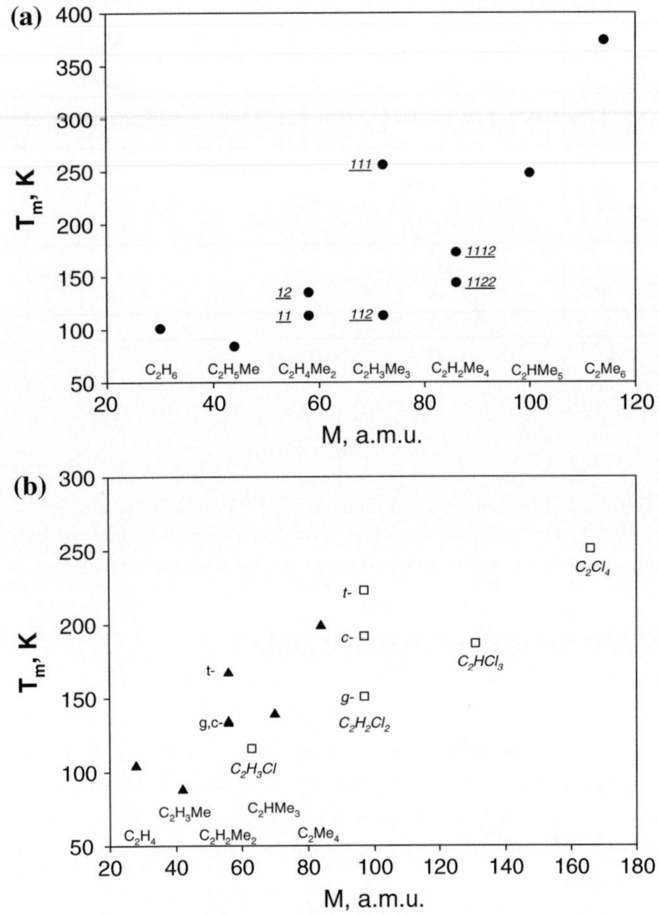

Fig. 8 Melting points T_m as a function of molecular mass M: (**a**) Me-substituted ethanes, (**b**) substituted ethylenes $C_2H_{4-n}X_n$ (*filled triangles*: X=Me, *open squares*: X=Cl).

(see Scheme 1 for notation of isomers). However, mp of trimethylbenzene isomers show an unusual 135 ≈ 124- < 123- succession. Fluorobenzenes have strongly depressed mp and bp, but the succession of the temperatures for their isomers mostly agree with general trends among substituted species.

Substituted benzene derivatives containing oxygen and/or nitrogen but without strongly H-bonding (e.g., hydroxyl, amino or imino) groups, have elevated mp and bp compared to their hydrocarbon analogues with similar masses and molecular dimensions (Table 2 and Fig. 9b; *Tb* data for high-boiling species are incomplete due to their decomposition). Thus, mp in $CgH_{6-n}(OMe)_n$ set lie 40–90 K higher than those of $C_6H_{6-n}Et_n$, whereas bp in OMe-set are elevated by 20–30 K. Melting points in phenols $C_6H_{6-n}(OH)_n$, additionally stabilised by H-bonds, reach 500–540 K. Both T_m and T_b values available for the isomers with neighbouring positions of OH groups (12-, 123 and 1234- in solids; 12- in liquids) are lower due to formation of H-bonded dimers whose hydroxyls cannot participate in a 3D network of H-bonds. Accordingly, their mp successions are altered to 12- < 13- < 14-, 123- < 124- < 135- and 1234- < 1235- < 1245- for $n=2$, 3 and 4, respectively.

Substituted Cyclohexanes $C_6H_{12-n}X_n$

Substituted cyclohexanes $C_6H_{12-n}X_n$ have more geometric isomers than substituted benzenes $C_6H_{6-n}X_n$, due to a possible axial (*ax*) or equatorial (*e*) orientation of the substituents X relative to the plane of the C_6 ring, as well as trans-, *cis*- and *gem*-isomers depending on positions of the substituents relative to each other. The T_b and T_m data available in this family are more limited, mostly covering monosubstituted and some disubstituted species. In Me_n-substituted cyclohexanes, mp is known up to $n=3$ and bp up to $n=4$ for many isomers.

Monosubstituted cyclohexanes $C_6H_{11}X$ often have lower T_m, but higher T_b as compared with the parent cyclohexane C_6H_{12} (Table 3); the reduction ?T_m of the melting temperature upon substitution is proportional to the van der Waals volume of the substituent X (see [10] for more details). Of the two known undeca-substituted species C_6HX_{11} (where X=F, Cl), the mp of the Cl-derivative is very much lower (by 201 K) than that of C_6Cl_{12}, but C_6HF_{11} melts only 6 K lower than C_6F_{12}. Another available example of undeca-substituted cyclohexane is $C_6HMe_{10}Et$ whose reported mp (458–469 K) is lower than the mp of C_6Me_{12} (568 K).

The available T_b and T_m data for $C_6H_{10}X_2$ (7 isomers, Table 4) and $C_6H_9Me_3$ (12 isomers, Scheme 2) generally resemble the corresponding patterns for substituted benzenes. Thus, T_m of disubstituted isomers with X substituents in different positions of the ring generally follow the "13- < 12- < 14" succesion, resembling the mp succession in $C_6H_4X_2$. Melting points of Me_3-derivatives, except for *gem-substituted* ones, correspond to "124- < 123" succession, and *all-cis*-1,3,5-isomer with the highest (C_3v) molecular symmetry has one of the highest T_m (227 K). However, these trends are modified by further geometrical isomers in the cyclohexane family, as can be seen from the compounds in the same series that have the lowest (1,*3cis*, *5trans*-$C_6H_9Me_3$, *Cs* symmetry, 166 K) and the highest mp (1,1,2-isomer, C_1 symmetry, 244 K).

Boiling points, known for all Me_3- and most of Me_4- cyclohexanes (Scheme 3), vary considerably among isomers (up to 20–25 K), forming "13- < 14- < 12-", "135- < 124- < 123-" and "1234- < 1245- < 1235-" successions, in similarity with the benzene family. The higher bp values, including those of *cis*-, *trans*- and *gem*- substituted isomers, correspond to non-uniform distributions of Me groups over the ring, either pointing to one side of the average ring plane (e.g., *all-cis* 1234- or 1245-species on Scheme 3) or grouping around the same region of the ring (112- and 1122-isomers). Similarly to hexa-substituted ethylenes C_2X_6 (see above), dodeca-substituted C_6X_{12} species (Table 3) are stabilised as solids and destabilised

Table 2 Melting (T_m) and boiling points (T_b) of substituted benzenes $C_6H_{6-n}X_n$

	T_m, K				T_b, K			
Non-associated molecules	X=Me	X=F	X=Cl	X=Br	X=Me	X=F	X=Cl	X=Br
1,2-$C_6H_4X_2$	248	239	256	280	417	366	452	498
1,3-$C_6H_4X_2$	225	214	248	266	412	356	446	492
1,4-$C_6H_4X_2$	286	260	327	360	411	361	447	493
1,2,3-$C_6H_3X_3$	248	–[a]	326	360	449	368	492	556
1,2,4-$C_6H_3X_3$	229	238	290	317	442	362	486	548
1,3,5-$C_6H_3X_3$	228	267	336	393	438	350	481	544
1,2,3,4-$C_6H_2X_4$	267	231	319	333	478	368	527	611
1,2,3,5-$C_6H_2X_4$	249	225	325	371	471	357	519	602
1,2,4,5-$C_6H_2X_4$	352	277	413	452	–	362	516	599
	T_m, K				T_b, K			
Associated molecules	X=OH	X=NO_2	X=OMe	X=Et[b]	X=OH	X=NO_2	X=OMe	X=Et[b]
1,2-$C_6H_4X_2$	377	392	296	242	518	592	480	456
1,3-$C_6H_4X_2$	382	363	221	189	553	570	490	454
1,4-$C_6H_4X_2$	444	447	329	230	559	572	485	457
1,2,3-$C_6H_3X_3$	406	398	319	–	573	–	510	–
1,2,4-$C_6H_3X_3$	413	333	293	195	–	–	522	491
1,3,5-$C_6H_3X_3$	492	396	326	207	–	588	529	488
1,2,3,4-$C_6H_2X_4$	433	391	362	285	–	–	–	525
1,2,3,5-$C_6H_2X_4$	439	400	319	252	–	–	544	520
1,2,4,5-$C_6H_2X_4$	505	462	376	242	–	–	–	521

[a]Data not available
[b]Reference data

as liquids. Having a very narrow T_b-T_m gap; these compounds sublime easily at ambient pressure with a partial decomposition.

Discussion

Molecular Mass as a Variable in $T_{m,b}(M^\alpha)$ Plots

It is a common wisdom that among similar molecular substances both T_m and T_b usually increase with increasing molecular mass (M). Even though M does not figure explicitly in any expression of the van der Waals (or, indeed, any other) bonding energy, the reasons for such trend are obvious. The energy of van der Waals attraction is proportional to the number of pairwise atom–atom interactions and hence the number of atoms in the molecule. For substances of similar composition, e.g., hydrocarbons, a roughly linear correlation has been described [7]. On the other hand, the energy of each interaction is proportional to the atomic polarisabilities, which are higher for larger (i.e., usually heavier) atoms.

One can hardly expect a clear-cut mathematical dependence of T_b (or T_m) from M on the basis of the above, purely qualitative considerations, as they relate *via* a number of generalisations which are only approximately true. Nevertheless, in a very simple model of a condensed phase that consists of idealised rigid molecular "bodies" (limited by van der Waals atomic spheres), the stabilising van der Waals energy per molecule $-\varepsilon_{\text{vdW}}$ is proportional to the accessible area of the molecular surface [7]. The absolute value of energy e_{vdW} therefore increases when the molecular van der Walls volume increases, giving rise to an approximate power law dependence of ε_{vdW} from a total molecular volume V calculated from atomic increments [3]. Since the steric molecular volume of organic compounds (which usually comprise a very limited selection of elements, viz. C, H, N, O or halogens) is additive on the same crude level of approximation [3, 7], ε_{vdW} may be empirically related to the molecular mass M by a power law

$$\varepsilon_{\text{vdW}} \approx KM^{\alpha}, \qquad (2)$$

where K is a scaling constant, and α has to be selected between 0.5 and 1 for different sets of organic molecules to achieve better linearisation. It is therefore remarkable that for noble gases, both T_b and T_m are proportional to $M^{1/2}$ (Figs. 1 and 3 respectively). For the time being, we accept as a working hypothesis that this correlation represents a general trend for van der Waals condensed phases. Note that all noble gases crystallise similarly in close-packing structures, unperturbed by any steric factors or interatomic interactions other than van der Waals forces.

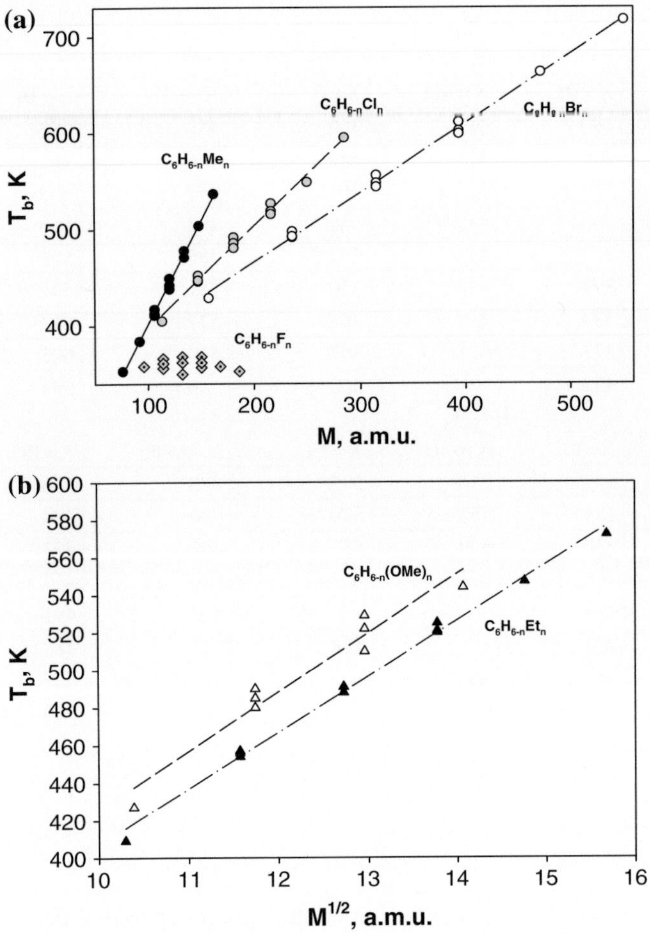

Fig. 9 (**a**) Boiling point T_b as a function of molecular mass M in substituted benzenes $C_6H_{6-n}X_n$ with sphere-like X, (**b**) boiling points T_b versus $M^{1/2}$ for X = Et (*filled triangles, dash-dot line*) and OMe (*open triangles, dashed line*); for T_b of isomers see Table 2.

Melting and boiling temperatures of a substance are the higher, the larger ε_{vdW}. Based on Eq. (2), one may use molecular mass, which is known precisely, instead of van der Waals volume of a molecule (which is a less definite variable) for the analysis of T_m and T_b trends described above. In particular, van der Waals liquids obey Trouton's rule,

$$N_A\ \varepsilon_{vdW} T_b^{-1} \approx 8R \qquad (3)$$

Table 4 Melting (T_m, K) and boiling points (T_b, K) of disubstituted cyclohexanes $C_6H_{10}X_2$

Isomers	X = Br, T_m	X = CN, T_m	X = Me, T_m	X = Me, T_b
1,1	242	337	239	392
1,2-*trans*	268	322	184	397
1,2-*cis*	283	337	223	403
1,3-*trans*	274	323	181	397
1,3-*cis*	385 (193)[a]	311	196	393
1,4-*trans*	385	413	236	393
1,4-*cis*	322	337	183	397

[a]Inconsistent data

(where N_A is Avogadro number, $N_A\varepsilon_{vdW}$ is a latent heat of vaporisation per mole, and R is the universal gas constant) [5]. Thus the T_b of such liquid is linearly related to its molecular cohesive energy ε_{vdW}, hence

$$T_b \approx K' \cdot M^{\alpha} \qquad (4)$$

Although no such simple rule is known for T_m, a comparison of T_b versus M^{α} and T_m versus M^{α} plots with the same α may help to reveal differences between melting and boiling trends in a same set of related substances.

General Trends in T_m and T_b

Through the analysis of T_b and T_m data in the sets of compounds with closely related molecular structure one can find the main factors that define stability of the corresponding condensed phases. In particular, mp and bp of symm-tetrasubstituted methanes CX_4 are determined mostly by van der Waals forces since their molecules of T_d symmetry have zero dipole moment ($\mu = 0$), zero quadrupole moment (Q = 0), and a scalar moment of inertia *I*. Mixed derivatives $CR_{4-n}X_n$ (where R = H, Me, and X = halogen) have nonzero molecular electric moments that notably influences on their properties (see Fig. 1). On the other hand, methane and some other *symm*-substituted CX_4, C_2X_6 and C_6X_{12} compounds have a roughly spherical molecular van der Waals shape whereas extended X substituents (e.g., CN or Et) make this shape less symmetrical. Deviations of asymmetric methane analogues

Table 3 Melting (T_m) and boiling points (T_b) of mono-, undeca- and dodeca-substituted cyclohexanes

	T_m, K					T_b, K				
Compound	X = H	X = Me	X = F	X = Cl	X = Br	X = H	X = Me	X = F	X = Cl	X = Br
$C_6H_{11}X$	*280*	147	285	230	216	*356*	374	373	415	437
C_6HX_{11}	*280*	–[a]	320	358	–	*356*	–	336	–	–
C_6X_{12}	*280*	568	326	559	–	*356*	–	326	–	–

Reference data for C_6H_{12} are shown in Italic
[a]Not available

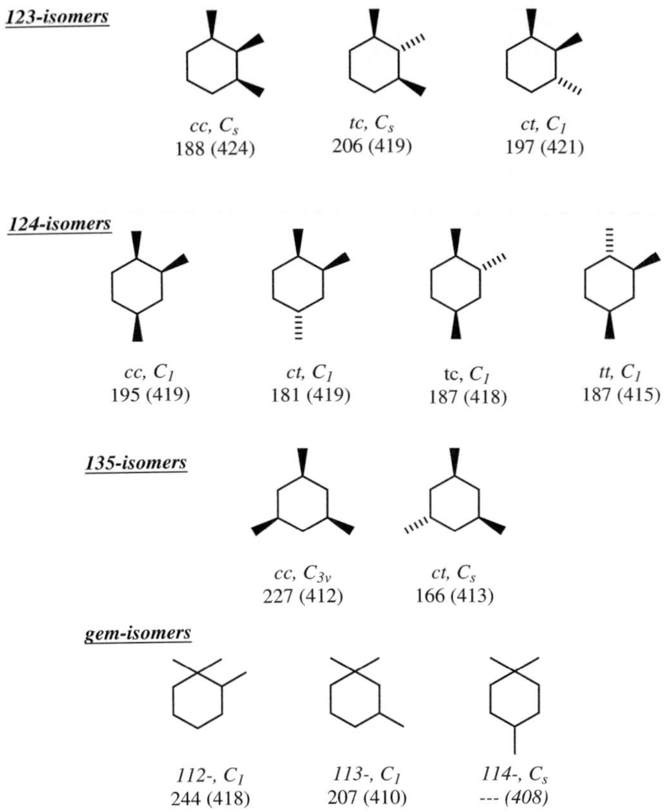

Scheme 2 Point groups, melting temperature, and boiling temperature (*in brackets*) of trimethylcyclohexane isomers (*c*=*cis*-, *t*=*trans*-).

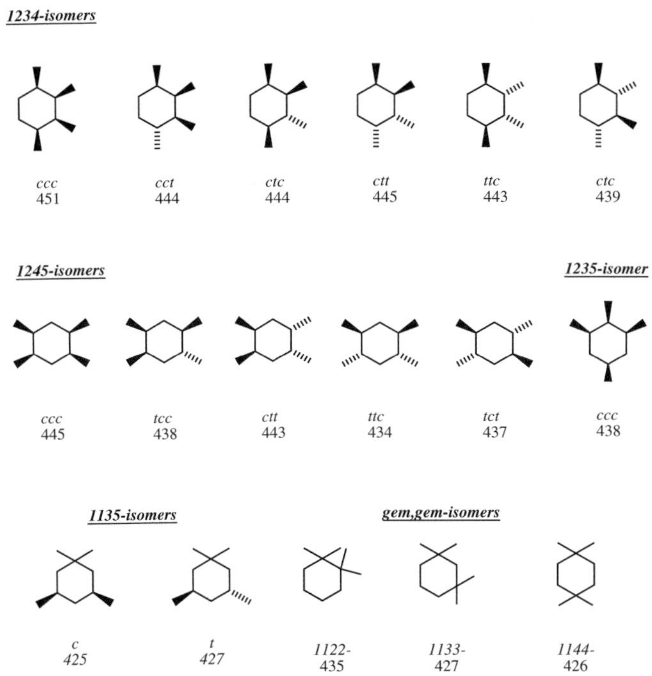

Scheme 3 Notations and boiling points for known isomers of tetramethylcyclohexanes.

from a simple linearised dependence of T_b (Fig. 1) and T_m (Fig. 3), shown in Figs. 2, 4, and 5 are therefore due to a balance of electrostatic forces, molecular van der Waals asymmetry and specific bonding (see below).

The corresponding $T(M^{\alpha})$ plots in Figs. 1, 2, 3, 4, 5, 6, 7, 8, and 9 reveal the following general trends:

I. Boiling point T_b increases smoothly and regularly with M; a difference of T_b among isomers is usually small (10–20 K)

II. Melting point T_m depends non-uniformly on M with a stronger difference of T_m among isomers (ca. 50–100 K)

III. In the same set of isomers different successions of T_m and T_b are observed.

Note that bulkier substituents usually bring higher T_b, but very often depress T_m in a set of closely related substances. For example, mp decreases strongly from benzene (279 K) to toluene (178 K) and from cyclohexane (280 K) to methylcyclohexane (146 K), whereas the bp of the same monosubstituted derivatives increases uniformly as a steric volume (and mass) of the molecule increases: C_6H_6 (353 K) → C_6H_5Me (384 K) and C_6H_{12} (356 K) → $C_6H_{11}Me$ (374 K).

Relative T_m and T_b values in all sets of isomers considered above are summarised in Table 5. Physical causes of the known dependence of T_m of organic compounds on molecular geometry [2, 3, 5] have been analysed from the symmetry viewpoint in our previous communication [10]. Variations of T_b at 760 torr between isomers, discussed here, are less common. Most intriguing are the different successions of T_b and T_m in the same sets of isomers (see Table 5)—a fact that has to be explained theoretically.

Hypothesis of van der Waals Symmetry

A relationship between bulk characteristics of substances and their molecular shape, is based on the long-known fundamental property of a molecule to occupy a finite volume in three-dimensional space. Considering molecules as bodies (rather then sets of point masses and charges), densely packed "bump-to-hollow" in a molecular crystal, was a key postulate proposed by Kitaigorodskii as the cornerstone of organic crystallography [3]. With this idea in mind we can attempt to explain the T_m and T_b trends, by comparing the symmetry of the local field created by all molecules in a condensed phase, with the symmetry of molecular van der Waals "bodies".

Crystals with close packing of molecules display a narrow range of such bulk characteristics as a packing coefficient (0.65–0.80), van der Waals energy per atom (5–10 kJ/mole) and molecular coordination number (10–14) [3, 5, 7].

Table 5 Melting point (T_m) and boiling point (T_b) orders among isomers of some organic compounds

	T_m	T_b
Ethanes $C_2H_{6-n}X_n$ (n=2–4)	11- < 12-	11- < 12-
	112- < 111-[a]	111- < 112-
	–[b]	1112- < 1122-
Ethylenes $C_2H_2X_2$	*gem-* < *cis-* < *trans-*	*gem-* < *trans-* < *cis-*
Benzenes	13- < 12- < 14- (*m-* < *o-* < *p-*)	13- < 14- < 12- (*m-* < *p-* < *o-*)
$C_6H_{6-n}X_n$, $n=2-4$	124- < 123- < 135-	135- < 124- < 123-
	1235- < 1234- < 1245-	1235- < 1245- < 1234
Cyclohexanes	13- < 12- < 14*tr*- ≈ 11	13- ≈ 14- < 12-
$C_6H_{12-n}X_n$, $n=2, 3$	13*cis*5*tr* < 124- < 123- < 13*cis*5*cis* < 112-	135- < 124- < 123-

Beilstein database [11]; for notation of the isomers see Scheme 1
[a]112- > 111- for X=F
[b]No definite succession observed

On a very qualitative level, dense molecular packing in such crystals may be deduced from the close packing of identical van der Waals spheres (e.g., in solid noble gases), whereas different van der Waals radii of the atoms and their tightening into molecules by interatomic bonds may be considered as a perturbations of packing (Fig. 10). A measure of such perturbation is a "contact" term in the local van der Waals field that acts on each molecule in a crystal. This assumption was actually a basis of our hypothesis of van der Waals symmetry [10].

A qualitative comparison of molecular shapes has been suggested two decades ago by Kitaigorodskii in his theoretical analysis of solid solutions of organic compounds [19]. To construct a more quantitative model, we have postulated in [10] that the local van der Waals field U that acts on a middle-sized molecule (of several dozen atoms) in a condensed phase, may be presented as

$$U = U_0 + W, \tag{5}$$

where U_0 is a van der Waals potential arising from all *distant* molecules (and possibly from outer fragments of the nearest ones), and W is a short range ("contact") term originating from the molecule's immediate environment (Fig. 11). We suppose that the averaged U_0 term, randomised by the influence of many molecules and fragments at distances larger than a close non-valence contact, is not sensitive to the molecular geometry whereas the perturbation term W is geometry sensitive.

As shown schematically in Fig. 11, the topology of W reflects the shape of the molecule composed additively of rigid van der Waals atomic spheres. On the other hand, spatial properties of the U_0 term are determined by the average structure of a condensed phase. The latter is different for molecular crystals and molecular liquids due to vacancies and "jumping" of molecules in a liquid state [4]. Trends in the structure-dependent bulk properties, like mp and bp, in a series of closely related compounds therefore reflect the similarity—or dissimilarity—between W and U_0, i.e., how well the molecular "bodies" fit to the averaged local potential U_0 in a given condensed phase.

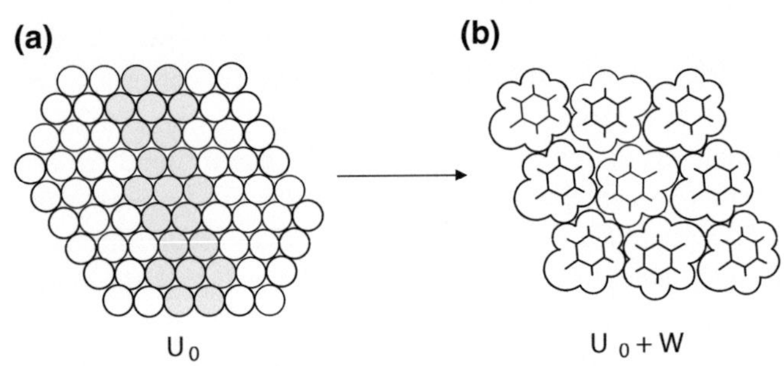

Fig. 10 Transformation of a close packing of van der Waals spheres (**a**) into a close packing of molecules (**b**) as a result of perturbation (schematic).

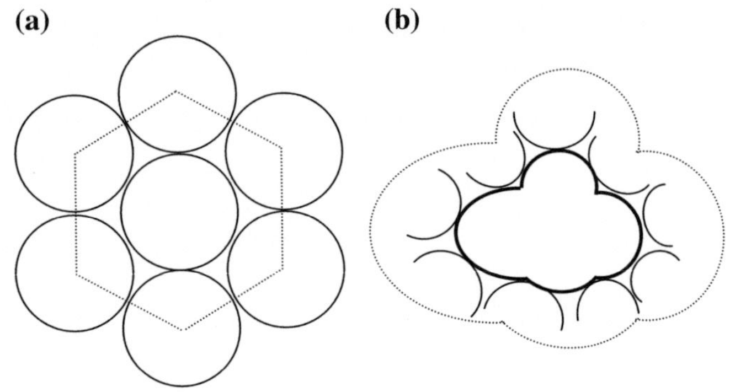

Fig. 11 Origin of the contact term W in (**a**) a dense packing of van der Waals spheres, (**b**) a close packing of low-symmetry molecules (schematic).

One can reasonably believe that spatial properties of U_0 are the same in a set of isomers of the same condensed state (i.e., all solid or all liquid). We also postulate that the W term, determined mostly by the shape of a molecule, remains unchanged upon melting if vacancies are randomly distributed over the average local molecular environment in a liquid [4, 5]. The model thus links a different succession in physical properties of isomers, or of other closely related substances, in crystal and liquid phases to variations of their U_0 term in Eq. (5).

To quantify deviations in van der Waals shape of different molecules from the main local field, a reference idealised point group, called *van der Waals group* $\widetilde{\mathbf{G}}(\mathbf{U_0})$, has been suggested [10]. Such a group corresponds to the approximate symmetry of the main term U_0 in Eq. (5). The perturbation introduced by W is proportional to the index of molecular asymmetry $\boldsymbol{a}_s$, a measure of the gap between the molecule's shape and the symmetry of the U_0 field as expressed by the van der Waals group $\widetilde{\mathbf{G}}(\mathbf{U_0})$:

$$a_s = (1/V)\sum \Delta V(\tilde{g}_i). \tag{6}$$

Here $\Delta V(\tilde{g}_i)$ is a difference in van der Waals volume between a molecule and its image after transformation by the symmetry element $(\tilde{g}_i) \in \widetilde{\mathbf{G}}$, V is the van der Waals volume of the molecule, and the sum is taken over all elements of $\widetilde{\mathbf{G}}$.

Molecular asymmetries in the main crystal field, calculated in [10] for several classes of organic compounds, displayed a good correlation with their melting points. In methane analogues EX_4 (see Figs. 3, 4, and 5), not considered in [10], van der Waals asymmetry increases with the increase of the E–X bond length, making larger tetrahedral molecules "less spherical". A general decrease of T_m from CEt_4 to $PbEt_4$, where a central atom is protected from additional bonding by the bulky ethyl substituents (Fig. 4), is therefore due to an increase of steric perturbation W in this row. The decrease of T_m for bulkier molecules is also observed in $CMe_{4-n}Et_n$ set (Fig. 5b). The non-uniform T_m versus $M^{1/2}$ patterns in other EX_4 substances shown in Figs. 3 and 4, is due to destabilisation of the crystal packing by increased van der Waals asymmetry from C to Si and further stabilisation by secondary bonding which grows from Si to Pb.

Van der Waals Symmetry in Crystals and Liquids

The idealised symmetry of the main crystal field U_0 may be taken, in the crudest approximation, as the O symmetry of van der Waals field in the face centered cubic (fcc) close packing, which is the actual atomic packing in crystalline noble gases [20] (see Figs. 10 and 11). In fact, such approximation is not as far-fetched as it seems, if one takes into account that (i) all known atomic van der Waals radii vary only by a factor of 2, and if we exclude hydrogen and iodine, only by 30 %; (ii) packing coefficients in molecular crystals do not differ from the ideal close-packing value of 0.74 by more than 8–12 %. From this viewpoint, molecular asymmetry corresponding to formula (6) in a crystalline phase is the sum of deviations of the molecule's van der Waals shape from the O_h group. Difference between T_b and T_m trends among isomers therefore corresponds to the molecular liquid being described by a different $\widetilde{\mathbf{G}}(\mathbf{U_0})$ group than the molecular crystal, providing different qualitative successions of the molecular $\boldsymbol{a}_s$ values in these two condensed phases.

The most prominent differences in mp and bp trends in Table 5, is a relative decrease of T_b for highly symmetric derivatives that have high T_m, e.g., *trans*-disubstituted ethylenes $C_2H_2X_2$, 1,1,1-trisubstituted ethanes $C_2H_3X_3$, 1, 4-benzenes, *trans*-1,4-cyclohexanes, etc. Their high-boiling isomers (*cis*-$C_2H_2X_2$, 1,2-benzenes, etc.) have lower molecular symmetry. This effect is stronger for 1,1,1-$C_2H_3X_3$ and 1,3,5-$C_6H_3X_3$ species possessing the highest molecular symmetry (C_3v and D_{3h}, respectively). These particular isomers usually have the highest T_m and lowest T_b in the corresponding sets. At the same time, such isomers as *gem*-$C_2H_2X_2$ or 1,2,3,5-$C_6H_2X_4$, having large van der Waals asymmetry in the crystal phase, are both low-melting and low-boiling. Different mp and bp values in pairs of isomers whose molecules have the same point symmetry, but different van der Waals shape (*gem*- and *cis*-$C_2H_2X_2$, 1,2 and 1,3-$C_6H_4X_2$, etc.) therefore reveal some systematic differences in their W term, both in solid and liquid phases.

The bp are generally higher for isomers with a smooth anisotropic van der Waals surface, viz. 1,2,3-$C_6H_3X_3$, 1,2,3,4-C6H2X4 and their *all-cis*-analogues in cyclohexane family. Most probably, this reflects a particular topology of the local field U in a liquid phase.

The existence of vacancies in a liquid phase [5] violates the dense packing and induces anisotropy in the local $U_0^{(L)}$ term. The corresponding change of van der Waals group $\widetilde{\mathbf{G}}(\mathbf{U_0})$ will transform the W term of highly symmetric molecules from a weak perturbation in a solid to a strong perturbation in a liquid. For example, a removal of one vertex from the cuboctahedron (the closest environment of a sphere in the fcc packing) reduces the local symmetry of the environment from O_h to C_2v and the coordination number from 12 to 11, whereas two vacancies in a cubooctahedral environment correspond to four "isomeric" arrangements with $N = 10$ (Fig. 12). A corresponding decrease in N by 16–17 % agrees well with the experimentally observed drop of density of organic compounds upon melting [5].

Fig. 12 Idealised closest molecular environment and their model van der Waals groups: (**a**) regular cuboctahedron, (**b**) a cuboctahedron without one vertex, (**c**)–(**f**) four "isomers" of a cuboctahedron without two vertices. The symbol '□' denotes a vacant site.

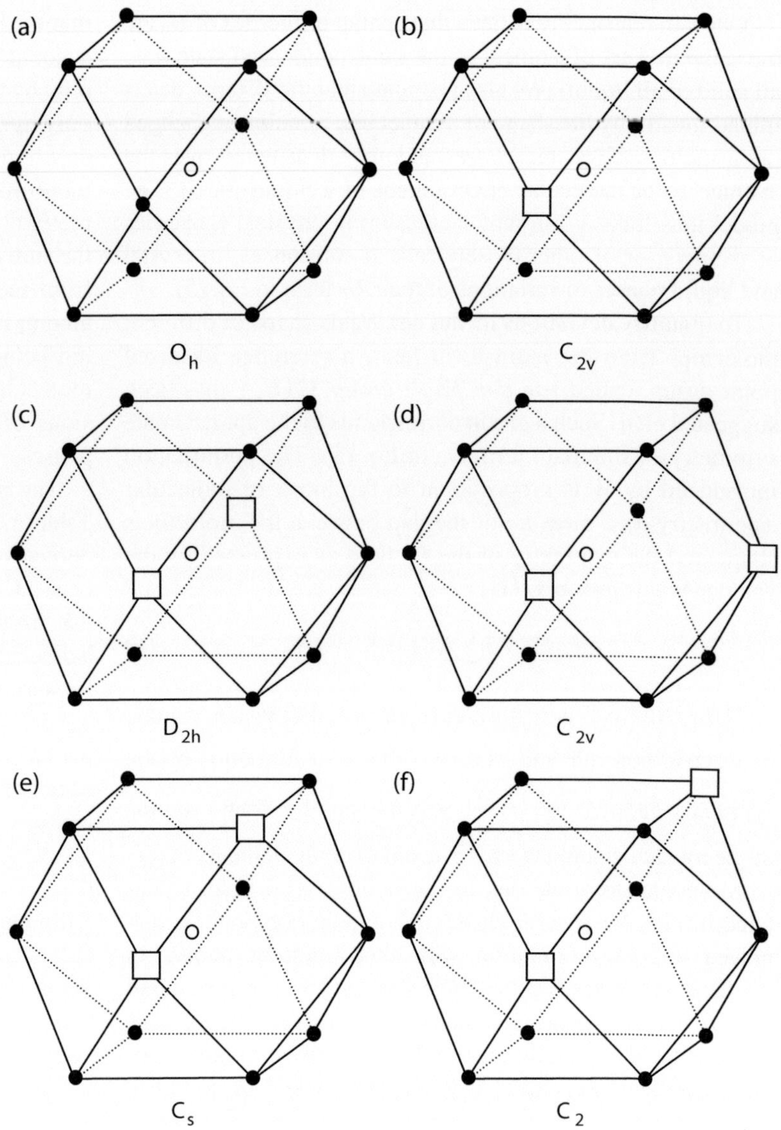

For a rigid local environment, molecular asymmetry in a model "liquid" becomes a weighted sum over all possible arrangements of vacancies [e.g., (b)–(f) in Fig. 12]:

$$< a_s > = \sum w_i a_s^{(i)}, \qquad (7)$$

where $\boldsymbol{a}_s^{(i)}$ is the asymmetry index of a molecule in the ith environment, and the weight w_i is its relative population. Thus, w_i values for the polyhedra in Fig. 12c–f should relate as 1:2:4:4 on simple geometric considerations. Likewise, a possible redistribution of 10 or 11 surrounding molecules in a liquid, will still correspond to the symmetry lower than cubic.

The neighbouring positions of vacancies in two of the distorted 10-vertex polyhedra [(d) and (e) in Fig. 12] qualitatively corresponds to a non-uniform distribution of substituents around the ring in high-boiling isomers of substituted benzenes and cyclohexanes. On the other hand, highly symmetrical molecules must have a smaller deviation from the cubic closest environment (Fig. 12a) but a larger misfit to any lower-symmetry van der Waals field. As a result, compounds built of such molecules will have high T_m but low T_b, in a qualitative agreement with the experimental observations.

The "gauge" in the symmetry of the van der Waals term U_0 in crystals and liquids helps to explain the observed high mp and low bp of substances composed of molecules with a rounded van der Waals shape—e.g., neopentane CMe_4 that has the highest melting point (256 K) and the lowest boiling point (283 K) in the whole family of C_5 hydrocarbons (see [10]). The same is true for the compounds with most symmetric sphere-like molecules (adamantane, C2Me6, C6F12 etc.) that sublime easily at 760 torr [2]. The shape of these molecules causes a weak perturbation of the average van der Waals field $U_0^{(C)}$ in a crystal but a strong perturbation of $U_0^{(L)}$. So the liquid phase does not exist at the temperature where thermal motion destroys the crystal state, and the substance passes directly into gaseous state.

Non-van der Waals Factors Influencing T_m and T_b

Differences of van der Waals field in organic compounds have to be compared with other factors that also may be in theory responsible for the observed trends of T_m and T_b. These factors are (i) Coloumb intermolecular interactions, (ii) secondary intermolecular bonding and (iii) geometry-sensitive entropy changes $T\Delta S$ in melting and boiling of the organic substances.

Electrostatic Forces

Coloumb intermolecular interactions caused by polar bonds in the molecules may modify T_m and T_b in two opposite ways. Firstly, a strong $X^{\delta-}...X^{\delta-}$ intermolecular electrostatic repulsion between substituents bearing partial negative charges (e.g., X=F or Cl) reduce both T_b and T_m (Figs. 2, 4, 6b, and 9a). The destabilising effect of close intermolecular F···F contacts have been deduced earlier from packing analysis of F-substituted organic compounds [21]. A less steep slope of T_b versus M^a plot for Cl- and Br-substituted derivatives in comparison with Me-substituted ones (Figs. 6, 7, and 9a) may also be due to a destabilising X···X repulsion of (more weakly) charged halogen atoms. Secondly, positively charged H atoms bonded to electronegative X (OH, NH_2, SH groups, etc.) give rise to a stabilising $X^{\delta-}-H^{\delta+}\cdots X^{\delta-}$ hydrogen bonding [5, 22]. A weaker electrostatic attraction may be responsible for other types of secondary interactions like $NO_2\cdots NO_2$ interactions in nitro-substituted organic compounds. Therefore, electronegative substituents not containing H atoms (OR, SR, COOR, NR_2, CN, etc., where R is a hydrocarbon moiety) elevate T_m and T_b in comparison with the related non-polar substances (Figs. 1 and 9b).

Secondary Bonding

Apart from H-bonds, other sorts of a weak attraction may exist between adjacent molecules having atoms with low coordination numbers (S···O=C, Hg···X, Au···Au, $I_2\cdots I^-$, etc.) [6, 23]. A special case of "additional" ligands weakly bonded to a non-saturated coordination centre is well known in inorganic chemistry. These secondary bonding interactions elevate both T_m and T_b. Higher mp and bp of tetrafluorides in the CF_4–GeF_4 series, where polarity of E–F bonds increases (see Figs. 2 and 4), is most probably due to such additional intermolecular E···F bonding to larger central atoms. A survey of the CSD reveals that Ge and Sn atoms are usually penta or hexacoordinated in the solid state (Fig. 13). A weak additional secondary bonding of Si is typical for tetracoordinated silicon derivatives [26] (where it is probably more than compensated by a strong van der Waals asymmetry, see Figs. 3 and 4). The increase of the melting temperature in EX_4 series with relatively small X substituents, like halogens, from E=Si to E=Pb, may also be due to the secondary E···X intermolecular coordination.

It is noteworthy that both stabilising (NO_2, CN) and destabilising (F) substituents usually do not alter a general order of T_m and T_b among isomers (except where strong H-bonds exist, e.g., in *ortho*-dihydroxo derivatives, discussed in *RESULTS*). In most cases, bp and mp of all isomers are uniformly shifted to higher or to lower temperatures by electrostatic interactions and/or secondary bonding. Moreover, similar T_m (T_b) trends are observed for the compounds with

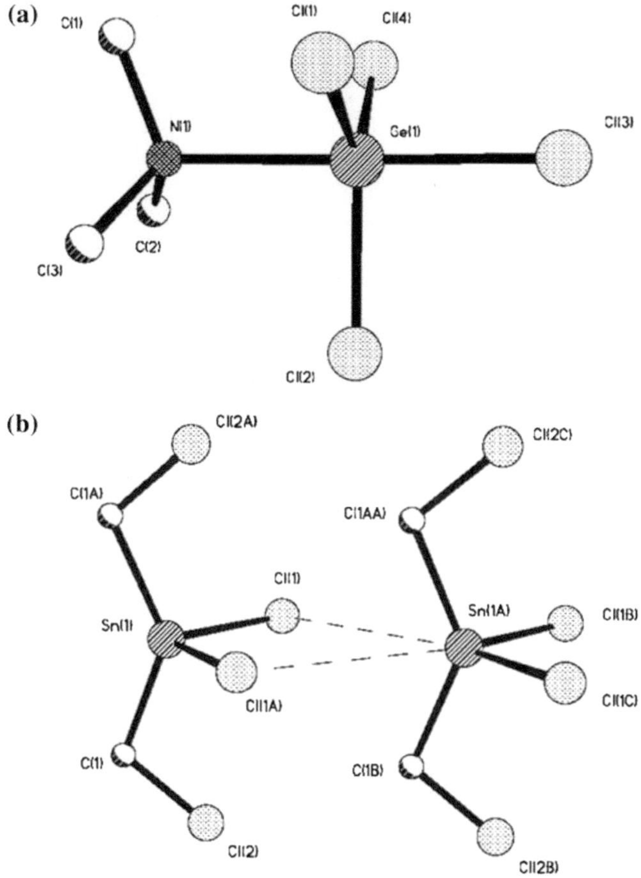

Fig. 13 Two examples of secondary bonding with a formally four-coordinated central atom of group 14 elements: (**a**) $GeCl_4 \cdot NMe_3$ [24], (**b**) $Sn(CH_2Cl)_2Cl_2$ [25]. Atomic coordinates have been taken from CSD.

opposite polarity of substituents (e.g., Me- and F-substituted isomers) as well as with different types of chemical bonding (benzene vs. cyclohexane derivatives). Therefore, both electrostatic and specific intermolecular interactions cannot be a primary reason for the observed regular geometry-dependence of mp and bp, although they may alter a general succession in some particular cases. Note that mp and bp variations are smaller (and in the case of bp may be negligible) for isomers containing X and Y substituents with opposite polarity but similar steric dimensions (e.g., Me and Cl). This fact strongly supports the principal role of van der Waals forces, in their molecular field, for all non-ionic organic substances.

Geometry-Dependent $T\Delta S$ Term

The discussion of T_m and T_b purely in terms of energy is valid only if discrete changes of entropy ΔS in the phase transitions are not very sensitive to the molecular geometry. It has been noticed [8] that for real and hypothetical polymorphs of several $C_{12}H_{12}$ isomers, the computed $T\Delta S$ terms are comparable with ΔH. However, a measurable difference in entropy of melting or boiling between different isomers with rather similar, compact and mostly rigid molecules discussed here, may be ruled out. This is indirectly supported by the lack of noticeable dependence of either mp or bp from the molecular moments of inertia. (Note that most I_{jj} values, e.g., in disubstitutes benzenes $C_6H_4X_2$, uniformly increase in the 12- < 13- < 14- succession, in disagreement with both T_m and T_b trends).

The latent heat ΔH of melting or vaporisation at the corresponding temperature evidently is

$$\Delta H = T\Delta S, \quad (8)$$

since the change of Gibbs potential $\Delta G = \Delta H - T\Delta S$ between two phases is zero at the point of equilibrium. A strict account for changes of entropy upon melting and boiling requires calculating statistical sums for organic substances in the liquid phase [4]. Since there is no accepted model of the liquid state to make such calculations straightforward, we compare here measured thermodynamic parameters [27] for several substituted benzenes with their T_m and T_b (Table 6). No direct correlation of T_m with either ΔH or ΔS at the melting temperature can be observed in various monosubstituted benzenes, as well as in the "anomalous" set of trimethylbenzenes (see above), whereas for $C_6H_4X_2$ isomers (X=Me, Cl) ΔH_m and T_m are of the same order. In a good agreement with the Trouton's rule [Eq. (3)], the ΔS values at boiling temperatures are almost constant (Table 6). Standard enthalpy of evaporation at 298 K, shown in the neighbouring column, therefore displays a good linear correlation with T_b (Fig. 14). Thus changes of entropy in Table 6 are not very sensitive to the molecular structure. We are therefore left with "static" van der Waals interactions, as the most probable reason for the variations of T_m and T_b in the sets of closely related compounds.

Table 6 Melting (T_m) and boiling (T_b) temperatures, enthalpy (ΔH_m) and entropy (ΔS_m) of melting, standard enthalpy of vaporisation at 298 K ($\Delta H_{vap}{}^0$) and entropy of vaporisation at T_b (ΔS_b) in benzene family

	T_m, K	T_b, K	ΔH_m, kJ/mol	ΔS_m, J/(mol K)	$\Delta H_{vap}{}^0$, kJ/mol	ΔS_b, J/(mol K)
C_6H_6	279	353	9.9	35.5	33.9	87.0
C_6H_5Me	178	384	6.6	37.2	38.0	86.5
C_6H_5F	231	358	11.3	47.0	34.6	87.2
C_6H_5Cl	228	405	9.6	41.9	41.0	86.9
C_6H_5Br	242	428	10.7	44.0	44.5	–[a]
C_6H_5I	242	461	9.8	40.3	47.7	–
$C_6H_5CF_3$	244	375	13.8	56.5	37.6	86.9
1,2-$C_6H_4Me_2$	248	417	13.6	54.9	43.5	86.8
1,3-$C_6H_4Me_2$	225	412	11.6	51.4	42.7	86.7
1,4-$C_6H_4Me_2$	286	411	17.1	59.7	40.0	86.9
1,2,3-$C_6H_3Me_3$	248	449	8.2	33.0	49.1	–
1,2,4-$C_6H_3Me_3$	229	442	12.6	56.0	47.9	–
1,3,5-$C_6H_3Me_3$	228	438	9.5	41.7	47.5	–
1,2-$C_6H_4Cl_2$	256	452	12.9	–	50.5	–
1,3-$C_6H_4Cl_2$	248	446	12.6	–	50.6	–
1,4-$C_6H_4Cl_2$	327	447	18.1	–	55.5	–

Data taken from [27]
[a]Data not available

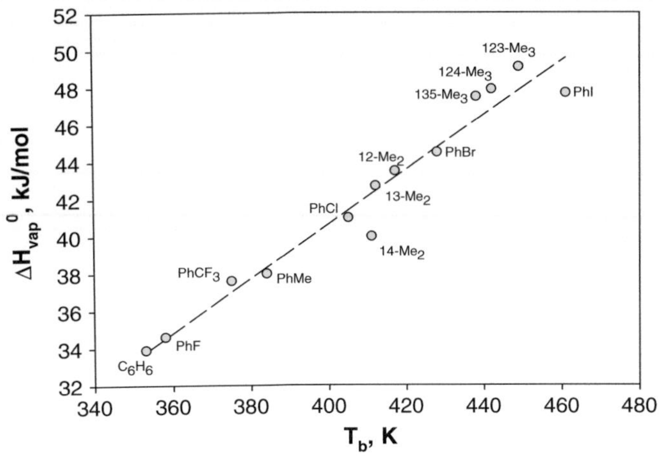

Fig. 14 Linear correlation of standard enthalpy of vaporisation at 298 K $\Delta H_{vap}{}^{0}$ with boiling point T_b in substituted benzenes [27].

Conclusions

The available melting and boiling temperatures of organic compounds and related substances strongly depend upon van der Waals volume and shape of their molecules. The relationship of T_m with a molecular structure is known for along time in physical organic chemistry whereas T_b shows different dependence upon molecular geometry. Both T_m and T_b trends may be explained on the basis of a model that represents a total van der Waals field U in a condensed phase, as a sum of the averaged long-range local field U_0 and a short-range "contact" term W determined by a molecular geometry. A symmetry of crystal's $U_0{}^{(C)}$ and liquid's $U_0{}^{(L)}$ local field is different due to vacancies in the nearest molecular environment in a liquid phase. Perturbations of U_0 by W are supposed to correlate with a discrepancy between the symmetry of the molecular steric shape and the symmetry of the van der Waals group $\widetilde{\mathbf{G}}(\mathbf{U}_0)$ of the U_0 term in a given condensed phase. Variations of such a misfit among isomers explain qualitatively the different successions of their T_m and T_b, including lower boiling temperatures of liquids composed of high-symmetric molecules. A more rigorous theoretical treatment, including model calculations of both U_0 and W terms, is required for a quantitative analysis of T_m and T_b data.

Acknowledgement Y.L.S is grateful to the Royal Society of London and the Russian Foundation for Basic Research (grant 06–03–08192) for a financial support. J.A.K.H. thanks the EPSRC for a Senior Research Fellowship.

References

1. Silverman RB (2004) The organic chemistry of drug design and drug action, 2nd edn. Academic Press, San Diego
2. Herbrandson HF, Nachod FC (1955) In: Braude EA, Nachod FC (eds) Determination of organic structures by physical methods. Academic Press, New York, p 3, and references therein
3. Kitaigorodskii AI (1973) Molecular crystals and molecules. Academic Press, New York
4. Gray CE, Gibbins KE (1984) Theory of molecular fluids, vol 1. Clarendon, Oxford
5. Israelachvili J (1991) Intermolecular & surface forces, 2nd edn. Academic Press, New York
6. Desiraju GR (1989) Crystal engineering: the design of organic solids. Elsevier, Amsterdam
7. Gavezotti A (1994) In: Bürgi H.-B, Dunitz JD (eds) Structure correlation, vol 2. WCH, Weinheim, p 509
8. Dunitz JD, Filippini G, Gavezzotti A (2000) Helv Chim Acta 83:2317
9. Boeze R et al (2001) Helv Chim Acta 84:1561
10. Slovokhotov YuL, Neretin IS, Howard JAK (2004) New J Chem 28:967
11. Beilstein Institut zur Förderung der Chemischen Wissenschaften licenced to Beilstein Chemie Daten und Software GmbH and MDL Information Systems GmbH, Version 4, December (2002)
12. 2002 Gesellschaft Deutscher Chemiker licensed to MDL Information Systems GmbH, Version 4, December (2002)
13. http://www.mimas.ac.uk/crossfire/docs.html
14. Allen FH, Taylor R (2004) Chem Soc Rev 33:463
15. http://www.systat.com/products/SigmaPlot
16. Hoppe R, Daehne W (1962) Naturwissenschaften 49:254
17. Hoffmann R, Hoppe RZ (1969) Anorg Allg Chem 368:271
18. Emsley J (1989) The elements. Clarendon, Oxford
19. Kitaigorodskii AI (1983) Smeshannye Kristally (Mixed crystals). Nauka, Moscow, p 128 (in Russian)
20. Vainshtein BK, Fridkin VM, Indenbom VL (eds) (1994) Modern crystallography, 2nd edn, vol 2. Springer-Verlag, Berlin
21. Thalladi VR et al (1998) J Amer Chem Soc 120:8702
22. Jeffrey JA (1997) An introduction to hydrogen bonding. Oxford University Press
23. Pyykkö P (1997) Chem Rev 97:597
24. Bilton MS, Webster MJ (1972) Chem Soc, Dalton Trans 722
25. Veith M, Agustin D, Huch VJ (2002) Organometal Chem 646:138
26. Burgi H-B, Shklover V (1994) In: Bürgi H-B, Dunitz J D (eds) Structure correlation, vol 1. WCH, Weinheim, p 303 and references therein
27. NIST Chemistry Web Book (2003) http://webbook.nist.gov/chemistry/

The Role of Symmetry in Building Up Zeolite Frameworks from Layered Zeolite Precursors Having Ferrierite and CAS Layers[a,b]

W.J. Roth[c,d] and D.L. Dorset[c]

Abstract

Layered zeolite precursors represent an emerging class of materials that expand the zeolite field in a new direction and have already produced new intriguing discoveries. One is a new fundamental insight that a zeolite framework can assemble by two pathways, directly in 3-D and via a layered precursor. Also new types of materials such as pillared and delaminated zeolites have been synthesized. Starting with the discovery of MCM-22 precursor in 1990s, other framework structures have now been obtained by topotactic condensation, including classical zeolites such as sodalite and ferrierite. The recent structural diversity observed with FER and CAS layers, which were found to produce two frameworks each, i.e. FER/CDO and CAS/NSI, respectively, are herein rationalized as resulting from the absence of in-plane mirror symmetry. A systematic treatment is envisioned, such as consideration of translational and pseudo-translational interactions in condensation of elementary shapes into close-packed assemblies, providing a blueprint for a general approach to analyze new precursors in the future. Other already documented instances of structural diversity include the presence of both types of packing in one preparation (CAS and NSI) and incomplete pairing of silanols connecting FER layers in ESR-12.

Keywords

Layered zeolite precursor • MCM-22 • Ferrierite • CDO • CAS • NSI

Introduction

Zeolites are crystalline solids comprising an extended network of corner-sharing MO_4 tetrahedra, where M is mostly Si and optionally elements like Al ([1–10], IZA Structure Commission web site). These periodic, so-called four-connected networks or frameworks often contain regular pores and/or channels of molecular dimensions; in short, they exhibit microporous structures ([2–5], IZA Structure Commission web site). Several among the about 200 known zeolites or related materials have been found valuable for practical applications and used widely as catalysts, cation exchangers, and adsorption media [9, 10]. This demonstrated usefulness has made zeolite synthesis and the discovery of new frameworks a very active area. One of the very successful and common strategies has been the use of appropriate organic molecules as structure directing agents (SDA's) which induce framework formation depending on composition and synthesis conditions. Many new zeolites have been discovered through the use of an SDA, most often empirically rather than based on prediction.

[a] *Structural Chemistry* 2010, 21(2):385–390.

[b] Dedicated to Professor Adam Bartecki on the occasion of his 90th birthday.

[c] Corporate Strategic Research, ExxonMobil Research and Engineering, Annandale, NJ 08801, USA

[d] 123 Bound Brook Ct, Sewell, NJ 08080, USA
e-mail: wies.roth@gmail.com

I. Hargittai, B. Hargittai (eds.), *Science of Crystal Structures: Highlights in Crystallography*,
DOI 10.1007/978-3-319-19827-9_26

As new zeolite synthesis is a major source of innovation, the underlying mechanism of framework formation has been of considerable interest but it is rather poorly understood. It was once envisioned as occurring by the condensation of rather complex structural building units but this remains unsubstantiated experimentally. It may be discounted for majority of cases, which are now believed to involve mainly small molecule/ion additions with assumed growth of the framework in 3-D.

Discussion

The Discovery of Layered Zeolite Precursor MCM-22P and Its Consequences

In contrast to the direct pathway implicit in zeolite synthesis, a new synthetic zeolite MCM-22 [11, 12] demonstrated its framework formation by an indirect route [11–13]. The hydrothermal synthesis with hexamethyleneimine (HMI) as the SDA afforded first a layered 2-D precursor, MCM-22P, with layers non-covalently but, rather, hydrogen-bonded through silanols [14] with possible HMI involvement, separated by *ca.* 2 Å. The completed four-connected MCM-22 zeolite framework was obtained upon calcination as the layers apparently came closer with concomitant condensation. The discovery of MCM-22P was a breakthrough that gradually expanded into a new direction and a separate area with broad implications and many novel findings [15].

The new indirect, layer precursor pathway turned out not to replace but only complement the conventional 3-D route. It was found that the MCM-22 structure could also be formed, under slightly different synthesis conditions, by the direct 3-D assembly route, represented by the MCM-49 synthesis [13].

The indirect layered precursor routes were subsequently identified for some traditional zeolites, well-known for years via minerals and synthetic solids, i.e., ferrierite [16] and sodalite [17]. A hypothesis was formulated [15] that many, if not most, zeolite structures may be accessible by both direct and indirect pathways, currently represented only by the 2-D layered precursors but in general one cannot rule out others, such as 1-D species (A reviewer kindly pointed out that Akporiaye and Price [4] and Shannon [5] had already demonstrated that most zeolites could be built by the stacking of layers. However, these layers are predominantly conceptual three-connected 2-D nets. The precursor layers discussed in this article are actual building blocks comprising thin crystals with internal zeolitic connectivity).

Layered precursors expanded the traditional zeolite area in new directions. The MCM-22 layers are separable and can adopt or be manipulated into a variety of spatial arrangements [18]. In theory their number is infinite, mostly of a random or disordered kind, but among this continuum of possibilities, MCM-22 demonstrated discrete structures. They were obtained under special conditions or made by design, have distinct characteristics and can be identified based on unique X-ray diffraction patterns. The examples of such discrete materials include pillared and delaminated zeolites, represented by MCM-36 [19] and MCM-56 [18, 20] or ITQ-2 [21, 22], respectively, and recently, the stabilized precursors like YNU-1 and others [23, 24].

General Comments on the Formation of Precursors and Their Conversion to Frameworks

New precursor findings are accidental, usually through the use of novel SDA or in new compositions. While zeolites are obtained when the framework can propagate uninterrupted in 3-D structures, a precursor ensues when conditions become conducive to termination of tetrahedra with silanols/siloxanes and stabilization with organic structure directing agents. Most precursors are found with alternating inorganic and organic layers with extensive hydrogen bonding.

The transformation to ordered zeolite framework can occur upon removal of the template and bridging of the layers. The silanols play a key role by generating interlamellar connections via Si–O–Si bonds determining the nature of the product. In general, these layers can come together in a non-congruent way, producing a disordered structure. The number of possibilities is infinite and the product is likely to be not only uncharacterizable but inferior, i.e., with blocked and/or irregular micropores. With congruent condensation the diversity is manageable, and often affords discrete ordered frameworks. In the case of MCM-22 only one such structure has been obtained (MWW).

As the area of precursors expands and new members are identified, they reveal new possibilities such as two distinct complete structures based on FER layers. Such findings can be usually readily rationalized in the hindsight and in general would be predictable if systematic structural and symmetry considerations were applied. Below, we discuss how some recent discoveries concerning layered precursors can be analyzed with the application of symmetry considerations and we probe some of the consequences as already found in real systems.

Layered Precursors with FER Sheets

The initially observed structural variability of MCM-22 family and its proposed continuum arose from spatial arrangement of sheets (distance and orientation) without the need to include the nature of the sheets themselves. Ferrierite and some newer precursors showed additional complexity [16,

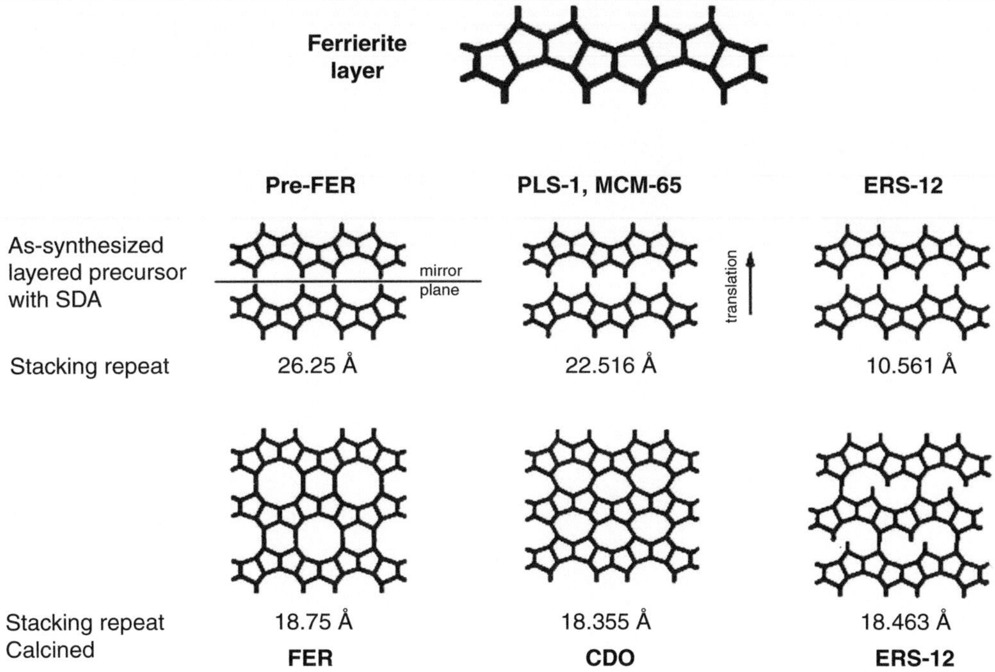

Fig. 1 Structural relationships among materials with ferrierite sheets in the 2-D projections.

17, 25–29] associated with silanol distribution on the surface of the layers, which in turn relates to internal symmetry of these layers.

Ferrierite has been one of several most extensively studied traditional zeolites, with no evidence of its possible formation via layered precursor ([1–3], IZA Structure Commission web site) going back many years. The accidental discovery of its layered form [16] justifies the speculation that many other zeolites may await evidence of their alternative formation via the precursor route [18].

The ferrierite layer, shown in Fig. 1, has been identified in a number of precursors with different SDA's (see Table 1). In contrast to MCM-22, two different completed frameworks have been obtained with these layers, one is FER (Ferrierite) [16], and the other CDO [25, 26]. This additional variability is seen as the result of the absence of lateral internal mirror plane symmetry within the precursor layers. This creates two different ordered stacking possibilities in terms of relationship between adjacent layers. In FER the layers are related by a mirror plane while in CDO they are related by translation (see Fig. 1). An alternative view could be that the layers are related by an offset in the horizontal direction. The outcomes appear to be the same, but as apparently shown by the CAS example below, the layer must show additional symmetry. It is noteworthy that horizontal offset implies a continuous, infinite number of possibilities but the indication to date is that positioning of the layers with respect to each other is governed by the SDA and its H-bonding interaction with the silanol population.

The example of ERS-12 [28] suggests that even initially ordered precursors may end up condensing partially in an incongruent manner. This, and the observation that some (apparently) calcined FER-related derivatives could not be structurally characterized (MCM-47 [27]), may reflect the complex and non-congruent transformation from layered precursor, possibly even within the same crystals.

Zeolites CAS and NSI

A second example of structurally distinct zeolites sharing the same layer are CAS ([2, 3], IZA Structure Commission web site) and NSI [29]. As with FER the two structures can be rationalized by the mirror plane and translation transformations relating the adjacent layers (see Fig. 2). The former has been known for a long time but its pure precursor has not been identified, i.e., no layered material showed congruent full conversion to CAS. Recently it was shown [30] that a previously known layered silicate EU-19 [31], which on calcination yielded disordered EU-20 [32] in the hands of Marler et al., gave a product, designated EU-20b, with 88 % CAS and 12 % NSI. The latter is the recently solved ordered microporous structure from the precursor made a few decades ago NU-6(1) [33]. It appears that in the case of EB-19/EB-20b combination one has both the mirror plane and translation related packing, CAS and NSI-like, in the same material. The structural characterization of NU-6(1) and NSI was hampered by the poor product quality obtained

Table 1 Cell parameters in A° and cell angles in degrees (only no 90° angles listed) for the ferrierite and CAS families of layered precursors and the corresponding framework structures

FER layer	**Mirror plane**	**Translation**			
Precursor	Pre-FER	PLS-1	MCM-65	MCM-47	ERS-12
SDA	4-amino-2,2,6,6-tetramethylpiperidine with fluoride	TMA (tetramethylammonium)	Quinuclidine/tetramethylammonium	Tetramethylene bis(N-methylpyrrolidinium dibromide)	TMA (tetramethylammonium)
Cell parameters	26.25	10.571	22.516	22.454	10.561
	14.05	14.0064	13.869	14.018	13.969
	7.44	7.422	7.367	7.386	7.415
Cell angle		98.014			97.981
Zeolite	FER	CDO (CDS-1)	CDO (MCM-65)	Disordered	ERS-12 (partial disorder)
Cell parameters	18.75	18.355	18.006		18.463
	14.05	13.779	13.489		13.917
	7.42	7.3674	7.373		7.413
CAS layer		**Mirror plane**		**Translation**	
Precursor			EU-19	NU-6(1)	
SDA			Piperazine	4,4′-Dipyridyl	
Cell parameters			22.36	27.7287	
			4.90	4.9731	
			13.57	13.935	
Cell angle			91.67	103.73	
Zeolite		CAS	EU-20b (0.88CAS.0.12NSI)	NSI	
Cell parameters		16.776	16.637	17.257	
		5.021	4.995	4.9881	
		13.828	13.817	13.848	
Cell angle				106.09	

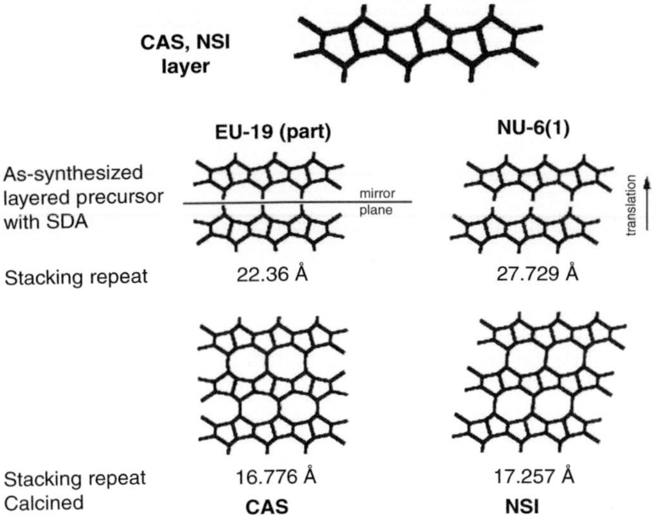

Fig. 2 Structural relationship between zeolite CAS and NSI in the 2-D projection.

after calcination, NU-6(2), probably resulting from tendency for non-congruent condensation.

In contrast to the ferrierite layer precursors the horizontal offset of neighboring layers does not operate in this case. The rationale may be that FER layers have different symmetry, i.e., perpendicular mirror planes while CAS does not. Again, this points to the involvement of symmetry, which would reveal itself in systematic analysis.

The material EU-20b illustrates another possibility that one could anticipate on purely structural basis, i.e., formation of mixtures with both types within the same preparation if not the same crystals. It may be a statistical phenomenon but most likely caused by changing the local synthetic environment or conditions. Like the zeolite structure templating in general, the answer to what directs disposition a nucleating layer with respect to its predecessor may be a key to a better predictability of a synthesis outcome.

The Role of Symmetry in Building Up Layered Frameworks

We discussed above two cases of zeolite layers, each producing two ordered zeolite frameworks. The underlying process of condensation of the layers through silanols from adjacent layers has been rationalized by applying translation and symmetry operations to match the surface silanol distribution.

These two specific cases, as well as MCM-22, fall into the category of the condensation of elementary shapes into a closely packed lattice. In a theoretical treatment of the periodic close packing of arbitrary shapes, Kitaigorodskii [34] invoked the interactions of translational and pseudotranslational symmetry elements. In terms of two-dimensional projections, 15 flat chain projections, could be envisioned [34]. These will incorporate pure translations: **1** and $\mathbf{2_1}$ axes, and the pseudo-translations: center of symmetry (**−1**) and mirror plane (**m**) elements.

The assembly of the structures built up from the FER type precursor layer involves either the pure translations **1** or the pseudo-translation **m**. This produces the two frameworks FER and CDO as shown in Fig. 1. As also pointed out above, an equivalent result is obtained by shift in the horizontal direction of the layers in 'FER configuration'. This can (probably) be explained by combining the perpendicular mirror plane with the translation. By contrast, the lack of mirror symmetry perpendicular to the CAS layer does not generate the same structure by horizontal offset.

So far the treatments for ordered structural variations among layered zeolite precursors have been used ad hoc to rationalize limited experimental observations showing apparently related structures. A more general application of symmetry and systematic derivation of all possibilities may give complete list of all possible variations. Such an enumeration is a laborious task, but may demonstrate value in theory and practice, e.g., by revealing possible interesting or useful structures. As previously mentioned (vide supra), the derivation of zeolite structures by stacking (conceptual) layers in the form of corrugated sheets via the application of various symmetry operators was proposed earlier [5]. Herein a similar process is applied to layers that are real entities.

We have shown that symmetry principles may govern and rationalize formation of different zeolite frameworks from corresponding layered precursors. By systematic enumeration of various contributions of symmetry operations a comprehensive set of possible structures may be obtained and used in future advancement.

References

1. Breck DW (1984) Zeolite molecular sieves. Robert E. Krieger Publishing, Malabar
2. Szostak R (1992) Handbook of molecular sieves. Kluwer, Dordrecht
3. Baerlocher C, Maier WM, Olson DH (2007) Atlas of zeolite framework types, 6th edn. Elsevier, Amsterdam
4. Akporiaye DE, Price GD (1989) Zeolites 9:23
5. Shannon MD (1993). In: von Ballmoos R, Higgins JB, Treacy MMJ (eds) Proceedings of the 9th international zeolite conference, Montreal 1992, vol 1, p 389. Butterworth-Heinemann, Stoneham, Massachusetts
6. Szostak R (1998) Molecular sieves, 2nd edn. Blackie Academic and Professional, New York
7. Xu R, Pang W, Yu J, Huo Q, Chen J (2007) Chemistry of zeolites and related porous materials. Wiley, Singapore
8. Barrer RM (1982) Hydrothermal chemistry of zeolites. Academic Press, New York
9. van Bekkum H, Flanigen EM, Jacobs PA, Jansen JC (eds) (2001) Introduction to zeolite science and practice vol 137. Stud Surf Sci Catal. Elsevier, New York

10. Cejka J, van Bekkum H, Corma A, Schuth F (eds) (2007) Introduction to zeolite science and practice, 3rd edn, vol 168. Stud Surf Sci Catal. Elsevier, New York
11. Leonowicz ME, Lawton JA, Lawton SL, Rubin MK (1994) Science 264:910 (Washington, DC)
12. Rubin MK, Chu P (1990) Composition of synthetic porous crystalline material, its synthesis and use. US Patent No. 4 954 325
13. Lawton SL, Fung AS, Kennedy GJ, Alemany LB, Chang CD, Hatzikos GH, Lissy DN, Rubin MK, Timken H-KC (1996) J Phys Chem 100:3788
14. Roth WJ (2006) Pol J Chem 80:703
15. Roth WJ (2007) In: Cejka J, van Bekkum H, Corma A, Schuth F (eds) Introduction to zeolite science and practice, 3rd edn, vol 168, p 221. Stud Surf Sci Catal. Elsevier, New York
16. Schreyeck L, Caullet P, Mougenel JC, Guth JL, Marler B. J Chem Soc, Chem Commun 1995, 2187; Micropor Mater 1996, 6:259
17. Moteki T, Chaikittisilp W, Shimojima A, Okubo T (2008) J Am Chem Soc 130:15780
18. Roth WJ (2005) In: Cejka J, Zilkova N, Nachtigall P (eds) Molecular sieves: from basic research to industrial applications, vol 158A, p 19. Stud Surf Sci Catal. Elsevier, New York
19. Roth WJ, Kresge CT, Vartuli JC, Leonowicz ME, Fung AS, McCullen SB (1995) In: Beyer HK, Karge HG, Kiricsi I, Nagy JB (eds) Catalysis by microporous materials, vol 94, p 301. Stud Surf Sci Catal. Elsevier, New York
20. Fung AS, Lawton SL, Roth WJ (1994) Synthetic layered MCM-56, its synthesis and use. US Patent No. 5 362 697
21. Corma A, Fornes V, Pergher SB, Maesen TLM, Buglass JG (1998) Nature 393:353
22. Climent MA, Corma A, Velty A (2004) Appl Catal A 263:155
23. Ruan J, Wu P, Slater B, Terasaki O (2005) Angew Chem Int Ed Engl 44:6719
24. Wu P, Ruan J, Wang L, Wu LL, Wang Y, Liu Y, Fan W, He M, Terasaki O, Tatsumi T (2008) J Am Chem Soc 130:8178
25. Dorset DL, Kennedy GJ (2004) J Phys Chem B 108:15216
26. Ikeda T, Akiyama Y, Oumi Y, Kawai A, Mizukami F (2004) Angew Chem Int Ed Engl 43:4892
27. Burton A, Accardi RJ, Lobo RF, Falconi M, Deem MW (2000) Chem Mater 12:2936
28. Millini R, Carluccio LC, Carati A, Bellussi G, Perego C, Cruciani G, Zanardi S (2004) Micropor Mesopor Mater 74:59
29. Zanardi S, Alberti A, Cruciani G, Corma A, Fornes V, Brunelli M (2004) Angew Chem Int Ed Engl 43:4933
30. Marler B, Camblor MA, Gies H (2006) Micropor Mesopor Mater 90:87
31. Blake AJ, Franklin KR, Lowe BM (1988) J Chem Soc, Dalton Trans 2513
32. Andrews SJ, Papiz MZ, McMeeking R, Blake AJ, Lowe BM, Franklin KR, Helliwell JR, Harding MM (1988) Acta Cryst B44:73
33. Whittam TV (1983) US Patent No 4 397 825
34. Kitaigorodskii AI (1961) Organic chemical crystallography. Consultant's Bureau, New York

Part IV

Data Mining and Management

The Correspondence Fedorov-Schoenflies: The Groups Pm, Pc, Cm, Cc[a]

Johann Jakob Burckhardt[b]

Abstract

In the discovery of the crystallographic space groups there occur some differences between Fedorov and Schoenflies. We try to clarify this point on the basis of their correspondence.

It may be well to copy the truth, but it is better to invent the truth, really much better…. (G. Verdi)

As we know, Fedorov and Schoenflies began to discover the crystallographic space groups from different viewpoints and independently. Not later than at the end of 1889 their correspondence began [1] in which they compared their results. It became evident that some results differed.

The letters nos. 223, 234, 503, and 541 show the source of a first difference. It is amazing that it happened in a relatively simple case concerning the groups Pm, Pc, Cm, and Cc. I will try to clarify the reason.

Fedorov:	5 s)	6 s)	1 h)	2 h)
Schoenflies:	C_s^1	C_s^3	C_s^2	C_s^4
International Tables (1935):	Pm	Cm	Pc	Cc

Let us look at these groups; we give their analytic expressions and the figures following the paper [3]. Please note that in paper [4] in the formulae for 1 h) and 2 h) the coordinates y and z are interchanged in relation to [3]. Moreover, in [3] Fedorov starts from the point (b,b,d), whereas in [4] he starts from (b,c,d).

We begin with the group:

1. $C_s^1 = \mathrm{Pm} = 5\mathrm{s}$

The equations of Fedorov are

$$y = b + \lambda, z = n^k b + \lambda_0, x = d + \lambda_1 \qquad (1)$$

written in orthogonal coordinates. Note that

$$n = -1, k = 0 \text{ or } 1 \qquad (2)$$

Here, λ, λ_0, λ_1 are repeating translations in the *y*, *z*, and *x* directions which we illustrate in the *yz*-plane in Fig. 1. We write Eq. (1) for $k=0$:

$$y = b + \lambda, z = b + \lambda_0 \qquad (3)$$

This equation is the expression of the lattice that is generated from the point (*b*, *b*) by the translations λ and λ_0, giving the points marked by a dot in Fig. 1.

For $k=1$, Eq. (1) assumes the form

$$y = b + \lambda, z = -b + \lambda_0. \qquad (4)$$

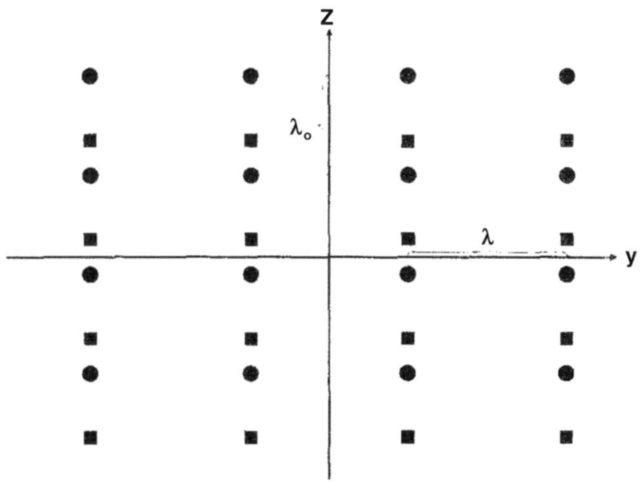

(4) $y = b + \lambda, z = -b + \lambda_o$

Fig. 1

---

[a] *Structural Chemistry* 1991, 2(7):645-649.

[b] Bergheimstraβe 4, CH-8032 Zürich, Switzerland (deceased)

I. Hargittai, B. Hargittai (eds.), *Science of Crystal Structures: Highlights in Crystallography*,
DOI 10.1007/978-3-319-19827-9_27

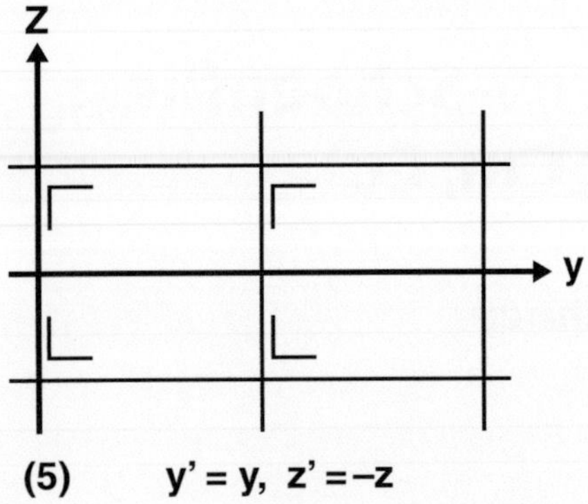

Fig. 2

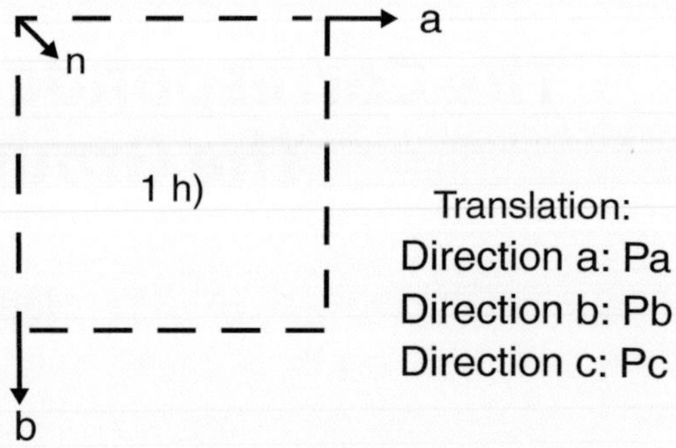

Fig. 4

This is the lattice symmetric to the first one, with the points marked by a square in Fig. 1.

In modern crystallography one writes for this symmetry (Fig. 2)

$$y' = y, z' = -z \tag{5}$$

Note: The Eq. (1) are the invention of Fedorov. They determine the *points of the lattice* and their symmetry, where "point" has to be taken in the sense of Euclid. Formula (1) does not state anything about the *intermediate space*. This point may be important for the crystallographic interpretation. I wonder whether this has been noticed in the existing literature? Is my interpretation of formula (1) correct?

The Eq. (5) are expressions concerning the *whole space* and thus differ in principle from the Eq. (1) of Fedorov. He has drawn Fig. 3 for the group 5 s): a square with solid lines. This implies that the drawing plane *yz* is a symmetry plane.

The following three groups are treated more briefly. An explicit exposition is found in [2, 5, 6].

2. $C_s^2 = \mathrm{Pc} = 1\mathrm{h})$

The equations of Fedorov are:

$$y = b + \lambda, z = n^k b + \lambda_0, x = d + k\frac{\lambda_1}{2} \tag{6}$$

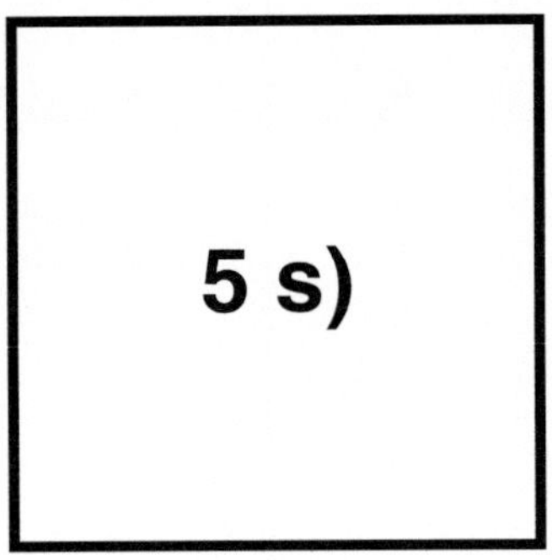

Fig. 3

These equations show that we have a glide reflection in the x-direction. The diagram of Fedorov (Fig. 4) shows a dashed square with arrows in the directions of the possible translations. Figure 4 thus corresponds to the symbols P11a, P11b, and P11c.

The following two groups are, alas, unfortunately marked in the literature and in the International Tables with a C, which means "centered rectangular lattice." But it must be considered as a rhombic lattice. When this was overlooked much trouble arose since the time of Schoenflies.

Figure 5 shows what happens in the *yz*-plane: the rhombus OABC has the centers A and C of the rectangles MNBO and OBPQ as vertices. Therefore the theorem: "In the rhombic lattice a succession of parallel mirror lines (mirror planes) m…m…m… stipulates a succession of parallel glide mirror lines (glide mirror planes) g…g…g… and vice versa."

In this way we are able to treat the groups Cm and Cc:

3. $C_s^3 = \mathrm{Cm} = 6\mathrm{s}$

The equations of Fedorov are:

$$y = b + f\frac{\lambda}{2}, z = n^k b + f\frac{\lambda_0}{2}, x = d + \lambda_1 \tag{6}$$

with $n = -1$ and $k = 0$ or 1.

$k=0$ gives

$$y = b + f\frac{\lambda}{2}, z = b + f\frac{\lambda_0}{2}, x = d + \lambda_1 \tag{7}$$

$k=1$ gives:

$$y = b + f\frac{\lambda}{2}, z = -b + f\frac{\lambda_0}{2}, x = d + \lambda_1 \tag{8}$$

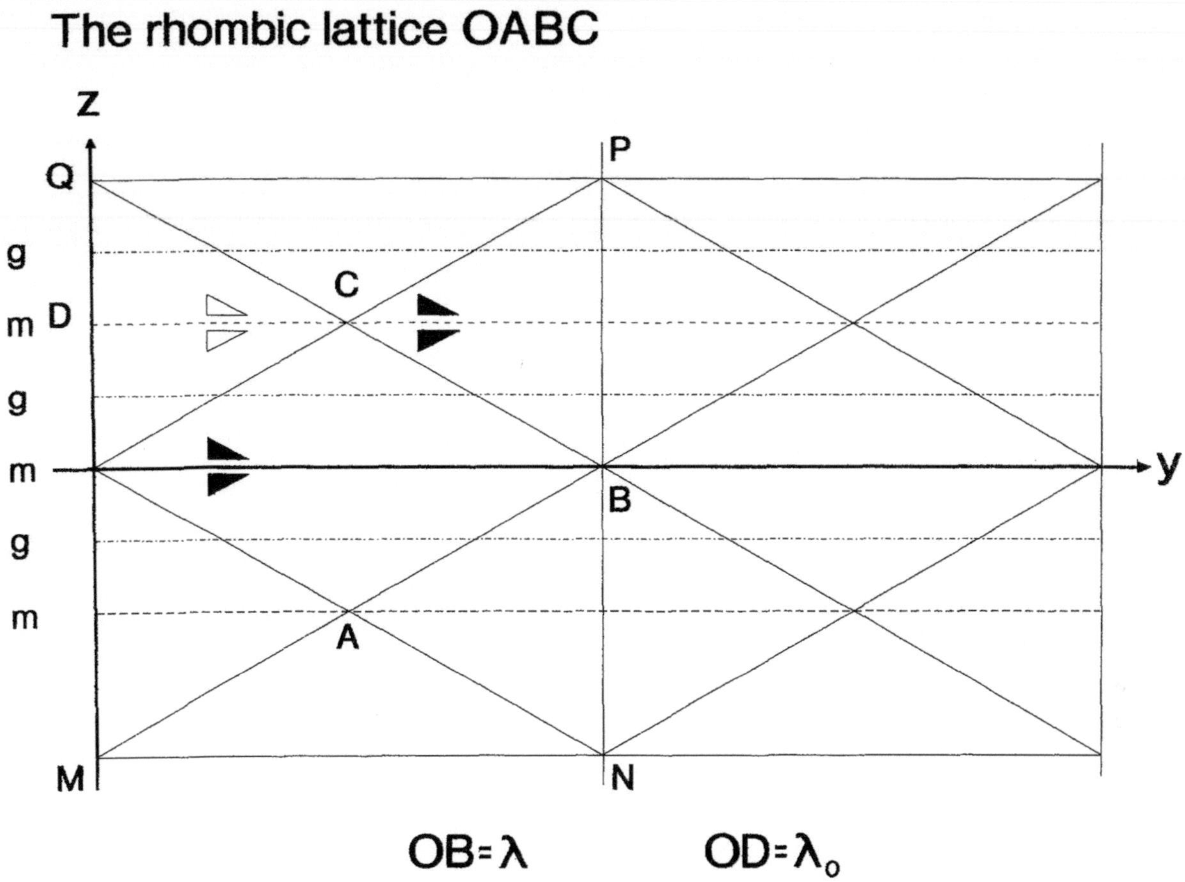

Fig. 5 The rhombic lattice OABC.

In Fig. 5 we illustrate the *yz*-plane, $x=0$. In space we obtain a series of mirror planes alternating with glide mirror planes. We see the graphic of Fedorov in Fig. 6: alternating mirror planes (solid square) and glide mirror planes (dashed square).

4. $C_s^4 = \mathrm{Cc} = 2\mathrm{h}$

The equations of Fedorov are

$$y = b + f\frac{\lambda}{2}, z = n^k b + f\frac{\lambda_0}{2}, x = d + k\frac{\lambda_1}{2} \tag{9}$$

In contrast to (6) a translation in the *x*-direction occurs. The diagram of Fedorov (Fig. 7) consists of a glide mirror plane (dashed square) and a glide mirror plane with arrow (dashed square with arrow in a vertex) which marks the translation in the x-direction.

Remark: After presenting in these examples the analytical as well as the geometric representation of the above groups by Fedorov, I may ask: (1) did Fedorov first look at the geometric representation, (2) or did he first look at the analytic representation, (3) or did he look more or less at both representations? I wonder whether anyone can answer these questions.

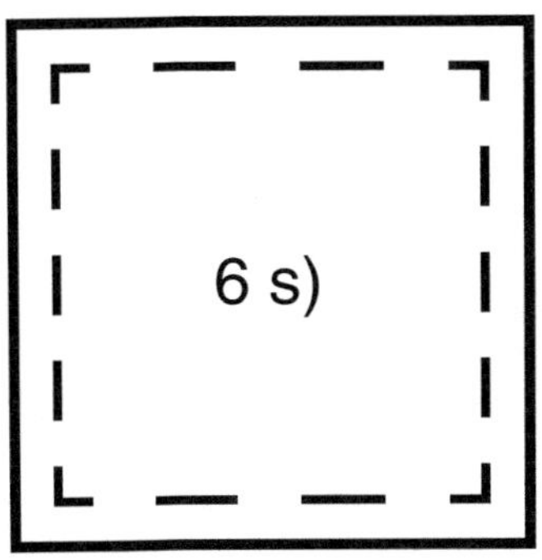

Fig. 6

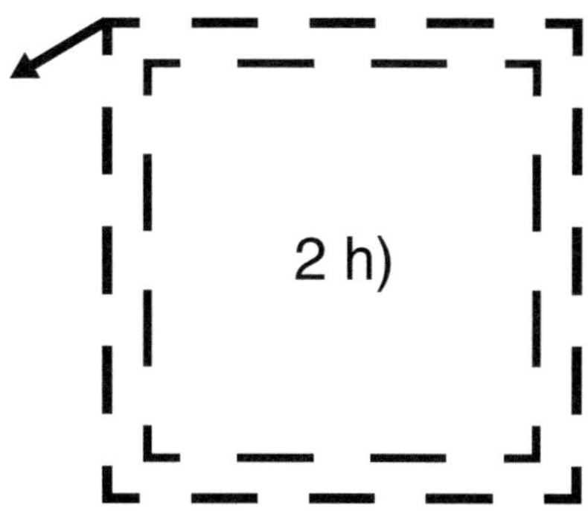

Fig. 7

Let us now look at the correspondence Fedorov-Schoenflies where these four groups play an important role. After Fedorov [4] had found the 230 crystallographic space groups (with some discrepancies), Schoenflies remarks in letter no. 541[1] that he noted from the Fedorov systems 5 s), 6 s), 1 h), and 2 h) only one of each [[7], p. 182 and [1] p. 96 and 182]. Concerning this point Fedorov writes in footnote 2 on page 4 of [4]: "Schoenflies missed in any case one of the pairs of systems 5 s) and 6 s), as well of the systems 1 h) and 2 h)."[2]

Moreover, Schoenflies writes in letter 541: "Concerning the systems 5 s, 6 s, 1 h and 2 h, I was not in a position to see to which of my systems they correspond."[3]

It is amazing that Schoenflies had difficulties with these very simple groups. I will try to clarify this. In letter 233 from Fedorov to Schoenflies we read:

> Instead of your System S(O) in my paper two different systems occur: 5 s) and 6 s). The essential difference is already to be seen in the fact that in the system 6 s) only the same kind of symmetry planes (that is planes which are in coincidence with the whole system) exist, whereas in the system 5 s), in addition to these planes, another kind of planes occurs, the 'Resultante Symmetrieebene'. The systems 1 h) and 2 h) which in your derivation are represented only by S(t) have the same distinction.[4]

At first we remark that in the above text of Fedorov we have to interchange 5 s) and 6 s) (see Figs. 3 and 6, and 4 and 7).

Why did Schoenflies overlook these two systems? Perhaps we find the answer in letter no. 543 from Schoenflies to Fedorov.

> Concerning my paper I did not distinguish the translation groups according to the operations which map them onto themselves, because otherwise I would have had to consider every special lattice as a distinct species. The problem is different when we look at a heap of molecules which can be derived from an *arbitrary* molecule as I have demonstrated in my last paper about the structure of crystals.[5]

Fedorov answered in letter no. 242 (before November 1890):

> Concerning the groups S and S(τ) I think that it is erroneous if you say "The problem is different when we look at a heap of molecules… As far as I see this seems to be also opinion." The way I have deduced the systems 5 s), 6 s) and 1 h), 2 h), the difference is not the result of any consideration about heaps of molecules but I exclusively considered the difference of the laws of symmetry for these systems. In the case of the identity of these laws I consider these systems as of the same kind (more or less "geometrically identical" in your nomenclature).[6]

From this letter we see that Fedorov considered exclusively lattices of points as we have emphasized at the beginning of this paper. Schoenflies, on the other hand, probably worked with heaps of molecules. It seems to me that he attributed properties of symmetry to the molecules which replace the lattice points of Fedorov. This is only possible for the symmorphic systems 5 s) and 6 s). In his book [[8] p. 405] Schoenflies mentions these groups as:

$$C_s^1 : x\ y\ z,\ x\ y\ \bar{z} \qquad C_s^2 : x\ y\ z,\ x\ y\ \bar{z}+\tau$$
$$C_s^3 : x\ y\ z,\ x\ y\ \bar{z}, \qquad C_s^4 : x\ y\ z,\ x\ y\ \bar{z}+\tau$$

and he says: "The groups C_s^1 and C_s^3 are those which arise from the point group S by multiplication with Γ_m, respectively, with Γ'_m. The difference of these groups, as well as of C_s^2 and C_s^4, is not seen from this table because it is based only on the group of translation."[7] For a detailed presentation of the correspondence Fedorov-Schoenflies see Erhard Scholz [[9], pp. 131–132].

Our demonstrations above show that Fedorov knew a perfect derivation of the 230 space groups. He used his own

---

[1] Page 4, letter no 541, line 24: "Heute komme ich nur dazu, wegen der auf S.4 Ihres Werkes enthaltenen Anmerkung Ihnen folgendes zur Beurteilung zu unterbreiten. Wenn ich Sie recht verstehe, so sind die von mir mit V_8^d und V_8^q bezeichneten Gruppen identisch, während ich von denjenigen Systemen symmetrischer Figuren, die Sie mit 5 s und 6 s resp. 1 h und 2 h bezeichnen, nur je eines aufführte."

[2] The paper [4] is written in Russian; I give a German translation: "Schoenflies liess sich entgehen, je eines aus den Paaren von Systemen 5 s) und 6 s) und ebenso aus den Paaren 1 h) und 2 h)."

[3] "Was die Systeme 5 s, 6 s, 1 h, 2 h betrifft, ist es mir noch nicht möglich gewesen zu sehen, welchen der von mir augestellten Systeme sie entsprechen. Sie würden mich daher sehr verbinden, wenn Sie mir dies gefälligst mitteilen."

[4] "Statt Ihres Systems S(O) figurieren in meiner Ableitung zwei verschiedene: 5 s) und 6 s). Der wesentliche Unterschied ist schon daraus zu ersehen, dass in dem System 6 s) nur die gleichen Symmetrieebenen (also solche, die sich zusammen mit dem System selbst decken) existieren, wlihrend in dem System 5 s) ausser solchen es noch die ungleichen, "die Resultante Symmetrieebene" geben. Die System 1 h) und 2 h), welche in Ihrer Ableitung nur in einer S(t) repräsentiert sind, haben einen entsprechenden Unterschied."

[5] "Was zunächst meine Arbeit selbst betrifft, so habe ich darin die Translationsgruppen nicht nach den Operationen unterschieden, welche sie in sich überführen, weil ich sonst jedes spezielle Gitter als eine besondere Art hätte rechnen müssen … Anders liegt die Sache, wenn man die Molekelhaufen ins Auge fasst, die aus einer *beliebigen* Molekel, wie in meiner letzten Arbeit über Krystallstruktur, abgeleitet werden können."

[6] "Bezüglich der Gruppen S und *S*(τ). Ich halte es für irrtümlich, wenn Sie sagen:" Anders liegt die Sache, wenn man die Molekelhaufen ins Auge fasst … Dies ist, soweit ich sehe, wohl auch Ihre Auffassung." Wenn ich die betreffenden Systeme 5 s) und 6 s), auch 1 h) und 2 h) abgeleitet habe, so ist dies kein Resultat irgendwelcher Ueberlegungen über die Molekelhaufen, sondern ausschiesslich das Konstatieren der Verschiedenheit des Symmetriegesetzes dieser Systeme; im Fall der Identität dieser Gesetze halte ich die Systeme für *gleichartig* (ungefähr 'geometrisch identisch' nach Ihrer Nomenklatur)."

[7] "Die Gruppen C_S^1 und C_S^3 sind diejenigen, welche durch Multiplikation der Punktgruppen S mit Γ_m resp. Γ'_m entstehen. Die Verschiedenheit dieser Gruppen, sowie der Gruppen C_S^2 und C_S^4 tritt in der Tabelle deshalb nicht hervor, weil sie allein auf der Translationsgruppe beruht."

analytic formulation as well as an excellent graphical representation, the latter being an example for later researchers. The basis of Fedorov's investigations was the concept of the point lattice and its symmetries and nothing else.

The working method of Schoenflies is not as evident as that of Fedorov. However, my exposition above shows clearly that he started from the concept of symmetry and the point groups based on it. To obtain the space groups he had to multiply them with lattices (as shown by the quotation above). In this way two of the four groups considered by us were missed. Schoenflies imagines matter to consist of molecules (or heaps of molecules) instead of mathematical points. Thus the problem arose, whether one should ascribe to them certain symmetries. Unfortunately Schoenflies did not use any of the methods of the lucid derivation of Fedorov; his book [8] lacks figures. The treatises of Fedorov, written in Russian or difficult-to-understand German, were hardly noticed in non-Russian-speaking countries. The monographs written in German (and later on in English) were based on Schoenflies with figures added that resemble those of Fedorov.

Postscript 1

After having sent a copy of my paper to Prof. I. I. Schafranovsky (Leningrad) he acquainted me with the two volumes of the book of Prof. S. A. Bogomolov [10]. In the first volume Prof. Bogomolov mentions letters 531, 541, and 543 of A. Schoenflies to Fedorov on pages 9–10 and remarks the missing of two of the above groups by Schoenflies. In the second volume on pages 16–17 Prof. Bogomolov gives the derivation of the four groups following Fedorov in the same notation as I have given above. There we also find the diagrams.

Postscript 2

In my diary I find that on October 27, 1948 I met then-Professor R. W. G. Wykoff. He told me from his memory of Schoenflies that he had the impression that Schoenflies had found the 230 groups essentially in an intuitive way.

Along with my exposition this gives a hint why Schoenflies missed two groups.

References

1. Burckhardt, J. J; Der Briefwechsel von E. S. von Fedorov und A. Schoenflies, 1898–1908. *Arch. History Exact Sci.,* **1971**, *7* 91–141.
2. Burckhardt, J. J. Die Bewegungsgruppen der Kristallographie, 2. Aufl. Birkhäuser Verlag; Basel, **1966**.
3. von Fedorov, F. S. Zusammenfassung der kristallographischen Resultate des Herrn Schoenflies und der meinigen. *Zeitschrift für Kristallographie und Mineralogie,* **1892**, *20*, 25–75.
4. von Fedorov, E. S. Die Symmetric der Figuren regelmässiger Systeme (Russ). *Verhandlungen der russisch-kaiserlichen mineralogischen Gesellschaft zu St. Petersburg* (= Zapiski imperator S. Peterb. Mineralog. Obščestva), Zweite Serie, **1891**, *28* 1–146. Einzelwerk: St. Petersburg 1890.
5. Internationale Tabellen zur Bestimmung von Kristallstrukturen. Borntraeger Verlag: Berlin, 1935, und weitere Auflagen.
6. M. Klemm, Symmetrien von Ornamenten und Kristalen. Springer-Verlag: Berlin, 1982.
7. A. Schoenflies, Ueber Gruppen von Transformationen des Raumes in sich. *Math. Annal.,* **1889**, *34*, 172–203.
8. A. Schoenflies, Krystallsysteme und Krystallstruktur. Teubner Verlag; Leipzig 1891.
9. E. Scholz, Symmetric, Gruppe, Dualität. Birkhäuser Verlag: Basel, 1989.
10. Bogomolov, S. A. *Derivation of the Regular Systems Following the Method of Fedorov,* 2 vol. Leningrad, 1932, 1934 (in Russian).

Prediction of H-Bonding Motifs for Pyrazoles and Oximes Using the Cambridge Structural Database[a]

Lourdes Infantes[b,c] and Sam Motherwell[b]

Abstract

H-bonding motifs for pyrazoles and oximes have been examined in the Cambridge Structural Database. The accessible surface of the N atoms has been found to be useful as a discriminator to divide structures into dimer and catemer motifs for both pyrazoles and oximes. Low accessibility favors dimers and tetramers and high values favor catemers and trimers. Total molecular volume shows some correlation for oximes, while high values favor dimers. Empirical rules were successfully applied to predict the motifs of eight new structures in the subsequent release of the CSD.

Keywords

1H-Pyrazoles • Oximes • Hydrogen-bonding motifs • Secondary structure • Supramolecular structure

Introduction

The objective of this work is to understand the factors that appear to govern the formation of hydrogen-bonding motifs in pyrazoles and oximes, which could possibly lead to the ability to predict motifs for unknown crystals. These systems were chosen because of the observation that they form very similar H-bonding structures [1–3] and also because we have found one interesting example of an isostructure. The following motifs have been observed (Fig. 1), namely dimers, trimers, tetramers, and infinite chains known as catemers.

The question of prediction of dimer or catemer hydrogen-bonding motifs has previously been considered in early studies for carboxylic acids [5] and amides [6]. A more recent survey of bimolecular hydrogen-bonded ring motifs includes these dimer forms [7, 8] and, rather surprisingly, the probability of observation of the carboxylic acid dimer ca. 33 % and amide dimer ca. 23 %. The probability is the ratio of the number of structures with the observed motifs, Nobs, divided by the number of structures containing carboxylic groups, and Nposs, i.e., the possibility to form the motif. The important point is that if we restrict the survey to only those entries where the carboxylic or the amide is the only strong H-bond donor/acceptor group, then the probabilities of dimer formation are carboxylic ca. 96 % and amide ca. 85 %. The lower figures of 33 and 23 % are because the survey is over the entire CSD and the presence of many other H-bond donor/acceptors in competition.

Examination of the catemer monocarboxylic acid structures R—COOH showed some tendency to have very small R groups, e.g., formic (FORMAC01) and acetic (ACETAC01). It is notable also that some small R groups form alternatives to the dimer motif, e.g., difluoro acetic acid (NAGVUM) forms an unusual nonplanar catemer, while chloroacetic acid (CLACET01) forms a rare almost planar tetramer. Another common factor among catemers is a sterically highly accessible -carbon atom, e.g., tetrolic acid (methyl-acetylenecarboxylic acid) (TETROL01). [This molecule also has a dimer polymorph (TETROL) probably showing that there is not a great energy difference between the catemer and dimer

[a] *Structural Chemistry* 2004, 15(3):173–184.

[b] Cambridge Crystallographic Data Centre, Cambridge, UK

[c] To whom correspondence should be addressed;
e-mail: xlourdes@iqfr.csic.es

I. Hargittai, B. Hargittai (eds.), *Science of Crystal Structures: Highlights in Crystallography*,
DOI 10.1007/978-3-319-19827-9_28

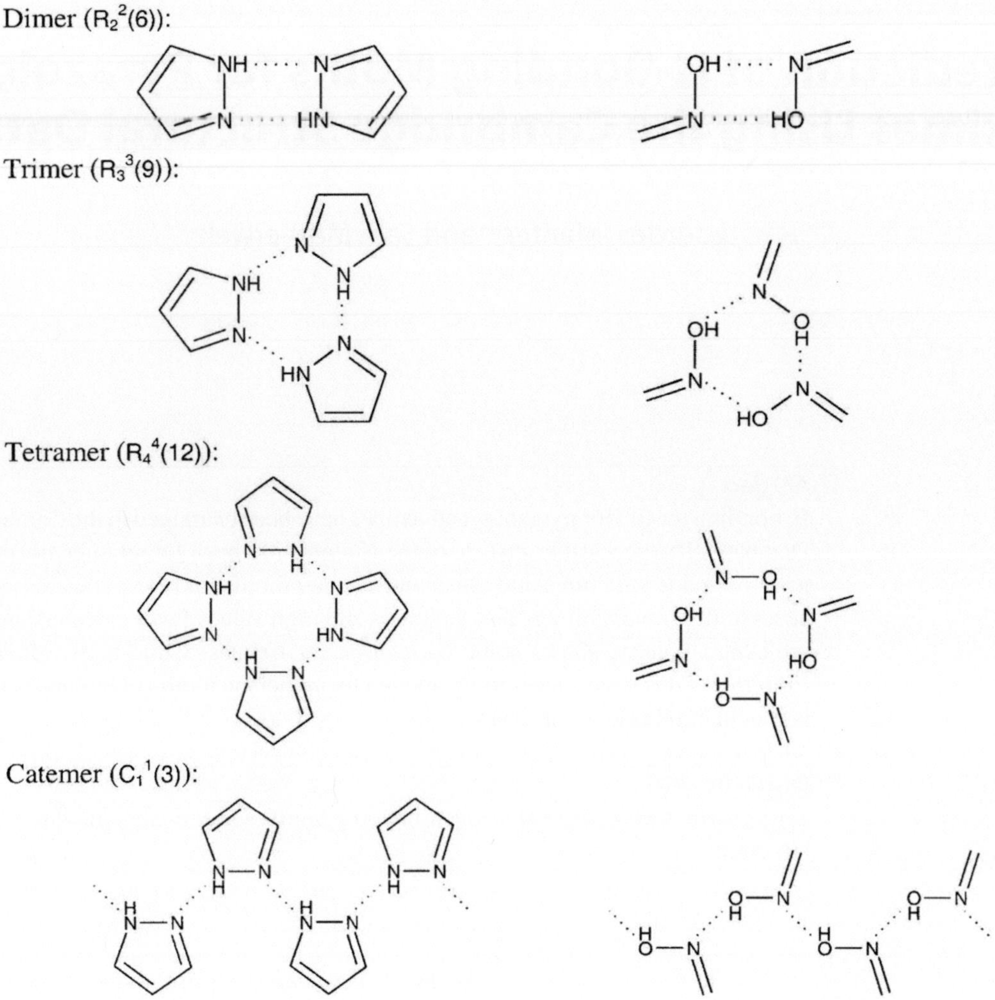

Fig. 1 Secondary structure of 1H-pyrazoles and oximes: dimer, trimer, tetramer, and catemer. The Etter/Bernstein graph set descriptors are given for each motif [4].

motifs.] However, it must be said that there are other catemer carboxylic acid structures where this explanation of steric accessibility does not hold, e.g., the unsubstituted cubane-carboxylic acid (JUNQAK), 4-phenyl-(JUNQEO), and 4-carboxamido- (JUNQIS) show dimer forms [9].

These observations give some support to the idea of high steric accessibility of donor/acceptor groups being a factor in catemer/dimer formation, which we now investigate in this paper, which is restricted to pyrazoles and oximes.

Methodology

The CSD search program ConQuest [10] was used to retrieve organic structures for pyrazoles and oximes from CSD release 5.23, April 2002. We specified that only one pyrazole or oxime group should be present and no other strong hydrogen bond donor/acceptor, and no metals, but organic halogens, Si, P, S, NO_2, ether groups and NR_3 groups with sp^2 nitrogen hybridization were allowed. We hoped that by excluding competing H-bonding groups, we might be better able to observe steric effects on the self-association motifs of pyrazoles and oximes. Forty-four pyrazole and 101 oxime structures were obtained. These were inspected visually using the CSD visualizers, Mercury and RPluto and the motifs assigned. We found that all the 44 pyrazole compounds and 99 of the 101 oximes form one of the secondary structures drawn in Fig. 1.

The program RPluto [11] has a facility to calculate the accessible surface of any given atom to a probe of given radius (see Fig. 3). The command VATM selects the named atom, or atom type, and randomly positions the probe point on the surface of the van der Waals sphere of radius Rvdw for that atom (Van der Waals radii from Bondi [12]). If the probe is given a radius Rprobe, then the points are scanned at a distance (Rvdw + Rprobe), with optional display on the screen.

We used the command "NOCRYST" for the free molecule without its crystal-packed neighbors, Rprobe = 1.1 Å, and 6000 random sample points (command: VATM atom-name N = 6000 P = 1.1 SURF NOCRYST). We express the accessible area of the atom in a normalized manner as the fraction, F, of the probe points, which do not intersect with any atom. Thus, a completely accessible free atom has F = 1.00, and a completely inaccessible atom has F = 0.0. For 6000 sample points, the estimated standard deviation of F is 0.005, so we quote just two decimal places in our results. We used H-atom positions as given in CSD normalized with a C—H distance of 1.085; if H is missing from the 3D coordinates, we calculated at the idealized positions (see HCAL command). Because of the uncertainty in H-atom positions in the crystal structures, we find a spread of the values of F of about 0.01, depending on renormalizing of C—H to 1.085 or 1.075 Å. When more than one independent molecule is present in the asymmetric unit, we have used the average value of F, as there was generally close agreement between the figures for the molecules.

We also used the RPluto feature for calculation of molecular volume, which explores an enclosing sphere around the molecule with 300,000 random points within the sphere (command MVOL). The fraction of points which lie in free space, not within Rvdw of any atom, multiplied by the volume of the sphere, enables one to estimate Mvol, the molecular volume. We recorded this for all molecules as a possible factor influencing H-bonding motifs.

Results and Discussion

The values of Mvol and F are given in Table 1 for pyrazoles and Table 2 for oximes. For pyrazoles, we probed atoms as follows, N having no H, the N of the NH group after removing the hydrogen atom, and the C atoms adjacent to the nitrogen atoms (C1,C3) (Fig. 2a); Sum-F is the sum of F over the two nitrogens. For oximes, we probed the terminal O (after removing the hydrogen atom), the N, and the C adjacent to N (C1) (Fig. 2b); Sum-F for oximes is the sum of F over the oxygen and the nitrogen atoms. The measurement over the carbon atoms was not found to correlate with the secondary structure groups and has not been included in the table. For oximes, the hydroxyl group has always a high accessible surface and does not distinguish between the motifs. After examination of all these parameters, we focused our studies on the accessible surface over the nitrogen atoms. Figure 3 is an example of two pyrazole compounds **(a)** HOQHUQ with high accessible surface F(N1) = 0.30 and F(N2) = 0.31, and **(b)** WILBAU with low accessible surface F(N1) = 0.18 and F(N2) = 0.19.

Table 1 Data for pyrazoles[a]

Motif	Refcode	F(N)	Sum-F	Mvol	Motif	Refcode	F(N)	Sum-F	Mvol
$R_2^2(6)$	CIQHIT	0.07	0.32	322.7	$R_3^3(9)$	DASXEA10	0.26	0.52	96.3
	EDIXOE	0.15	0.41	396.8		FITQAA	0.26	0.57	96.8
	FULKUS	0.19	0.50	129.4		GOQXIT	0.29	0.59	81.3
	HEHTUJ	0.27	0.57	103.0		HEHVAR	0.25	0.56	102.6
	LADBEX	0.19	0.39	223.1		HOQHUQ	0.30	0.61	80.3
	NIBFIN	0.22	0.52	214.2		LABHEB	0.25	0.54	174.4
	OBIZAA	0.21	0.45	155.9		PAMTAY	0.23	0.52	151.9
	TEHQAY	0.17	0.48	183.7		RIKNOO	0.25	0.54	85.9
	TIJJUR	0.19	0.47	215.4		TIWKIT	0.27	0.57	134.0
	VEHCOA	0.17	0.49	165.8		VILPEL	0.23	0.50	197.0
	WILBAU	0.18	0.37	217.8		WIKZUL	0.29	0.59	85.9
	WILBEY	0.19	0.43	227.1	$C_1^1(3)$	FITQEE	0.26	0.54	112.9
	YEYQOI	0.16	0.35	183.2		GOQXOZ	0.26	0.53	114.1
	YULNUO	0.17	0.34	194.5		INDAZL	0.28	0.59	106.9
$R_4^4(12)$	DIRKOE	0.23	0.47	124.8		LETNAZ	0.26	0.52	118.4
	LADBIB	0.21	0.42	203.6		NEZMOU	0.17	0.47	270.9
	LIYGOP	0.19	0.45	169.1		NIBFIN	0.22	0.52	214.2
	MEPHPY	0.21	0.47	150.8		NOPRUF	0.29	0.61	194.9
	QAMQEA	0.17	0.44	145.2		PUMLUE	0.24	0.53	208.2
	QOFWOX	0.16	0.48	179.0		PYRZOL02	0.29	0.61	63.4
	QOFWUD	0.16	0.47	212.4		TELXUD	0.28	0.59	177.7
	QOFXAK	0.25	0.56	219.1		YAXYIF	0.17	0.39	267.6
	XAHBUD	0.17	0.48	211.0					

[a]Accessible surface over the nitrogen atom with lowest steric accessibility, F(N), sum over both nitrogen atoms, Sum-F, and molecular volume, Mvol. Motif is the Etter graph set symbol for the H-bond network

Table 2 Data for oximes[a]

Motif	Refcode	F(N)	Sum-F	Mvol	Motif	Refcode	F(N)	Sum-F	Mvol
$R_2^2(6)$	ACAMOX	0.17	0.63	104.0		MEPHOX	0.15	0.63	181.2
	AEBHXM10	0.16	0.58	153.4		MRPLNB	0.16	0.59	210.7
	BAPMEK	0.16	0.61	172.3		NITBAL02	0.22	0.76	138.4
	BOCBEA	0.22	0.66	190.4		NUVVAB	0.19	0.63	172.8
	BOCBIE	0.17	0.62	172.9		OXBAZP	0.18	0.65	294.5
	BUMDOX	0.15	0.62	251.5		PMNBOX	0.19	0.66	241.2
	CABXOS10	0.16	0.63	176.8		PORVOH	0.29	0.78	208.0
	CABZUA	0.16	0.58	170.2		PORYAW	0.17	0.54	179.8
	CAMPHX10	0.19	0.67	168.3		QEJCAJ	0.19	0.59	208.4
	CARVOX01	0.18	0.65	169.6		SEKKIC	0.16	0.64	186.2
	CBALOS01	0.21	0.75	128.8		SOLAOX	0.21	0.66	234.2
	CDECOX	0.14	0.55	180.6		SOSDOT	0.16	0.53	171.0
	CDODOX	0.15	0.59	213.6		SOSDUZ	0.17	0.53	185.5
	CISJET	0.09	0.63	238.9		SOSFAH	0.18	0.53	184.8
	CNBZHM	0.18	0.66	150.3		TANMAW	0.16	0.71	161.8
	CTETOX	0.15	0.59	247.5		TMCHOX	0.21	0.67	179.6
	CUNDOX	0.18	0.56	197.3		TMPCHO	0.21	0.62	234.9
	CYCOXA	0.21	0.63	146.9		TUHWEY	0.21	0.76	184.9
	CYNONA	0.20	0.62	163.0		TUHWIC	0.2	0.74	202.3
	DEBAOX	0.12	0.56	189.6		TUHWOI	0.21	0.76	200.8
	DMTOXA	0.23	0.72	203.9		UCULIN	0.17	0.62	149.9
	DMTOXB	0.23	0.71	203.5		VEJROR	0.19	0.74	169.4
	DORKAW	0.15	0.60	187.5		VIMCID	0.17	0.66	203.4
	DOTJUR	0.14	0.49	251.4		VUCJAE	0.11	0.52	218.7
	DOTKAY	0.16	0.51	233.2		WOVROO	0.17	0.71	186.2
	EDMBOX	0.15	0.59	177.2		YABRAU	0.25	0.73	178.6
	EPDSOX	0.18	0.65	261.1		YORWUX	0.16	0.60	207.6
	FAKRIS	0.19	0.66	130.6		YORXAE	0.21	0.75	216.6
	FAPXEZ	0.19	0.74	231.8		YUPGOF	0.12	0.47	173.6
	FAVSIE	0.14	0.62	202.5		ZUKSIH	0.25	0.73	196.2
	FAVSOK	0.14	0.62	207.1	$R_4^4(12)$	ABACOX10	0.22	0.64	149.3
	GADLEC	0.19	0.59	191.3		CAHDAQ	0.20	0.75	114.6
	GIZSOX	0.15	0.62	145.2	$R_3^3(9)$	ACEOXM	0.25	0.72	76.6
	HEVJIB	0.15	0.49	256.5		BAYDUA	0.19	0.66	163.1
	HIQLAU	0.14	0.56	288.0		MZACHX	0.26	0.76	85.0
	HUFXEL	0.15	0.60	212.9		ZZZVPO01	0.24	0.72	114.1
	IBUNIC	0.19	0.65	175.4	$C_1^1(3)$	AANIIOX	0.29	0.71	138.7
	KEVSAF01	0.13	0.54	313.0		BDSOXI	0.22	0.68	140.3
	KITMIJ	0.16	0.63	241.7		CBALOA10	0.29	0.71	127.5
	KOKCIW	0.17	0.62	186.3		DOXZOF	0.29	0.75	121.5
	KOPFUQ	0.12	0.48	228.6		FAKROY01	0.23	0.65	130.2
	KUNDOM	0.22	0.69	163.0		FURDOX10	0.29	0.71	95.0
	LCARVX	0.18	0.66	168.8		IBUNEY	0.22	0.76	159.3
	LOGFES	0.19	0.67	186.5		JIHVEB	0.21	0.69	82.2
	LOGFIW	0.19	0.68	185.5		POTLUF	0.29	0.75	161.2
	LONPUZ	0.10	0.56	271.5		SOJGED01	0.30	0.69	119.4
	LOVLUD	0.16	0.57	274.1		WOVRII	0.30	0.70	159.9
	LOVMAK	0.20	0.59	292.1		XEDTOP	0.29	0.69	105.1
	MABZOX	0.20	0.74	159.3		YUYCOK	0.24	0.77	119.2
	MCHIMB	0.24	0.70	314.1					

[a]Accessible surface over the nitrogen atom, F(N), sum over the nitrogen and oxygen atoms, Sum-F, and molecular volume, Mvol. Motif is the Etter graph set symbol for the H-bond network

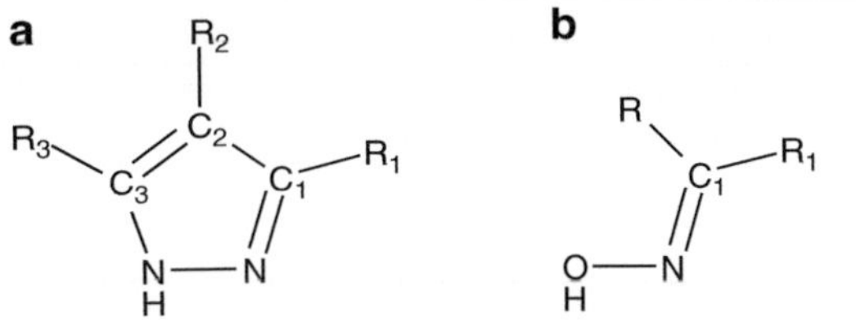

Fig. 2 Diagram of (**a**) 1H-pyrazole and (**b**) oxime molecules.

Pyrazoles were previously studied by Infantes [2] looking for empirical rules to predict the type of motif; a rule was suggested depending on a combination of weights assigned to the substituents R1 and R3 (Fig. 2a), the weights being related to the bulk of the substituent. Bulky substituents adjacent to either side of the—NH-N=group had a definite promoting effect toward dimer or tetramer formation. Conversely, small substituents tended to form trimer or catemer motifs. An attempt to rationalize the fact of the segregation into two groups dimer/tetramer and trimer/catemer was make by Foces-Foces *et al.* [3], who concluded that dimer and nonplanar tetramer motifs are less sensitive to steric effects and the selection was based on steric and polarizability effects. We should clarify here that, with one exception, the tetramer motif is nonplanar, having –4 symmetry or pseudo-symmetry between the molecules.

The steric factors that govern the formation of hydrogen-bonding motifs in the solid state for pyrazoles and oximes are directly related to the idea of aggregation of molecules in solution, prior to joining the crystal surface. If we propose a discriminatory rule, based on the accessible surface of the nitrogen atoms, it means that those structures with low accessibility in these atoms will have to form dimer or tetramer aggregates but, in no case, trimers or chains. Nevertheless, the opposite is not certain. Structures with high accessibility will favor trimers and catemers, but they will also be able to form either dimers or tetramers.

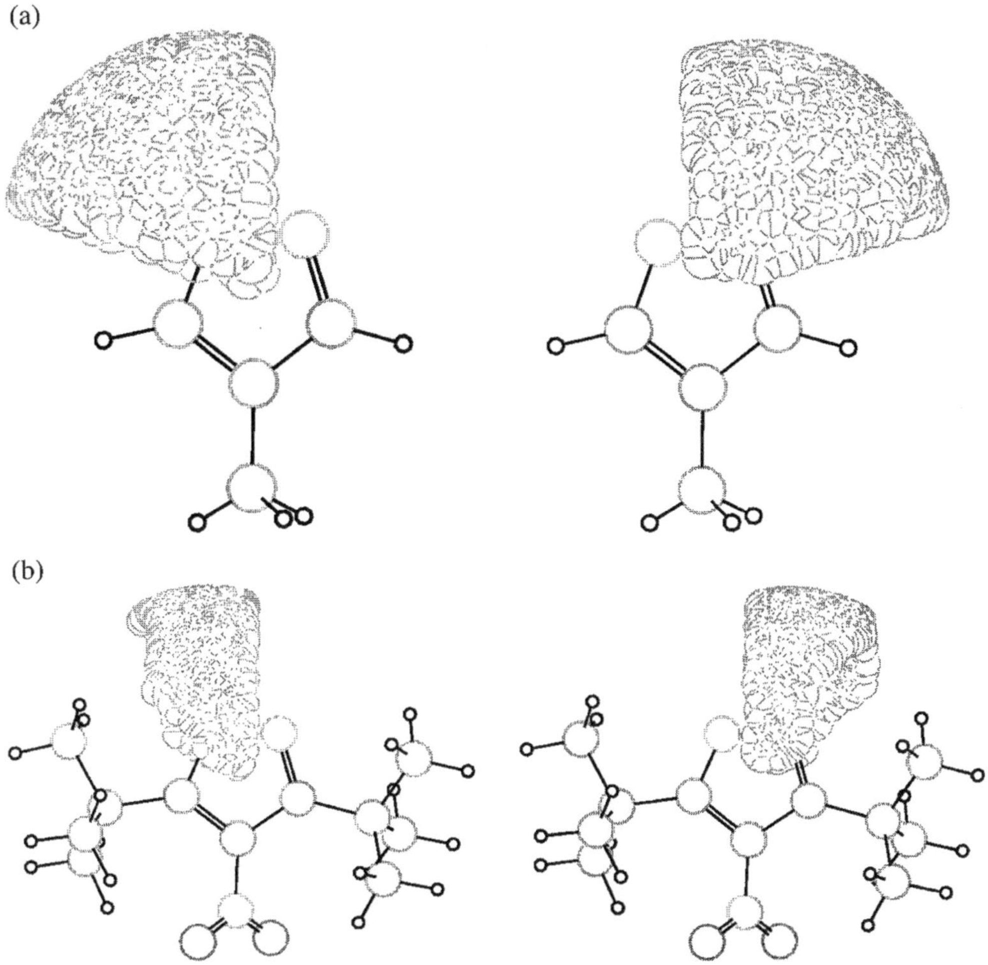

Fig. 3 Display of the available surface probe over the nitrogen atoms for the compounds: (**a**) 4-methylpyrazole (HOQHUQ), high accessibility, F=0.30, 0.31; (**b**) 3,5-di-*t*-butyl-4-nitropyrazole (WILBAU), low accessibility, F=0.18, 0.19.

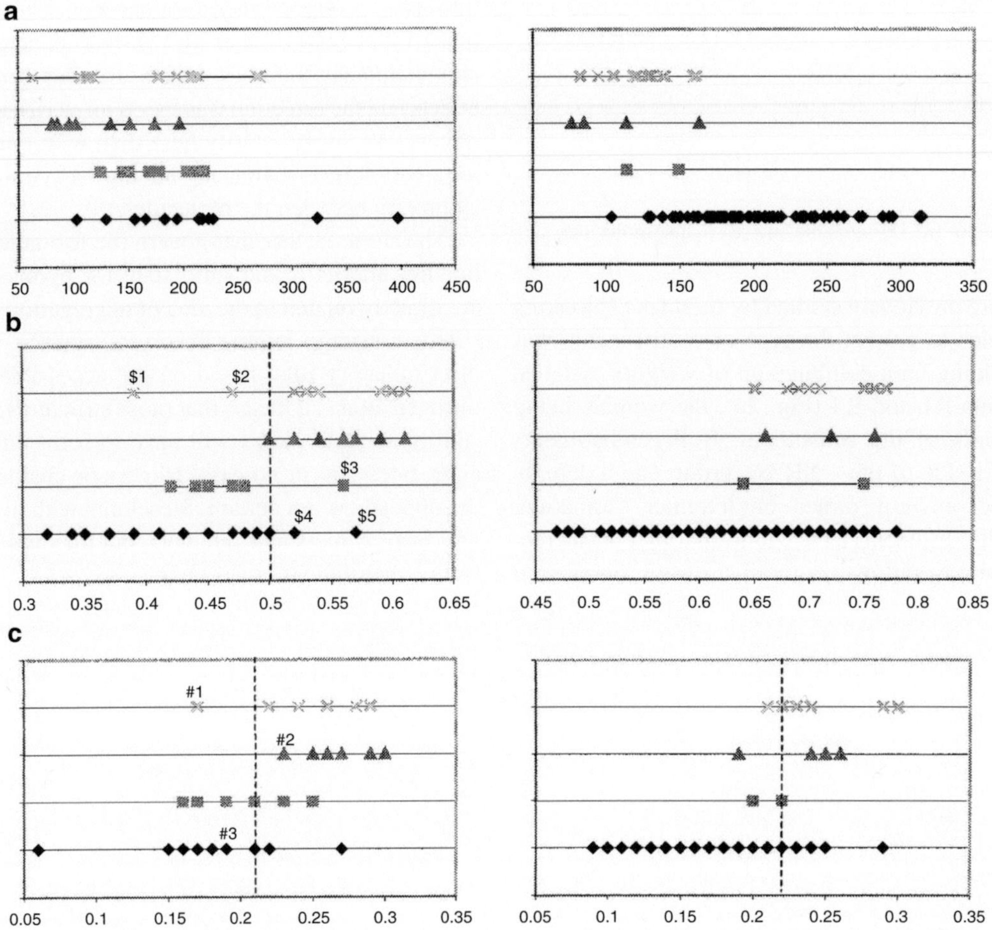

Fig. 4 Diagram showing the distribution of the accessible surface and the molecular volume for pyrazoles and oximes. ◆ Dimer, ■ Tetramer, ▲. Trimer and × Catemer. (**a**) Molecular volume (Mvol) of 1H-pyrazoles (*left*) and oximes (*right*). (**b**) Sum of the accessible surface (Sum-F) for both nitrogen atoms for pyrazoles (*left*) and Sum-F for nitrogen and oxygen atoms for oximes (*right*). The structures marked $ are discussed in the text. (**c**) Accessible surface [F(N)] of the nitrogen atom with lowest steric accessibility for pyrazoles (*left*), and F(N) for the nitrogen atom for oximes (*right*). The structures marked # are discussed in the text.

Pyrazole Results

The sample of 44 pyrazoles was classified into 13 dimers, 11 trimers, 9 tetramers, 10 catemers, and the remaining unclassified structure NIBFIN shows both dimer and catemer motifs in the same crystal structure.

The distribution of the molecular volume in pyrazoles (Fig. 4a), does not differentiate between the hydrogen-bond motifs (dimer: 103–397 Å^3, tetramer: 125–219.1 Å^3; trimer: 80–197 Å^3; and catemer: 63–271 Å^3). We also measured the accessible surface for the adjacent carbon atoms, C1 and C3, but found no correlation with motifs and have not studied these further.

However, the motifs show correlation with Sum-F, the sum of the accessible surface for the nitrogen atoms (Fig. 4b). In an attempt to establish an empirical rule, we can draw a line at the value Sum-F equal to 0.5, which divides the two groups (dimer, tetramer: lower F value; trimer, catemer: higher F value). The selection of this value is completely arbitrary to obtain the best agreement for the segregation into two groups. In this way we have five exceptions ($1–5) and two compounds on the borderline of 0.5, namely, FULKUS (dimer) and VILPEL (trimer). The second one is the most nonplanar trimer in our sample and it is formed from three independent molecules held together through a weak hydrogen bond with long N…N distances (2.940, 2.995, and 3.024 Å).

Examination of the exceptions to this rule of Sum $F < 0.5$ shows the following:

$1, YAXYIF, and $2, NEZMOU, the rule expects a dimer or tetramer motif, but we observe catemers; examination of the structures does not suggest any explanation.

$3, QOFXAK, is a unique arrangement; all other tetramers show a nonplanar –4 symmetry between their molecules, which is planar, formed by four independent molecules.

$4, NIBFIN, dimer and catemer motifs are observed in the same crystal. The discriminatory rule would predict a catemer (Sum-F=0.52) but the dimers are also present in the crystal structure, joining the chains in layers parallel to the plane (10–1).

$5, HEHTUJ, is an example of desmotropy, i.e., molecules are tautomers, crystallizing from different solvents. The dimer, HEHTUJ (3-methyl-4-nitropyrazole), is the exception, and the trimer, HEHVAR (5-methyl-4-nitropyrazole), obeys the rule.

Another attempt to find a discriminatory rule is presented in Fig. 4c, where we choose the nitrogen atom with the lower value of F(N), i.e., the one with higher steric impediment. This is to test the idea that if even one of the nitrogens is sterically impeded, then perhaps the dimer/tetramer motifs will be favored. The diagram does not show any advantage over the Sum-F dividing rule in Fig. 4b in terms of number of exceptions to the rule. However, after studying the molecular structures, we observe that if we draw a line at F(N)=0.21 (dashed line in Fig. 4c), we obtain a trend toward the dimer/tetramer motifs at F<0.21. The trend occurs even though the other nitrogen is highly accessible.

If we look at the structures that were on the borderline Sum-F=0.5 in Fig. 4b, we see that FULKUS (#3) now has value 0.19, in agreement with the idea that dimers should have lower F. The other borderline structure, the trimer VILPEL (#2) has F(N) of 0.23, in agreement with the idea that more accessible N favors catemer/trimer motifs. The exceptions noted in Fig. 4b, YAXYIF ($1) and NEZMOU ($2), both catemers, remain unexplained with values of 0.17, 0.17, at point #1 in Fig. 4c. We note that all other catemers and trimers are above the line at 0.21.

Oxime Results

The oxime compounds show similarities with the pyrazoles, but it is more difficult to establish a simple discriminatory F value empirical rule. The total sample of 101 structures was classified visually into the motifs drawn in Fig. 1, giving 99 classified and the remaining 2 unclassified, utilizing an ether group as acceptor for the hydrogen of the OH. From these 99 structures, 80 form dimer motifs, which is a rather unbalanced distribution in favor of dimers. Among the remaining 19 structures, there are only two examples of tetramer arrangement, which display a nonplanar −4 symmetry between the molecules that form the motif; we include these in the dimer count because of the similarities with pyrazole tetramers. There are four compounds which form trimers and 13 which form catemers.

For oximes, the molecular volume seems to be correlated with the secondary structure (Fig. 4a), in that bulky molecules have a tendency to form dimers; in this sample we have exclusively dimers above 164 Å^3. The accessible surface probe F(N) for the nitrogen atom (Fig. 4c) shows a tendency for dimers to be found at lower F(N), exclusively below 0.18. We show in Fig. 5 the frequency of F values for the motif classes (dimers/tetramers) and (catemers/trimers) for the pyrazoles and oximes. In this distribution we see evidence of two overlapping distributions, perhaps clearer in the oximes than the pyrazoles. With such a small amount of available data, the ripples in the curves are not significant, but we present this visualization to possibly give insight as to trends. As the number of the CSD entries increases [10] these frequency distributions should show smoother curves. If we attempt to find a discriminatory value of F, we always have exceptions to the rule because of overlapping distributions. A suggested value as discriminator to get the best fit according to the

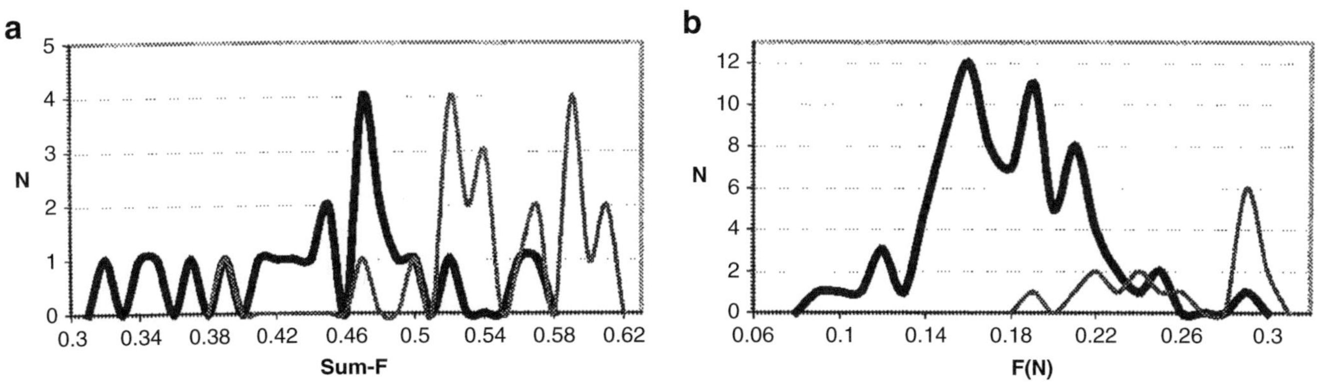

Fig. 5 Diagram showing number of occurrences N of the accessible surface for the nitrogen atoms: (**a**) for pyrazoles, Sum-F, and (**b**) for oximes, F(N). This is a deliberately smoothed version of a histogram in an attempt to visualize the broader distributions. –, Dimer and tetramer; –, trimer and catemer.

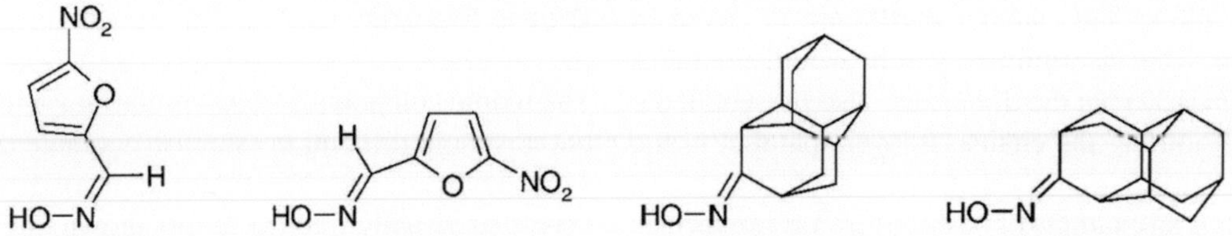

SOJGED01	YUYCOK	DMTOXA	DMTOXB
Catemer	Catemer	Dimer	Dimer
F(N)=0.30	F(N)=0.24	F(N)=0.23	F(N)=0.23

Fig. 6 Examples of stereoisomeric oximes giving the same motif in the crystal structure.

frequency distribution in Fig. 5, is F(N) = 0.22 (dashed line in Fig. 4c), which gives eight compounds breaking the rule (six dimers, one trimer, and one catemer) and six on the line (three dimers, one tetramer, and two catemers). We cannot make reliable predictions statistically with such a small sample, but there is a suggestion that low F values below about F(N) = 0.18 will form dimer/tetramers. The trimer that breaks the rule, BAYDUA, corresponds to the most nonplanar distortion of the trimers and it shows the largest O... N distances. The catemer, which breaks the rule, JIHVEB, shows a bifurcated H bond involving the nitro substituent in *trans* position with respect to the hydroxyl group.

E/Z Oximes

The accessible surface of the nitrogen atom of oximes depends greatly on the configuration about the double bond N=C. Different stereoisomeric forms (*E*/*Z*) can present the same or different motif in the crystal structure (Figs. 6 and 7) The steric impediment on the nitrogen changes, depending what substituent is in *trans* position with respect to the hydroxyl group. More bulky substituents increase the likelihood of the dimer motif.

The subset of 23 oximes with R H (Fig. 2) is interesting. Eight present the *trans* conformation [(OH—N=C—H) = 180°], and 15 the *cis*. In all of these, *trans* structures there is a high accessible surface on the nitrogen [F(N) > 0.29], seven form catemers and one a dimer. Thus, we can conclude that CH-oximes with E configuration (*trans* conformation) are more likely to form catemer or trimer motifs in the crystal.

Comparison of Pyrazole/Oxime Structures

Pyrazoles have a five-membered planar ring that can present proton tautomerism and can form three different substituents (R1-3) (Fig. 2). They have a high tendency to crystallize with more than one independent molecule. Only 16 structures of the 44 analyzed have Z′ = 1 (36 %) and we find structures with Z′ = 0.5, 1.5, 2, 3, 4, and 6. In comparison, oximes are three-membered chains with two possible stereoisomeric forms (*E*/*Z*), and form two substituents (R and R1). Of the oximes, 72 % have only one independent molecule in the cell; 21 % have *Z*′ = 2 and there are few with *Z*′ = 0.5, 3, 4, and 6. The combination R1, R3 = H, Me in pyrazoles and R1 = Me in oximes forms catemer/trimer motifs, whatever the nature of R2 or R, e.g., pyrazoles FITQAA and HEHVAR, oximes ACEOXM, FAKROY01, and MZACHX. The combination R1, R3 = H, Ph in pyrazoles forms catemer/trimer motifs and R1 = Ph in oximes form dimer/tetramer motifs, regardless of whether we have R2 or R, e.g., pyrazole PAMTAY and oximes

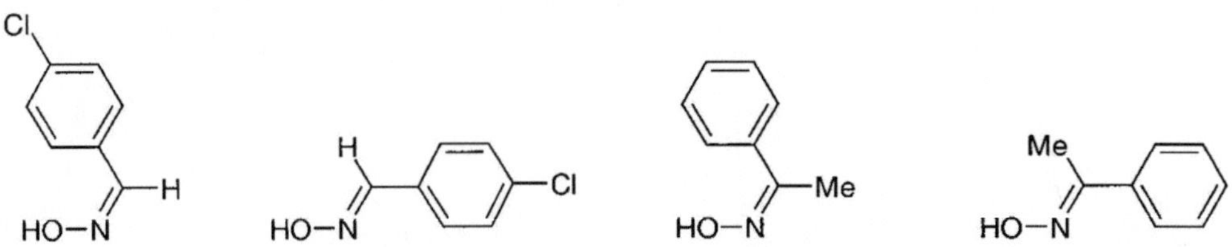

CBALOA10	CBALOS01	FAKROY01	FAKRIS
Catemer	Dimer	Catemer	Dimer
F(N)=0.29	F(N)=0.21	F(N)=0.23	F(N)=0.19

Fig. 7 Examples of stereoisomeric oximes giving different motifs in the crystal structure.

Table 3 CSD reference code[a]

1 H-pyrazole									Oxime (R=H)								
KUFVAI	d	$P2_1/c$	15.1	5.6	19.5	90	111	90	SALOXM	d	$P2_1/c$	10.4	5.1	13.6	90	113	90
RIZYEE	c	*Cc*	11.1	12.5	11.8	90	118	90	WOMXIF	c	$P2_1/c$	16.4	8.4	9.0	90	106	90
									Oxime (R Me)								
RIKNOO	t	*P*-1	10.1	12.5	13.1	100	104	111	JIHVEB	c	$P2_12_12_1$	4.8	8.0	11.2	90	90	90
RIZYEE	c	*Cc*	11.1	12.5	11.8	90	118	90	KOWXOJ	c	$P2_1/c$	8.1	8.1	10.6	90	96	90

[a]Motif (*d* dimer, *t* trimer and *c* catemer), space group and cell parameters *a*, *b*, *c*, (Å), alpha, beta gamma (°) for some comparable pairs of pyrazole/oxime molecules

CAHDAQ, FAKRIS, and LOVLUD. Although there are differences in molecular structure, we have tried to find approximate isostructures using pyrazoles and oximes comparable at the molecular level. We examined R2 and R3 = H in pyrazoles and R = H or Me in oximes (Fig. 2 and Table 3).

However, none of these form isostructures. When R is a methyl group, we obtain greater differences in the packing in the secondary structure and even in the molecular structure-{[N=C—(C—N)$_{\text{pyridine}}$]} = 176.8° and 177.8° for the two independent molecules in RIZYEE and 135.4° for KOWXOJ. The pair KUFVAI/SALOXM fits very well at secondary structure level, but the packing of the dimers is different (Fig. 8). However, looking more broadly at the projections along the *b* axis, we see the similar architectural feature of the two crystals, namely, the H-bonded dimers forming a stacking *b* axis direction of ca. 5 Å arranged differently with phenyl-phenyl hydrophobic surfaces parallel to (100) planes. We can say that these structures show the same principles of construction or architectonics and are homostructural.

We found one interesting example of isostructure between pyrazole and oxime compounds (Fig. 9). These are not in our sample because there is more than one oxime or pyrazole group present. 3,3′-Bipyrazole (HEPJER) and ethanedial dioxime (GLOXIM) crystallize in the space group $P2_1/c$ with cell parameters [*a*, *b*, *c* (Å), (°)]: 5.465, 5.491, 11.203, and 116.3 and 3.868, 4.414, 10.949, and 91.2, respectively. This pair shows the same H-bonding scheme in layers parallel to the plane (100). The structures are very similar (as shown later in the projections in Fig. 9).

During the progress of this work, we noticed several mimetic structures. For example, there is a stacking similarity fortrimerstructuresofthecompounds:3(5),4-dimethylpyrazole (FITQAA) and acetoxime (ACEOXM). FITQAA crystallizes with six independent molecules that form two independent trimers and no symmetry relations other than inversion

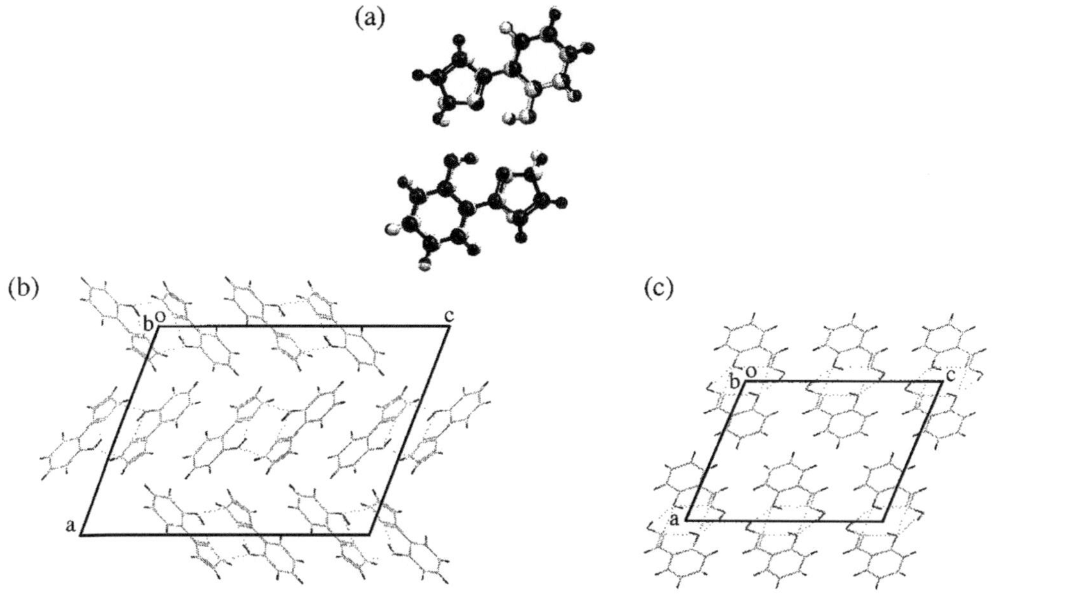

Fig. 8 (**a**) Overlap of the dimers for KUFVAI (*dark*) and SALOXM (*light*). (**b**) KUFVAI projected down *b* axis. (**c**) SALOXM projected down *b* axis.

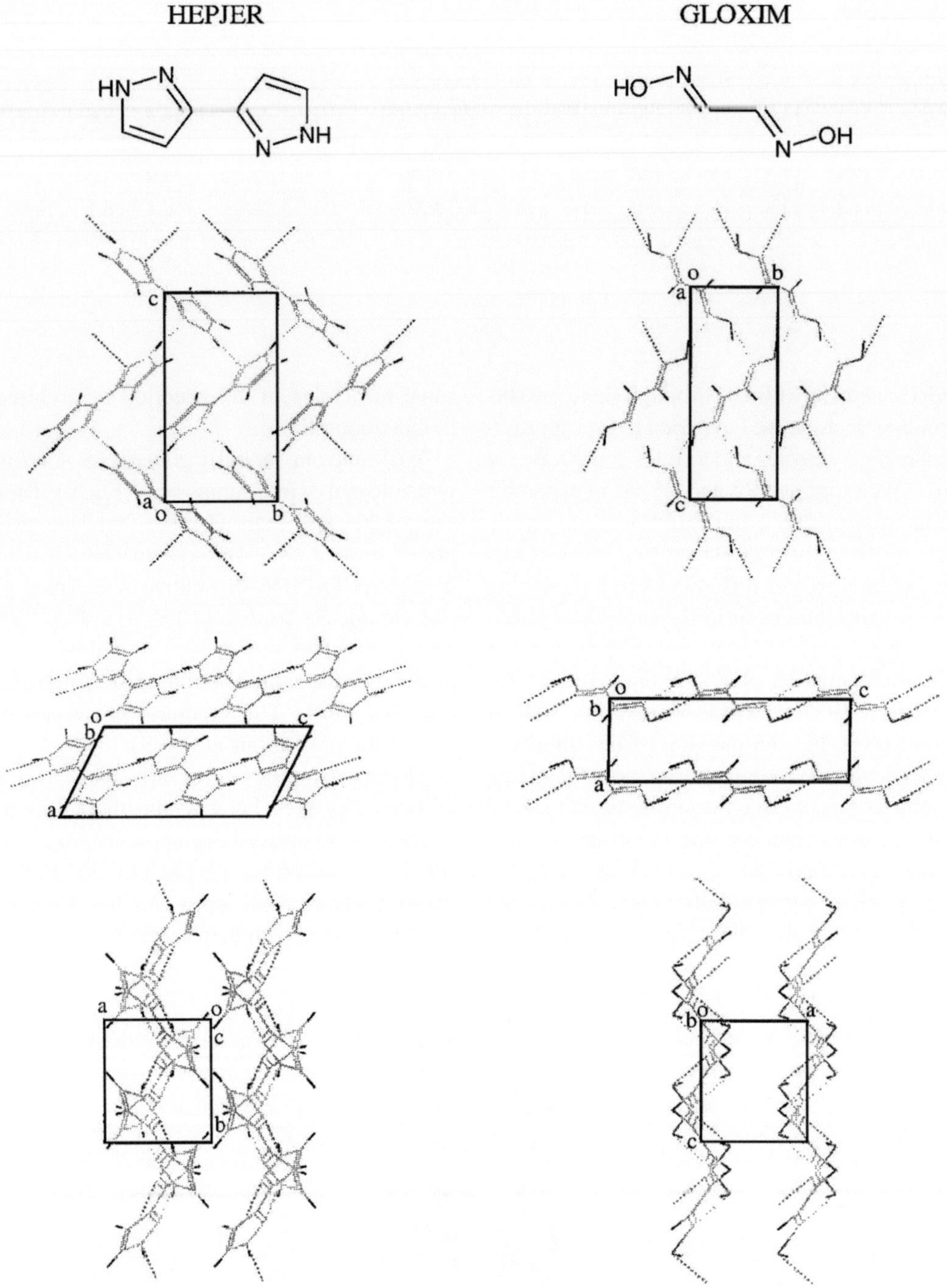

Fig. 9 View of the projections for the compounds 3,3′-bipyrazole (HEPJER) and ethanedial dioxime (GLOXIM).

centers in the crystal (space group *P*-1). ACEOXM crystallizes with one independent molecule in a high symmetry space group, $P6_3/m$, and the molecules in the trimers are related by a threefold axis. In both, the trimers pack forming sheets parallel to the planes (100) and (001) for FITQAA and ACEOXM, respectively, and with hexagonal topology (Fig. 10). The cell parameters for these two structures are [lengths in Å, angles in degrees], (14.4, 12.0, 12.0, 120, 106, and 87) and (10.6, 10.6, 7.0, 90, 90, and 120). If we consider only the planar layers of molecules, we find almost exact hexagonal two-dimensional lattices with area parameters (FITQAA, 12.0, 12.0, and 120, and ACEOXM, 10.6, 10.6, and 120). These are comparable and the differences are due to the different size of the molecules.

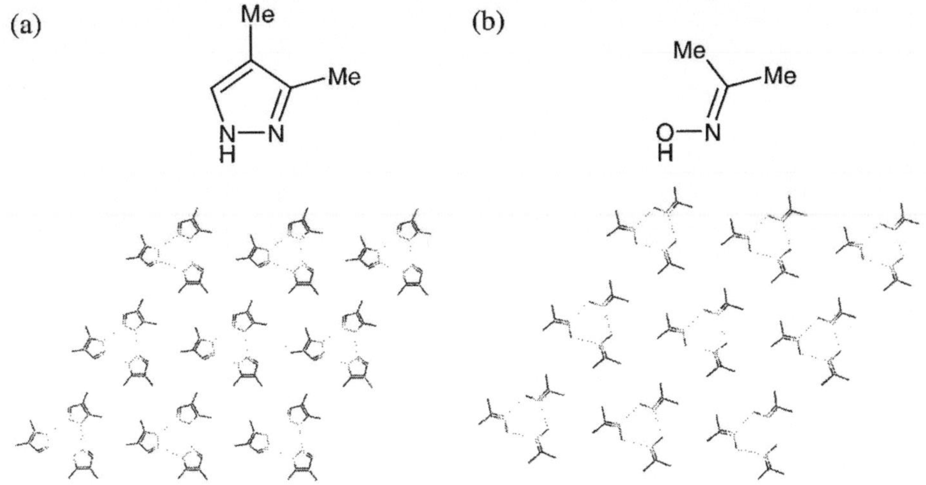

Fig. 10 View of one sheet of trimers for the compounds (**a**) FITQAA and (**b**) ACEOXM. Hydrogen atoms are omitted.

A Predictive Test

To be useful, an empirical rule has to explain the given sample observations and then to predict the secondary structure of new compounds. Before we finished this work, a new version of the CSD became available (CSD release 5.24, November 2002); this includes one new pyrazole and seven new oximes. The F(N) values are given in Table 4.

Using the empirical rules shown in Table 5, all the new compounds should form dimers or tetramers with the exception of BAGWIQ, which should form trimers or catemers. This prediction is correct; BAGWIQ shows catemers and all others show dimers.

Energetic Considerations

It is not to be expected that an empirical rule should function with 100 % reliability, since the physical basis for the rule is to concentrate only on the N atom available for H bonding, and all other factors should be ignored, such as C—H...O bonding and polarizible groups. The energetic basis for the formation of the crystal structures is that the crystal-packing energy should be as low as possible, which requires not only the satisfaction of all H-bond donor atoms by acceptors in a preferred geometric manner, but also the close packing of the rest of the molecule. In this limited sample, with only one H-bonding group, the hydrophobic parts of the molecule must be in close contact with each other. Even approximate empirical calculations of the packing energy show that the H-bonding contribution is not necessarily a major component, e.g., if the NH···N bond contributes about 3 kcal-mol^{-1}, we have only a contribution of about 6 kcal-mol^{-1} to the crystal-packing energy of around 50 kcal-mol^{-1}. Nevertheless it is the norm for the H-bond donors to be satisfied in all these structures and the motif is the result of finding the region of lowest energy in the total packing. The rationalization of the effectiveness of the rule is that as the steric availability of the N decreases, it becomes more difficult to geometrically form the catemer motifs in the sort of chain geometry with low energy; thus the other options of dimers, trimers, or tetramers are formed. As the total molecular volume increases, one might expect the packing requirements of the hydrophobic part of the molecule to predominate, which seems to favor dimers, as was observed for the oxime sample.

Conclusions

It has been shown that for pyrazoles and oximes there is a strong relation between the motif and the accessible surface over the atoms involved in the H bond. There is a tendency for the trimer or catemer motif to be favored by high-accessibility donor/acceptor atoms (Fig. 4). A simple dividing rule can be established at the values of Sum-F = 0.50 for pyrazoles and F(N) = 0.22 for oximes, which performs well for pyrazoles, but less so for oximes.

Table 4 CSD reference code[a]

1H-pyrazoles				Oximes								
CAMFUS	d	0.20	0.21	ADOJUY	d	0.15	BAGWIQ	c	0.30	OFIWOP	d	0.18
				ADOKAF	d	0.15	BAGWOW	d	0.21	UFEPAW	d	0.14
				ADOKEJ	d	0.14						

[a]Motif (*d* dimer and *c* catemer), and available surface on the nitrogen atoms for new compounds, F(N)

Table 5 Summary of the empirical available surface rule for pyrazoles and oximes

	Dimer or tetramer	Trimer or catemer
1H-pyrazoles	Sum-F<0.5	Sum-F>0.5
Oximes	F(N)<0.22	F(N)>0.22

It was also found that for oximes the molecular volume parameter has a tendency for large volumes to favor dimer motifs.

The application of this rule to a small number of "new" compounds, not known at the time of the formulation of the rule, was successful in this case. Although the samples were small and the absolute statistical probabilities are low, it is hoped that as the CSD grows in size the predictive power of such rules will increase and help in designing crystals with a required H-bonding motif.

Acknowledgments Lourdes Infantes acknowledges the Ministerio de Educacion y Cultura of Spain for a postdoctoral fellowship.

References

1. Kerr, L. C. PhD Thesis, University of Cambridge, 2001.
2. Infantes, L.; Foces-Foces, C.; Claramunt, R. M.; López, C.; Jagerovic, N.; Elguero, J. *Heterocycles* **1999,** *50,* 227.
3. Foces-Foces, C.; Alkorta, I.; Elguero, J. *Acta Crystallogr.* **2000,** *B56,* 1018.
4. Bernstein, J.; Davis, R. E.; Shimoni, L.; Chang, N-L. *Angew. Chem. Interm. Ed. Engl.* **1995,** *34,* 1555.
5. Leiserowitz, L.; Schmidt, G. M. J. *J. Chem. Soc. A* **1969,** p. 2372.
6. Leiserowitz, L. *Acta Crystallogr.* **1976,** *B32,* p. 775.
7. Allen, F. H.; Raithby, P. R.; Shields, G. P.; Taylor, R. *Chem. Commun.* **1998,** p. 1043.
8. Allen, F. H.; Motherwell, W. D. S.; Raithby, P. R.; Shields, G. P.; Taylor, R. *New J. Chem.* **1999,** *23,* 25.
9. Kuduva, S. S.; Craig, D. C.; Nangia, A.; Desiraju, G. R. *J. Amer. Chem. Soc.* **1999,** *121,* 1936.
10. Allen, F. H. *Acta Crystallogr.* **2002,** *B58,* 380; Bruno, I. J.; Cole, J. C.; Edgington, P. R.; Kessler, M.; Macrae, C. F.; Pearson, J.; Taylor, R. *Acta Crystallogr.* **2002,** *B58,* 389.
11. RPluto, A program for crystal structure visualisation, Cambridge Crystallographic Data Centre, Cambridge, 2002 (http://www.ccdc.cam.ac.uk, freely downloadable for noncommercial use)
12. Bondi, A. *J. Phys. Chem.* **1964,** *68,* 441.

Hierarchies of Intermolecular Potentials and Forces: Progress Towards a Quantitative Evaluation[a]

Angelo Gavezzotti[b]

Abstract

Conventional views of intermolecular cohesion, based on the traditional categories of hydrogen bonding, of aromatic interactions, of dipolar or quadrupolar contacts, and of the broad, gray zone of 'van der Waals' liaison, often define strength hierarchies on the basis of qualitative categories like approximate molecular orientations or distances between atomic nuclei in molecules. When interaction energies are quantitatively evaluated between molecular pairs, in a more justifiable partitioning scheme, often a completely different picture emerges. Examples are given for selected molecular dimers and organic crystals, using a new semiempirical scheme, the PIXEL method, which also allows a separate evaluation of coulombic, polarization, dispersion and repulsion energy terms.

Keywords

Molecular dimers • Organic crystals • Hydrogen bonding • PIXEL method

Introduction: Hierarchy and Significance of Intermolecular Bonds

All chemical cohesion results from nuclei and electron clouds attracting or repelling one another. Chemists usually distinguish between an intramolecular and an intermolecular [1] case, but the laws of our physics do not see a difference between intramolecular and intermolecular bonding. In its simplest form, the Hellmann-Feynman electrostatic theorem states [2, 3] that in the Born-Oppenheimer approximation nuclei see a static, smeared out electron cloud, and that the forces acting at nuclei are just coulombic forces exerted by other nuclei and by the electron cloud. In more plain words, this can be taken as meaning that any chemical bond or bonding can be explained in pure coulombic terms, as a matter of electrons being delocalized between nuclei, so that more ± attraction than +/+ and −/− repulsion develops. The quantum mechanical Hamiltonian contains just 1/R terms and there are no strange, hidden quantum mechanical forces around.

In fact, the Lewis electron pair theory of chemical bonding (*c.a.* 1920) was a distant echo of this. In more modern terms, Bader's Atoms in Molecules theory (AIM; *c.a.* 1980 [4]) defines atomic basins and bonds in terms of the topological properties of the electron density $\rho(r)$, its gradient $\nabla\rho(r)$, and its Laplacian $\nabla^2 \rho(r)$. Points where $\nabla\rho(r)=0$ are called critical points. Types of critical point depend on the signature (the algebraic sum of signs of eigenvalues of matrix of second derivatives): if the signature is (3, −3), all curvatures are negative and one has a local maximum in the density which corresponds to nuclear positions; if the signature is (3, −1), two curvatures are negative and one is positive, corresponding to saddle points occurring between each pair of nuclei linked by a chemical bond. Given the rigorousness of the proceedings, this recognition of chemical bonding is not a speculation, but a truth in terms of the first principles of the physics we know—and a truth that does not distinguish between intra- and intermolecular occurrences.

That apparently settles the matter. But sometimes truth, rather than providing an advancement in our ability to understand and control the outer world, is just an ornament

[a]*Structural Chemistry* 2005, 16(3):177–185.

[b]Dipartimento di Chimica Strutturale e Stereochimica Inorganica, Università di Milano, via Venezian 21, 20133 Milan, Italy
e-mail: angelo.gavezzotti@unimi.it

I. Hargittai, B. Hargittai (eds.), *Science of Crystal Structures: Highlights in Crystallography*, DOI 10.1007/978-3-319-19827-9_29

of beautiful appearance but of limited use. In intramolecular matters, AIM belongs to the first category [5], while in intermolecular matters it sometimes tends to melt into an indistinct tool in the hands of less critical users. Firstly, our wavefunctions are never exact and one becomes quickly entangled with problems of basis set adequacy and superposition, and in the treatment of electron correlation. Furthermore, given the diffuseness of the wavefunction, intermolecular saddle points are found all too often, and everything seems bound to everything, unless the recognition of the saddle point is corroborated by an appreciation of the actual value of the density there. A saddle point with insignificant weight is a bond of zero strength—an oxymoron; and one has to be critical [6].

Enter, then a different concept: the concept of chemical significance. As Pauling considered the matter, "We shall say that there is a chemical bond between two atoms or two groups of atoms in case that the forces acting between them are such as to lead to the formation of an aggregate with sufficient stability to make it convenient for the chemist to consider it as an independent chemical species [7]." Elaborating a little more, we conveniently say that two atoms are joined by a chemical bond when: (1) the electron density of one is significantly perturbed by the approach of the other; (2) as a consequence, the energy of the system is lowered and a restraining force acts against separation of the nuclei; (3) hence, the nuclei are constrained at a distance which is significantly smaller than the sum of the bare atomic radii. As corollaries, we may consider that (a) what other atoms are doing is less important (a true bond is a 1 ··· 2 effect, without 1 ··· 3 and 1 ··· 4 dependence); (b) the bond line is identified as the line joining the nuclei, and thermal motion is along that line; (c) when a bond is broken, chemical and structural consequences are important, that is, a large destabilization follows and other bonds must immediately be formed to, so to speak, repair the damage. This last corollary is the heart of chemistry—bond breaking and bond making, the cut and paste job that transforms petrol into shopping bags or water and carbon dioxide into sugar.

Is it convenient to identify bonds between atoms in intermolecular interactions? Such a way of thinking is familiar to the chemist from the intramolecular context, and lends itself to the easiest of the categorizations: a bond is identified when two atomic nuclei are approximately arranged in a favorable way for some kind of electronic interaction to occur, that is, mostly, when they are somewhat close to one another. The indeterminacy and possible subjectivity of these geometrical criteria are evident. Sometimes, more solid arguments are invoked, like the depletion of electronic interaction in some intramolecular bond as a consequence of the formation of the postulated intermolecular bond, as detected for example by quantum chemical calculations or by spectroscopic methods. The lengthening of a O—H bond on formation of an H···O hydrogen bond is one such example. Valid and adequate as these arguments may be, they still leave undetermined the question of hierarchy, or relative significance of the intermolecular bond in the overall molecular recognition process.

In a somewhat different perspective, we start by observing that when two molecules are being attracted towards mutual recognition, an overall electrical force acts between the two separate electron clouds, which then modify upon coming into closer contact until they merge into the new wavefunction of the molecular pair. The extent of the modification upon merging depends on the chemical nature of the partners: it is largest for a strong hydrogen bond as in a ($HCOO^-$ H_3O^+) pair, and smallest for two strictly closed-shell methane molecules (we need be reminded at this stage that a hydrogen bond is considered 'strong' when its energy is in the 80–130 kJ mol^{-1} range [8], which implies that anything below that is considered as moderate to weak). Given the exact wavefunction one could calculate the electron density and the Hellmann-Feynman forces; since this is never the case, a different approach is taken and one usually subdivides the interaction into a coulombic plus polarization part, a dispersion part, and a repulsion contribution.

In any case, the two molecules reach equilibrium when repulsion balances attraction. Since these forces involve the entire electrical system of the molecule, all nuclei and all of the electron density, the positions of the nuclei of outer atoms should normally be seen as the result, rather than the cause of this equilibrium and the definition of intermolecular atom–atom bonds has to be justified on a rather exceptional basis. After all, this is the deep root of the concept of close packing, and the only molecular system which clearly and significantly violates the close packing principle is water (to the point that the crystal is less dense than the liquid, a fact that applies to no other substance, to this author's knowledge). Although they may qualify as bonds according to some geometrical, quantum chemical or spectroscopic definition, very few of these intermolecular atom–atom bonds qualify as *significant* chemical bonds, that is, (1) surviving stretching, and (2) of predictable formation. The medium-weak O—H···O or N—H···O hydrogen bond between organic acids and amides (20–40 kJ mol^{-1}) does, but, to be more specific, calculations show that even in molecular systems with hydrogen bonding capability, these bonds may not be first in the hierarchy of intermolecular cohesion forces.

We enter an era in which a large part of frontier chemistry deals with intermolecular interactions, for such diverse fields as the control of the polymorphism of pharmaceuticals [9] or protein folding. There is a need for the replacement of qualitative thinking on the implied forces by reasonably accurate but affordable methods of quantitative evaluation. We describe in this paper some calculations that may provide a quantitative basis to the problems of hierarchy and of relative significance in intermolecular bonding.

Methods

The PIXEL Procedure

The calculation and analysis of intermolecular forces by theoretical methods has of course a long history, and a complete review would be far beyond the scope of the present work. Nevertheless, in perspective, mention should be made of former attempts at calculating and partitioning these energies using test dimer systems: extensive quantum chemical calculations by Radom and coworkers [10], Scheiner et al. [11], Hobza et al. [12], Kim et al. [13]; Stone's IMPT (intermolecular perturbation theory) [14], and SAPT (symmetry-adapted perturbation theory) [15]. All these calculations involve heavy computing resources, and the results are sometimes quite dependent on the choice of the basis set and of the method for the representation of electron correlation, i.e., the dispersion term, MP2 being the cheapest, and DFT perhaps the most expensive. The problem of the treatment of the basis set superposition error (BSSE) also hampers a clear and unequivocal assessment of the implied interaction energies. Many efforts have also been devoted to the development of computational methods for the analysis of crystalline systems, using distributed multipolar models [16] or energy decompositions according to the variation-perturbation scheme [17]. Periodic orbital calculations are embodied in the widely used CRYSTAL computer program package [18] (this last reference is also a review on the available methods for the calculation of intermolecular potentials).

We use in this paper a recently developed formulation called the SCDS-PIXEL approach [19, 20] (a more appropriate name would be VOXEL, since the methods involves volume elements rather than surface elements, but the former denomination has already gained some common use and is now too late to change it). The method is based upon the calculation of a molecular electron density by standard quantum chemical programs, and on the description of each molecule in an intermolecular energy calculation by the positions of the atomic nuclei plus some 10,000 charge sites (e-pixels), each one of them being derived from a partial condensation of the charge distribution represented by that electron density. These positions and sites are repeated in space in a rigid manner to form a molecular cluster (e.g., by space group symmetry operations in a crystal). Then, the coulombic energy is a sum of coulombic terms over e-pixels and nuclei (a parameter-less calculation); the polarization energy is a sum of linear polarization terms over e-pixels endowed with a local polarizability derived from additive atomic polarizabilities (the calculation involves one disposable screening parameter); the dispersion energy is a sum over pixel–pixel London-type terms (involving a second disposable screening parameter); and the repulsion energy is proportional to the overlap integral between electron densities, requiring two more disposable parameters. The values of these parameters, as well as the atomic polarizabilities and atomic radii adopted, are described in previous work [21].

The strong points of the PIXEL formulation are the delocalization of the electrical description of the molecule, the small number of disposable parameters, the many-body character of polarization, the subdivision of total energies into clearly identifiable coulombic or dispersion terms, and the considerably high accuracy against its ready applicability; calculations on dimers take seconds, calculations on crystals a few minutes to 1 hour on standard PC's. The main weak point is the assumption of a completely rigid approach between closed-shell repeated electron densities, which neglects any electronic rearrangement that might occur after molecular contact and represents the corresponding stabilization by just a linear polarization term. For reasonably closed-shell contacts, like, say, between hydrocarbon molecules, the assumption is acceptable although it forbids a dynamical evaluation of polarizability. For more complicated cases—typically, the hydrogen bond—the assumption introduces an intrinsic error which grows larger with the extent of charge transfer and-or covalency in the system. Nevertheless, thanks to careful parameterization, the theory has proved effective also for the treatment of the medium-weak hydrogen bond (the cyclic dimer of formic acid being probably the upper limit of covalency the theory can deal with). It is also clear that the PIXEL energy terms are not 'the' coulombic, polarization and dispersion terms; when intermolecular interactions are dissected into separate contributions, each method defines its own partitioned energies. But comparisons of the different terms over similar systems, for example crystal polymorphs, can be much more revealing than comparisons of total energies.

In spite of all these caveats, the performance of the PIXEL method is often outstanding for its extremely reduced price in computational effort: for example, the PIXEL results [21] for the hydrogen-bonding dimer found in the urea crystal (requiring seconds of CPU time) are almost identical to the results obtained in accurate variation-perturbation (VP) quantum chemical calculations [17] at the complete basis set limit: coulombic, −53.7 against −59.0; exchange repulsion, 41.0 against 38.9; dispersion plus polarization, −26.4 against −24.3 for the sum of the other contributing terms in VP, for total interaction energies of −39.2 for PIXEL and −44.4 kJ mol^{-1} for VP.

Analysis of Crystal Packing

Instead of looking for short atom–atom distances and presuming 'bonds,' the new view in terms of whole molecules shifts the focus of the analysis from the positions of the nuclei (which in themselves are physically irrelevant) to the electron density (which is the physically relevant quantity), or, in other words, from atom–atom chemical bonds to molecule–molecule chemical bonding. Molecular crystals of course provide an ideal ground for such studies, as they portray molecular recognition in action. The analysis of crystal packing proceeds by identifying a reference molecule in the crystal (called REF), and then all the molecular pairs formed by this molecule with surrounding molecules (labeled A, B, C, D, …), which contribute a significant amount of energy to the total lattice energy. These pairs are called structure determinants, and are identified by a symmetry operator joining the two molecules in the pair, by a distance between centers of mass and, when applicable, by an angle between main molecular planes. For such determinants, the coulombic, dispersion and repulsion interaction energies can be easily calculated, as they all result from pair-wise summations. The polarization energy is a many-body term because the electric field felt by any one molecule in the crystal depends on the vector sum of all fields from all surrounding molecules in the crystal, and as such cannot be partitioned over a molecular pair. One can choose between two alternatives: the first is to calculate just the electric field due to molecule A, B, C, … on the reference molecule, and the resulting energy of polarization of the reference molecule. Since, however, the calculation actually refers to a hypothetical gas-phase dimer extracted from the crystal, the second alternative is to calculate the complete polarization energy, that is, the sum of the REF → A and A → REF polarizations. This second alternative is probably more adequate especially when total energies have to be compared with those calculated for other, hypothetical dimers not appearing in the crystal. In both alternatives, the total energy over the determinants is not equal to the total lattice energy.

It goes without saying that the overall equilibrium crystal structure results from the combined action of intermolecular forces originating from all the molecules in the packing sphere, i.e., none of the structure determinant dimers need be a separate energy minimum. Nevertheless, ranking total energies of structure determinants is *ipso facto* an assessment of the hierarchy of strength of the observed recognition modes; these energies may also be compared with optimized energies of hypothetical dimers, for more understanding. Furthermore, ranking the partitioned energies provides information of the kind of forces at work (mainly, coulombic, directly related to the charge distribution, versus dispersion, related to the bulk electron density). These separate terms can be in turn related to chemical constitution and molecular structure, and a knowledge of their relative importance can be in principle used for manipulation of either constitution or structure towards a desired crystal structure—quantitative crystal engineering.

Three Crystal Structures

We illustrate in this section the potential and reliability of the molecule–molecule analysis of crystal packing using PIXEL calculations of partitioned energies, in comparison with more traditional atom–atom or anyway qualitative electrostatic considerations. We chose three symbolic crystal structures, naphthalene [22], a pure hydrocarbon (space group $P2_1/a$, $Z=2$), naphthoquinone [23], as a polar substituted derivative (space group $P2_1/c$, $Z=4$), and 2-naphthoic acid [24], as a hydrogen-bonding derivative (space group $P2_1/n$, $Z=4$). All calculations were carried out using the comprehensive program package OPiX [25].

Naphthalene

The structure determinants are shown in Figs. 1 and 2, while the corresponding energies are reported in Table 1. The largest interaction (structure determinant A) is between molecules at a high interplanar angle, 52°; structure determinant C is at the same interplanar angle, but at a higher distance and with a much smaller interaction energy. Determinant B is between parallel molecular planes, but highly offset, and the interaction energy is much smaller than the one between perfectly overlapping rings (dimer PS in Table 1 is the optimized parallel stacked dimer with exactly overlapping molecular planes). Dispersion energy is seen to dominate all these interactions, and the sequence of increasing distance between centers of mass follows almost exactly the sequence of descending dispersion contributions. The PIXEL dispersion energies for aromatic hydrocarbons have been shown [21] to be reliable, following closely the values obtained by intermolecular perturbation theory [1] calculations and being, if any, underestimated. Coulombic terms are reduced, in the PIXEL view, to minor modulators of intermolecular arrangement, far from providing any substantial driving force for cohesion. Besides, the same coulombic contribution obtains in dimers with nearly perpendicular (A) or with parallel (B) molecular planes, thus negating any major relevance of C—H···π electrostatic interactions in this crystal. The C—H···π interaction is hard to define anyway, because H···C distances span a continuum and the arrangements of aromatic rings against C—H bonds show no detectable pattern. Neither do in this picture apply simple electrostatic concepts, e.g., molecular quadrupoles. Once the basic molecular interlocking motif in the first coordination shell (12 molecules) has been satisfied, other, second-nearest neighbors (Fig. 2) are energetically much less important (Table 1, determinants D–F).

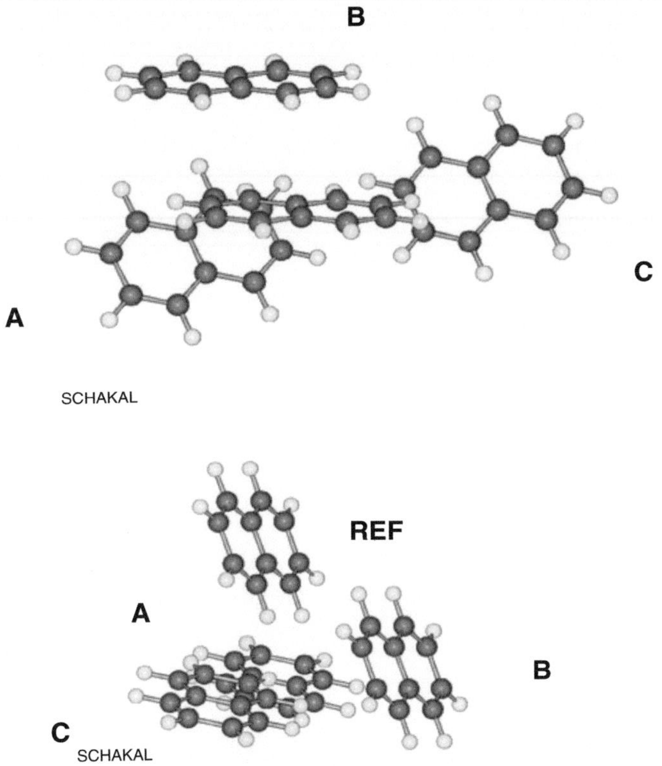

Fig. 1 Two views of structure determinants *A*, *B*, *C* in the naphthalene crystal. *Above*: the reference molecule is seen edge-on at the center of the picture, straddling the *A* and *C* molecules. *Below*: shows the large offset in the REF-B stacking interaction. Symmetry codes: *A*, screw axis, 1/2−*x*, 1/2+*y*, −*z*; *B*, translation, *x*, 1+*y*, *z*; C, screw axis, 1/2−*x*, 1/2+*y*, 1−*z*.

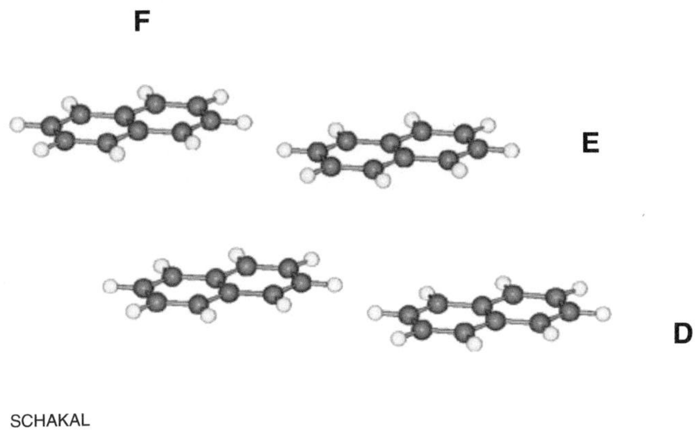

Fig. 2 The *D*, *E* and *F* structure determinants in the naphthalene crystal. All molecules are related by pure translation and parallel: *D*, *x*, *y*, 1+*z*; E, 1+x, *y*, 1+*z*; F, 1+*x*, *y*, *z*.

For further comparison, Table 1 also reports the optimized energy for the T dimer with perpendicular molecular planes, and no offset (overlapping centers of mass). In its crystal the naphthalene molecule disregards both the perfect parallel stacked and the T recognition modes, choosing instead a structure that largely offsets the first (B) and in which one molecule straddles two neighbor semiperpendicular molecules (A and C). Table 1 also shows that a parallel, perfectly stacked arrangement would be largely more favorable than any one of the dimeric arrangements found in the crystal; but this PS arrangement would leave only rim hydrogens exposed for further propagation into a 3-dimensional structure, with consequent destabilization, or, at least, lack of stabilization. It has been shown [26] that a naphthalene crystal structure with all parallel planar molecules would result in a very large loss of stabilizing coulombic interactions. Thus the coulombic contribution, although minor, does make a difference, but only in a bulk sense, and it would be extremely difficult, if not impossible, to use coulombic energy criteria to predict the detail of molecule–molecule arrangements in the naphthalene crystal. In fact, the crystal structure of naphthalene is a complex compromise between many demands, in a dosage that defies any identification of local aspects, and is totally unpredictable on the basis of simple geometrical or electrostatic reasoning. The only way to crystal structure prediction is an accurate calculation of the total lattice energy. A quite similar analysis applies to the benzene crystal [27].

Naphthoquinone

The structure determinants are illustrated in Figs. 3, 4, 5, and 6, while their energies are collected in Table 2. As is immediately evident (Fig. 3), naphthoquinone takes full advantage of the very favorable, dispersion-enhanced, parallel stacked recognition mode. Molecules are related by a center of inversion, but, contrary to naive thinking, the stacking arrangement is not stabilized by dipolar interactions (although $\mu = 1.30$ D), because the corresponding coulombic energies are very small. The A and B determinants alone take up more than one third of the total cohesive energy of the crystal.

Table 1 Structure determinants in the naphthalene crystal (A–E) in comparison with the optimized energies of some ideal dimers (PS and T)

Label[a]	Distance[b]	Interplanar angle	E_{coul}[c]	E_{pol}	E_{disp}	E_{rep}	E_{tot}	E_{qq}[d]
4 A	5.08	52	−5.1	−2.9	−23.5	13.2	−18.3	−1.2
2 B	5.97	0	−4.5	−1.1	−14.0	6.4	−13.2	−2.6
4 C	7.84	52	−1.9	−0.9	−7.7	5.0	−5.6	−0.4
2 F	8.65	0	−1.1	−0.6	−6.0	3.7	−4.0	−0.1
2 D	8.00	0	−0.2	0.0	−1.1	0.0	−1.4	−0.2
2 E	8.21	0	0.2	0.0	−0.6	0.0	−0.4	0.2
PS	3.50	0	−0.4	−5.0	−51.1	25.0	−31.6	7.9
T	4.88	90	−13.0	−7.9	−30.7	32.0	−19.6	−3.5
Lattice	–	–	−19.6	−8.8	−87.9	46.5	−69.8[e]	−6.4

[a]A-E, labels as in Figs. 1 and 2, the number is the multiplicity of the dimer in the crystal structure. Idealized dimers: *PS* parallel stacked, perfectly overlapping molecular planes; *T* perpendicular molecular planes
[b]Distance between centers of mass, Å
[c]PIXEL partitioned energies, kJ mol^{-1}
[d]Coulombic energy calculated with point-charge atom–atom parameters
[e]Sublimation enthalpy 72 kJ mol^{-1}

Why this is possible in naphthoquinone, differently from naphthalene? Lateral inter-stack interactions in naphthoquinone are not a problem, thanks to the possibility of favorable adhesion between positively charged C—H moieties and oxygen basins. In fact, determinants C, E, F show just that (Fig. 4): the interaction is sustained by its coulombic component, and short H···O distances result (2.47–2.52 Å), marginally shorter than the sum of accepted average interatomic radii in crystals, 2.68 Å [28]. Whether one would like to call these interatomic arrangements 'C—H···O hydrogen bonds' depends on the rather subjective choice of drawing a connecting bonding line between the implied atoms, or on a quantum mechanical calculation of AIM bond paths. In the PIXEL view, in partial favor of this localized bond picture stands the fact that there is a relatively large overlap of electron density, as appears from the implied repulsion energies. Determinant D (Fig. 5), with a very similar molecular arrangement, shows a H···O distance, 2.87 Å, marginally longer than the previously mentioned sum of atomic radii; in the literature, such distances are sometimes called bonds, sometimes not. Determinant G, only marginally less stabilizing than determinants E—F, has a similar approach of a C—H bond towards a C=O bond but with a significantly longer H···O distance, 3.35 Å. The 'C—H···O bond' is thus a very easily stretchable bond. In any case, the analysis of the hierarchy of intermolecular interactions shows that the C—H···O bond, if at all definable, is far from being decisive in determining the crystal structure of naphthoquinone. After the formation of the stacking modes A and B, there is a continuum of possible C—H···O distances and interactions. A computer experiment [26] shows that in 100 computational polymorphs of naphthoquinone, the number and length of short C—H···O contacts does not correlate with either the coulombic or the total lattice energies.

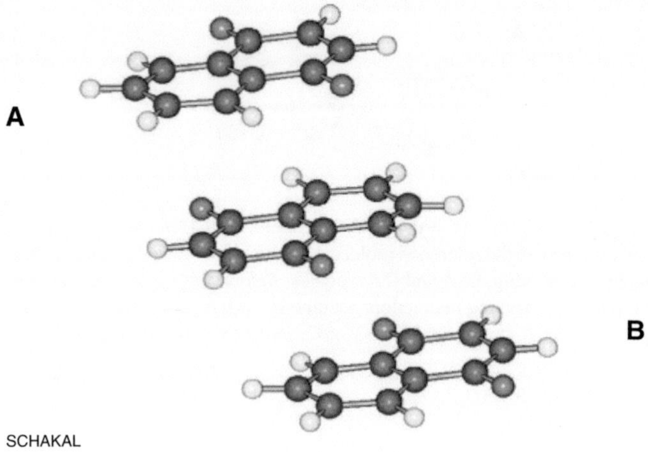

Fig. 3 The *A* and *B* structure determinants in the naphthoquinone crystal. Symmetry codes: A, $-x, -y, -z$; B, $1-x, -y, -z$.

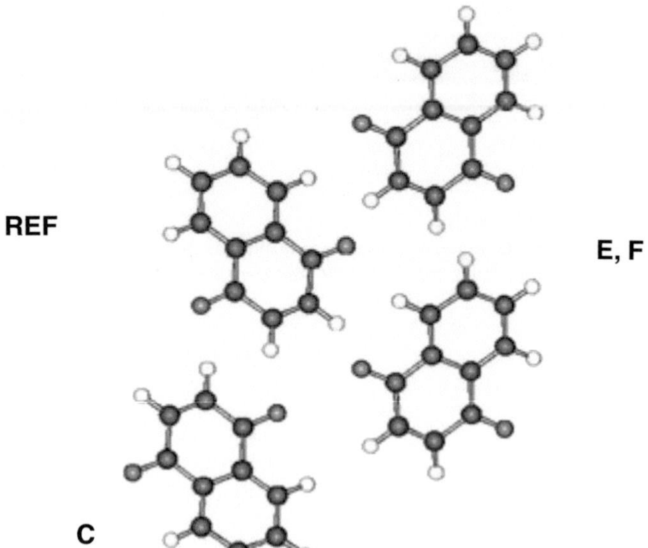

Fig. 4 The determinants *C*, *E* and *F* in the naphthoquinone crystal. Short O···H distances range from 2.47 to 2.52 Å (see Table 2). Symmetry codes: $-x, -y-1, -z$; $1-x, \pm 1/2+y, 1/2-z$.

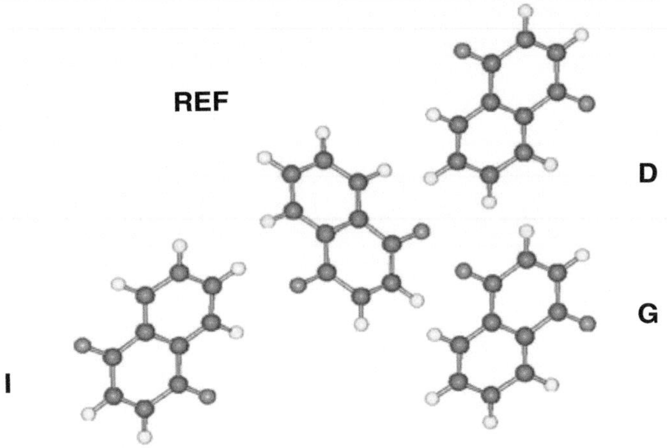

Fig. 5 The determinants *D*, *G* and *I* in the naphthoquinone crystal. Short O···H distances with *D* and *I* range from 2.52 to 2.87 Å, the shortest with *G* is 3.35 Å (see Table 2). Symmetry codes: *D*, *x*, $-1/2-y$, $1/2+z$; *G*, *x*, $1/2-y$, $1/2+z$; I, $-x$, $y-1/2$, $-z-1/2$.

The inspection of second-nearest neighbors (Fig. 6) offers some interesting considerations. The long-range electrostatic component of the interaction energy allows for some net repulsive partners, with a confrontation of molecular moieties of the same sign. The crystalline edifice takes that small amount of strain without apparent difficulty. In fact, the crystal structure of a fully charged species like the glycine zwitterion shows largely destabilizing near-neighbor partners [21, 29]. In crystals, this fact is more and more unavoidable as the net charge of the constituting molecules increase (think of sodium chloride).

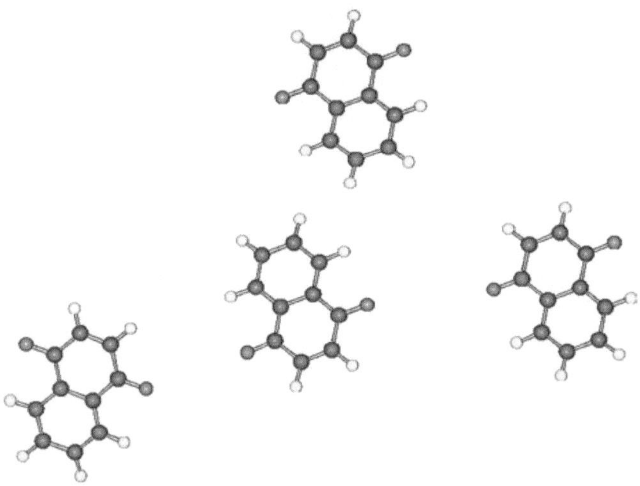

Fig. 6 Second neighbors in the naphthoquinone crystal; all coulombic energies are destabilizing (+1.0 to 1.2 kJ mol^{-1}). One short H···H contact (2.16 Å) appears.

2-Naphthoic Acid

The bond lengths in the carboxylic group of the molecular structure of 2-naphthoic acid in its crystal are poorly determined due to partial disorder, a not uncommon occurrence in crystals of carboxylic acids. Therefore, the C—O bond length was reset at 1.30 Å, the C=O bond length at 1.25 Å, and the O—H distance at 1.0 Å.

The by far dominating structure determinant is, as expected, the hydrogen bonded cyclic dimer portrayed in Fig. 7 (determinant A, see energies in Table 3), described by the PIXEL approach as a strongly coulombic interaction, with a substantial polarization component, which is supposed to represent also the not easily quantifiable covalent character of the bond, as well as any missing stabilization deriving from a dynamic rearrangement of the electron density upon bonding. While the treatment of this particular interaction is certainly the most problematic for the PIXEL approach, using rigid monomer densities, the results can be considered still satisfactory, as, for example, the hydrogen-bonding dimer energy is similar to the experimental dimerization energy of most carboxylic acids, as already discussed [21].

Next in amount of stabilizing energy come two offset stacked dimers, B and C (Fig. 8), where the dispersion component dominates. A further offset stacked dimer is D (not shown), where the two polarized carboxylic groups overlap with, rather surprisingly, no net coulombic stabilization or repulsion.

The hydrogen-bonded dimer is nicely complemented by two in-plane molecular interactions, those with determinants E and F (see again Fig. 7). Determinant E shows two short C—H···O distances, 2.64 Å, with the expected coulombic stabilization. However, the shortest C—H···O distance, 2.52 Å, occurs in determinant F, where the total coulombic energy is much smaller, due to the concomitant confrontation of two hydrophobic rims with H···H distances of 2.6–2.8 Å. All considered, it is probably safer to consider this determinant as stabilized by dispersion than by the coulombic C—H···O interaction. This in-plane interlocking results in a sort of 'Watson-Crick' bonding scheme: the hydrogen-bonded REF-A dimer is flanked by two molecules, whereby a CO—CO—CH triad points at a CH—CH—CO triad, thus forming three short H···O contacts. In this case the consideration of C—H···O bonds may be of intuitive help in the description of the bonding situation; to that, one could counter that the molecular interlocking is also nicely explained by close-packing, shape-fitting arguments. In any case, even the sum of the REF-E and REF-F energies, −14.4 kJ mol^{-1}, is still smaller than the smaller of the energies in the stacked dimers.

Table 2 Structure determinants in the crystal of naphthoquinone

Label[a]	Distance[b]	Short O···H distances	Interplanar angle	E_{coul}[c]	E_{pol}	E_{disp}	E_{rep}	E_{tot}	E_{qq}[d]
1 A	4.07	None	0	2.4	2.5	−45.7	13.9	−36.7	+1.0
1 B	4.38	None	0	−1.5	−1.7	−37.1	10.0	−30.3	+0.5
1 C	8.13	2.52	0	−14.8	−3.3	−8.6	11.6	−15.0	−4.7
1 E, 1 F	7.40	2.47, 2.49	42	−13.2	−3.6	−10.2	14.9	−12.0	−3.9
2 D	6.65	2.87, 2.71	42	−2.2	−1.6	−12.8	5.8	−10.9	−0.2
2 G	7.45	3.35	42	−4.7	−1.4	−10.0	5.5	−10.6	−1.7
2 H	7.76	None	0	−1.8	−0.8	−8.8	3.1	−8.2	−0.6
2 I	7.90	2.52, 2.89	42	−6.9	−2.6	−8.1	9.6	−8.1	−2.2
lattice	–	–	–	−36.8	−11.0	−102.7	57.2	−93.3[e]	−19.0

[a]A-I, labels as in Figs. 3, 4, 5, and 6, the number is the multiplicity of the dimer in the crystal structure
[b]Distance between centers of mass, Å
[c]PIXEL partitioned energies, kJ mol^{-1}
[d]Coulombic energy calculated with point-charge atom–atom parameters
[e]Sublimation enthalpy 91 kJ mol^{-1}

With all the favorable stacking interactions in determinants B, C and D, and the favorable in-plane interactions in determinants A, E and F, there is still some lateral interaction that cannot be favorably satisfied between parallel molecules, and that concerns the molecular tail opposite to the carboxylic group. The C—H···H—C confrontation there would be too unfavorable, so the problem is solved by determinant G, the screw axis, which brings about a deep corrugation in the whole structure (Fig. 9). The C—H groups of the tail find an acceptable way of contact by pointing at a neighbor benzene ring, again with a predominantly dispersive contribution. These interactions are better considered as expedients for avoided destabilization than as active stabilization agents of the crystal structure.

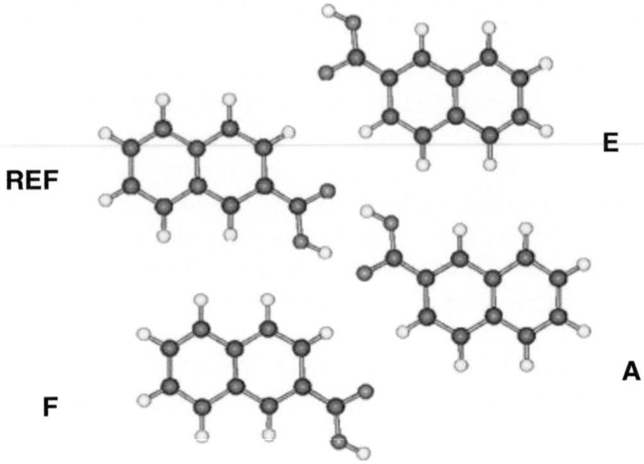

Fig. 7 The in-plane determinants in the crystal structure of 2-naphthoic acid.

Conclusion

One sees with some surprise that in the crystal structure of naphthoic acid an energy requirement of only a few kJ mol^{-1}, like the G determinant, brings about a major structural consequence like the deep corrugation of the whole pattern. Crystal stability results from a fine balance of attractive and repulsive forces, and major structural consequences sometimes result from minor differences in total energy. On the other hand, separate energy terms like repulsion and coulomb energies show large variations with minor structural changes; this is a consequence of the order of magnitude of electrical interactions at short range (recall that two 1 electron charges at 1 Å distance repel each other by 1389 kJ mol^{-1}, and that the field due to a 1 electron charge at the distance of 1 Å is 1.44 10^{11} V m^{-1}) This is why qualitative structural analyses, and attempts at prediction based on localized weak forces, are mostly unsuccessful. The PIXEL calculation gives a faithful portrait of these effects, and we have shown how such calculations can be used for some quantitative views of molecular packing in organic crystals. We stress the importance of a quantitative evaluation of intermolecular energies in terms of whole molecular electron densities. In a pragmatic view, the definition of intermolecular bonds is considered less important, in the sense that these bond entities seldom convey useful, convenient understanding or predictive power.

Table 3 Structure determinants in the crystal of 2-naphthoic acid[a–d]

Label	Distance	E_{coul}	E_{pol}	E_{disp}	E_{rep}	E_{tot}	E_{qq}
1 A	9.68	−146.0	−78.2	−13.5	183.8	−53.8	−85.1
2 B	5.00	−6.0	−3.1	−37.6	19.6	−27.1	−0.4
2 C	5.63	−4.6	−1.0	−16.6	4.5	−17.7	−3.4
1 D	7.33	−1.0	−2.4	−10.5	4.4	−9.5	−0.2
1 E	8.29	−7.9	−3.0	−5.4	5.8	−10.4	−5.5
2 F	7.53	−1.9	−1.7	−7.5	7.1	−4.0	−0.1
2 F	7.53	−1.9	−1.7	−7.5	7.1	−4.0	−0.1
2 G	8.87	−1.1	−0.7	−6.7	3.1	−5.5	−0.3
Lattice	–	−92.4	−49.7	−96.6	140.7	−98.0[e]	−48.0

[a]A-I, labels as in Figs. 7, 8, and 9, the number is the multiplicity of the dimer in the crystal structure
[b]Distance between centers of mass, Å
[c]PIXEL partitioned energies, kJ mol^{-1}
[d]Coulombic energy calculated with point-charge atom–atom parameters
[e]Sublimation enthalpy 91 kJ mol^{-1}

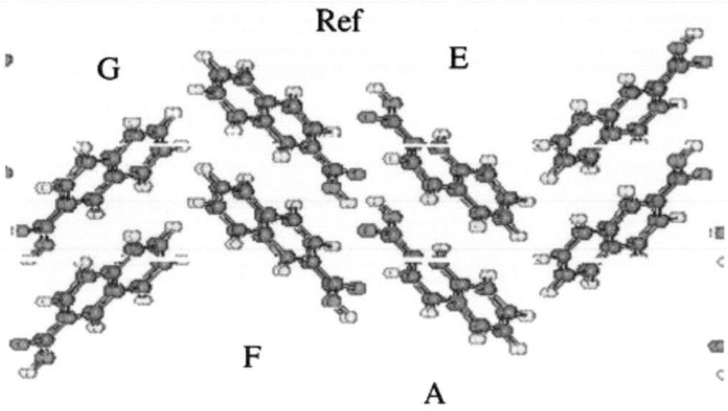

Fig. 9 The corrugated crystal structure of 2-naphthoic acid: *a* cell axis horizontal, *b* cell axis vertical. Labels *A*, *E*, *F* as in Fig. 7.

Acknowledgment The molecular pictures were drawn by program SCHAKAL [30].

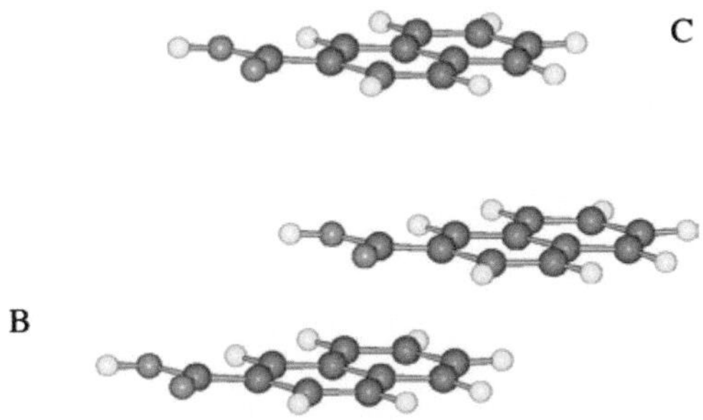

Fig. 8 The stacking determinants in the crystal structure of 2-naphthoic acid. Symmetry codes: *B*, *x*, 1 + *y*, *z*; *C*, *x*, *y*, 1 + *z*.

The comparison of coulombic energies calculated by the exact PIXEL method against those calculated by localized point-charge models shows large discrepancies due to the fact that the latter do not account for penetration energies; the sign of the point-charge coulombic energies is sometimes even opposite to the correct one. This is a further warning against inferences based on localized coulombic arguments.

We do not claim to have found a unique and straightforward way to crystal structure prediction or control; far from that. Since the interplay of weak intermolecular forces is so complicated, there is still no guarantee that even a quantitative view may give at once the correct answer. What is sure is that without quantitative views, correct answers will never be reached.

References

1. Stone, A. J. *The Theory of Intermolecular Forces*; Clarendon Press: Oxford, 1996.
2. Levine, I. N. *Quantum Chemistry*; Prentice Hall, p. 469 ff; 5th Edn.
3. Hunt, K. L. C. *J. Chem. Phys.* **1990,** *92*, 1180.
4. (a) Bader, R. F. W. *Atoms in Molecules: A Quantum Theory*, Oxford University Press, 1990; (b) Bader, R. F. W.; Anderson, S. G.; Duke, A. J. *J. Am. Chem. Soc.* **1979,** *101*, 1389.
5. Matta, C. F.; Hernandez-Trujillo, J.; Tang, T.-H.; Bader, R. F. W. *Chem. Eur. J.* **2003,** *9*, 1940.
6. Gatti, C.; May, E.; Destro, R.; Cargoni, F. *J. Phys. Chem.* **2002,** *A106*, 2707–2720.
7. Pauling, L. *The Nature of the Chemical Bond*; Cornell University Press, 1939; 2nd Edn, Chapter 1.
8. Gilli, G.; Gilli, P. *J. Mol. Struct.* **2000,** *552*, 1.
9. Bernstein, J. *Polymorphism in Organic Crystals*; Oxford University Press: Oxford; 2002.
10. Wetmore, S. D.; Schofield, R.; Smith, D. M.; Radom, L. *J. Phys. Chem.* **2001,** *105*, 8718.
11. Gu, Y.; Kar, T.; Scheiner, S. *J. Am. Chem. Soc.* **1999,** *121*, 9411.
12. Hobza, P.; Spirko, V.; Havlas, Z.; Buchhold, K.; Reimann, B.; Barth, H.-D.; Brutschy, B. *Chem. Phys. Letters* **1999,** *299*, 180.
13. Tarakeshwar, P.; Choi, H. S.; Kim, K. S. *J. Am. Chem. Soc.* **2001,** *123*, 3323.
14. Price, S. L.; Stone, A. J.; Lucas, J.; Rowland, R. S.; Thornley, A. E. *J. Am. Chem. Soc.* **1994,** *116*, 4910.
15. Bukowski, R.; Szalewicz, K.; Chabalowski, C. F. *J. Phys. Chem.* **A1999,** *103*, 7322.
16. Mitchell, J. B. O.; Price, S. L. *J. Phys. Chem.* **2000,** *A104*, 10958.
17. Gora, R. W.; Bartkowiak, W.; Roszak, S.; Leszczynski, J. *J. Chem. Phys.* **2002,** *117*, 1031.
18. Pisani, C. In *Handbook Molecular Physics*; Wilson, S.; McWeeny, R., Eds.; Chichester: Wiley, 2001.

19. Gavezzotti, A. *J. Phys. Chem.* **2002,** *B106*, 4145.
20. Gavezzotti, A. *J. Phys. Chem.* **2003,** *B107*, 2344.
21. Gavezzotti, A. *Cryst. Eng. Comm.* **2003,** *5*, 429.
22. Brock, C. P.; Dunitz, J. D. *Acta Crystallogr.* **1982,** *B38*, 2218.
23. Gaultier, C.; Hauw, C. *Acta Crystallogr.* **1965,** *18*, 179.
24. Trotter, J. *Acta Crystallogr.* **1961,** *14*, 101.
25. Gavezzotti, A. *OPiX, A computer program package for the calculation of intermolecular interactions and crystal energies*, University of Milano, 2003. The package is available upon request at angelo.gavezzotti@unimi.it for a nominal fee.
26. Gavezzotti, A. *CrystEngComm* **2003,** *5*, 439–446.
27. Dunitz, J. D.; Gavezzotti, A. *Angew. Chem.*, submitted.
28. Rowland, R. S.; Taylor, R. *J. Phys. Chem.* **1996,** *100*, 7384.
29. Volkov, A.; Coppens, P. *J. Comput. Chem.* **2004,** *25*, 921–934.
30. Keller, E. *SCHAKAL92, A program for the graphic representation of molecular and crystallographic models*, University of Freiburg, 1993.

Where Are Genes in Paulingite? Mathematical Principles of Formation of Inorganic Materials on the Atomic Level[a]

V.Ya. Shevchenko[b] and S.V. Krivovichev[b,c]

Abstract

By means of the concept of universal optimum and general principles of inorganic gene, structures of paulingite-related zeolites and minerals have been constructed. The structures of zeolite Rho and paulingite are considered as members of zeolite family generated by the work of 4-colored cellular automaton (CA). The ideal symmetry of the members of the family is cubic, space group $Im\bar{3}m$, $a = 15 + 10n$ [Å], where n is a number of the CA cycle ($n = 0$ for zeolite Rho and 2 for paulingite). A new hypothetical zeolite of the family with $n = 1$ is predicted and named ISC-1 (Institute of Silicate Chemistry-1); atomic coordinates and theoretical X-ray powder diffraction pattern have been calculated. It appears to be very probable that universal optimum contains all necessary information for its material realization (sharp configurations) and construction of materials with certain type of interaction potential (even with restrictions existing in theory for this potential).

Keywords

Universal optimum • Inorganic gene • Paulingite • Cellular automaton • Zeolite

Nobody enters here, if he is not a geometrician.
Platon
Le chimiste est le seul scientifique, qui cree l, objet de ses etudes.
Marcelline Berthelot

Introduction

Over the last 20–30 years, we have been witnessing (and participating in) the development of new approaches to the problem of the structure of matter and considerable expansion of the borders of classic crystallography.

First of all, this expansion was induced by the advances in experimental techniques (in particular, high-resolution microscopy) applied to small natural objects: biological, organic, inorganic, and their combinations. To explain the experimental data obtained, there is a need to understand the very basic principles of construction of matter as a part of the material world around us. Indeed, how many atoms are needed to form a substance of macro-world as we know it? For instance, metallic iron. We know the series of atomic Fe clusters consisting of N atoms forming a "magic" sequence of numbers. What are the properties of these clusters? Do their structure correspond to the structure of macro-iron? These problems are under active debate and investigation at the moment. Perhaps, at the most advanced degree, development of these ideas has been determined by Mackay who first published on this issue in 1967 [1]. The most important parts of this idea are atomistics (i.e., everything starts from atoms) and hierarchy (i.e., matter is organized on the levels of increasing complexity). At the core of this idea is the concept of information that determines an inorganic structure in

[a]*Structural Chemistry* 2008, 19(4):571–577.

[b]Institute of Silicate Chemistry, Russian Academy of Sciences, Nab. Makarova, 2, 199034 St. Petersburg, Russia
e-mail: exchem@isc.nw.ru

[c]Department of Crystallography, St. Petersburg State University, University Emb. 7/9, 199034 St. Petersburg, Russia

I. Hargittai, B. Hargittai (eds.), *Science of Crystal Structures: Highlights in Crystallography*,
DOI 10.1007/978-3-319-19827-9_30

the same way as genetic code determines biological structures. The ideas of Mackay were in correspondence with the works of W. Pearson who in 1972 suggested describing structures of intermetallics by means of codes that characterize stacking sequences of atomic layers (for the sake of justice, we note that analogous approaches were developed by the Soviet scientist G. S. Zhdanov back in 1946). Mackay further developed the idea of description of structures of inorganic materials on the basis of another important concept of 'curvature' in the sense of 'grammar' of inorganic and biological structures. In this grammar, letters are atoms, and words are clusters of connected atoms (building blocks, structural complexes, etc.). What determines the structures of these clusters and the structure of matter comprising these clusters? Today we are able to provide some partial answers to this question using mathematics (geometry) and information theory. It happens that mutual arrangement of N points (atoms) interacting with each other in n-dimensional space is uniquely determined and, in turn, determines so-called spherical code that can be used to save and to transmit information. Authors of Ref. [2] named these configurations "universal optima". A special condition is the requirement that interaction potential has to be continuous and differentiable. All known universal optima are listed in Table 1. It is worthy to note that the sharp configurations—independently from the type of interaction potential—are the most closely packed arrangements. Thus, the principle of dense filling of space suggested by Laves in 1967 for metallic alloys has now been proved mathematically, rigorously.

Table 1 Known universal optima. Energy-minimizing point configurations

n	N	t	Description
2	N	$\cos(2\pi/N)$	N-gon
n	$N \le n+1$	$-1/(N-1)$	Simplex
n	$2n$	0	Cross-polytope
3	12	$1/\sqrt{5}$	Icosahedron
4	120	$(1+\sqrt{5})/4$	Regular 600-cell
5	16	1/5	Klebsch graph
6	27	1/4	Schlaefli graph
7	56	1/3	Equilateral lines
8	240	1/2	Root lattice E_8
21	112	1/9	Isotropic subspaces
21	162	1/7	Sharply regular graph
22	100	1/11	Higman-Simps graph
22	275	1/6	McLaughlin graph
22	891	1/4	Isotropic subspaces
23	552	1/5	Equilateral lines
23	4600	1/3	Kissing configurations
24	196560	1/2	Leech lattice
$q\frac{q^3+1}{q+1}$	$(q+1)(q^3+1)$	$1/q^2$	Isotropic subspaces

n space dimension, N number of points, t scalar product

Experimental proof of the results of [2] had been performed in [3] using advanced computer modeling. Two more universal optima have been found (for 40 points in 10-dimensional space and for 64 points in 14-dimensional space). First five configurations in Table 1 are vertices of certain regular polytopes (these are polytypes with all faces being simplices). Next seven cases were obtained from the root lattices E_8 in R^8 and Leech lattice in R^{24}. 240-point configurations and 196550 points have minimal vectors in these lattices. In terms of sphere packings, these are contact configurations with kissing points in corresponding packings. Each configuration has a spherical cap in each point with its radius being as maximal as possible but without overlaps. Kissing points of this cap provide spherical code in the space of corresponding dimensionality.

Mathematical problem of contact numbers of spheres is a partial case of the problem of a spherical code. For each sphere, there is a corresponding spherical cap on the central sphere and a kissing point. These points are called an alphabet. The problem is to transfer kissing points from certain caps to the other sphere. The set of caps from which signals are transmitted form a word. If transmission is somehow distorted, the caps should be relatively large. In that case, the point should first appear in the cap where it has previously been, so it can be reconstructed since the caps do not intersect. The word can be reconstructed accordingly. The more spherical caps there are on the certain sphere, the more different information can be encoded.

In the framework of information theory, instead of continuous wave in 1-dimensional space, one can speak about vector set in n-dimensional space.

If connection channel has no memory and with the increase of the word length n, the number of decoding steps do not increase exponentially (2^n) but as n^2. Using hierarchical principles of coding-decoding, one may decrease computational complexity till ~n. Coding as well as decoding are done step-by-step toward higher levels of hierarchy. On the lower level, simple symbols (alphabet) form blocks (words). On the second level, each block (word) is considered as a new symbol, and blocks group into higher-level blocks (sentences). By placement of certain points in n-dimensional space and looking at their arrangements, we determined further directions of self-assembly using purely geometrical operations.

Thus, the information theory and structure can be merged in an unified theory. All sharp configurations are m-designs, i.e. they are not simply spherical codes, but also the set of roots of some polynom (Hermit polynom) of the degree m.

The science of structure of matter is based upon self-assembly of atoms first into sharp configurations, and then, according to the principle of a cellular automaton, into more hierarchically complex structures, into nanoparticles and then into macromolecules and macroobjects.

The structural diversity of the nanoworld is determined by the fact that finite packings may represent fragments of some "parent" structures that do not necessarily exist in 3-dimensional Euclidean space. We shall call these structures fundamental. The reciprocal transition into space with dimension $n \le 3$ is by the 'cut-and-project' procedure.

The chemical or physical nature of the subunits is of secondary importance (whether these are atoms or protein molecules). Geometry determines and forms a structural model (universal optimum) before the atoms or other subunits and their interaction potentials are chosen. Thus, in this philosophy, geometry is more than a description principle. It is a reflection of the properties of space where nanoobjects exist.

In order to understand mystical appearance of regular polytopes in the problems of applied mathematics, one has to replace direct calculations of differential geometry by more simple and general approach of symplectic and contact geometry. General principle here is as follows: one has to embed geometrical objects from 'configurational' space V into 'phase' space $V \times F$, where peculiarities either disappear or become simpler (in theory of differential equations and quantum theory, this approach is called 'microlocal'). This concept transforms simple facts of differential geometry into general theorems of symplectic and contact geometry with more wide field of application. In this way, perspectives arise for very wide exploration of differential-geometric intuition (V.I. Arnold).

In the past few years, a large number of nanoparticles had been found that have structures not describable in terms of classical crystallography. To express their admiration, scientists use such terms as "unprecedented", "unique", "mysterious". To explain their structure, we have used local (microlocal) approach. General principle is the same: atom configurations should be considered as geometrical objects and embedded from 3-dimensional Euclidean space into another space—of higher dimensionality and (probably) different topology. However, in local terms, this space shall still be represented as a direct product $V \times F$. In terms of modern geometry, this procedure corresponds to the construction of a fiber space, where V is the basis of fibering and F is the structural group of the fiber.

In general case, fibering is a manifold. Any manifold can be isometrically embedded into Euclidean space of corresponding (or even higher) dimensionality, where we have a possibility to choose a fundamental structure.

These principal results allow us, instead of search of concrete structure responsible in E^3 for the minimum of some potential, to consider a whole class of derivative structures by the prior knowledge that they all have been obtained from some universally optimal non-Euclidean structures and thus close to the local minimum for the whole class of interaction potentials.

The real type of atoms (or any other subunits) involved will influence preferential formation of derivative structures by the atoms of particular type.

Accordingly, we may consider a complex inorganic structure, e.g. paulingite,[1] a crystalline zeolite with several hundreds of atoms in its unit cell. Where are genes for this structure? Which information does it contain? How may we describe it? Can we describe complex inorganic structure (periodic or aperiodic) as a 1-dimensional chain with its "gene" being the basis of construction and mutation and suitable for the formation of a whole range of structures? In this case, the particular structure can be chosen by the free energy minimum but different minima should not coincide.

We shall try to use these "inorgenes" in genetic algorithm of search of structures that fulfill the requirements of the lowest energy level. This will be required to calculate the gene that corresponds to the corresponding structure. As in biology, there could be genes that may not generate a 'living' structure.

Paulingite: Structural Architecture

Paulingite ($K_6Ca_{16}[Al_{38}Si_{130}O_{336}](H_2O)_{113}$) is one of the most structurally complex minerals. Its crystal chemistry was the subject of a large number of studies in mineralogical crystallography [4, 5]. At the present time, there are several mineral species of the paulingite group that have the same root name and different suffices according to the dominating extra-framework cation: paulingite-K, paulingite-Na, and paulingite-Ca. A synthetic analogue of paulingite is known under the name 'zeolite ECR-17' that had been used in chemical technology [6]. However, the fundamental interest in paulingite is not in chemical diversity of extra-framework cations but in structural complexity of its tetrahedral framework. Unit cell of paulingite contains 768 tetrahedral anions TO_4 ($T = Al^{3+}$, Si^{4+}) linked into complex 3-dimensional architecture (Fig. 1a).

Minerals of the paulingite group have a cubic symmetry and crystallize in the space group $Im\bar{3}m$, with the a parameter in the range of 35.00–35.43 Å. Topology of zeolite framework in paulingite is usually described in terms of polyhedral building units [7]. Symbolic (or nodal) description is usually used, when tetrahedra are symbolized by vertices and vertices corresponding to the linked tetrahedra are linked by an edge. Graphs obtained along this procedure are

[1] We have chosen paulingite because the question 'Where are genes in paulingite?' had been asked first in the paper of A.L. Mackay "Generalised crystallography" Journal of Molecular Structure (Teochem) 336 (1995) 293–303.

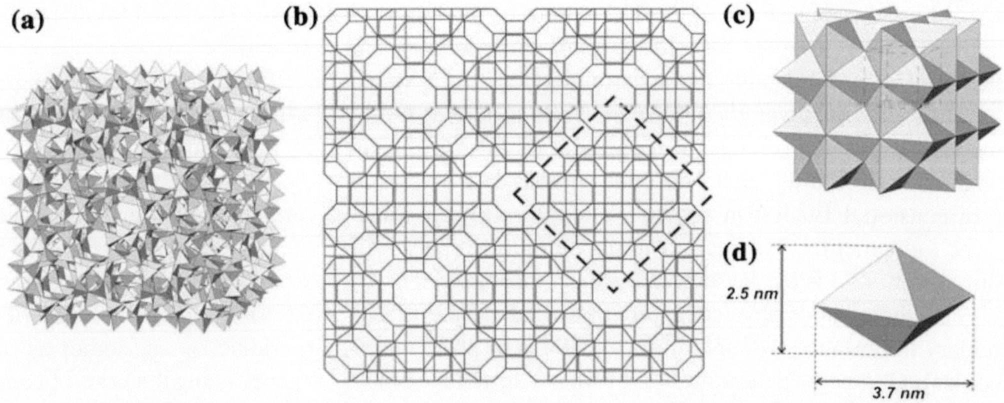

Fig. 1 Tetrahedral framework in paulingite (**a**) and its 3-dimensional graph (**b** striation denotes bipyramidal nanocomplex). Arrangement of the bipyramidal nanocomplexes (**c**) corresponds to the space partition into tetragonal bipyramids (= flattened octahedra) (**d**).

4-connected nets that are subject of a separate area of mathematical crystallography [8]. Figure 1b shows graph of the paulingite framework. In contrast to previous works [4–7], we prefer to describe this framework not in terms of polyhedral units but as a result of linkage of bipyramidal (octahedral) complexes outlined in Fig. 1b by striated line. These complexes are centered in the middle points of edges and in the centers of the cubic cell (e.g. in points (0; 0; 0.5) and (0.5; 0.5; 0)). In total, they form space partition into flattened octahedra with dimensions of $3.7 \times 3.7 \times 2.5$ nm^3 (Fig. 1c, d).

Let us consider the structure of complexes in more detail. In Fig. 2, each 4-membered ring (4-MR) is replaced by a flat square. The complex can be split into five levels, each formed by a closed contour of vertically standing squares (Fig. 2b). The highest level consists of a double octagonal ring (D8-MR), the second is formed by 24 squares, the third and central level contains 40-square contour and internal D8-MR. The fourth and fifth levels repeat the first and the second, respectively. The levels are linked by inclined squares so that the whole construction reminds of hierarchical construction of Middle-Age castles that also had many-level architecture.

It is noteworthy that, in general, construction of the complexes corresponds to the principles of sharp configurations. At the first level, atoms form tetrahedra—3-dimensional simplices (see second row in Table 1), then tetrahedra are linked into 4-MRs and 8-MRs (first row in Table 1). General architecture corresponds to an octahedron that is a cross-polytope of 3-dimensional space (third row in Table 1).

Paulingite: Cellular Automata Modeling

Projection of the nanocomplex along its axis is shown in Fig. 3. Obviously, the 2nd and 3rd levels are similar to the 1st level and can be obtained from the latter by some simple geometrical operations. As an instrument, we may use a cellular automaton (CA), a discrete deterministic system that develops in time and space according to simple set of transition rules [9]. The use of CAs in structural crystallography was first proposed by Mackay [10] and recently was further developed in [11].

Figure 3a shows schematical construction of paulingite nanocomplex in terms of vertical squares. The CA lattice is a dihedral partition of plane into squares and octagons. Initial condition (zero-cycle of the CA) is a regular octagon. In order to simplify the model further without the loss of generality, we may transform the initial CA lattice into a square lattice by a series of simple substitutions (Fig. 3b). The closed 8-MR will correspond to a black cell, 8-MR with one or less bold edges to a white cell, 8-MRs with three and five bold edges to blue and red cells, respectively. As a result, we shall obtain 4-colored CA shown in Fig. 3c.

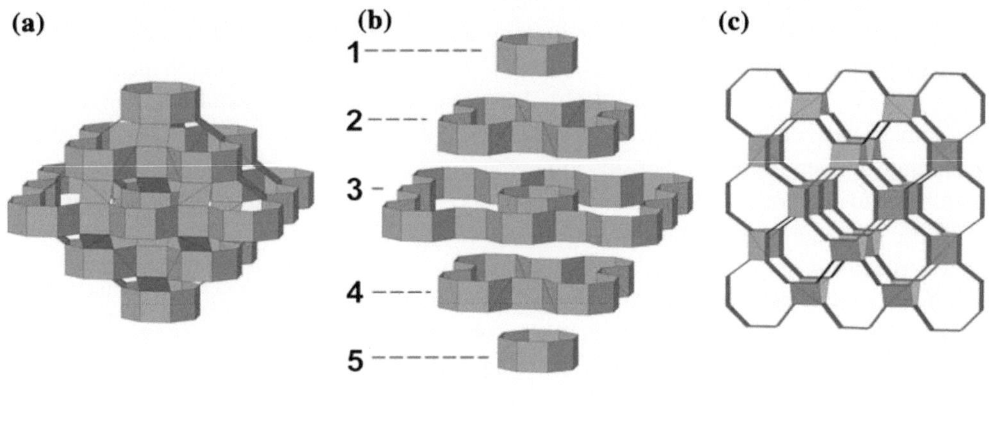

Fig. 2 Bipyramidal nanocomplex as construction of 4-MRs (shown as *squares*): side view (**a**), stratified structure (**b**), top view (**c**).

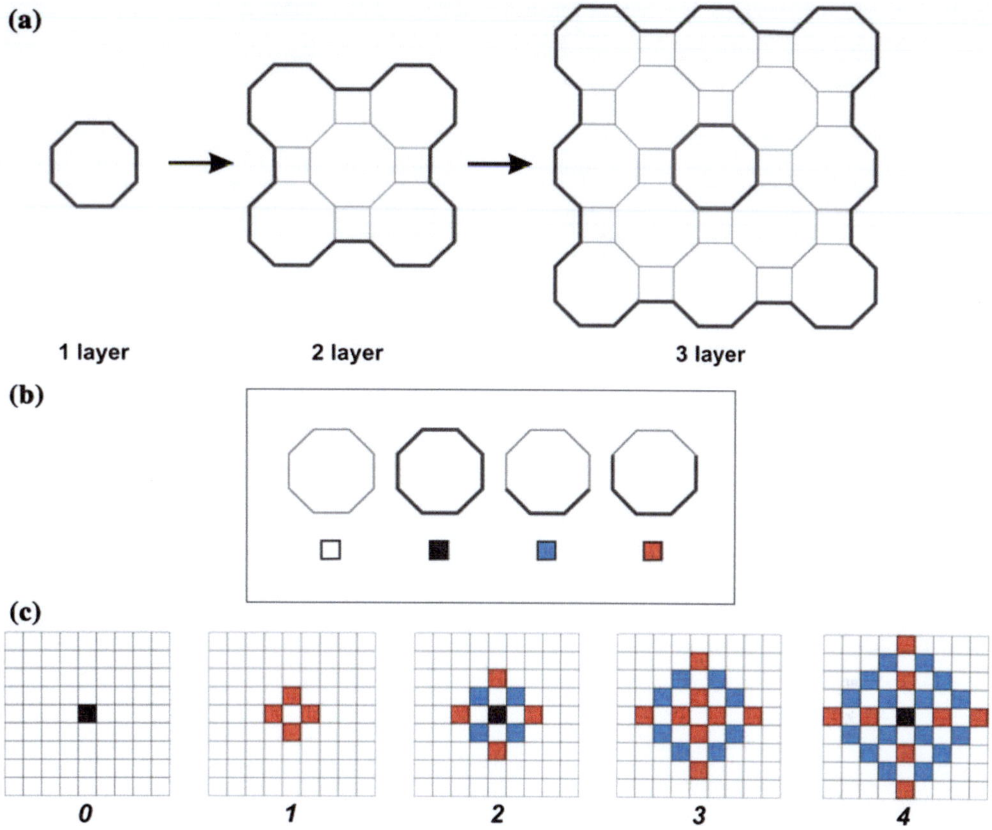

Fig. 3 Levels of paulingite bipyramidal nanocomplex as a result of work of the CA: schematic structure of contours in different layers (**a**), transition to the square lattice (**b**), 4 cycles of work of 4-colored CA (**c**).

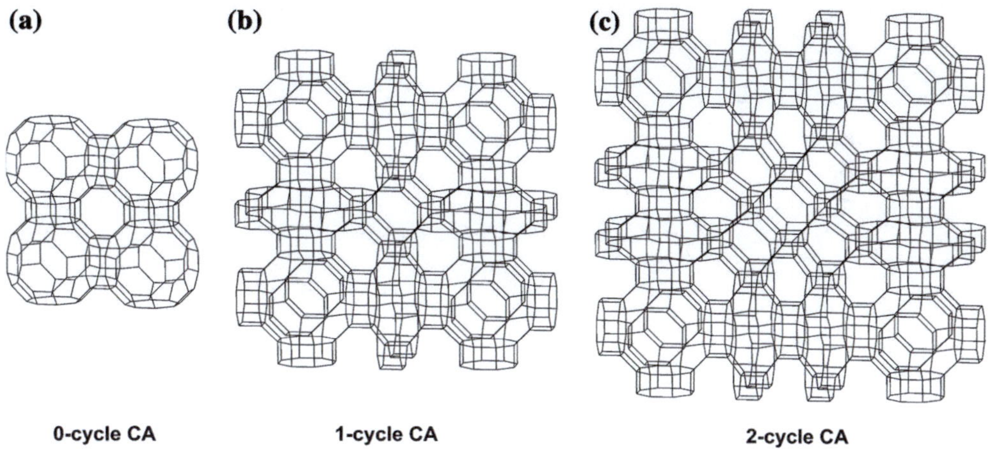

Fig. 4 Graphs of zeolite frameworks consisting of nanocomplexes formed as a result of work of paulingite CA: zeolite Rho RHO (**a**), hypothetical zeolite ISC-1 (**b**) and paulingite (**c**).

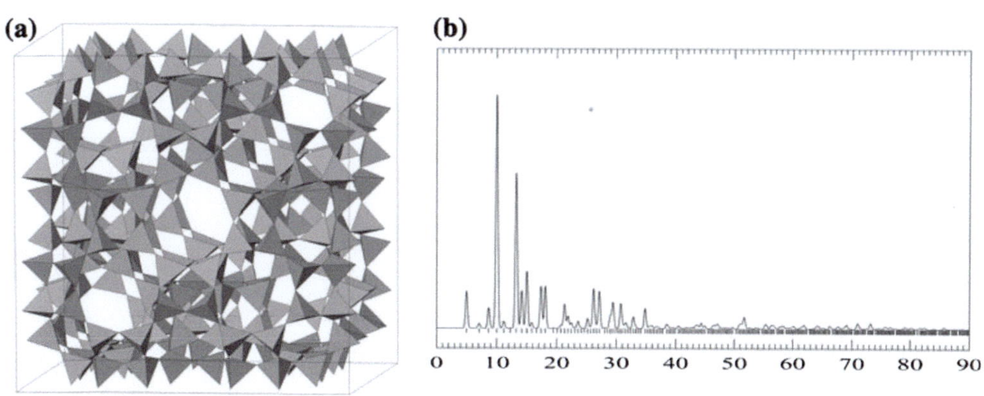

Fig. 5 Tetrahedral framework of the hypothetical material ISC-1 (**a**) and its calculated X-ray powder diffraction pattern (for $\lambda = 1.54178$ Å) (**b**).

Table 2 Atom coordinates and geometrical parameters of tetrahedral framework in hypothetical zeolite ISC-1 ($a=25$ Å, sp. gr. $Im\overline{3}m$)

Atom coordinates

Atom	*x*	*y*	*z*
Si1	0.2400	0.0632	0.1520
Si2	0.3640	0.0632	0.1520
Si3	0.4368	0.0632	0.25
O1	0.2291	0	0.1317
O2	0.2256	0.1000	0.1000
O3	0.3014	0.0698	0.1710
O4	0.2015	0.0765	0.2015
O5	0.3756	0	0.1370
O6	0.3755	0.0980	0.0980
O7	0.4260	0	0.2661
O8	0.4020	0.0820	0.1999

Geometry of tetrahedra

Bond	Bond length (Å)	Angle	Angle value (°)
Si1–O1	1.682	O1–Si1–O2	104.64
Si1–O2	1.633	O1–Si1–O3	109.81
Si1–O3	1.615	O1–Si1–O4	109.29
Si1–O4	1.603	O2–Si1–O3	112.74
		O2–Si1–O4	111.42
		O3–Si1–O4	108.81
Si2–O3	1.644	O3–Si2–O5	109.25
Si2–O5	1.650	O3–Si2–O6	110.67
Si2–O6	1.631	O3–Si2–O8	108.66
Si2–O8	1.599	O5–Si2–O6	106.95
		O5–Si2–O8	110.26
		O6–Si2–O8	111.01
Si3–O7	1.652 2x	O7–Si3–O7	111.74
Si3–O8	1.596 2x	O7–Si3–O8	112.53 2x
		O7–Si3–O8	106.40 2x
		O8–Si3–O8	107.21

Intertetrahedral angles

Angle	Angle value (°)	Angle	Angle value (°)
Si1–O1–Si1	139.89	Si2–O5–Si2	146.55
Si1–O2–Si1	148.00	Si2–O6–Si2	148.38
Si1–O3–Si2	144.04	Si3–O7–Si3	145.95
Si1–O4–Si1	152.21	Si2–O8–Si3	145.65

It is of interest that infinite work of this CA will result in formation of a periodic 3-dimensional framework with an axial structure. Along its axis, with the period of 1 nm, will appear 8-MRs. We do not see any principal reasons why this structure cannot be formed under natural or experimental conditions. However, if they really form, their identifications would be a complex and separate task.

It is quite probable that paulingite is the by-product of the work of the CA described above that has been stopped on the 2nd cycle. In order to check this hypothesis, it is of interest to consider related structures

The zero-cycle of the CA shown in Fig. 3 corresponds to the structure of zeolite Rho [12–14] that is depicted in Fig. 4a. As paulingite, zeolite Rho crystallizes in cubic symmetry, space group $Im\overline{3}m$ with $a \sim 15$ Å. In this zeolite, D8-MRs are linked directly by 4-MRs. Thus, whereas paulingite corresponds to the 2nd cycle of the CA (Fig. 4c), zeolite Rho corresponds to the zero cycle of the CA.

Table 3 Theoretical X-ray diffraction pattern of hypothetical zeolite ISC-1 calculated from the data given in Table 2 ($\lambda=1.54178$ Å, $2\theta_{max}=30°$; 0-intensity lines are listed as well)

h	*k*	l	*d* (Å)	2θ (°)	*I/Io*
0	1	1	17.6777	5.00	16
0	0	2	12.5000	7.07	2
1	1	2	10.2062	8.66	9
0	2	2	8.8388	10.01	100
0	1	3	7.9057	11.19	3
2	2	2	7.2169	12.26	0
1	2	3	6.6815	13.25	66
0	0	4	6.2500	14.17	16
1	1	4	5.8926	15.03	0
0	3	3	5.8926	15.03	24
0	2	4	5.5902	15.85	2
2	3	3	5.3300	16.63	0
2	2	4	5.1031	17.38	18
0	1	5	4.9029	18.09	13
1	3	4	4.9029	18.09	5
1	2	5	4.5644	19.45	0
0	4	4	4.4194	20.09	0
0	3	5	4.2875	20.72	1
3	3	4	4.2875	20.72	1
0	0	6	4.1667	21.32	10
2	4	4	4.1667	21.32	0
1	1	6	4.0555	21.92	0
2	3	5	4.0555	21.92	5
0	2	6	3.9528	22.49	2
1	4	5	3.8576	23.05	0
2	2	6	3.7689	23.61	3
1	3	6	3.6860	24.14	0
4	4	4	3.6084	24.67	0
0	1	7	3.5355	25.19	1
0	5	5	3.5355	25.19	3
3	4	5	3.5355	25.19	0
0	4	6	3.4669	25.70	1
1	2	7	3.4021	26.19	12
2	5	5	3.4021	26.19	0
3	3	6	3.4021	26.19	5
2	4	6	3.3408	26.68	1
0	3	7	3.2827	27.16	16
1	5	6	3.1750	28.10	0
2	3	7	3.1750	28.10	0
0	0	8	3.1250	28.56	0
1	1	8	3.0773	29.02	2
1	4	7	3.0773	29.02	3
4	5	5	3.0773	29.02	0
0	2	8	3.0317	29.46	11
4	4	6	3.0317	29.46	0
3	5	6	2.9881	29.90	0

ISC-1: A New Theoretical Zeolite Framework

We could not find in literature any data concerning zeolite intermediate between paulingite and zeolite Rho. However, the model developed above allows to predict its hypothetical structure (Fig. 4b). It has a cubic unit cell, $Im\bar{3}m$ with $a \sim 25$ Å. We were able to construct model of its tetrahedral framework (Fig. 5a). Atom coordinates and basic geometrical parameters are given in Table 2. We denote this zeolite as ISC-1 (Institute of Silicate Chemistry-1). Theoretical X-ray diffraction pattern of ISC-1 is shown in Fig. 5b (diffraction lines till $2\theta_{max} = 30°$ are listed in Table 3).

Thus, zeolite Rho, hypothetical material ISC-1, and paulingite form a family of zeolites that result from the CA shown in Fig. 3. It can be proposed that more complex zeolites of the family can be synthesized that correspond to higher cycles of the CA work. Ideal symmetry of these zeolites will be cubic, space group $Im\bar{3}m$, $a = 15 + 10n$ [Å], where n is a number of the CA cycle.

Conclusions

1. By means of the concept of universal optimum and general principles of inorganic gene, structures of paulingite-related zeolites and minerals have been constructed.
2. A new hypothetical zeolite of the paulingite family has been predicted and named ISC-1 (Institute of Silicate Chemistry-1).
3. It appears to be very probable that universal optimum contains all necessary information for its material realization (sharp configurations) and construction of materials with certain type of interaction potential (even with restrictions existing in theory for this potential).

References

1. Mackay AL (1967) Nature 216:159
2. Cohn H, Kumar AJ (2007) Am Math Soc 20:99
3. Ballinger B, Bleknerman G, Cohn H, Giansiracusa N, Kelly E, Schurmann A. Experimental study of energy-minimizing point configurations on spheres. arxiv:math.MG/0611451
4. Gordon EK, Samson S, Kamb WB (1966) Science 154:1004
5. Passaglia E, Gualtieri AF, Marchi E (2001) Eur J Miner 13:113
6. Vaughan DEW, Strohmaier KG (1999) Microporous Mesoporous Mater 28:233
7. Baerlocher Ch, Meier WM, Olson DH (2001) Atlas of zeolite framework types. Amsterdam, Elsevier
8. Friedrichs OD, Dress AWM, Huson D, Klinowski J, Mackay AL (1999) Nature 400:644
9. Toffoli T, Margolus N (1987) Cellular automata machines: a new environment for modeling. Boston, MIT Press
10. Mackay AL (1976) Phys Bull 495
11. Krivovichev SV (2004) Acta Crystallogr A60:257
12. Robson HE, Shoemaker DP, Ogivie RA, Manor PC (1973) Adv Chem Ser 121:106
13. McCusker LB (1984) Zeolites 4:51
14. Parise JB, Gier TE, Corbin DR, Cox DE (1984) J Phys Chem 88:1635

Supramolecular Architectures of Metal–Organic Host–Guest Compounds[a]

Alessia Bacchi[b] and Mauro Carcelli[b]

Abstract

The so-called "wheel and axle" compounds play an important role in the developing of supramolecular chemistry and there are a lot of studies dealing with their inclusion properties and host–guest chemistry. Classical "wheel and axle" molecules have a long, thin, central part (the axle) with two bulky ends (the wheels). Here, new compounds with host properties are described, that we called "wheel and axle metal–organic" diols (WAMOD), that are decorated with OH groups in the wheel. In particular, their clathration properties are discussed in connection with their crystal structures. A modular strategy is applied to obtain WAMODs: coordination chemistry is used together with soft interactions (H-bond, π–π interaction), with the aim to realize a dynamic framework that is able to reversibly capture and release a guest. The coordination bond is robust and permits to obtain WAMODs with different arrangements of the axle; the hydrogen bonds and/or the other soft interactions are responsible for the thin adjustments in the crystal packing that allow reversible adsorption/desorption of the guest.

Keywords

Crystal engineering • Inclusion compounds • Host–guest chemistry • Metal–organic compounds

The "Wheel and Axle Metal–Organic" Diols

Metal–organic frameworks are a class of materials with new and interesting properties [1–3]. In particular, porous metal organic systems are attracting considerable interest because of their potential use as sensors, catalysts and, in general, in host–guest chemistry [4, 5]. The first examples of "wheel and axle" compounds were reported by Toda and Agaki [6], who discussed the marked tendency of some tetraphenyldiols to give solvates with different organic molecules. Since then, a lot of studies were carried out with the aim to correlate the modifications of the molecular architecture of the host with the properties of the resulting materials; a part of this effort was recently reviewed [7]. In general, there are two types of "wheel and axle" compounds. One is formed by a large molecule (wheel) crossed by another, elongated molecule (axle); this arrangement is used, for example, with rotaxanes [8]. However, we are interested in "wheel and axle" compounds that are constituted by two bulky and relatively rigid end groups (wheels) and a linear rigid link (axle) (Fig. 1); from now on the expression is used to characterize compounds with this type of structure. The axle is generally an organic molecule, but it may also be obtained by means of supramolecular interactions, as in the case of the dimerization of (triphenylmethyl) benzoic acid [9] or 4-tritylbenzamide [10]. Sometimes, the end groups are decorated with hydrogen bonding groups, as in the case of

[a] *Structural Chemistry* 2012, 23(4):945–954.

[b] Dipartimento di Chimica Generale ed Inorganica, Chimica Analitica, Chimica Fisica, Parco Area delle Scienze 17/A, 43124 Parma, Italy
e-mail: alessia.bacchi@unipr.it; mauro.carcelli@unipr.it

I. Hargittai, B. Hargittai (eds.), *Science of Crystal Structures: Highlights in Crystallography*,
DOI 10.1007/978-3-319-19827-9_31

"wheel and axle" diols [11]. It is generally recognized that the dumbbell shape of "wheel and axle" compounds frustrates an efficient crystal packing and it is probably for this reason that these compounds have marked inclusion properties [5].

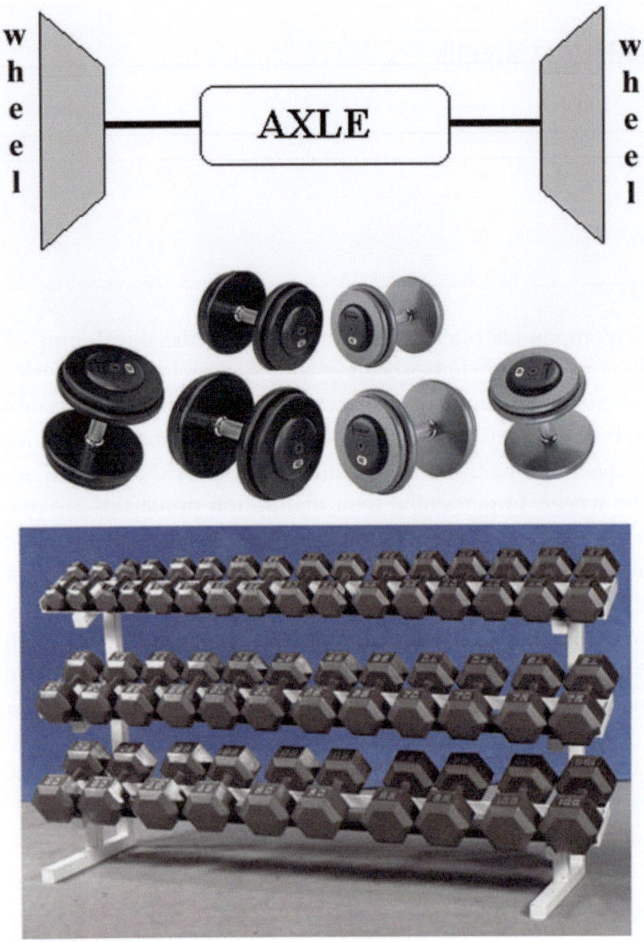

Fig. 1 *Top* sketch of the skeleton of a "wheel and axle" molecule. The shape reminds that of dumbbells (*middle*), that are difficult to pack efficiently (*bottom*).

A considerable innovation in the field is represented by the use of a metal to obtain "wheel and axle metal– organic" hosts and, in particular, "wheel and axle metal– organic" diols (WAMODs). The metal can participate to the definition of the spacer by using metal–organic bonds [12–17] or coordination bonds [18–25]. The coordination preferences of the metal were used to model the axle, and in this case the results obtained with calixarenes are particularly worth of note [26], but also redox and photophysical [27] properties of the metal ions are interesting for the design of new materials.

"Wheel and axle" compounds and WAMODs show interesting sorption properties and, from the beginning, a lot of studies focused on the analysis of the adsorption/desorption process and the relative rearrangement of the crystal packing. Our attention focused on WAMODs (Fig. 2). We can say that the crystal packing is "guest mediated", when the presence of the guest realizes the condition of maximum efficiency, and so the result is a clathrate. On the contrary, the packing is "self-mediated" when the host interacts only with other molecules of the host. Even if it is in general assumed that shape and symmetry of the molecule are more important than molecular functionality in the design and function of hosts as the "wheel and axle" compounds [28], we thought that the decoration of the wheels, for example with OH groups, could favor the switch between "guest mediated" and "self-mediated" packing, and, in this sense, it could influence the adsorption/desorption process.

Following the Kitagawa's view [29], if the attention is appointed to the sorption properties, there are three different types of microporous frameworks: first-generation materials have a microporous framework that collapses on removing the guest; second-generation materials exhibit a permanent porosity, also in the absence of included guests (for example, the metal–organic frameworks [2]), and third-generation hosts can rearrange reversibly in response to the presence of guests. Accordingly to this classification, a host that supports networks that can switch between "guest mediated" and

Fig. 2 Wheel and axle metalloorganic diols (WAMODs): the wheels carry two OH groups (*top*). *Bottom* realization of WAAMOD shape for [Pd(LOH)$_2$Cl$_2$].

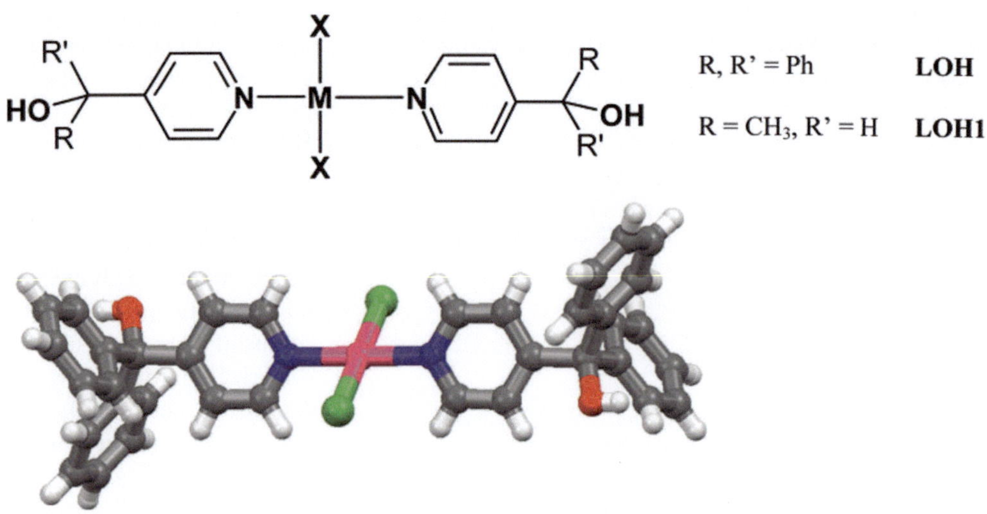

"self-mediated" packing in response to an external stimulus, represents a third generation host. In this sense, we are seeing that WAMODs can be considered third-generation hosts.

Building of the Metal–Organic Host

For the realization of a WAMOD host, some functionalities are necessary: (1) the group coordinating the metal, (2) some bulky groups for the wheel, and eventually, (3) the group that can donate/receive H-bonds for the decoration of the wheels. α-(4-pyridyl)benzhydrol (LOH, Fig. 2) satisfies these three criteria and it was used for the realization of new WAMODs (Fig. 2). Another ligand (1*S*)-1-(4-pyridinyl)ethanol (LOH1, Fig. 2), was successively used to validate this choice; it is expected to have poor clathrating properties, because of the reduced bulkiness of the wheels. As for the metals, Ag(I), Pd(II), Zn(II), and Pt(II) are used to synthesise the complexes.

Ag(I) Complexes as WAMODs: 1D Coordination Polymers

LOH reacts with $[Ag(CH_3CN)_4]BF_4$ and $AgPF_6$ giving the complexes [*trans*-$Ag(LOH)_2]BF_4$ and [*trans*-$Ag(LOH)_2]$ PF_6 [18]. In both cases the silver ions are coordinated by two pyridine rings giving a linear, centrosymmetric $Ag(LOH)_2^+$ cation. The anions $(BF_4^-$ or $PF_6^-)$ bridge the cations by means of charge assisted hydrogen bonds involving the OH groups: the resulting structure is based on zig-zag chains built by –OH···F–anion–F···HO– interactions (Fig. 3). The geometry and symmetry around the anion differentiate the two complexes: it is tetrahedral, with a pseudo twofold axis on BF_4^-, or octahedral, with a centre of symmetry on PF_6^-. In the former, the geometry is locked by mutual interactions of the aromatic groups, while in the latter there are no interactions between the triarylcarbinol ligands and the geometry is dictated only by the centre of symmetry. By using a nomenclature already introduced, it is possible to describe the final packing as "anion mediated". It is interesting to note that the final structure can also be described as 1D coordination polymer of "wheel and axle" cations and anions. Unfortunately and contrary to other examples reported in the literature [30], these "wheel and axle" based 1D polymers do not show sorption properties.

Pd(II) Complexes as WAMODs

The Structure

The complexes obtained by reaction of palladium(II) salts and LOH are square planar with a *trans* disposition of the ligands, and the general formula $[Pd(LOH)_2X_2]$ (X = anion) (Fig. 4). To study the influence on the crystal packing and on the sorption properties, many different complexes are synthesised and characterised by varying X (X = Cl, I, CH_3COO). In addition to the non solvates $[Pd(LOH)_2X_2]$, many solvates $[Pd(LOH)_2X_2]$nG with hydrogen bond acceptor guest molecules are isolated (X = Cl, n = 2, G = acetone, THF, DMSO, DMF; X = I, n = 2, G = DMSO, DMF; n = 3/2, G = 1,4-dioxane; X = CH_3COO, n = 1, G = THF). As already mentioned, these compounds are examples of "guest mediated" networks. It is

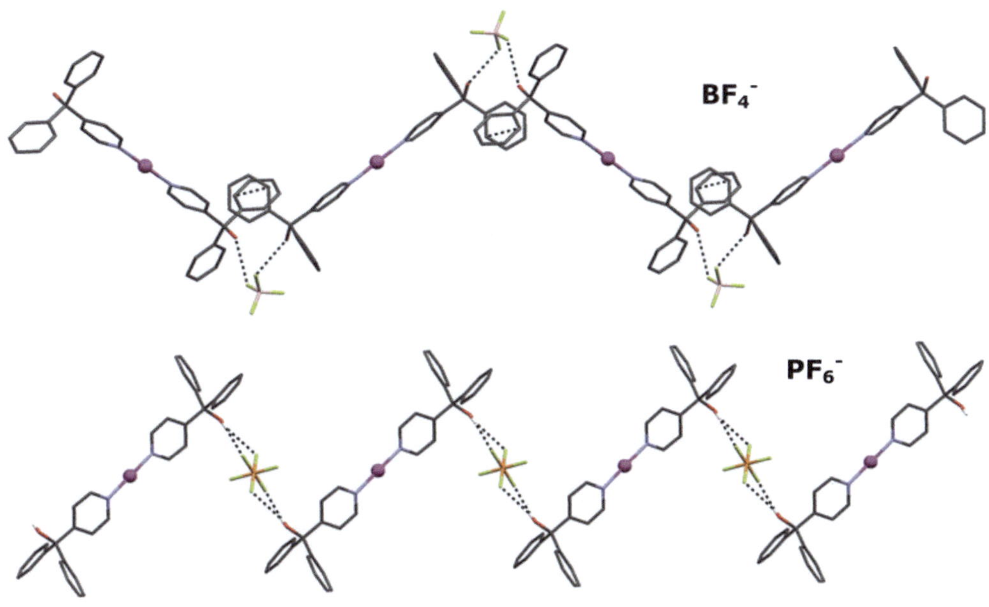

Fig. 3 Zig-zag chains in the crystal organization of $[Ag(LOH)_2]BF_4$ and $[Ag(LOH)_2]PF_6$.

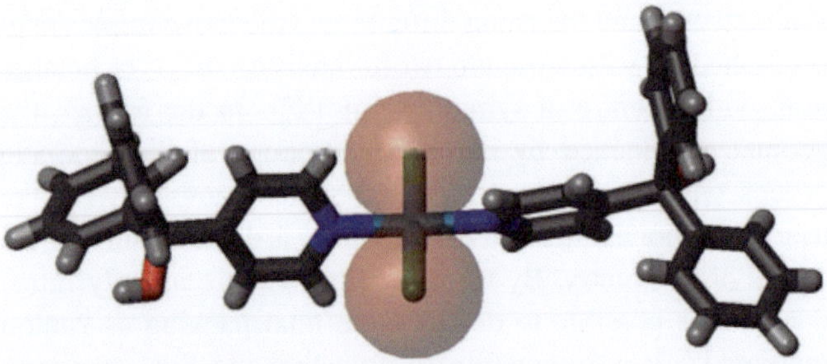

Fig. 4 General structure of $Pd(LOH)_2X_2$ molecules; X anions are indicated by the *red transparent balls*.

possible to obtain the solvate by crystallization from solution or by exposure of the non solvate crystalline powder to the vapor of the guest. On the contrary, it is possible to obtain the non solvate by removing the guest by heating. The processes are characterised by TGA and X-ray diffraction powder at room and at variable temperature [20–24].

From these studies emerge that, effectively, the WAMOD $[Pd(LOH)_2X_2]$ is a good unit to realize crystalline materials that adsorb and release reversibly small organic guests. It is also worth of note that these processes occur with retention of crystallinity [24].

An analysis of the crystal packing of the non solvate and of the solvate compounds is necessary to try to understand the mechanism of solvation/desolvation.

In the $[Pd(LOH)_2Cl_2]$2G complexes, solvation occurs by hydrogen bonding between the OH groups of the WAMOD and the acceptor atom on the guest (Fig. 5). All solvates are organized in layers with similar metrics. The layers consist in arrays of parallel columns made by stacked $[Pd(LOH)_2Cl_2]$2G units, that are inclined by $\theta = 54°$ relative to the column axis. Average Pd···Pd separation is $a' = 7.6$ Å within a column and $b' = 18.1$ Å between adjacent columns ($\alpha' = 90°$). The stacking of complex units within a column occurs by –OH···G interactions (Fig. 6). The guest molecules are included in the space between the chains within the layer; this is a guest mediated network. The layers may then be assembled in a parallel or antiparallel (herringbone) orientation.

The structures of the non-solvate $Pd(LOH)_2Cl_2$ is based on –OH···Cl hydrogen bonds which again generate columns of parallel units, arranged in layers (Fig. 6). The WAMODs in the columns are rotated of 28° respect to the solvate forms within the layer plane (Scheme 1), so that the complex long axis makes an angle of 26° with the column axis; the Pd···Pd separations are only slightly altered ($a' = 10.7$ Å, $b' = 17.4$ Å, $\alpha' = 90°$).

$[Pd(LOH)_2I_2]$ shows the same configuration of the ligands, the same OH (halogen) chains and the same layer metrics ($a' = 11.15$ Å, $b' = 18.13$ Å, $\alpha' = 82°$, $\theta = 26°$) as $[Pd(LOH)_2Cl_2]$, but the layers are stacked in parallel, while in $[Pd(LOH)_2Cl_2]$ there is an herringbone arrangement [20].

$[Pd(LOH)_2I_2]$2DMSO, $[Pd(LOH)_2I_2]$2DMF, $[Pd(LOH)_2I_2]$3/2(1,4-dioxane) are isolated by recrystallization from the appropriate solvents [20]. $[Pd(LOH)_2I_2]$2DMSO and $[Pd(LOH)_2I_2]$2DMF crystallize with very similar unit cells

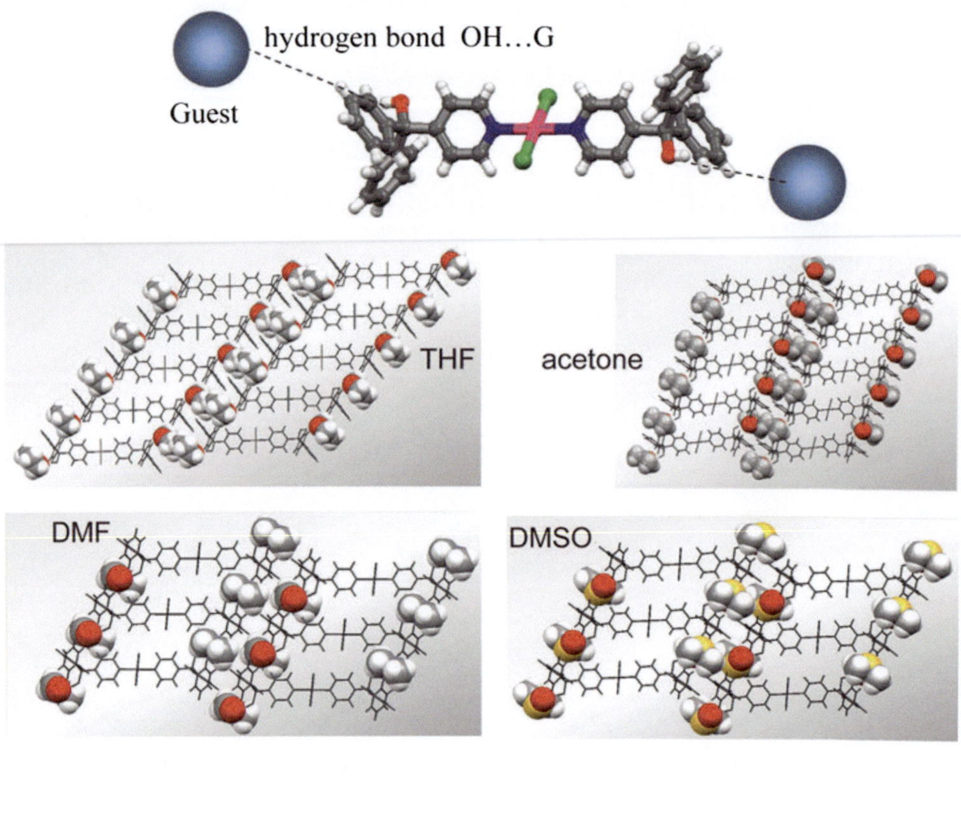

Fig. 5 Inclusion propensity of WAMOD complexes in $[Pd(LOH)_2Cl_2]$2G (G=THF, acetone, DMSO, DMF), organized in layers. The guests interact with the complex by OH···G hydrogen bonds.

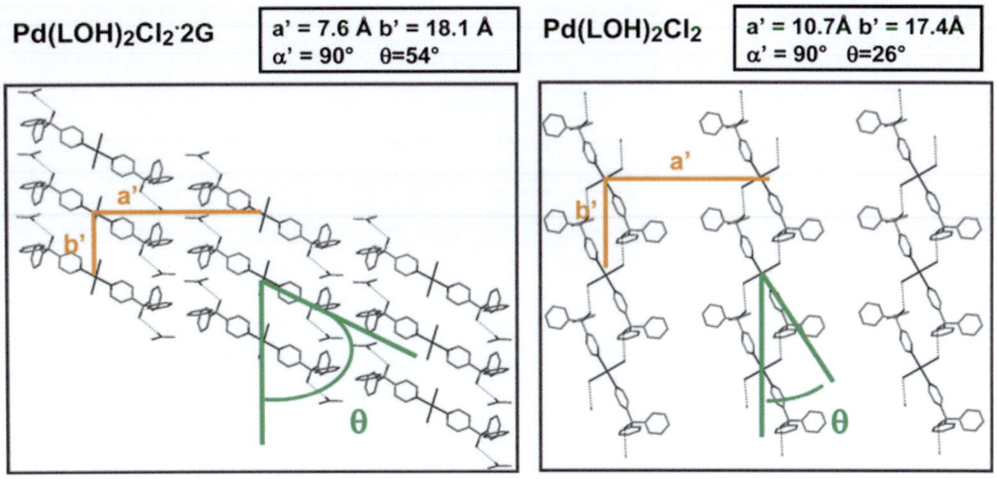

Fig. 6 Layer organization and metrics for the non solvate $[Pd(LOH)_2Cl_2]$ (*right*) and for the host–guest system $[Pd(LOH)_2Cl_2]$2 acetone (*left*).

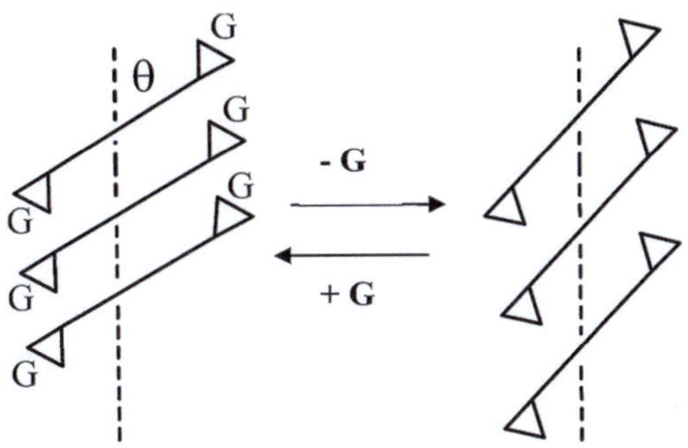

Scheme 1 Guest mediated (*left*) and self-mediated (*right*) networks.

and show the same packing arrangement. Each $[Pd(LOH)_2I_2]$ unit interacts with two guest molecules by –OH···G (G = DMSO, DMF) hydrogen bonds and the stacked $[Pd(LOH)_2I_2]$2G units form columns. The complex molecules are arranged in layers as in chloride solvates (average $a' = 6.75$ Å, $b' = 18.2$ Å, $\alpha' = 99°$). However, in both cases the guests have slided away from the site between the columns within a layer, being displaced to a position between the layers. Evidently, the layer arrangement of $Pd(LOH)_2X_2$ possesses two main solvation sites: site A between the columns within a layer, and site B between the aromatic

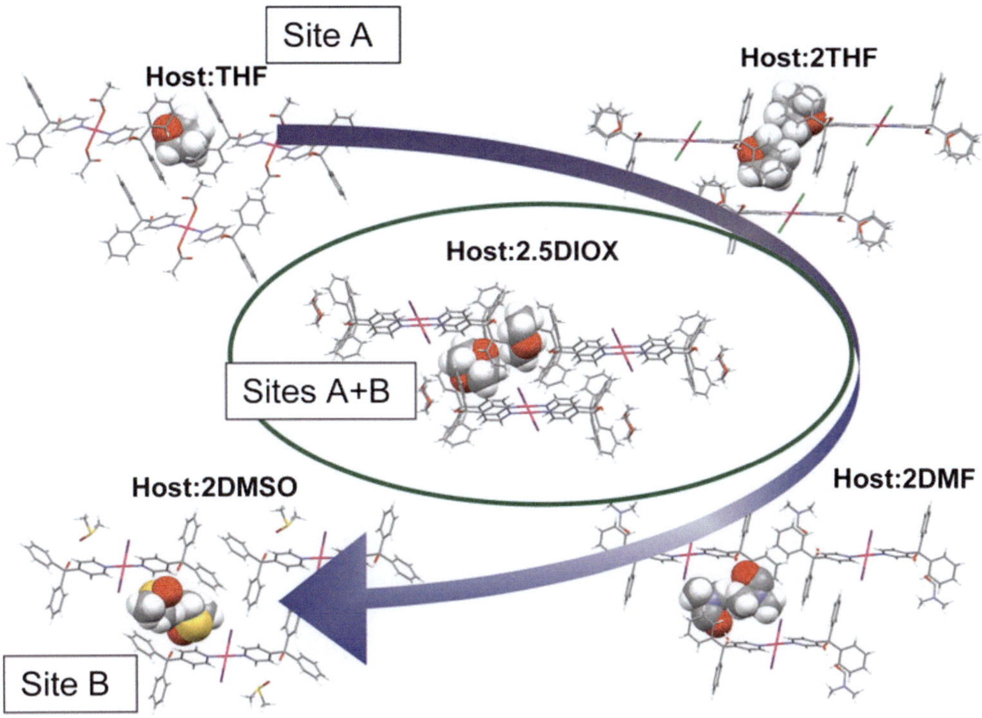

Fig. 7 Imaginary snapshot sequence of solvent migration path through the solvate lattice, by guest sliding between *sites A* and *B*, with the assistance of conformational rearrangement of aromatic rings. The structure of $[Pd(LOH)_2I_2]$3/2(1,4dioxane) (*central inset*) suggests a structural continuity for the process, as it can be viewed as the combination of $[Pd(LOH)_2(CH_3COO)_2]$THF (*top left*) and $[Pd(LOH)_2I_2]$ 2DMSO (*bottom left*).

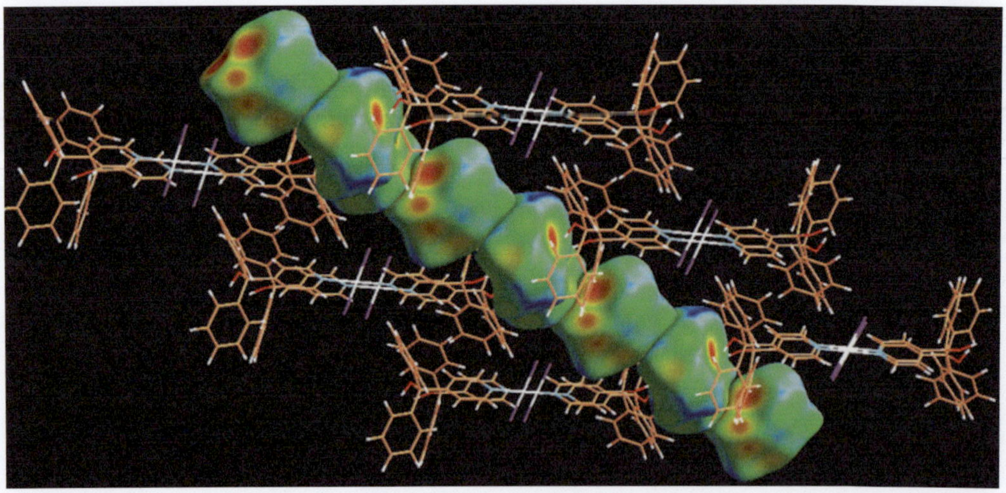

Fig. 8 Hirshfeld surface of solvent domains for $[Pd(LOH)_2I_2]3/2(1,4\text{-dioxane})$: guest *continuum* through the layers. Hirshfeld surface is *colored* according to *de* values.

groups and the anions of two adjacent layers (Fig. 7). This is confirmed by the analysis of the structure of $[Pd(LOH)_2I_2]3/2(1,4\text{-dioxane})$, where both solvation sites A and B are occupied. This is an interesting novelty and, with the aid of the study of the Hirshfeld surfaces [31, 32], it plays an important role in the following discussion of the solvent migration through the lattice.

In $[Pd(LOH)_2(CH_3COO)_2]$ [21] and in non solvate $[Pd(LOH)_2X_2]$ the overall organization of the complex units is similar, but $–OH{\cdots}O(CO)CH_3$ hydrogen bonds replace $–OH{\cdots}$halogen. However, the metrics of the layer ($a'=7.6$ Å, $b'=15.9$ Å, $\alpha'=86°$) is similar to the one observed in the solvates and the molecular orientation within the chains ($\theta=36°$) is intermediate between nonsolvate and solvate forms. This arrangement leaves some space between the chains within the layers.

$[Pd(LOH)_2(CH_3COO)_2]$THF is quasi isostructural with $[Pd(LOH)_2(CH_3COO)_2]$ [21]. Interestingly, there is no host–guest hydrogen bonding, but THF is clathrated in channels running along axis *a*. In particular, clathration is assisted by inter-chain stacking of THF between two parallel aromatic rings belonging to adjacent chains, with distances comparable to the ones typical of the intra-chain stacking of disolvates.

Sorption/Desorption of Organic Guests

On the basis of the structural analysis carried out on non solvate e solvate compounds, we proposed that guest release/uptake is essentially based on a "venetian blinds" mechanism [21], that is a concerted rotation of the $[Pd(LOH)_2X_2]$ units by ~30° in the layer plane (Scheme 1), in concomitance with the solvent migration through the lattice (Fig. 7). In particular, the structural analysis of $[Pd(LOH)_2(CH_3COO)_2]$ and of $[Pd(LOH)_2\ (CH_3COO)_2]$ THF suggests that the first step of the solid/gas solvation process could be the clathration of one molecule of guest between the aryl rings. Clathration successively triggers the collective reorientation of the host molecules. By studying the succession of the Hirshfeld surfaces of the guests in all the collected structures (Fig. 8), it is possible to identify a possible path for solvent migration through the lattice that involves guest sliding between sites A and B, with the assistance of conformational rearrangement of aromatic rings [20]. The aromatic rings act as "gate" which are "closed" in the first and last pictures and "open" in the two intermediate steps.

As a whole, the process is reversible because uptake or release of the guests require a minimum displacements for the single WAMOD and a modest reorientation for their molecular long axes.

Validation of the WAMOD Concept

The [*trans*-$Pd(LOH)_2X_2$] complexes show reversible inclusion properties due to their favorable shape, that allows reversible interaction with the guest with modest reorganization of the crystal packing. The modification of the steric demands of the wheels or of the geometry of the axle by using a less bulky ligand ((1*S*)-1-(4-pyridinyl) ethanol = LOH1, Fig. 2) or by changing the metal (Zn(II) instead of Pd(II)), respectively, leads to complexes without inclusion properties.

Zn(II) has coordination preferences different from Pd(II) and this can be used to modify the geometry of the axle. LOH reacts with ZnX_2 (X = Cl, Br) giving complexes with formula $[Zn(LOH)_2X_2]$; repeated attempts to obtain solvates were unsuccessful [22]. In $[Zn(LOH)_2X_2]$ the metal ion is coordinated by two pyridine rings in *trans* each other and the halide anions complete the coordination, as in Pd(II) complexes, but the Zn(II) cation is tetrahedrally coordinated. The modification of the geometry of the axle has an important effect on the crystal packing: the "bent" WAMOD units form a quadrangular networks based on –OH···X hydrogen bonds, which would be impossible to obtain with a square planar stereochemistry on the metal and LOH, due to reciprocal hindrance of the aromatic rings in the middle of the supramolecular squares (Fig. 9, right).

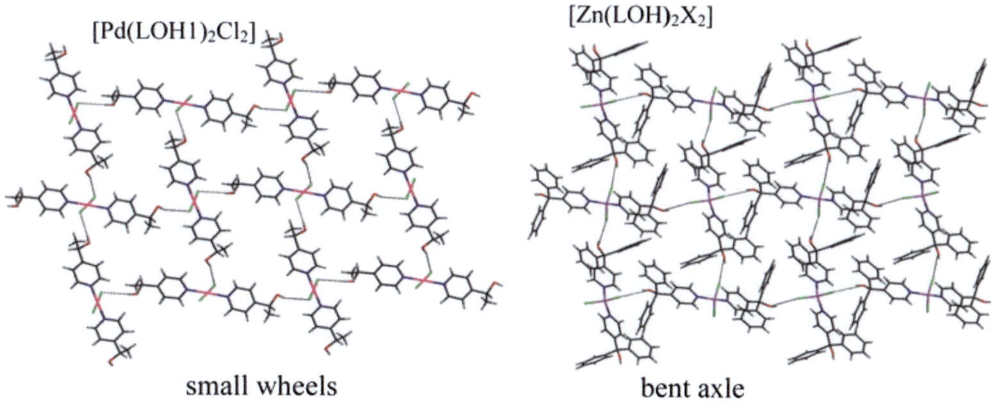

Fig. 9 Square patterns in [Pd(LOH1)$_2$Cl$_2$] (contracted wheel hindrance, *left*) and [Zn(LOH$_2$)Cl$_2$] (bent axle, *right*): these compounds do not show clathrating ability.

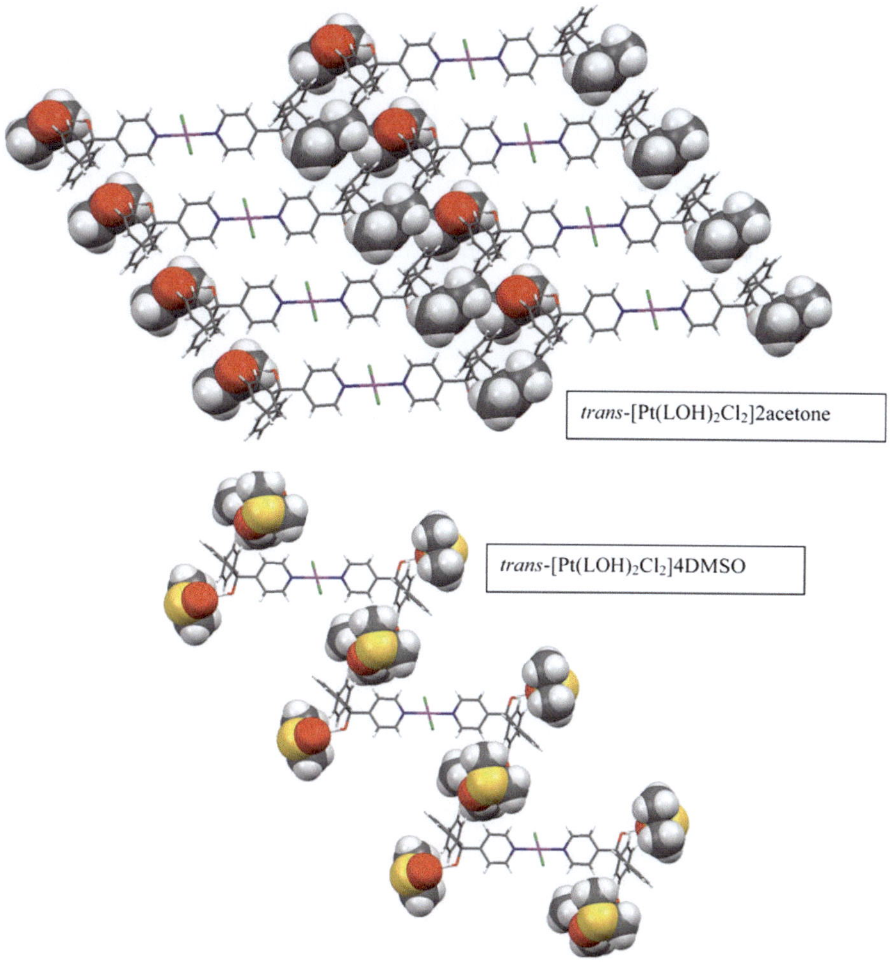

Fig. 10 Crystal packing of *trans*-[Pt(LOH)$_2$Cl$_2$]2G compounds (in figure G = acetone) (*top*) and of *trans*-[Pt(LOH)$_2$Cl$_2$]4DMSO (*bottom*). Guest molecules are evidenced in spacefill mode. The arrangement of *trans*[Pt(LOH)$_2$Cl$_2$]2G is isostructural with the related palladium(II) complexes with G in site A, while in *trans*[Pt(LOH)$_2$Cl$_2$]4DMSO two extra guest molecules are placed in solvation site B, between the layers.

In LOH1, the two aromatic rings of LOH are replaced by one methyl and one hydrogen. The less cumbersome ligand reacts with Pd(II) salts giving [*trans*-Pd(LOH1)$_2$X$_2$] (X=Cl, Br, I) complexes [23]. If the single metal–organic unit is without surprises, as in [*trans*-Pd(LOH)$_2$X$_2$], the packing organization is new for Pd(II) compounds. In fact it is based on a quadrangular pattern realised by means of –OH···X hydrogen bonds, in analogy to the one observed for [Zn(LOH)$_2$X$_2$] (Fig. 9, left). In [Pd(LOH1)$_2$X$_2$], the modest steric demands of the terminal substituents, that can be easily accommodated without the need to pucker the hydrogen bonded layers, permit the planar arrangement of the supramolecular squares. As for Zn(II) complexes, [Pd(LOH1)$_2$X$_2$] do not have clathrating properties.

The Pt(II) Complexes: From WAMODs to Supramolecular Metallamacrocycles

As a logical continuation of our studies on LOH coordination chemistry, we have reacted LOH with K_2PtCl_4. After the resolution of some synthetic troubles, *cis*- and *trans*-[Pt(LOH)$_2$Cl$_2$] complexes are isolated [25]. In particular, with *trans* configuration we have isolated the non solvate [Pt(LOH)$_2$Cl$_2$], and the two solvates [Pt(LOH)$_2$Cl$_2$]2acetone and [Pt(LOH)$_2$Cl$_2$]2DMSO (Fig. 10). They are isostructural to the respective palladium complexes, confirming the strength of the structural motifs: they present a layer arrangement of the complex molecule, with arrays of guests hydrogen bonded to the terminal carbinol OH groups. Interestingly, from DMSO solution [Pt(LOH)$_2$Cl$_2$]4DMSO was also isolated. The alteration of the stoichiometry with the insertion of a higher amount of guest destroys the layered pattern, since two DMSO molecules form the usual *trans*-[Pt(LOH)$_2$Cl$_2$]2DMSO unit, but the others are intercalated between the complexes (Fig. 10, bottom).

The *cis*-Pt(LOH)$_2$Cl$_2$ unit was found in one non solvate complex and in diverse solvates (acetone, 1/2CHCl$_3$, CH$_3$OH, 2CH$_3$OH, 3CH$_3$COOH, 2.5THF); in all the structures, it presents substantially the same conformation (Fig. 11) [25].

In the non-solvate [*cis*-Pt(LOH)$_2$Cl$_2$], there are two non equivalent molecules in the asymmetric unit. The new relevant structural motif is that the two non equivalent metal–organic units form two centrosymmetric supramolecular rectangles by means of OH···OH hydrogen bonds: in each complex one OH acts as donor and the other as acceptor (Fig. 12). The two non equivalent rectangles are hydrogen bonded: the two OH groups involved as acceptors in one rectangle behave as hydrogen bond donors towards two chloride ligands belonging to two adjacent rectangles. In this way, a chain along [110] direction is formed. Concomitantly, in the *ab* plane, layers of supramolecular rectangles are made by chains linked by OH…Cl hydrogen bonds (Fig. 12). Finally, the layers are stacked along [001] through phenyl–phenyl contacts.

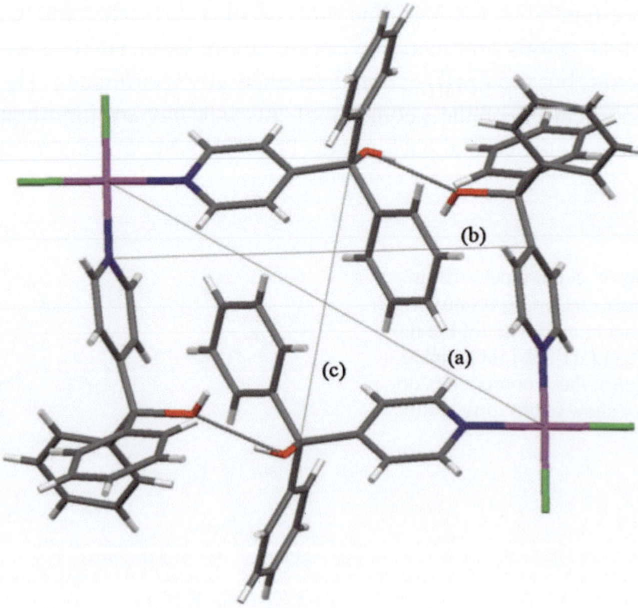

Fig. 11 Supramolecular rectangles in the structure of *cis*- [Pt(LOH)$_2$Cl$_2$]nG. Rectangle dimensions are found as: (**a**) 13.4 Å, (**b**) 11.5 Å, (**c**) 7.8 Å in $n=0$ and they are similar for the solvate compounds.

The insertion of small molecules into the packing produces effects that depend essentially on the hydrogen bonding nature of the guests, but the supramolecular rectangles, with dimensions more or less conserved (Fig. 11), remain the main structural motif in these *cis*-Pt(II) complexes. In the acetone solvate the layer structure made by cross-linked herringbone chains of supramoleculars rectangles observed in the non solvate compound is maintained; the acetone molecules are nested between these layers. On the contrary, the layers are destroyed by a moderate hydrogen bond donor as CHCl$_3$ or by methanol (Fig. 13): one OH···Cl hydrogen bond is substituted with a CH···Cl interaction ([*cis-Pt*(LOH)$_2$Cl$_2$]1/2CHCl$_3$) or by MeOH···Cl and one OH···(OH)Me hydrogen bonds ([*cis*- Pt(LOH)$_2$Cl$_2$]MeOH).

Conclusions

Crystal engineering can find in coordination chemistry a big support. Here we have reported some of the results obtained with pyridylbenzhydrol ligands and Ag(I), Pd(II), Zn(II), and Pt(II) metal ions. In particular, a bistable network is obtained in the presence of a significant steric hindrance around the carbinol groups and of a linear stereochemistry of the axle, so to obtain reliable "metal–organic wheel and axle" diols. The favorable shape makes *trans*-Pd(LOH)$_2$Cl$_2$ able to

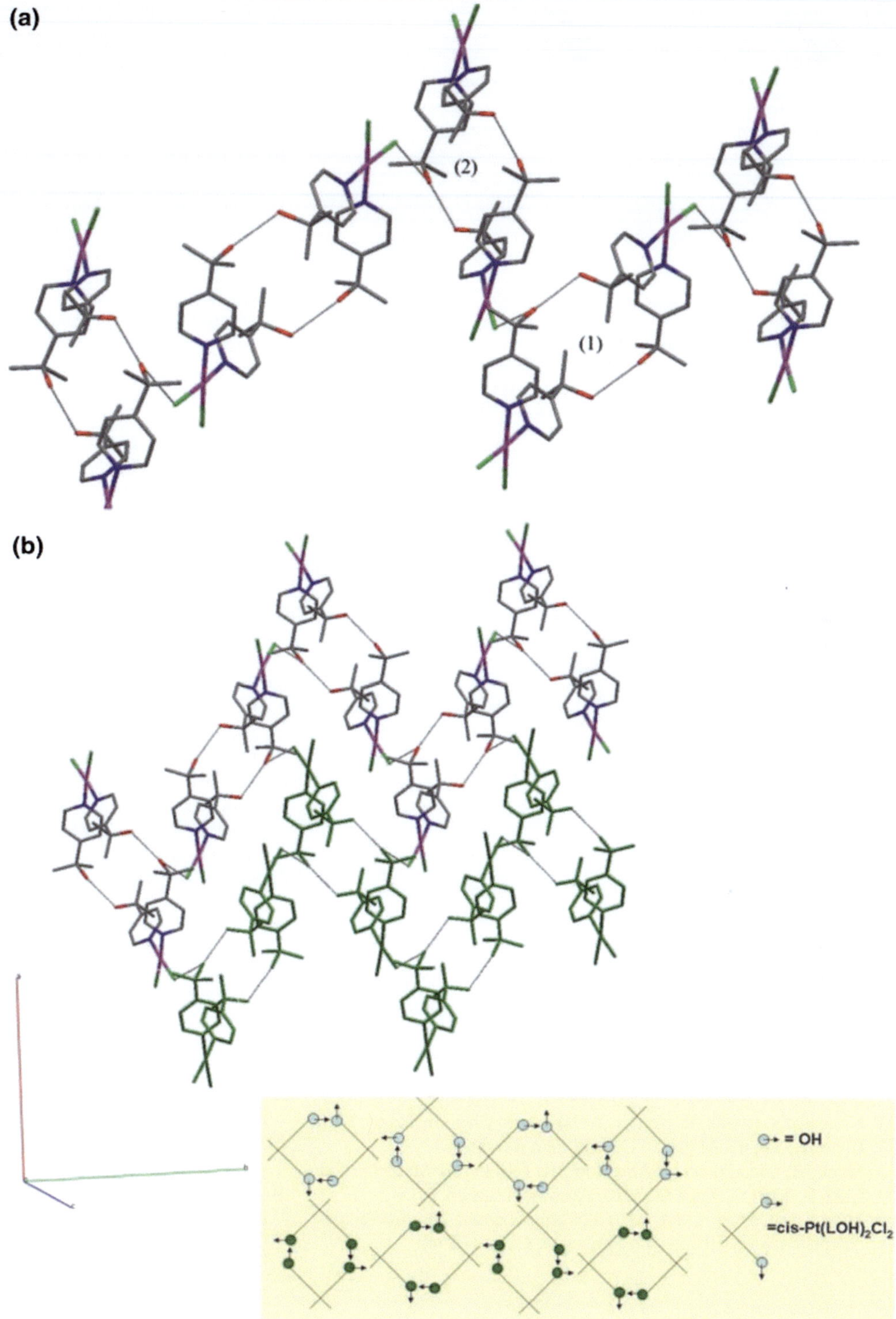

Fig. 12 Packing arrangement of *cis*-[Pt(LOH)$_2$Cl$_2$]: (**a**) chains of supramolecular rectangles linked by OH···Cl hydrogen bonds; (**b**) layers of supramolecular rectangles made by the chains displayed in (**a**) linked by OH···Cl hydrogen bonds. Two chains are represented, one *colored* in *green* for clarity. *Shaded inset*: scheme showing the hydrogen bond connectivity of the layers. Carbinol aryls and hydrogens omitted for clarity.

adsorb and release reversibly small guests. During the process there is a switch between "self-mediated" and "guest mediated" packing, with OH···anion bonds that become OH···guest hydrogen bonds. On the other hand, in [*cis-Pt*(LOH)$_2$Cl$_2$] the OHs that decorate the wheels are behind the formation of the supramolecular metallamacrocycles: the realization of new synthons that combine the robustness of the coordination bonds with the flexibility of hydrogen bonds seems a promising way to obtain new materials, with interesting properties.

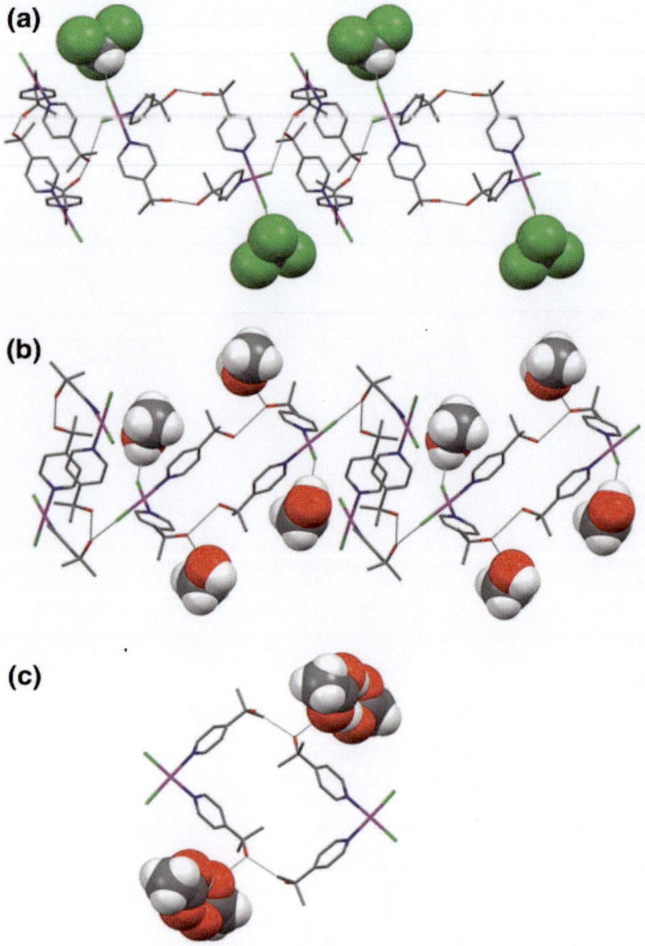

Fig. 13 Organization of supramolecular rectangles and their interaction with the guests in *cis*-[Pt(LOH)Cl_2]G for (**a**) G = 1/2$CHCl_3$; (**b**) G = CH_3OH; (**c**) G = 3CH_3COOH. Guest molecules are evidenced in space-filling mode; hydrogen bonds are *dashed*.

References

1. Kitagawa S, Kitaura R, Noro S (2004) Angew Chem Int Ed 43:2334
2. Long JR, Yaghi OM (2009) Coord Chem Rev 38:1213
3. Meek ST, Greathouse JA, Allendorf MD (2011) Adv Mater 23:249
4. Chen B, Xiang S, Qian G (2010) Acc Chem Res 43:1115
5. Toda F (1996) In: Lehn JM (ed) Comprehensive supramolecular chemistry, vol 6, Chapter 15. Pergamon, Oxford, p 465
6. Toda F, Agaki K (1968) Tetrahedron Lett 33:3695
7. Soldatov DV (2006) J Chem Crystallogr 36:747
8. Arduini A, Bussolati R, Credi A, Pochini A, Secchi A, Silvi S, Venturi M (2008) Tetrahedron 64:8279
9. Jetti Ram KR, Kuduva SS, Reddy DS, Xue F, Mak TCW, Nangia A, Desiraju GR (1998) Tetrahedron Lett 39:913
10. Malla Reddy C, Sreenivas Reddy L, Aitipamula S, Nangia A, Lamb C, Mak TCW (2005) CrystEngComm 7:44
11. Toda F, Ward DL, Hart H (1981) Tetrahedron Lett 22:3865
12. Bruce MI, Grundy KR, Liddell MJ, Snow MR, Tiekink ERT (1988) J Organomet Chem 344:C49
13. Che CM, Chao HY, Miskowski VM, Li Y, Cheung KK (2001) J Am Chem Soc 123:4985
14. Lu W, Xiang HF, Zhu N, Che CM (2002) Organometallics 21:2343
15. Chao HY, Lu W, Li YQ, Chan MCW, Che CM, Cheung KK, Zhu NY (2002) J Am Chem Soc 124:14696
16. Vicente J, Chicote MT, Alvarez-Falcon MM, Fox MA, Bautista D (2003) Organometallics 22:4792
17. Bruce MI, Halet JF, Le Guennic B, Skelton BW, Smith ME, White AH (2003) Inorg Chim Acta 350:175
18. Bacchi A, Bosetti E, Carcelli M, Pelagatti P, Rogolino D (2004) Eur J Inorg Chem 10:1985
19. Bacchi A, Bosetti E, Carcelli M, Pelagatti P, Rogolino D (2004) CrystEngComm 6:177
20. Bacchi A, Bosetti E, Carcelli M (2005) CrystEngComm 7:527
21. Bacchi A, Bosetti E, Carcelli M, Pelagatti P, Rogolino D, Pelizzi G (2005) Inorg Chem 44:431
22. Bacchi A, Bosetti E, Carcelli M (2007) Cryst Eng Comm 9:313
23. Bacchi A, Carcelli M, Chiodo T, Mezzadri F (2008) Cryst Eng Comm 10:1916
24. Bacchi A, Carcelli M (2010) Influence of size and shape on inclusion properties of transition metal-based wheel-and-axle diols. In: Comba P (ed) Structure and function. Springer, Dordrecht, pp 235–253
25. Bacchi A, Carcelli M, Pelagatti P, Rispoli G, Rogolino D (2012) Cryst Growth Des 12:387
26. Maharaj F, Bishop R, Craig DC, Jensen P, Scudder ML, Kumar N (2009) Cryst Growth Des 9:1334
27. Coropceanu EB, Croitor L, Botoshansky MM, Siminel AV, Fonari MS (2011) Polyhedron 30:2592
28. Weber E (1996) Shape and symmetry in the design of new hosts. In: MacNicol DD, Toda F, Bishop R (eds) Comprehensive supramolecular chemistry, vol 6, Chapter 17. Pergamon, Oxford, pp 535–592
29. Kitagawa S, Uemura K (2005) Chem Soc Rev 34:109
30. Soldatov DV, Moudrakovski IL, Ratcliffe CI, Dutrisac R, Ripmeester JA (2003) Chem Mater 15:4810
31. Hirshfeld FL (1977) Theor Chim Acta 44:129
32. McKinnon JJ, Spackman MA, Mitchell AS (2004) Acta Crystallogr B60:627

What Is a Chemical Substance and How Is It Formed?[a,b,c]

Vladimir Ya. Shevchenko[d]

Abstract

The objects of nature are structurally generalized, especially the objects undergoing the stage of formation from atoms to macrosized species. No exceptions from a large number of the examined structures of inorganic and organic compounds, algae, plants, viruses, and other objects have been found out. For the majority of substances, the build-up starts from fundamental configurations, then clusters are formed, whereby the structure is assembled. The hierarchical construction is the basic principle irrespective of whether the whole structure is periodic or aperiodic. The spatial domain, in which the substance is formed, is tens to hundreds of nanometers in size, the type of the space geometry being inessential because in the small all the geometries are equivalent.

Keywords

Nanoworld • Fundamental configurations • Multidimensional space • Self-assembly • Structural code • Icosahedrons • Convergence • Fundamental nanocluster

Nulla cogente natura, sed concurs quodau fortuito atomorum.
Cicero (106–43 BC) De Natura Deorum.
Not by teleology, by the chance association of atoms.

Introduction

This study does not claim at all to be the exhaustive and exact scientifically substantiated answer to the questions put in its title. This is an attempt to define chemistry as a basic science, which, as both physics and biology, has its own underlying objective.

As a matter of fact, in the beginning of this century, several journals, Science and Nature being among those, have published the lists of the outstanding problems of the twenty-first century for mathematics, physics, and biology. Among various problems for physics, the problem of the origin of the Universe is stated for physics and for biology it is the origin of life. The problems posed for chemistry are absolutely incommensurable with the scope of physics and biology. In the Ref. [1], the problem of synthesis of complex chemical compounds and growing nanostructures has been specified. In the journal Nature, two problems are presented, namely, "Are there any limitations in rational chemical synthesis?" and "Is it possible to construct an information 'gene' for inorganic (organic) substances?". It makes no sense to compare these problem statements with those assigned to physics and biology.

Interestingly, some mathematical problems, in particular, the Hodge problem and the Poincare problem that are associated with the space partition and its dimensionality also appeared in the Science journal. It should be noted that mathematics in its modern conception is not a fruit of the

[a] *Structural Chemistry* 2012, 23(4):1089–1101.

[b] This article is dedicated to Prof. Dr. Sten Andersson on the occasion of his 80th birthday.

[c] This paper has been done in conjunction with Professor Vladislav A. Blatov and Professor Grigoryi D. Ilyushin.

[d] Institute of Silicate Chemistry, Russian Academy of Sciences, Makarova Emb. 2, 199034 St. Petersburg, Russia
e-mail: shevchenko@isc.nw.ru

I. Hargittai, B. Hargittai (eds.), *Science of Crystal Structures: Highlights in Crystallography*,
DOI 10.1007/978-3-319-19827-9_32

human intellect, but it does exist as an independent basic science (queen of all sciences). Long ago, Plato said that mathematicians do not devise the laws but discover these. In his well known work "The unreasonable effectiveness of mathematics in natural sciences" [2], E. Wigner gave quite a number of examples of how mathematics unexpectedly simply describes complex phenomena of the material world.

An important inference following from this study is that such an adequacy of an observation and its description occurs in the cases when the space, in which these phenomena take place, is properly chosen. E. Wigner exemplifies this statement with the quantum theory of general relativity theory.

Both theories involve different mathematical concepts, namely, the concept of the 4D Riemann space and the concept of the infinite-dimensional Hilbert space. The relativity theory is applicable to macroscopic bodies, whereas the quantum theory goes into the microscopic phenomena domain. The attempts to create a unified theory so far have not succeeded, though intuitively it is clear that this is possible at least because the space in which our world exists is unified.

A very big issue of the nature of the space and its dimensionality, which has been investigated by many outstanding scientists for a very long time, falls beyond the scope of this study. However, we make use of the important inferences from these investigations. First of all, the space is structured and the position which may be occupied in it by a material object (particle) is mathematically exactly defined in terms of the concept of the finite simple groups. In 1830, Hessel derived 32 crystal classes, which are conventionally referred to as the finite groups of the proper and improper rotations in a 3D space containing only the first-, second-, third-, fourth-, and sixth-order elements. The crystal property discovered 50 years before by one of the founders of crystallography abbot R. J. Haüy was known to Hessel. This property was named the law of rational indices. The discovery of this law was based on even more interesting discovery by Bishop Nicolas Steno, that is, the regular position of atoms in the spatial lattice sites.

These are the pre-Roentgen studies that formed the basis for the notion of the symmetry, which indicates the space structuring. The notion of symmetry made it possible to use mathematical description of space using the tool of the simple finite groups. In the 1970s, more than 100 experts under the guidance of D. Gorenstein established a specific team aimed at presentation of a complete classification of the simple finite groups consisting of three infinite countable families and 26 so-called sporadic groups with special properties, including the group of the highest order that was named "monster." The order of this group amounts to about 10^{53}, which, probably, indicates the total number of compounds (including chemical ones) in nature. Note that the total number of protons in the Universe is about 10^{80}).

Why does symmetry underlie almost all structures and regularities of the living and nonliving nature?

It is likely that the source concept here is that of the finite set of species and the equality of the primary entities from which larger units are constructed at a certain level of the matter organization. There also should be the equality in geometry, that is, the equality of the relative position of equal parts. It is these reasons that govern the most important basic property of a substance, namely, its structure. By definition, the structure (composition, arrangement, order) is a system of stable connections of an object that provide its integrity and self-identity, i.e., the conservation of its basic properties during various inner and outer stresses.

The interatomic interaction gives rise to a geometrically simple structure, in which the atoms occupy fixed positions spaced at a certain distance from one another. Even consideration of this geometry per se contributes much to understanding of the regularities of the atomic structure of matter. Formally, the geometrical consideration involves investigation into the stable grouping of atoms, coordination polyhedra, complexes, clusters, etc. Their form and symmetry are studied, as well as the relation of symmetry to the spatial structure of a substance. It should be noted that, when the substance is formed, commonly the principle of the maximum filling of space is implemented as a tendency to the maximum number of atoms per unit volume.

As M. Planck wrote in 1910, the physical law of the highest priority for all that system is the principle of the least action comprising all four world coordinates distributed absolutely symmetrically.

It is noteworthy that the space geometry can be multidimensional as well, which follows from the generality of the laws known to date. All physical laws (and corresponding laws of geometry) known so far totally and necessarily extend to any number of dimensions: these laws do not imply anything that would distinguish the four dimensions. However, mathematics, the group theory in particular, acquaints us with such formations whose structure is absolutely different and depends on the number of dimensions.

One more important circumstance is that there exists closest relation between chemical and structural formulae of substances and their symmetry, and that the law of multiple proportions in chemistry is a mere consequence of the more general symmetry law. The product of the multiplicity of points S_i and the relative number of the points n_i is a constant quantity for all discontinua of the given kind of symmetry. This product is equal to the relative number of asymmetric points "n" of a fixed kind in the same unit cell. Using this relationship, one can a priori calculate (predict) all possible chemical formulas for binary, ternary, and so forth compounds. Commonly, the more complex the chemical formula

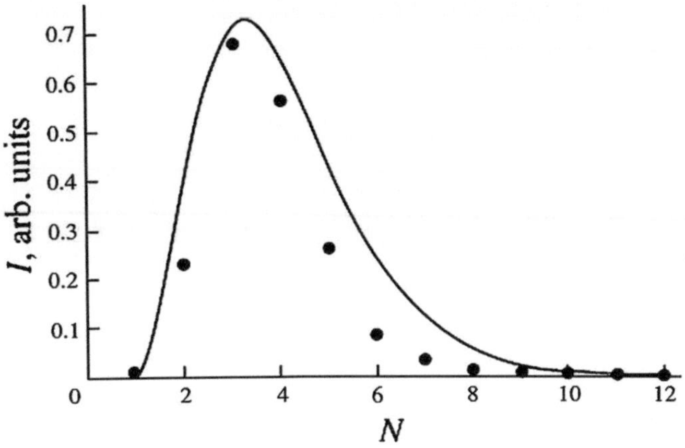

Fig. 1 Points denote the numbers of inorganic crystal structures (I) with 1, 2, 3, …. N different elements. $N=3$ has a maximum of 19,000. The line follows the Planck distribution of energy in an absolutely black body [3].

of a substance, the lower the symmetry of its structure. A. Mackay made one more interesting observation. Figure 1 shows the dependence of the number of substances (structures) on the number of components that is constructed according to the data of the International Database for inorganic substances [3].

It can be seen that rigorous restrictions are imposed on the complexity, though seemingly the number of substances composed of a large number of elements should have increased as N! or Fibonacci series.

A. Mackay draws an analogy with the blackbody radiation whose theory M. Planck resolved by introducing the notion of the energy quantum.

The energy distribution curve according to the Planck theory is also depicted in Fig. 1. Is this coincidence occasional? Of course not, the space is structured and quantized as it follows from consideration of the theory of simple finite groups, so that atoms can occupy only certain sites in space.

In conclusion of this section, it is appropriate to cite a saying of highly reputed English mathematician G. H. Hardy:

> But the most important point is that pure geometry is the space–time reality of the physical world rather than merely some pictures [4].

On the Definition of "Chemical Substance"

If we turn to various modern handbooks and textbooks, we can find the following definitions of the substance.

The substance is a form of matter that has a rest mass, in contrast to a field.

The substance is a variety of matter, which contains the number of chemical particles (atoms) from 10^{10} to 10^{15} and more.

In chemistry, an individual chemical substance is understood as a set of atoms bonded to one another by a certain law.

The attributes that characterize a chemical substance are its qualitative composition (atoms of elements), quantitative composition (the number of atoms of each element), chemical bond between the atoms, and their relative position. The collection of all these attributes is necessary and sufficient to characterize the essence of a chemical substance. Therefore, the structural formula or chemical structure is necessary and sufficient to characterize a chemical substance.

The notion "structure" is the evolving and most beneficial background for an analysis of this notion that is presented by structural chemistry.

This notion has been evolving for hundreds of years and hereafter we will try to analyze this process.

In 1899, M. Planck, based on considerations of dimensionality, derived the values of length $l_{pl}=10^{-33}$ cm, time $t_{pl}=10^{-43}$ s, and mass $m_{pl}=10^{-5}$ g from three fundamental constants, that are, speed of light "c," gravitational constant "G," and Planck constant "$\hbar$". A comparison with Heisenberg's uncertainty relationship brings about a conclusion (presently generally accepted) that measuring length and time shorter than l_{pl} and t_{pl}, respectively, is impossible. These quantities imply the limit of applicability of the classical gravitation theory and our notion of the space–time.

However, whereas the quantities l_{pl} and t_{pl} cannot be measured because of their small values, Planck's mass of tens of micrograms can be readily determined. For example, the least weight of ordinary substances can be measured using a Sartorius balance (Germany) with the minimum measurable weight of 10^{-6} g.

Nevertheless, no traces of particles of Planck's mass have been detected. The energy corresponding to that mass is so enormous (10^{19} GeV) that neither at present nor in the future its realization appears feasible. There exists an opinion that the value of Planck's mass is merely a result of a play with numbers. A controversy is evident. Planck's mass is a conventional boundary between macro and micro, i.e., between the gravitation theory and quantum mechanics. This is one of a few units in quantum mechanics in terms of which the objects in our world worth to be measured. On the assumption that this particle consists of atoms and its size, according to quantum theory, corresponds to the Compton wavelength $l_c \sim 10^{-33}$ cm, this is the gravitational radius (Schwarzschild radius) in the classical general relativity theory. Assuming the existence of Planck's mass as a boundary, $l_c = l_{gr}$. A substance of that size should have maximum density ($\sim 10^{90}$ g/cm^3). Second controversy is obvious.

Using the notion of the boundary character of this particle, let us try to determine the number of atoms in it. For example, such iron particle contains 2.3×10^{16} iron atoms. Certain conformity with the above definition is seen. Though, obviously, by definition matter is something that occupies space and can be perceived in some way by the sense organs, the basic feature of matter is its feasible conversion to energy and vice versa. In chemistry, a chemical substance is a form (piece) of matter with constant chemical and characteristic properties. It cannot be divided into components physically without breaking chemical bonds.

Therefore, the interatomic interaction is the basic principle of the substance formation. Thus, Planck's mass relates to the quantum limit and cannot be attributed to a real mass of a chemical substance. The coincidence of the number of atoms in the particle and compound is a lucky occurrence, as well as the (quantitative) coincidence of the mass and weight on the Earth. Does a spatial domain exist where the boundary between micro- and macroworld lies? Estimates obtained in Ref. [5] give the values for a body whose localization width is on the scale of its own size. Such a body occurs at the boundary between the micro- and macroworld.

For example, calculations for a zirconia particle give its size of 10^{-7} cm, that is, 1 nm. Allowing for the mean atom size of about 1 Å, about 1,000 atoms can be contained in a volume of 1 nm^3 and about billion atoms in a particle of 100 nm size.

For the number of atoms of 10^{15} mentioned in the definition of the substance in one of the handbooks, the particle size would amount to about hundreds of nanometers.

It follows from the aforesaid that, within the above definitions, the size of space in which the substance is formed makes up from one to hundreds of nanometers. Strictly speaking, this is just the "nano" domain.

Space Dividing and Fundamental Configurations

The arrangement of objects in space, in which any point of space belongs at least to one of the objects and they pairwise do not have common inner points is referred to as space partition. This operation makes the basis for construction of two, three, or more structures, which often can be readily perceived and visualized by identification underlying their skeleton.

The necessary condition for the regular space partition into cells is the requirement for symmetry, that is, such property of space S that the G-discrete symmetry group should operate transitively (i.e., any cell could be superimposed on any other one by transformation pertaining to group G). The cell belonging to such partition is referred to as a fundamental domain. To date, an excellent mathematical tool has been developed, which allows these partitions to be carried out and analyzed [6]. The whole history of structural chemistry is associated with one or another model of such tilings. By the way of example, let us consider the structure of metals and intermetallides.

The atoms tending to the closest packed arrangement in space is the basic geometrical factor, whereby the metal structure is commonly explained. At the same time, this principle is usually applied to the crystal structure on the whole (the closest packing model) without analysis of the mechanisms of the microstructure build-up from nanosized objects. Obviously, the local atomic configurations (nanoclusters) should also obey this principle. However, in this case, the atomic arrangement differs from that in a compact phase. In the most general form, the optimal arrangements of points (identical atoms) in spaces of various dimensionalities (fundamental configurations) were considered by Cohn and Kumar [7]. In particular, it was demonstrated that a fundamental configuration for the 3D space is the icosahedron (Fig. 2). The idea of fundamental can

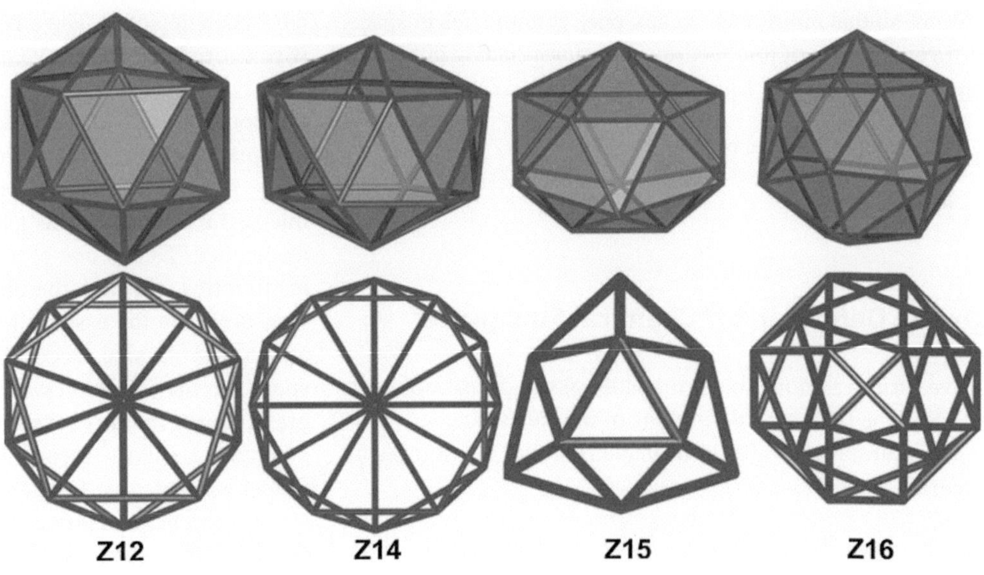

Fig. 2 Frank–Kasper polyhedra: icosahedron, the 14-vertex polyhedron, the 15-vertex polyhedron, the polyhedron Friaufa (16-vertex). Polyhedra characterize dense packing of identical atoms (icosahedron) or different types of atoms of the two (the other polyhedra) [7].

be naturally extended to systems of various atoms, as well as to nanosized atomic configurations falling beyond the first coordination sphere of the atom (coordination polyhedron). An exact mathematical solution has not been obtained for this case, however, fundamental configurations can be found out from analysis of the intermetallide structure.

The most well-known coordination polyhedra in intermetallides are the Frank–Kasper polyhedra (Fig. 2), the Z14, Z15, and Z16 polyhedra being considered as fundamental configurations corresponding to the optimal arrangements of atoms of two different types. Among these polyhedra, the most prevailing one is the Friauf's (Z16) polyhedron; therefore, it

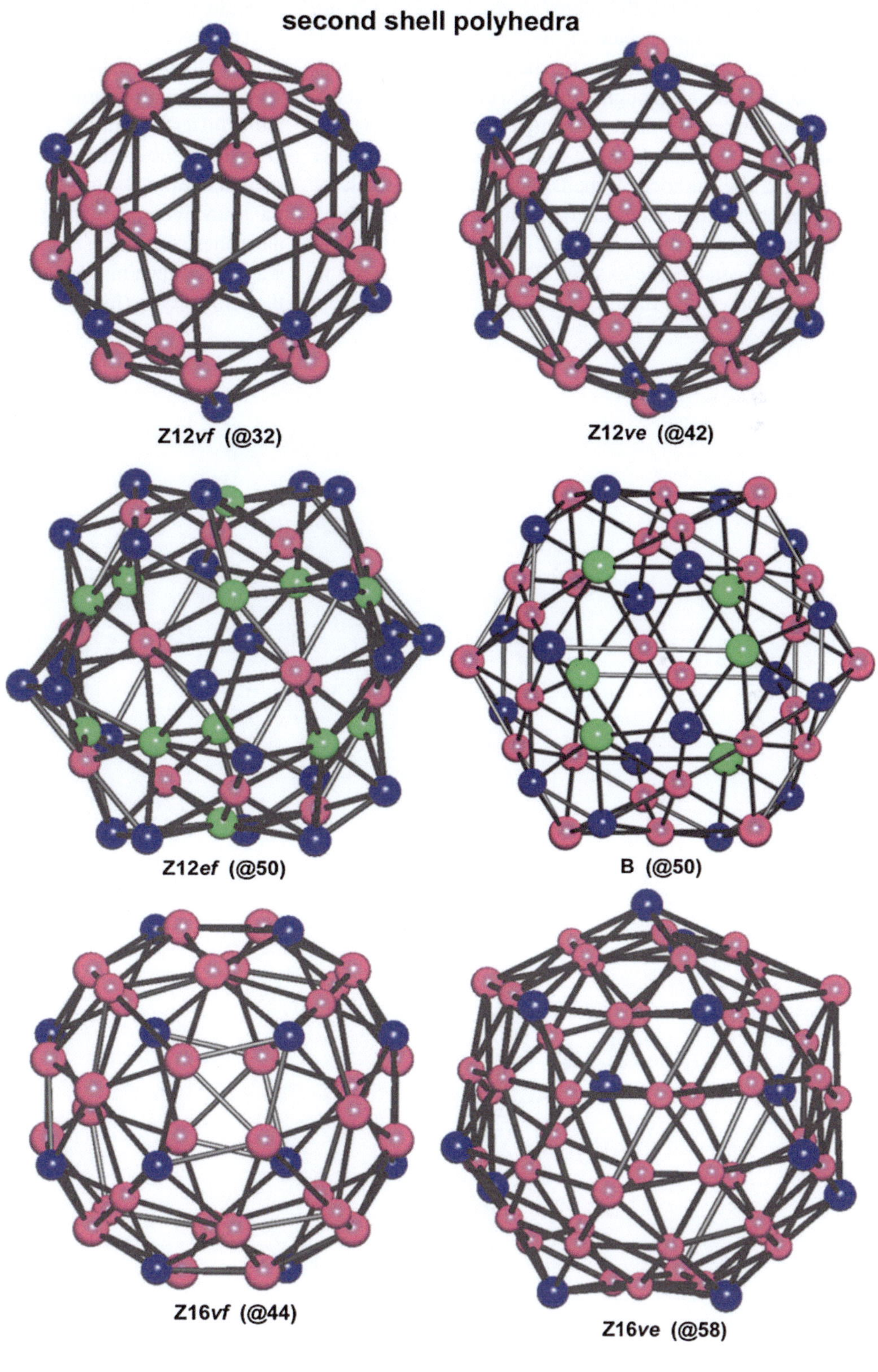

Fig. 3 The second shell bilayer clusters which are realized in the intermetallides. The first polyhedral shell—an icosahedron (Z12) or a polyhedron Friaufa (Z16). The letters *v*, *e*, and *f* shells in the symbols mean that the atoms of these shells are projected, respectively, on the top, the centers of the edges or center of the inner faces of the polyhedron. Shells Z12vf and Z12ve match Bergman and Mackay clusters. The shell of the same name cluster B corresponds to the central icosahedron, found in $NaCd_2$ and Mg_2Al_3 [8].

can be considered an analog of the icosahedral configuration for a system consisting of atoms of different sizes.

At present, it is accepted to describe intermetallides in terms of coordination polyhedra, that are, single-layer clusters (a central atom plus first coordination sphere). At the same time, the crystallization problems inevitably bring about the necessary consideration of more complex local configurations, in which the atoms in the second layer are closely packed on the surface of the inner polyhedral core. These multilayer configurations can be also treated as fundamental configurations, though in Ref. [7] they are not considered due to their structural complexity. We will refer to them as fundamental configurations of the second, third, etc. order according to the number of shells in them. An analysis of multilayer configurations in intermetallides [8] brings about typical shells that are characteristic of the growth mechanisms of the metal particles (Fig. 3), which are similar to nanoclusters based on Friauf's icosahedron. Thus, apart from the known Bergman and Mackay nanoclusters based on the icosahedron, we discovered nanoclusters with the inner Friauf's polyhedron, in which the second shells are constructed by the same principles (Fig. 3). At the same time, new variants of the structure of the second shells of the icosahedral two-layer fundamental configurations have been revealed, which are not found in the Bergman and Mackay nanoclusters (Figs. 3 and 4).

The further growth of the nanophase may proceed in two ways: either by build-up of new shells and forming third and higher-order fundamental configurations, or by packing nanoclusters with the subsequent macrophase formation. Our study shows that it is the first- and second-order fundamental configurations that are most prevailing ones, whereas the four-layer nanoclusters are extremely rare. Therefore, the nanophase transition to the macrophase occurs by the second way, the known nanocluster types often forming packings of the known type (Figs. 5, 6, and 7).

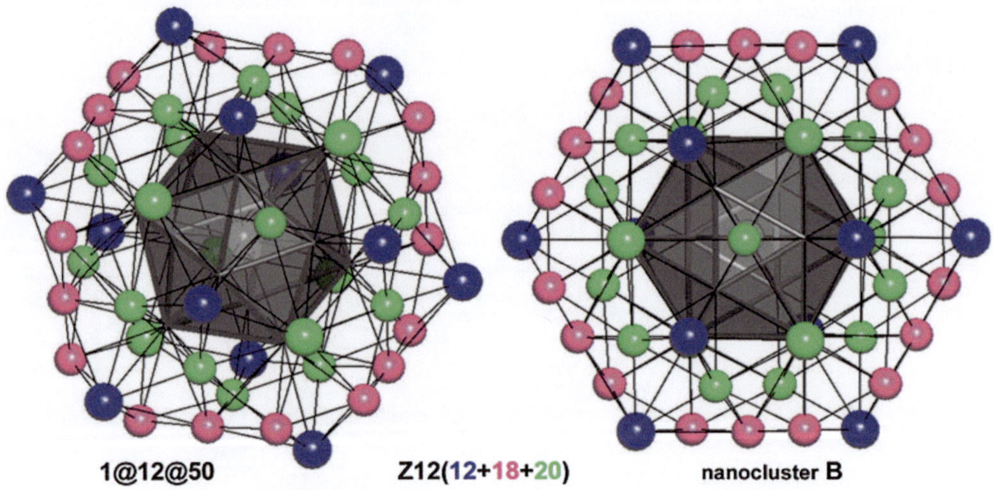

Fig. 4 Nanocluster B. The atoms of the second shell, projected onto vertices, edges, and faces of the inner icosahedron, respectively, are shown in *blue, pink,* and *green* [8].

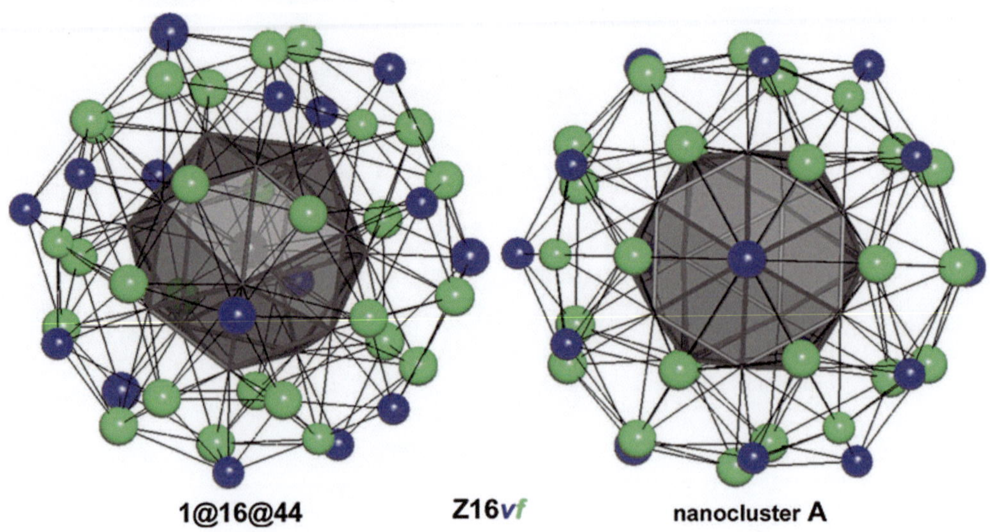

Fig. 5 Nanocluster A from structure of $NaCd_2$—a topological analog of the Bergman cluster, but based on the polyhedron Friaufa (16-vertex) [8].

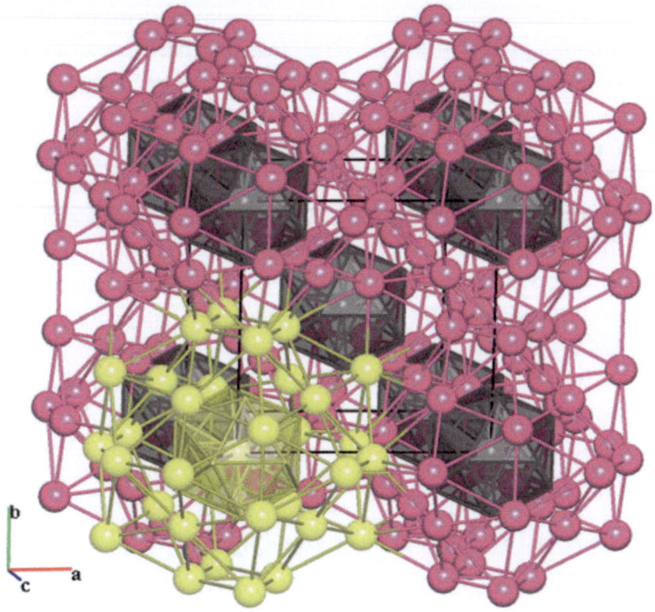

Fig. 6 Packing of nanoclusters Bergman (Z12vf) in the K_9NaTl_{13}. crystal structure with the body-centered unit. One of the nanoclusters is highlighted in *yellow* [8].

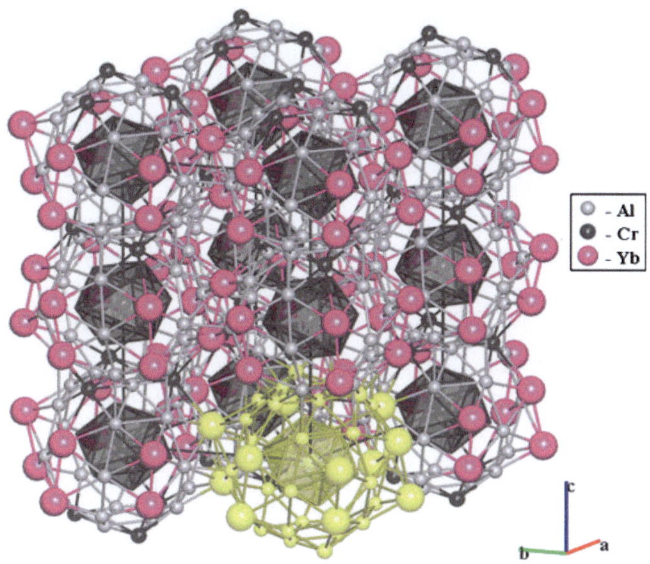

Fig. 7 The primitive hexagonal packing of nanoclusters Mackay(Z12ve) of $Yb_6Cr_{5.76}Al_{41.24}$. One of the nanoclusters is highlighted in *yellow* [8].

So, the nanophase build-up can be characterized by the following sequence: (1) formation of the first-order fundamental configurations of the polyhedral shape (icosahedron, Friauf's polyhedron, or some other polyhedra); (2) formation of the multilayer fundamental configurations (commonly, second or third-order ones); (3) closest packing of fundamental configurations.

For substances of other types, numerous nice packings can be outlined. For the 2D structures, these are Kepler, Penrose, Gummelt, and Truchet spherical coatings; for the 3D structures, besides icosahedra, these are Bergman, Mackay, Kuo, Tsai, Blatov, etc., clusters; for the helicoidal structures, these are a set of helix-forming polyhedra, including biological Bernal's "deltahedron," etc. To date, all this diversity of the fundamental configurations completely describes the total variety of (chemical) substances [9].

Calculations of Chemical Composition

The chemical composition of a compound can be calculated from the composition of nanoclusters, their stoichiometric ratio, and type of their bonding with one another.

The equation for calculation of α atoms of chemical element A in the composition of an intermetallide is presented in the general form:

$$\alpha = \sum_{I=1}^{N} K_I \left[\sum_{i=0}^{n(I)-1} \alpha_n^{(I)} + \sum_{j-1}^{m_I} \frac{\alpha_{n(I)j}^{(I)}}{j} \right], \quad (1)$$

where N is the number of different nanoclusters in a model; k_I is the relative number of the nanoclusters of the given type; $n(I)$ is the number of layers in the Ith nanocluster; $\alpha_{ij}^{(I)}$ is the number of atoms A in the ith layer of the Ith nanocluster that belong concurrently to j nanoclusters (including the Ith one); m_I is the maximum number of nanoclusters to which atoms A present in the composition of the Ith nanocluster can concurrently belong. Matrix α is always corner-shaped (the nonzero elements therein are contained only in the bottom row and first column), because the atoms of the inner shells belong only to the given nanocluster ($\alpha_{ij}=0$ at $j>1$). In the special case of the nanocluster *packing* (when they do not have common atoms), matrix α is the column vector; for the trivial clusters consisting of a single atom, the matrix contains only one element.

Thus, for example, the $TbNi_2$ phase is formed from the single-type nanoclusters composed of 1@12@50=Ni@Ni_6Tb_6@$Ni_{36}Tb_{14}$ and spacer Ni atoms (trivial clusters) with the 1:1 ratio. Each nanocluster B has 12 contacts with the other similar nanoclusters, the contacts forming two equal groups. In the first group, when each of the six contacts is formed, the nanoclusters have three common Ni atoms that concurrently belong to two nanoclusters and two Tb atoms shared between three nanoclusters. In the second group, in each of the six contacts, the nanoclusters have only one common Ni atom. Therefore, in this example, A_1=Ni, A_2=Tb;

$N=2$; $I=1$ and 2 correspond to nanocluster B and the spacer; $k_1=k_2=1$; $n_1=2$, $n_2=0$; $m_1=3$, $m_2=1$; and matrices $\alpha^{(1)}$ and $\alpha^{(2)}$ have the form for A_1, and $\alpha^{(1)} = \begin{pmatrix} 1 & 0 & 0 \\ 6 & 0 & 0 \\ 12 & 24 & 0 \end{pmatrix}$ and $\alpha^{(2)} =$

(1) for A_2. $\alpha^{(1)} = \begin{pmatrix} 1 & 0 & 0 \\ 6 & 0 & 0 \\ 8 & 0 & 6 \end{pmatrix}$ and $\alpha^{(2)} = (0)$

The rows in matrices $\alpha^{(1)}$ correspond to the zeroth, first, and second layer of the nanocluster (the zeroth layer formally relates to the central atom); the sum of all the elements in the row is equal to the number of atoms of the given type (A_1 or A_2) in the layer. The sum of the elements of the first, second, or third columns is equal to the number of atoms of the given type that belong to the nanocluster completely, by 1/2, or by 1/3. Calculation of the number of the Ni and Tb atoms in the composition of the compound (α_1 and α_2) gives

$$\alpha_1 = 1\cdot((1+6)+(12/1+24/2+0/3))+1\cdot(1)=32;$$
$$\alpha_2 = 1\cdot((0+6)+(8/1+0/2+6/3))+1\cdot(0)=16,$$

which corresponds to the stoichiometric composition $Ni_{32}Tb_{16}=Ni_2Tb$.

The form of the calculation equation is independent of the periodicity in a structure and whether it is finite or infinite.

Let us consider some examples of the intermetallide structures of different complexity that feature diamond symmetry (group Fd-3m). We will demonstrate that, despite quite different crystal-chemical complexity, these structures are constructed by the same mechanism involving topologically similar supraclusters.

1. *Structure* $ZrZn_2$ pertains to the Laves phase group, whose nanocluster cell is 3-atoms cluster in the shape of triangle $ZrZn_2$. Since the neighboring nanocluster cells do not have common atoms, the chemical composition of a substance coincides with the composition of the nanocluster cell.
2. *Structure* $ZrZn_{22}$ consists of primary nanoclusters $ZrZn_{16}$ (A) and $ZnZn_{12}$ (B) aggregated in triangular nanocluster cells AB_2 of the same topology as that in $ZrZn_2$. Its composition is $ZrZn_{39}$, which can be represented as $[ZrZn_5]Zn_{34}$, where the inner atoms of the nanocluster cell are indicated in the square brackets. Each outer atom is shared between two different nanocluster cells; hence, the structure composition can be calculated as $[ZrZn_5]Zn_{34/2}=[ZrZn_5]Zn_{17}=ZrZn_{22}$.
3. *Structure* $NaCd_2$ is built from two-layer primary nanoclusters (Figs. 4 and 5) $[Na(Na_4Cd_{12})]Na_{24}Cd_{20}$ (A) and $[Cd(Cd_6Na_6)]Na_{18}Cd_{32}$ (B), which are aggregated in nanocluster cells AB_2. In this case, six Na atoms of the nanocluster A are shared between two nanoclusters, 12 Cd atoms and 12 Na atoms of the nanocluster B are also shared between two nanoclusters, and six Na atoms of the nanocluster B are shared between three nanoclusters. The composition of this compound is calculated basing on the nanocluster cell $\{[Na(Na_4Cd_{12})]Na_{12}Na_{12/2}Cd_{20}\}$ $\{[Cd(Cd_6Na_6)]Na_{12/2}$ $Na_{6//3}Cd_{20}Cd_{12/2/}\}_2=\{Na_{23}Cd_{32}\}$ $\{Cd_{33}Na_{14}\}_2=Na_{51}Cd_{98}$, whence it is seen that this composition differs from the idealized formula $NaCd_2$.

Structural Generality of the Objects of Nature in the Nanostate

Small particles (nanoparticles) are capable of interacting with one another as actively as at the atomic level, retaining the structure motives, i.e., they interact coherently. By mixing nanoparticles, small molecules, and polymers, it is possible to control self-organization of the nanoparticles and form various structures without any changes in the chemical structure of the components involved.

Thereby, very complex hierarchical structures can be produced. This is characteristic of both inorganic and biological objects, e.g., proteins.

This property of the nanostate was observed as long ago as in 1999 [10], and it was referred to as convergence. It is known that the DNA molecule is used for self-organization of nanoparticles to be glue to each other.

Rather interesting particles are observed in the intermetallide family.

We analyzed databases and revealed a number of metal clusters with high nucleation, which are nanosized. For example, in Ref. [11], apparently the largest metal polynuclear complex $[Mn_{84}O_{72}(O_2CMe)_{78}(OMe)_{24}(MeOH)_{12}(H_2O)_{42}(OH)_6]$ was synthesized and its structure is corroborated by X-ray diffraction analysis. This complex consists of 84 Mn atoms and it is ring-shaped (torus) with 1.9 nm I.D., 4.2 nm O.D., and 1.2 nm thick (Fig. 8). In a crystal, these rings are packed in infinite stacks, forming tunnels.

The second in size high-nuclear manganese complex $[Mn_{44}O_8(O_2CMe)_{52}(pd)_{24}(py)_8]$ [12] comprises nearly twice as less Mn atoms (44); however, it features an unusual topological structure, being an aggregate of four loops (Fig. 9). Both the above-described complexes are monomolecular magnets.

The formation of the loops of various shapes is characteristic of biological structures; however, it is inherent feature of the so-called supramolecules to take most exotic shapes.

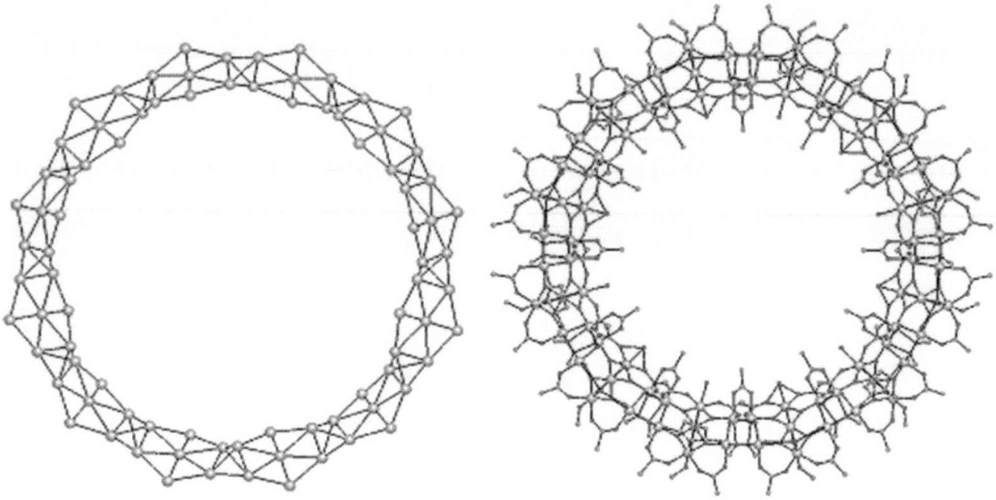

Fig. 8 The structure of nano-wheels $[Mn_{84}O_{72}(O_2CMe)_{78}(OMe)_{24}(MeOH)_{12}(H_2O)_{42}(OH)_6]$: (*left*) skeleton Mn_{84}; (*right*) polynuclear complexes [11].

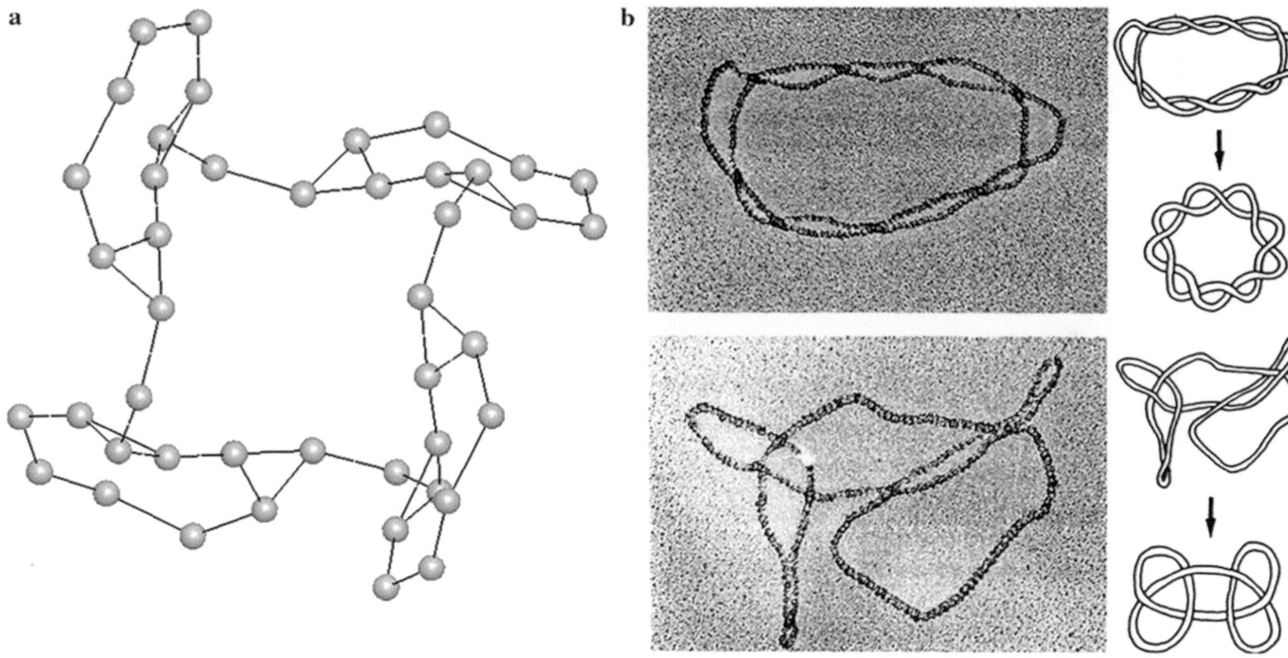

Fig. 9 (**a**) Four-knotted skeleton of Mn44 in the complex $[Mn_{44}O_8(O_2CMe)_{52}(pd)_{24}(py)_8]$ [12]. (**b**) (*Left*) Electron microscope pictures of knotted DNA strands; (*right*) schematic diagrams of the real and ideal knots.

This is exemplified in the nanosized molecules $\{M_{36}\}$ and $\{M_{150}\}$. After formation of the outer wheel, the $\{M_{36}\}$ molecule again transfers to a solution to prepare the buildup of a new molecule (Fig. 10) [13].

In biology, examples of the cell clustering in spherical shells with unexpected coordination numbers, 5, 6 and 7 were found. Specifically, the algae Volvox, one of the elementary multicellular organisms, is a spherical shell, mainly composed of heptagonal, hexagonal, and pentagonal cells (that are the cells having seven, six, or five adjacent cells with three cells converging to each "vertex" (Fig. 11). There occur specimens with the tetragonal and octagonal cells, but biologists noted that, if such atypical cells (with less than five and more than seven edges) are absent, the number of pentagonal cells are exactly by a factor of 12 larger than that of the heptagonal cells (altogether, there can be several hundreds and even thousands of the cells). This statement follows from the known Euler formula.

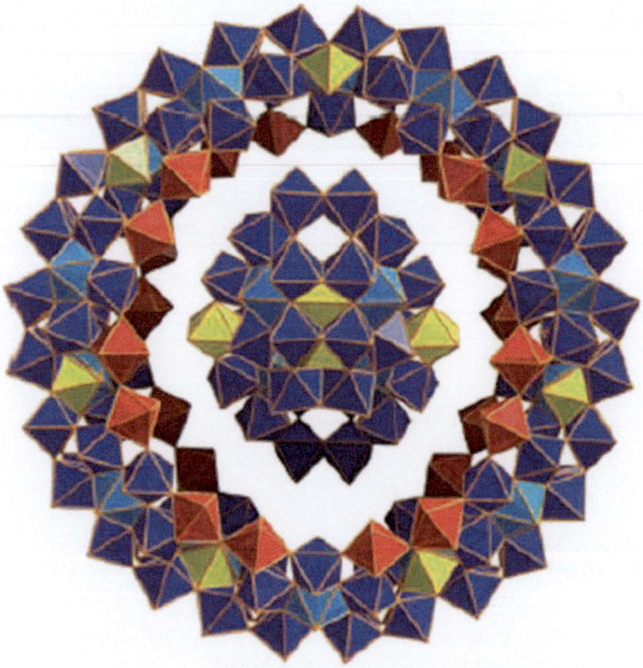

Fig. 10 A polyhedral representation of the nanoscale {Mo36} ⊂ {Mo150} wheel. The inner side of the ring is the template to the outer ring. The polyhedral building blocks are colored as follows: {Mo1} (*yellow*), {Mo2} (*red*), {Mo8} (*blue* with a *light blue* pentagonal central group) [13].

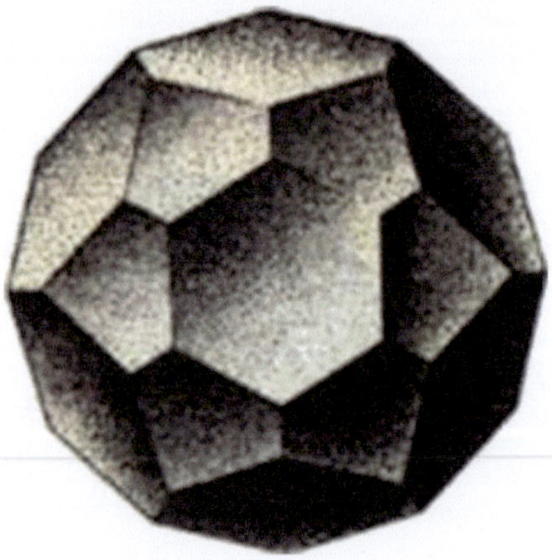

Fig. 11 Algae Volvox.

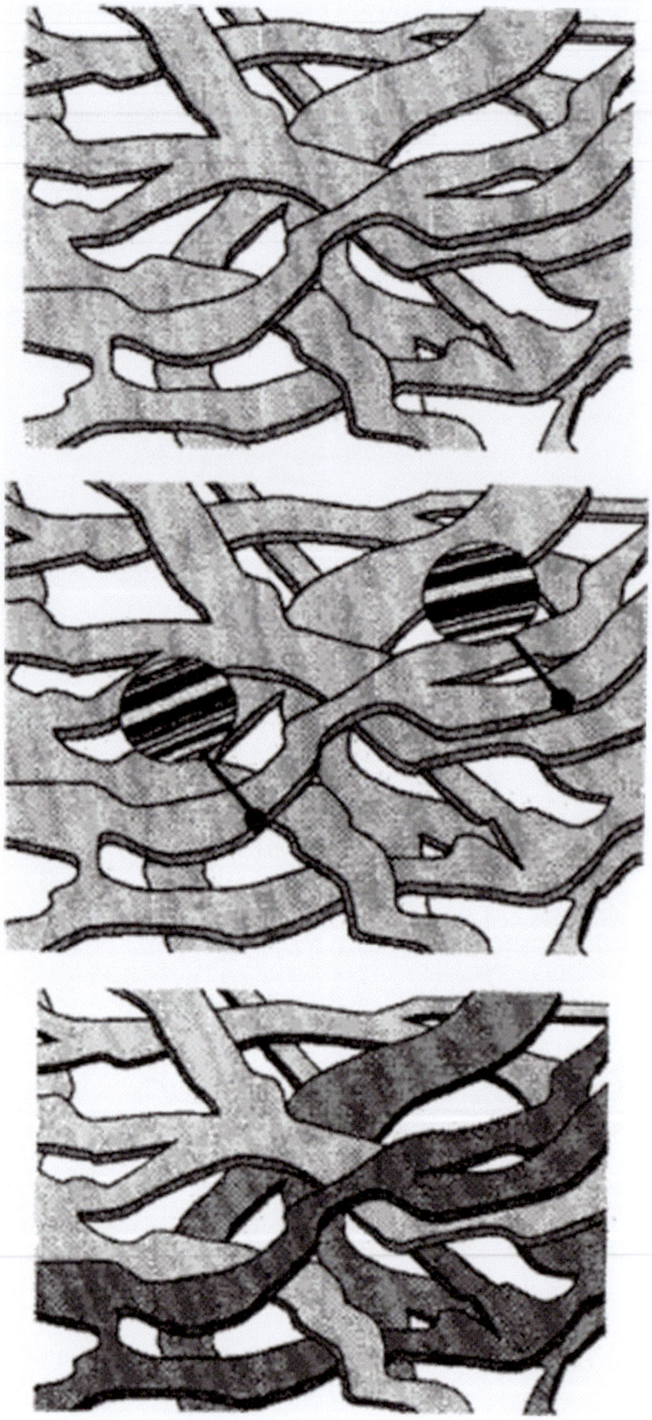

Fig. 12 Representation of mineral crystal layers, a possible primitive gene? (*Middle*) Variations in genetic "barcodes." (*Bottom*) A consortium of chemically different primitive genetic materials.

A cluster of 13 aluminum atoms behaves as an iodine atom. This property can be used for various applications, though it is necessary to study the process of consolidation of atomic clusters to other forms.

In 1949, J.D. Bernal assumed that (mineral) clays may be a source of the formation of new organic (biological) life molecules (Fig. 12). The present generally accepted viewpoint, though, is based on the simultaneous synthesis of DNA (RNA) from saccharides and amino acids.

Viruses play a special role in the hierarchical structure of the nature. The general mathematical theory of the structure of virus capsids was set forth by S. Andersson. A fundamental configuration, namely, the truncated dodecahedron made the basis of the theory. All the capsid structures, from the minute to the largest ones (altogether, there are about 102 viruses), have been described by means of the fundamental algebraic theorems combined with the exponential scale. The chemical virus particles consist of concentrated protein gradients. Mathematically, this gradient is described by means of the Gaussian distribution function and summing these functions over the capsid. This calls for calculations in the finite space. Since all these items are interrelated, a topology is needed for the description of Hardy's deviations as part of the topology.

This mathematical tool provides a link between the structural code and genetic code.

The genetic code directs the production of the structural code of the truncated dodecahedron via the relationship between the DNA nucleotide sequence and the amino acid sequence in proteins. The mutations are commonly accumulated inside the nucleotides that are the form and structure of protein.

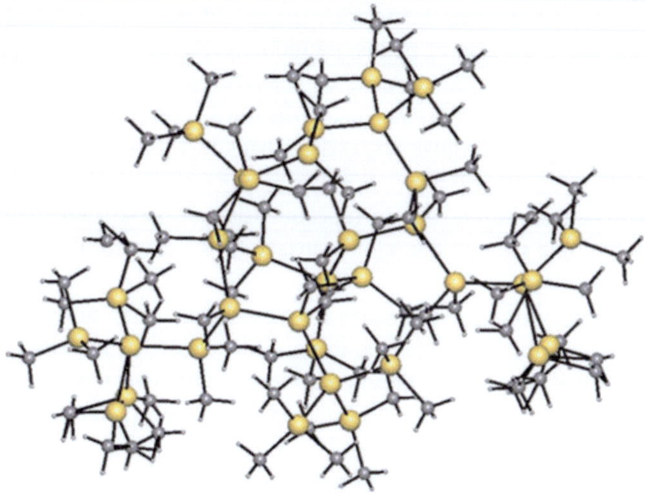

Fig. 13 Molecule 1,1,1,3,3,4,5,5,7,7,8,8,10,10,11,12,12,14,14,14-icosamethyl-2,2,6,6,9,9,13,13-octakis(trimethylsilyl)-4,11-bis(1′,1′,3′,3′,3′-pentamethyl-2′,2′-bis(trimethylsilyl)trisilanyl) tetradecasilane (two trimethylsilyl groups are disordered over two positions) [14].

On the "Silicon" Life Forms

One more aspect of the interrelation between the living and nonliving material is the problem of carbon and silicon compounds and molecules in the objects of Nature.

Of course, there is no absolute analogy between silicon and carbon. For example, the chains consisting of the Si–Si bonds are less typical for silicon compounds, no silicon analog of benzene is known, etc. However, instead of silicon chains, the macromolecules of the living substance quite likely may be constructed based on the Si–O–Si and Si–N–Si bonds.

There exists a "terrestrial" theory of the silicon life as well. Carbon was incorporated in the chemical structure of primordial organisms as a result of secondary processes, replacing silicon. The source of carbon was calcium carbonate, which thereby became a constituent of the composition of organisms, competing with silicon in skeleton formation.

Silica gel formed in the presence of an organic compound, after its removal, acquires enhanced selective adsorption capacity with respect to that compound. That is, in contrast to other inorganic gels occurring in nature, silica gel exhibits "memory" in terms of cybernetics.

The process of fossilization or silicification of organisms is not yet clear to a full extent and cannot be reproduced in laboratories. Commonly, only skeletal formations and plant tissues undergo fossilization.

This is a very slow process and it involves substitution of the organic material by silica, the organic structure being exactly reproduced.

Analysis of 1,299 compounds from the Cambridge Database of structures containing the Si–Si bonds reveals that silicon most commonly (in 655 cases) forms simplest diatomic fragments Si_2, however, 98 molecules have silicon skeletons consisting of 10 and more silicon atoms. The record holder is the 32-atom skeleton, which is contained in molecule 1,1,1,3,3,4,5,5,7,7,8,8,10,10,11,12,12,14,14,14-icosamethyl-2,2,6,6,9,9,13,13-octakis(trimethylsilyl)-4,11-bis (1′,1′,3′,3′,3′-pentamethyl-2′,2′-bis(trimethylsilyl)trisilanyl) tetradecasilane (Figs. 13 and 14) [14]. Therefore, silicon is capable of forming rather complex molecular fragments that are topologically similar to saturated hydrocarbons.

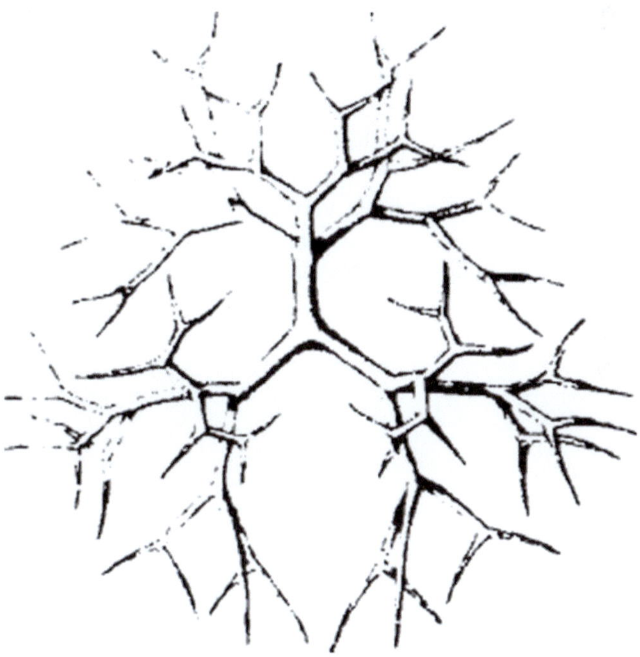

Fig. 14 *Thalassoxanthium cervicorae* (Spumellaria). A single cell three-branching crystalloidal silica body, whose three limbs meet at the same angle and repeatedly form forked branches [15].

These structural forms can be part of not only a living substance but also of plants, including protozoa.

The diatoms (Diatomeae or Bacillariophyta) are among the most abundant organisms on the Earth, comprising about 15,000 species, whose origin dates back to 120 million years ago.

The maximum diatom reproduction rate is so high that they could occupy the whole surface of the terrestrial globe in 16.8 days.

The diatomaceous species are remarkable for assimilating silicon compounds and building their shells of amazingly beautiful shape from these compounds. The diversity of the shell shapes makes the basis for classification of the diatomaceous species.

The chemical composition of the diatomaceous species is very peculiar and drastically differs from that of other algae.

Apart from oxygen and hydrogen, silicon is the predominant element in the diatoms, prevailing over carbon in its content. Therefore, the diatoms are real terrestrial representatives of the silicon life.

The main silicon content is concentrated in the 0.02–2.25 μm diatomaceous shell whose density is 2.07 g/cm^3. The shell is built of two asymphytous halves, which are inserted into each other as a box and lid. The diatomaceous species of the Centricae class have a light shell and those of the Pennatae class have a heavier skeleton. The benthos species are distinguished by the most massive shell.

The diatom's shell has a very intricate structure: it consists of pores, stria, areolae, ribs, and other structural elements, which make up infinite versions of quaint patterns.

Fig. 15 Delicate transparent shells of unicellular algae diatoms made of the same material as the glass (silicon oxide).

The diatoms can use only low-molecular silica species (that can form yellow silicomolybdate complex with ammonium molybdate). Tetraethoxysilane can be a source of silicon for cultivating diatoms in artificial media (Fig. 15).

How Does the Matter Think?

In his well-known lecture "Generalized Crystallography" given in Yugoslavia in 1975, A. Mackay asked a question "How does the matter think?" For a long time, this question

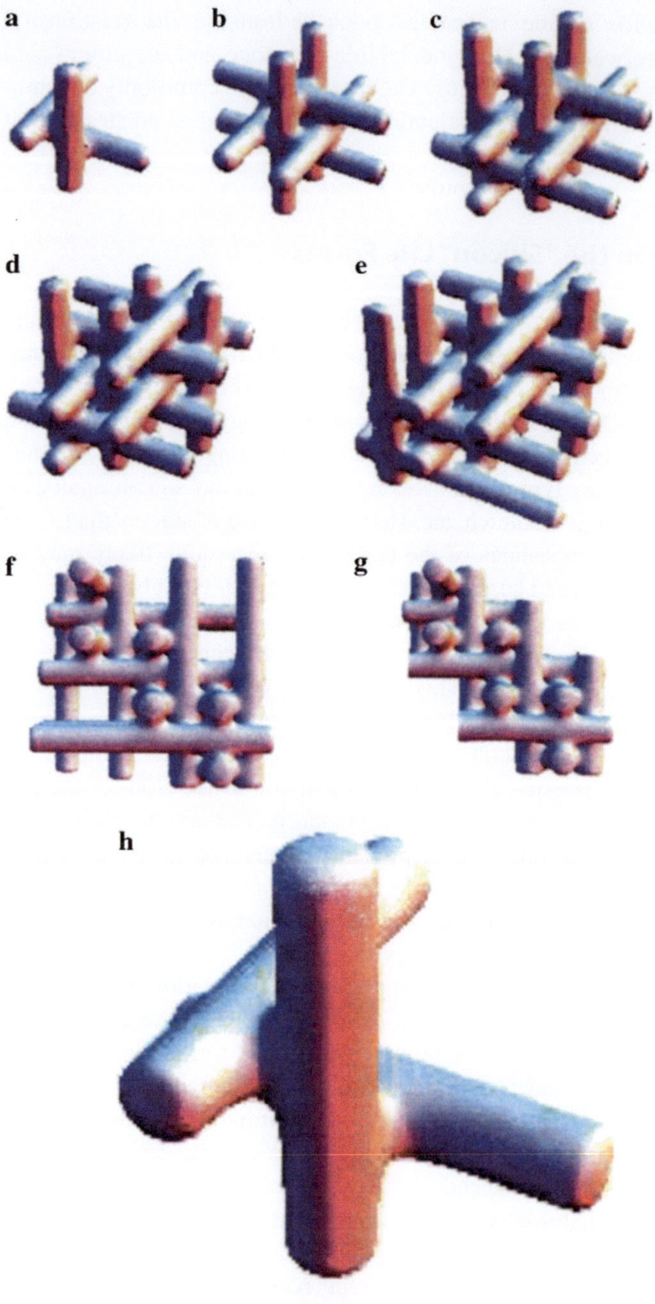

Fig. 16 Rod packings models with $n=1$ (**a**), $n=2$ (**b**), $n=3$ (**c**), $n=4$ (**d**), $n=5$ (**e**) and $n=6$ (**f, g**), cf. relative size of the triple rod junction for $n=1$ (**a**) and $n=6$ (**h**) [16].

could not be considered, because other questions of structural chemistry, such as "Are there genes in paulingite?", "What is the structure?", "Where does the structure arise from and how does it develop?" (morphogenesis, growth and form, etc.), were not decided.

Owing to the efforts of the outstanding structural chemists L. C. Pauling, J. D. Bernal, A. Mackay, S. Andersson, B. Vainshtein, and many hundreds of the other creative scientist, part of these questions was studied, and now we can consider the question of how the matter think. In our view, this question primarily pertains to chemistry, in a sense that the interatomic interaction (as well as the interaction of other material particles) is stated as the basis of structural chemistry.

Using the developed mathematical methods, S. Andersson [16] studied the formation of neurons, axons, and dendrites in the brain substance from the infinite type rods. Hermitian wave pulses are used to describe the cognition mechanisms.

The rod packings are well adapted for build-up of very complex structures. It was demonstrated how the successive transfer can proceed from 10^{11} neurons, which correspond to the age of one to three, to 10^{14} dendrites and axons characteristic of the age of about 10–14 (Fig. 16).

S. Andersson proposed a model of mentation in a brain as a kind of mechanical motion. We think and we speak. The mentation is accompanied by a simultaneous command for motion. It is such activity that is the basis of mentation. The wave pulses move back and promote contacts with some structures and packings in the memory. Information arrives to the rod packings via the wave pulses [16].

Summary

1. From the foregoing consideration, it follows that the objects of nature are structurally generalized, especially the objects undergoing the stage of formation from atoms to macrosized species. No exceptions from a large number of the examined structures of inorganic and organic compounds, algae, plants, viruses, and other objects have been found out.

   For the majority of substances, the build-up starts from fundamental configurations, and then clusters are formed, whereby the structure is assembled. The hierarchical construction is the basic principle irrespective of whether the whole structure is periodic or aperiodic.

   The spatial domain, in which the substance is formed, is tens to hundreds of nanometers in size, the type of the space geometry being inessential because in the small (after F. Klein) all the geometries are equivalent.
2. The chemical composition and the structure are unambiguously interrelated. It is necessary to know the structure symmetry and the construction sequence. As a rule, the chemical composition can be determined reasoning from the knowledge of the structure of the initial clusters (100–200 atoms). Several thousands of atoms may be needed for aperiodic structures.
3. The structure of the basic configuration is the basis for development of the whole structure of the material and includes a "design" the necessary information to do so.
4. It is obvious that the chemistry in the twenty-first century will find not only answers the question "how is substance formed?", but also to the question "How does the matter think?"

References

1. Philip Ball, 10 Unsolved Mysteries in Chemistry, Scientific American Magazine, October 10, 2011, pp 48–55
2. Wigner EP (1960) The unreasonable effectiveness of mathematics in the natural sciences. Commun Pure Appl Math 13(1):1–14
3. Mackay AL (2001) On complexity. Crystallogr Rep 40(4): 524–526
4. Hardy GH (1940) A mathematician's apology. Cambridge University Press, Cambridge, p 153
5. Kadomtsev BD (1999) Dynamics and information. Editorial Board of UFN, Moscow, p 427
6. Conway JH, Sloane NJA (1988) Sphere packing, lattices and groups. Springer, Berlin
7. Cohn H, Kumar A (2007) Universally optimal distribution of points on spheres. J Am Math Soc 20:99–148
8. Shevchenko V Ya, Blatov VA, Ilyushin GD (2009) Intermetallic compounds of the $NaCd_2$ family perceived as assemblies of nanoclusters. J Struct Chem 20:975–982
9. Shevchenko VYa (2011) Structural chemistry of the nanoworld—a new page of inorganic chemistry. Glass Phys Chem 37(5): 467–484
10. Shevchenko VYa, Tereschenko GF (2000) Research, development and innovations in field of ceramic and glass materials. Herald Russ Acad Sci 70(1):82–87
11. Tasiopoulos AJ, Vinslava A, Wernsdorfer W, Abboud KA, Christou G (2004) Single-molecule magnets: a Mn84 torus and its supramolecular nanotubes. Angew Chem Int Ed 43(16):2117–2121
12. Moushi EE, Lamprodopoulos G, Wernsdorfer W, Nastopoulos V, Christou G, Tasiopoulos AJ (2010) Inducing single-molecule magnetism in a family of loop-of-loops aggregates: heterometallic $Mn_{40}Na_4$ clusters and the homometallic Mn_{44} analogue. J Am Chem Soc 132:16146–16155
13. Miras HN, Cooper GJT, Long D, Bögge H, Müller A, Streb C (2009) Leroy Cronin unveiling the transient template in the self-assembly of a molecular oxide nano-wheel. Science 327:72–74
14. Krempner C, Köckerling M (2008) Nanoscale double-core oligosilane dendrimers: synthesis, structure and electronic properties. Organometallics 27(3):346–352
15. Haeckel E (1917) Kristal seelen. Alfred Kroner Verlag, Leipzig
16. Andersson S (2011) Anatomy of thought, mechanism and mathematics. J Inorg Gen Chem 637(7):840–845

Investigation of Structural Relations Among the Compounds in the ICSD Using Geometry Based Comparison Techniques[a]

M. Sultania[b], J.C. Schön[b], D. Fischer[b], and M. Jansen[b]

Abstract

We investigate the structural relations among the inorganic crystalline compounds listed in the ICSD using a geometry based similarity criterion. Using MYSQL-language based codes together with the structure comparison algorithm CMPZ, we classify the ordered crystalline structures contained in the ICSD into structure families, and explore the relations among these structures and structure families.

Keywords

Inorganic crystal structures • Crystalline solid state compounds • Structure classification • Structure comparison • Database analysis

Introduction

Atoms brought together interact, and, in responding to the force field they experience, approach characteristic equilibrium structures. Each of these stable chemical configurations corresponds to a locally ergodic region on the energy landscape of the chemical system [1], and is typically associated with a local minimum on this hypersurface of the potential energy. Projecting all already known or still undiscovered compounds onto such surfaces constitutes the foundation of the energy landscape concept of solids [2–4]. Already from a rough combinatorial estimate, e.g., by multiplying the number of possible combinations of about 80 stable elements with the different stoichiometries and polymorphs expected to exist in each system, it is obvious that the number of possible chemical compounds is truly gigantic, and for all practical purposes infinite. Although only a minute slice of this space of potential structures has been realized experimentally or addressed computationally, already the compounds and structures documented thus far pose great challenges in rationalizing and classifying them, and deducing the underlying principles of their structural systematics.

Soon after X-ray structure determinations had become a routine tool in solid state science, first classification schemes for crystal structures became available. Among the pioneers, there are P. P. Ewald and C. Hermann [5] suggesting a nomenclature for structure types, first used in "Strukturberichte", Victor Moritz Goldschmidt [6] providing basic factors influencing the evolution of a specific structure, and Linus Pauling [7] formulating the rules named after him. The box of descriptive tools has been advanced continuously till the present [8–15]. Exemplary milestones along this way are, for example, the ideas contributed by Fritz Laves [16], Sten Andersson's [17, 18] rod packing/twinning operations and minimal surfaces, or the development of space group–subgroup relations as a stringent concept for relating crystal structures [19]. All the respective approaches serve different purposes, and exhibit distinct weaknesses and strengths.

In the context of our long-term project aimed at developing a concept for rational solid state synthesis planning [20, 21], we explore the potential energy landscape for stable structures candidates, as a first step. These random walk-based searches, employing the energy as the only cost function, are performed without applying any symmetry-based restrictions, because this would undermine the ultimate

[a] *Structural Chemistry* 2012, 23(4):1121–1129.

[b] Max-Planck-Institute for Solid State Research, Heisenbergstr. 1, Stuttgart, Germany
e-mail: M.Jansen@fkf.mpg.de

I. Hargittai, B. Hargittai (eds.), *Science of Crystal Structures: Highlights in Crystallography*,
DOI 10.1007/978-3-319-19827-9_33

objective of being able to identify all stable configurations. After performing the usually considerable number of exploration walks for a given system, one easily ends up with some tens of thousands of structure candidates that need to be documented in terms of the crystallographic conventions, scanned for duplicates, and compared to structures already contained in databases. For this purpose we have developed a sequence of algorithms, symmetry finder (SFND) [22], space group searcher (RGS) [23], and compare cell (CMPZ) [24], that automatically assign conventional unit cells and space groups to the structures formed, and compare them with each other and with structures drawn from databases, respectively. Having arrived at this stage, the structure candidates might serve as targets for synthesis, and can be analyzed and categorized from a crystal chemical point of view, using the concepts discussed above. A first step during the latter undertaking usually is to group them into families, using distinct criteria like symmetry, topology or geometry. But such classifications are not only of relevance for theoretical investigations: in synthetic chemistry, such comparisons yield information about the plausibility of newly synthesized compounds and their expected chemical and physical properties.

In this work, we have performed systematic comparisons of the structures listed in the Inorganic Crystal Structure Database (ICSD) [25, 26] using the structure comparison algorithm CMPZ [24] implemented in the structure analysis program KPLOT [27]. Besides containing various informations about the crystal structures, such as crystal lattice, space group, atom positions, etc., the ICSD provides for many structures assignments of structure types like NaCl, Al_2MgO_4 etc., where information about bond-topologies, coordination polyhedra, symmetries, and relative atom positions of the various structures are taken into account. In contrast, the CMPZ algorithm performs a quantifiable geometry-based comparison by mapping the point patterns of the two structures into each other, and thus allows us to decide, whether two structures are the same with respect to their geometry within a given distance tolerance, irrespective of their symmetries.

Methodology

In this study, we automatically compare and classify the large numbers of crystalline structures listed in the ICSD into structure families. For this, we extract the structures from the ICSD employing search code written in the MYSQL-database language, and subsequently compare their geometries in a pairwise fashion using the CMPZ algorithm implemented in KPLOT. Based on the success of such a comparison, we group the structures into families. The structure, whose total distance from all other structures in the group is minimal and/or which is similar to the largest number of structures, is selected as the "father" of the family, and the other members are denoted as the "sons". Here, one should note that such a quantitative algorithmic definition of what constitutes a structure father will often result in a prototype different from the standard ones that are typically assigned based on history or structural idealizations. By combining two (or more) of the types of atoms present in a structure, we can also determine, whether a given structure is geometrically related to one with fewer types of atoms, and in this way construct geometrical structure trees connecting families with different numbers of types of atoms.

The CMPZ Algorithm [24]

The CMPZ algorithm compares two periodic geometric point patterns and decides, whether the two crystalline structures are the same within a given distance tolerance. The underlying concept is the affine mapping of structure A (represented by a unit cell of A or a multiple thereof) onto and into the periodic structure B. For this, we show that such a mapping and its inverse exist by demonstrating that every atom within the unit cell of A gets mapped to an atom in the unit cell (or a multiple thereof) of structure B while making sure that atoms of the same type are mapped to atoms of the same type, and the set of atoms contained inside the image of A in B is also mapped uniquely to atoms inside structure A by the inverse transformation. In order to be able to compare structures consisting of atom types differing with respect to their size, we usually rescale the structures such that they exhibit the same number density of atoms.

The algorithm is implemented in the program KPLOT, and is called using the command CMPZ, with parameters tols, tolg, tolw, tolv. The default values of these parameters are tols = 0.5, tolg = 0.1, tolw = 4.0, tolv = 0.5. Here, tols is the maximally allowed distance deviation of the corresponding atoms in structures A and B after applying the mapping (in Angstrom), tolg describes the allowed tolerances of the cell lengths a, b, c (1 + tolg), tolw is the maximal permitted angle deviation (in degrees), and tolv denotes the maximal allowed variation of the cell volume V ($1 \pm$ tolv). The comparison algorithm then follows the following strategy: after scaling to the same number density, it is checked, how the volume of the larger cell can be transformed to the one of the smaller cell. If necessary, the smaller cell (or both) is appropriately enlarged. Next, for each of the possible transformations, the two cells are shifted such that one of the atoms of the asymmetric unit of one cell has the same location as one atom of the same type in the asymmetric unit of the second cell. In the next step, it is checked, whether, within the tolerance tols, this agreement is also found for all other atoms including the ones generated due to the symmetries of the structures. If this is the case, the corresponding transformation matrix and the displacement are printed, together with information about the amount of misfit. If no affine mapping fulfilling these conditions is found, a failure notice is produced, indicating that the comparison between the structures has failed. For the purpose of this study, the parameters of the CMPZ command were always set to the default values.

Procedure for the Analysis of the Database

We employed the MYSQL-database language to extract the structures and the information about them from the ICSD. In a first step, we grouped the structures in the ICSD by the number of different atom types present in their chemical formula denoted by the so-called generic AB_FORMs (A_x, A_xB_y, $A_xB_yC_z$,…, $A_xB_yC_zD_kE_lF_mG_nH_o$). Then, we sorted all structures having the same generic AB_FORM according to their specific AB_FORMs, i.e., with specific values for (x, y, …). From these groups, we generated sets of structures with the same composition by combining all groups that exhibited the same overall stoichiometry, i.e., which fulfilled the criterion $x1/x2=y1/y2=\ldots$, where the structures considered were of type Ax_1By_1 …, Ax_2By_2 …, etc. Next, we eliminated all those structures, which contained atoms with occupation factor unequal to one. This was necessary, since the comparison algorithm by itself does not distinguish between full and partial occupation in a structure.[1] Subsequently, all structures that are duplicates were removed, in order to improve the statistics, since some compounds had been measured and entered into the ICSD several times. Finally, for every structure a KPLOT readable file was generated, and its structure prototype as registered in the ICSD (if available) was extracted.[2]

After these preliminary tasks, all structures with the same composition were pairwise compared, using CMPZ, and the structures were sorted by the number of successful comparisons. Next, the structures were assigned to families with structure fathers in the following way: the structure, which had the highest number of successful comparisons was selected as a structure father, and all the structures, with which it had been successfully compared were included as sons in its family, and removed from the list of structures for the composition under consideration. The procedure was repeated, until all structures had been selected as a structure father or assigned to a family.

[1]Of course, one can perform additional analyses of such structures by specially preparing such structures before applying CMPZ. However, this would go beyond the questions addressed in this study which focuses on perfectly ordered crystalline compounds and does not include solid solutions, or compounds with rotating complex ions, oscillating atoms/atom groups, etc.

[2]Note that these ICSD assigned structure types often do not contain the same number of elements as the structure they are assigned to, leading to complications when comparing the results of our analysis with the ICSD assignments.

As a further refinement, we selected all the structure fathers for a given AB_FORM, i.e., for a given number of atom types in their stoichiometric description, and generated new "structures" with fewer atom types by combining types of atoms. For example, by combining atom types A and B in a structure with composition $A_1B_2C_3$, we get a new structure with composition $A'_3B'_3$ (which can be reduced to $A'_1B'_1$, of course), where type A' refers to the former atoms of types A and B, and type B' to the former atoms of type C. In this fashion, it is possible to generate a tree-like structure, which, on a geometrical basis, allows us to connect structures where one atom position is occupied by one type of atoms in structure 1 and two types of atoms in structure 2, at least as long as we are dealing with ordered structures and not solid solutions.

Results

Preliminary Results

Out of the 100,243 structures listed in the ICSD (2008 issue), 50,552 structures fulfill the criterion that the atoms show full occupancy on their positions in the crystal lattice, and after eliminating duplicates, 21,925 remain. Table 1 represents the categorization of these structures into the eight generic AB_FORMs found in the ICSD for ordered structures, where we list the number of structures and structure families as function of the number of different atom types present, including those families that only contain the structure father and no sons.

Next, we subdivided the structures listed for a given generic AB_FORM into groups with the same composition, and performed the pairwise comparisons with CMPZ for each of these groups. As an example of such a comparison, we consider the two compounds $K(AlF_4)$, with specific AB_FORM=$A_2B_2C_8$ (ICSD call code 16413), and $Tl(AlF_4)$, with specific AB_FORM=$A_1B_1C_4$ (ICSD call code 202454), whose structures have the same overall composition $A_1B_1C_4$, but are listed in the ICSD with different numbers of atoms in the unit cell, and different symmetries, *P*4/*mbm* and *P*4/*mmm*, respectively. After reading in the two structures in KPLOT format, the comparison can easily be performed, yielding the following result:

```
Solution 1 (2->1)
Transformation matrix              Displacement
-1.00000     -1.00000         0.00000     0.00000
1.00000      -1.00000         0.00000     0.00000
0.00000       0.00000         1.00000     0.00000
Misfit of atoms: 0.2213 and of the cell: 0.0077
Comparison successful.
Mapping of elements:
1: Al K F
2: Al Tl F
```

Table 1 Number of structure fathers and sons for the eight generic AB_FORMs present in the ICSD

AB_FORM	No. of structures in ICSD before analysis	Total no. of fathers	No. of fathers having at least one son	Total no. of sons	No. of fathers having no son
Ax	348	38	18	310	20
AxBy	6,184	2,996	301	3,188	2,695
AxByCz	10,963	5,306	737	5,657	4,569
AxByCzDk	3,714	1,373	479	2,341	894
AxByCzDkEl	677	289	104	388	185
AxByCzDkElFm	64	36	11	28	25
AxByCzDkElFmGn	6	5	1	1	4
AxByCzDkElFmGnHo	1	1	0	0	1

Note that the number of structure fathers equals the number of structure families, and that the total number of structures for a given AB_FORM is the sum of the fathers and sons

In Fig. 1, we show for a couple of typical families the spread in the comparison misfits with respect to the structure father (located at the origin of the plot, of course). We note that four different types of scatter plots appear to be present: a more or less even spread of misfits both with regard to atom positions and cell parameters, structure families with no misfits at all, and families with misfits that only exhibit contributions from the cell parameters or the atom positions, respectively. We note that if no misfit for atom positions and/or cell parameters is observed within a family, this usually indicates that (a) the symmetry of the structures imposes serious restrictions on the structures, and/or (b) all possible small geometrical deviations in the atom positions and/or cell parameters of the real compounds from the perfectly symmetrical (idealized) structure had been eliminated during the structure determination process. Furthermore, we observe that typically the misfit in the atom positions does not reach the limiting value prescribed for the CMPZ algorithm, which implies that most of the structure families are reasonably distinct from one another.[3] On the other hand, we note that if there exist many structures with the same or very similar relative atom positions and space group symmetry but noticeable variability in the cell parameters, e.g., the c/a ratio, these structures are going to be considered geometrically different, of course, and assigned to different structure families in our classification scheme, although they might be treated as belonging to the same structure type in chemical or crystallographical discussions.

[3]Recall that due to the finite tolerances, the similarity based on the geometric structure comparison cannot be a full equivalence relation: In principle, it can happen that structure A is similar to B and B similar to C, but A not similar to C.

Families of Structures

In Table 2, we give the complete list of structure fathers for the elements. We find that 18 families exist with at least two members and 20 families with only a father and no sons. Several fascinating observations can be made just from these results. For one, it is not always the "most ideal" structure commonly used e.g., for simple visualization or pedagogical purposes, which corresponds to the structure father that lies in the centre of the structure family. Furthermore, a number of element structures that are commonly assigned to a particular highly symmetric "structure type", actually exhibit lower symmetry due to distortions away from this high-symmetry structure. An example is the "fcc-structure" type (symmetry *Fm-3m*), which is found in a family containing many structures with lower symmetry (*I4/mmm*, *R-3m*), and where the structure father (Indium, call code 53091) exhibits a lower symmetry (*I4/mmm*).

In Table 3, we show the results of an analysis of the 153 different compositions for which compounds exist with the generic AB_FORM AxByCzDk, in particular, how many different structure families are found for each such composition, and how many compounds are observed for a particular composition. In order to avoid excessive data, we only show the most diverse compositions, i.e., those with the most structure families, and the most common compositions, i.e., those with the largest number of compounds, respectively.

In Table 4, we focus on one particular composition, $A_1B_1C_2D_6$, and list all structure fathers, their prototype assignment according to the ICSD, how many sons they have each, and the most common prototype assignment for these sons in the ICSD. We note that most of the structures belong to only a handful of structure families, which can be derived from the perovskite structure followed by silicate analogue structures. Furthermore, we see that certain prototypes

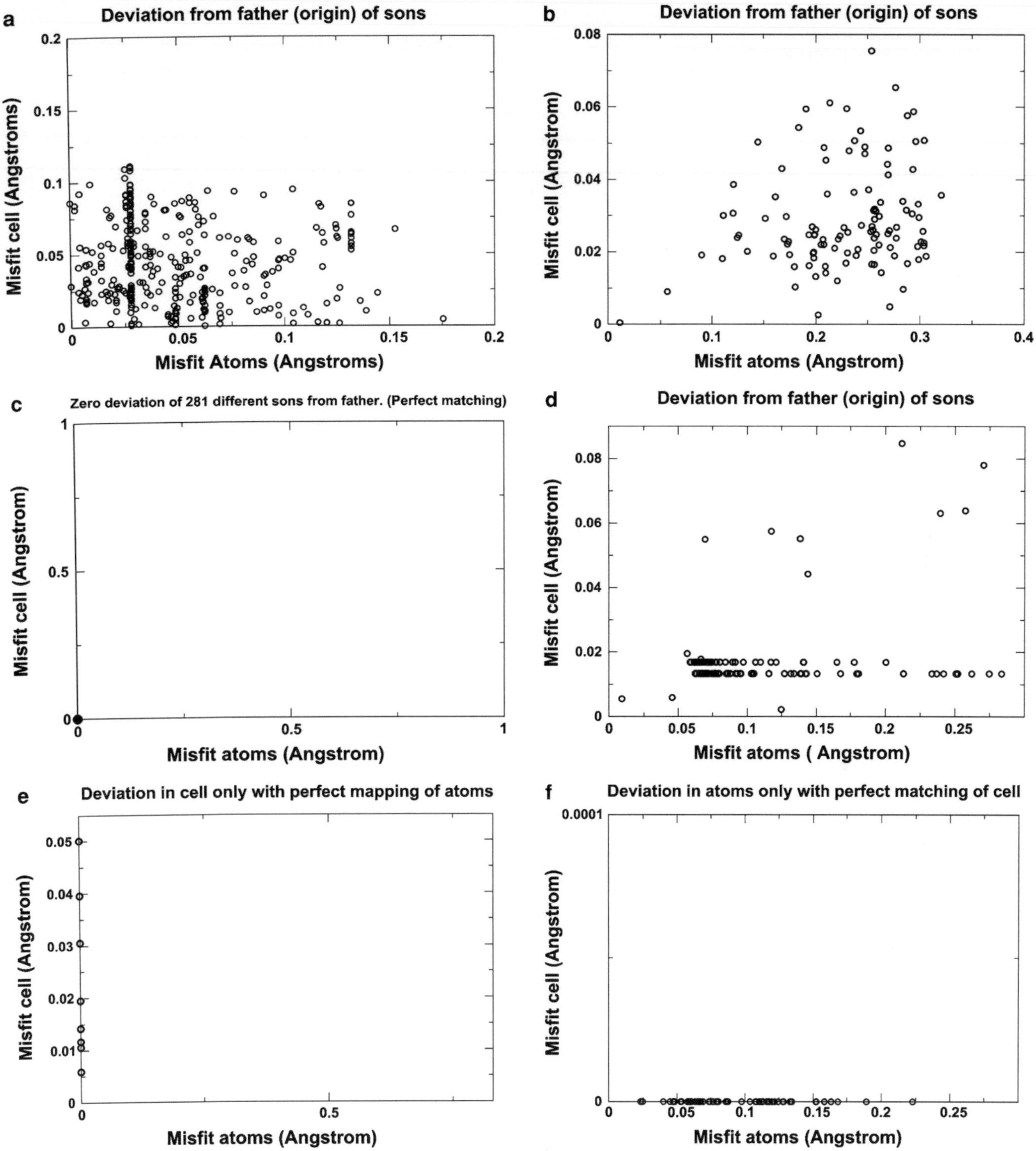

Fig. 1 Distributions of misfits with respect to the structure father for several structure families randomly selected from the results of the database analysis. The structure fathers chosen were: (**a**) $CaCo_2Ge_2$ (*I4/ mmm*, call code=406), (**b**) YbIrGe (*Pnma*, call code=414433), (**c**) $LiCu_2Si$ (*Fm-3m*, call code=15128), (**d**) $CuIr_2S_4$ ($I4_1/amd$, call code=75532), (**e**) $FeNiPt_2$ (*P4/mmm*, call code=42564), and (**f**) $Li_3Na_3In_2F_{12}$ (*Ia-3d*, call code=416926). Note that the structure father itself is always located at the origin of the plot, and that, by definition, the misfits in cell parameters and atom positions must always be larger or equal to zero.

Table 2 List of the structure families for element compounds (AB_FORM=A_x) listed in the ICSD, together with the corresponding structure fathers and sons

ICSD call code of father	Formula	Space group	*x*	No. of sons	Sons' space groups
53091	In	*I4/mmm*	2	93	*I4/mmm, Fm-3m, R-3m*
52412	Sc (HP)	*P4/nmm*	4	77	*Im-3m*
43426	He (HP)	$P6_3/mmc$	2	71	$P6_3/mmc$
76592	Nd	$P6_3/mmc$	4	16	$P6_3/mmc$
98120	P (HP)	*R-3m*	2	16	*R-3m*
150731	*β*-Se	*Fd-3m*	8	9	*Fd-3m*
109024	Sr (HP)	$I4_1/amd$	4	6	$I4_1/amd$
52497	Tb (HP)	*R-3m*	9	4	*R-3m*
16955	Si (HP)	*Ia-3*	16	4	*Ia-3*
16056	U	*Cmcm*	4	3	*Cmcm*
23836	P (black)	*Cmca*	8	2	*Cmca*
18154	C_{12}	*Cmca*	8	2	*Cmca*
57164	S (HP)	*R-3m*	3	2	*R-3m*
108326	Cr (layer)	*Pm-3n*	8	1	*Pm-3m*
104296	Hg (LT)	*R-3m*	3	1	*R-3m*
22251	Se (HP)	$P3_121$	3	1	$P3_121$
27840	S_8	*Fddd*	128	1	*Fddd*
109035	Si (HP)	*P6/mmm*	1	1	*P6/mmm*
2091	S_8	$P2_1/c$	32	0	#
15819	N_2 (HP)	$P2_13$	8	0	#
16570	Ge (HP)	$P4_32_12$	12	0	#
20502	Gd	*Fm-3m*	4	0	#
29123	C	*R-3m*	2	0	#
23058	Te	$P3_121$	3	0	#
41775	*β*-Mn	$P4_132$	20	0	#
42743	*α*-Mn	*I-43m*	58	0	#
43499	Am	*Fm-3m*	4	0	#
43539	Ga (HP)	*Cmcm*	4	0	#
52410	Sc (HP)	*I4/mmm*	2	0	#
52521	Ti (HP)	*P6/mmm*	3	0	#
56721	Si_{136}	*Fd-3m*	136	0	#
56897	Hg (As)	*P6/mmm*	3	0	#
30101	Si (HP)	$P6_3/mmc$	4	0	#
52230	C (graphite)	$P6_3/mmc$	4	0	#
105489	Np	*Pnma*	8	0	#
154318	P_4	*C2/m*	16	0	#
156481	O_8	*C2/m*	16	0	#
412326	S_8	*Fddd*	128	0	#

Note that the space group of a son does not have to agree with the one of the structure father

assigned in the ICSD appear to cover several of the structure families we have determined. The reason for this lies in the geometrical criterion we employ instead of the topology and symmetry-based considerations commonly used. Of course, the particular compound itself, which had been chosen as the prototype in the ICSD, will only belong to one of these families, typically the largest one. We also note in this context that sometimes the prototype assigned contains either fewer (here: 3) or more (here: 5) different elements than the structures we analyze (here: 4). The reason for this non-intuitive kind of structure type assignment is usually a mixture of tradition, history and pedagogical thinking, since trying to visualize a complicated multi-atom-type structure is often more easily done, if one starts from a well-known structure built up of fewer elements. However, this often leads to problems when comparing and classifying structures based on hard geometrical criteria as done in this study: the strong distortions that frequently occur when replacing some of the atoms in a structure by atoms of different size—distortions one is willing to accept for the sake of the argument or better visualizability of a complicated structure—make an affine mapping impossible, and thus lead to the result that the *n*-atom-type structure is not geometrically comparable to the $(n-1)$-atom structure.

Table 3 Excerpt from the list of the 153 compositions that are found in the ICSD for compounds with four elements (AB_FORM=AxByCzDk), where the number of structure fathers, sons and total number of compounds are given

*n** (*x,y,z,k*)	Total members in each group (Sum of fathers and sons)[a]	Total no. of fathers[b]	No. of fathers with at least one son	No. of fathers without any son	Total no. of sons
1 (1,1,2,6)	532	196	27	169	336
2 (1,1,1,4)	216	95	40	55	121
1 (1,1,2,4)	257	80	25	55	177
1 (1,1,2,5)	218	60	17	43	158
2 (1,1,2,7)	101	34	14	20	67
2 (1,2,2,8)	67	30	15	15	37
2 (1,1,1,6)	59	27	13	14	32
1 (1,1,2,8)	79	25	13	12	54
1 (1,2,3,8)	72	18	10	8	54
Rest (no. of fathers with no sons)	503	503	0	503	0
Rest (no. of fathers with son)	1,610	305	305	0	1,305
Total	3,714	1,373	479	894	2,341

*Note that the number "*n*" indicates the minimal number of formula units, for which a compound of the particular composition is found

We only present the compositions with the highest diversity i.e., [a]the ones with the largest number of compounds and [b]with the most structure families

Relations Between Families of Structures of Different AB_FORMs

In the final part of our analysis, we investigated possible geometrical relations between structures belonging to different AB_FORMs. To do so, we took the structure fathers found for the AB_FORM with *n* types of atoms, combined two of the atom types, and compared the result with the structure fathers of the AB_FORM with $(n-1)$ atom types, etc. For example, we find that the compounds YbAuSb ($P6_3mc$, call code=57240) and AlHoNi (*P*-62*m*, call code=57917), which are father compounds in the class of structures containing three elements, are geometrically similar to e.g., $ThSi_2$ (*P*6/*mmm*, call code=15449) and $BaCl_2$ (*P*-62*m*, call code=2190), respectively, where one needs to combine atoms Au and Sb, and atoms Al and Ho, respectively.

Quite generally, we find that it is quite rare that an ordered structure father with *n* atom types is actually geometrically similar to a structure with $(n-1)$ atom types. Out of the 28 fathers with the $A_2B_2C_2$ AB_FORM, only four, DyFeSi (*P*4/*nmm*, call code=57233), NiZrP ($P6_3/mmc$, call code=76678), UPSe (*P*4/*nmm*, call code=42169), and YbAuSb ($P6_3mc$, call code=57240), were "related" to a structure with two elements, and it appears that this ratio is even lower when one considers all the structures, including the many compositions, where only very small structure families exist. These low numbers appear rather surprising at first, but the reason might be that many of the "known" cases of expanding a structure by adding another type of cation, for example, actually result in either solid solutions (not considered here), or ordered structures that are perhaps approximately topologically and intuitively similar to the aristotype structure, but are too distorted to fulfill the criterion of geometric similarity.

Discussion

We have shown at a number of examples how an automated analysis of the fully ordered crystalline structures listed in the ICSD allows us to gain a quantitative overview over the large and often difficult to appraise field of inorganic solid state compounds. Besides presenting some averaged data regarding the distribution of structure families, we have restricted ourselves to a number of selected examples for more specific analyses since showing these data for all structure families would require a database on its own. We note that due to the way a structure father is defined, it is not necessary for a structure father to correspond to the aristotype of a group of related structures.

Of course, there are limitations to the approach we have employed. One technical problem we encountered were incomplete or incorrect entries in the ICSD; while these can sometimes be identified due to the unusual or contradictory results they cause during our database searches, and can then be analyzed and corrected on a case by case basis by checking the data in the original publications, this poses a considerable problem when trying to analyze not just a handful but tens of thousands of structures. In that case, rechecking the full database would be equivalent to re-creating the database, and thus one must put some trust in the data provided. As a consequence, one has to realize that the numbers and classifications found as a result of the database analysis might contain some outliers. Nevertheless, the overall insights into the crystal chemical connections among the many structures found in the database are still valid, and the methodology we present is applicable in general.

Another limitation is due to the quality and accuracy of the structure data derived from the original experimental

Table 4 List of the structure fathers of the compounds with composition $A_1B_1C_2D_6$

ICSD call code of father	Formula	Structure type of father (ICSD) (Family it belongs to)	No. of sons	Predominant structure type of sons (ICSD)
95517	$Sr_2(FeWO_6)$	Ba_2LaRuO_6 (95517)	212	K_2NaAlF_6
9671	$NaFeSi_2O_6$	$CaMg(SiO_3)_2$ (9671)	31	$CaMg(SiO_3)_2$
42153	$Cs_2Na(TiF_6)$	Ba_2NiTeO_6 (42153)	13	Ba_2NiTeO_6
154107	$NdBaCo_2O_6$	$PrBaCuFeO_{5+x}$ (n.a.)	6	$PrBaCuFeO_{5+x}$
81880	$La_2(MgIrO_6)$	Ba_2LaRuO_6 (95517)	6	Na_3AlF_6
71968	$AgIn(P_2Se_6)$	$AgScP_2S_6$ (71698)	4	$AgScP_2S_6$
30998	$CaSn(BO_3)_2$	$CaMg(CO_3)_2$ (30998)	4	$CaMg(CO_3)_2$
33624	Ba_2YSbO_6	$NaBaLiNiF_6$ (n.a.)	4	$NaBaLiNiF_6$
90152	$K(SbP_2Se_6)$	Ca_3UO_6 (n.a.)	3	Ca_3UO_6
280906	$LiNa_2(AlF_6)$	Na_3AlF_6 (n.a.)	2	Na_3AlF_6
86274	Rb_2NaHoF_6	Sr_2NiWO_6 (95517)	2	Sr_2NiWO_6
30974	$BaMg(CO_3)_2$	$BaMg(CO_3)_2$ (30974)	2	$BaMg(CO_3)_2$
80108	$NaScGe_2O_6$	$MgSiO_3$-LT (n.a.)	2	$MgSiO_3$-LT
22503	$LiGa(Si_2O_6)$	$CaMg(SiO_3)_2$ (9671)	2	$CaMg(SiO_3)_2$
172591	Sr_2FeTaO_6	Ba_2LaRuO_6 (95517)	1	Ba_2LaRuO_6
152890	K_2LiAlH_6	Na_3AlF_6 (n.a.)	1	Sr_2NiWO_6
88158	Ba_2LuIrO_6	Ba_2LaRuO_6 (95517)	1	$Cs_2Au_2Cl_6$
172169	Sr_2InNbO_6	Na_3AlF_6 (n.a.)	1	Na_3AlF_6
94079	Sr_2FeMoO_6	$Cs_2Au_2Cl_6$ (n.a.)	1	$Cs_2Au_2Cl_6$
88157	Sr_2LuIrO_6	Ba_2LaRuO_6 (95517)	1	Ba_2LaRuO_6
71488	$Li_2Zr(TeO_6)$	Ni_3TeO_6 (n.a.)	1	Ni_3TeO_6
152487	$LiTi(Si_2O_6)$	$BiMnO_3$ (n.a.)	1	$BiMnO_3$
100713	Bi_2NiMnO_6	$KAg(NO_3)_2$ (100713)	1	$KAg(NO_3)_2$
200128	$RbAg(NO_3)_2$	$CaMg(SiO_3)_2$ (9671)	1	$CaMg(SiO_3)_2$
15489	$LiScSi_2O_6$	$CaMg(SiO_3)_2$ (9671)	1	$CaMg(SiO_3)_2$
79786	$NaAl(Si_2O_6)$	Pb_2MgWO_6 (79786)	1	Pb_2MgWO_6
96292	$Pb_2Mg(WO_6)$	$CaMg(SiO_3)_2$ (9671)	1	$CaMg(SiO_3)_2$

The ICSD-prototype of the father (in brackets, the ICSD call code of the structure family is shown, to which the prototype itself belongs), the number of sons of the father, and the predominant structure type among the sons as assigned by the ICSD. Note that several ICSD prototypes belong to compounds with three or five elements, and thus are not part of any of the structure families. To save space, we do not list those fathers, which form a family just by themselves

measurements. These are determined by the quality of the crystals used, the sophistication of the measurement apparatus available at the time of the experiment, and the amount of work put into the actual measurement. Clearly, a lack of high-quality data will make it very difficult to separate true structural differences from noise in the set of structures investigated.

Besides the technical issue of using a similarity criterion that is not an equivalence relation, and thus, in principle, can lead to fuzzy edges of the structure families determined, there is the crystal chemical question of bringing the traditional (history based), the intuitive, the symmetry oriented (based on isotypicity), and the topological (bond-distance or coordination polyhedron based) points of view into agreement with the geometry-based approach we have presented here. We have chosen the geometry-based procedure since in many ways it reflects the underlying variability of real structures and represents a quantitative approach with a quantifiable criterion of similarity between structures. What we see in the diffraction experiments is just a time average over the distribution of electrons and nuclei in the crystalline solid—at any given instant in time, the arrangement of the nuclei of the atoms will not even exhibit any periodicity, not to mention other symmetries or well-defined bond-lengths or coordination polyhedra. Similarly, any global exploration of the energy landscape results in structures without any perfect symmetry, which thus need to be assigned space groups and similarities within some tolerances. Furthermore, many structures transform into geometrically closely related structures that exhibit quite different symmetries, however. In this context, we note that using e.g., topological criteria for comparing structures also suffers from the issue of finite tolerances since defining the bonds and coordination polyhedra on which the bond networks are based, also involves judgment calls as to what constitutes a bond between two atoms or when one can speak of two atoms as being neighbors. Nevertheless, it is clear that using database analysis techniques like the ones we have presented here, together with structural analysis algorithms and programs like CMPZ and KPLOT, respectively, can provide new insights and much-needed guidance through the jungle of both experimentally known and theoretically proposed crystalline structures.

Acknowledgments We would like to thank Dr. A. Hannemann and the FIZ Karlsruhe for arranging the use of the ICSD-database in a form searchable via MYSQL.

References

1. Schön JC, Jansen M (2001) Z. Kristallogr. 216: 307–325, 361–383
2. Schön JC, Jansen M (1996) Angew Chem Int Ed Engl 35: 1286–1304
3. Jansen M (2002) Angew Chem Int Ed Engl 41:3746–3766
4. Schön JC, Jansen M (2009) Int J Mater Res 100:135–153
5. Ewald PP, Hermann C (1931) Strukturbericht 1913–1928, Ergänzungsband. Akademische Verlagsgesellschaft, Leipzig
6. Goldschmidt VM (1926) Naturwissenschaften 14:477–485
7. Pauling L (1929) J Am Chem Soc 51:1010–1026
8. Niggli P (1928) Kristallographische und strukturtheoretische Grundbegriffe. Akademische Verlagsgesellschaft, Leipzig
9. Schwarzenbach D (1963) Schweiz Mineral Petrogr Mitt 43:41–48
10. Hellner E (1966) Acta Crystallogr 21(Suppl.):A265
11. Pearson WB (1967) A handbook of lattice spacings and structures of metals and alloys, vol 2. Pergamon, Oxford
12. Andrews LC, Bernstein HJ, Pelletier GA (1980) Acta Crystallogr A36:248–252
13. Lima De Faria JL, Hellner E, Liebau F, Mackovicky E, Parthe E (1990) Acta Crystallogr 46:1–11
14. Gelato LM, Parthe E (1987) J Appl Crystallogr 20:139–143
15. Blatov VA, Shevchenko AP, Serezhkin VN (2000) J Appl Crystallogr 33:1193
16. Laves F (1967) Phase stability in metals and alloys. McGraw-Hill, New York
17. Andersson S, Hyde BG (1989) Inorganic crystal structures. Wiley Interscience, New York
18. Andersson S (1983) Angew Chem Int Ed Engl 22:69–81
19. Bärninghausen H (1980) MATCH 9:139–176
20. Jansen M, Doll K, Schön JC (2010) Acta Crystallogr A66: 518–534
21. Schön JC, Doll K, Jansen M (2010) Phys Status Solidi B 247: 23–39
22. Hundt R, Schön JC, Hannemann A, Jansen M (1999) J Appl Crystallogr 32:413–416
23. Hannemann A, Hundt R, Schön JC, Jansen M (1998) J Appl Crystallogr 31:922–928
24. Hundt R, Schön JC, Jansen M (2006) J Appl Crystallogr 39:6–16
25. Bergerhoff G, Hundt R, Sievers R, Brown ID (1983) J Chem Inf Comput Sci 23:66–69
26. ICSD-Fiz-Karlsruhe (2008) Inorganic crystal structure database. http://icsdweb.fiz-karlsruhe.de
27. Hundt R (1979) KPLOT: a program for plotting and investigation of crystal structures (version 9.6.12 November 2011)

Index

A
Agranat, I., 205–211
Alpha-helix, 129–131, 233
Aqueous solutions, 54, 63, 68, 179, 182
Atomic structure, 5, 27, 38, 143–153, 310

B
Bacchi, A., 299–308
Batsanov, A.S., 239–253
Bernal, I., 179–185
Blatov, V.A., 117–124, 309
Blundell, T.L., 231–236
Boiling point, 239–253
Bolanos-Garcia, V., 231–236
Burckhardt, J.J., 263–267

C
Capsid, 113–116, 318, 319
Carcelli, M., 299–308
CAS (zeolites), 255–259, 264
CDO (zeolites), 257–259
Cellular automaton (CA), 293–297
Centaurs, 114
Chirgadze, D.Y., 231–236
Circle packing, 156, 157, 161
Clusters, 11, 43, 44, 46–51, 61, 85, 104, 113, 114, 118, 120, 122, 123, 144–152, 165–170, 224, 240, 283, 291, 292, 310, 313–316, 318, 321
Cohen, S., 205–211
Convergence, 316
Crystal engineering, 210, 222, 284, 306
Crystalline solid state compounds, 330
Crystal structures, 7, 19–33, 38–41, 50, 56, 91–93, 95, 96, 98, 113, 118–124, 155, 184, 185, 188, 194, 201, 202, 205–227, 232–236, 240, 271, 274–276, 279, 284–289, 311, 312, 315, 323, 324
  determination, 23–41, 91, 93, 211

D
Database analysis, 327, 329, 331
de Boissieu, M., 143–153
Deloudi, S., 165–170
DNA repair systems, 235
Dorset, D.L., 91–98, 255–259
Dunlap, R.A., 173–176

E
Electron crystallography, 91

F
Ferrierite, 255–259
Finney, J.L., 53–68
Fischer, D., 323–331
Fluorine interactions, 187, 188
Fluoro derivatives, 242
Fullerenes, 87, 101–111
Fyfe, M.C.T., 213–227

G
Gas electron diffraction, 23–33, 46
Gavezzotti, A., 281–289
General crystallography, 69, 71
Glusker, J.P., 187–202

H
Hargittai, B., 137–142
Hargittai, I., 125–131, 133–142
Harmer, N.J., 231–236
Hauptman, H.A., 6, 7, 19–22, 27, 30, 31
Helices, 33, 81–90, 115, 116, 129, 235
Hemihedrism, 180, 183
Hemolysin, 70, 71, 75–76
Hexagonal lattice, 69, 81, 82, 87
Hierarchic structures, 40, 44, 118
Hierarchy, 37, 44, 114, 281–282, 284, 286, 291, 292
Higher-dimensional approach, 165
Host-guest chemistry, 299
Howard, J.A.K., 239–253
Hydrogen bonding, 32, 33, 60, 62, 63, 128, 129, 190, 191, 193, 194, 198, 200–202, 214, 218–222, 251, 256, 269, 270, 273, 274, 282–284, 287, 299, 301, 302, 304, 306, 307
Hydrophobic interactions, 54, 60, 62, 65–67
Hyperpolarizability, 206–207

I
Icosahedral shell clusters, 43–51
Icosahedrons, 44–47, 49–51, 85, 86, 103, 109, 114, 120, 122, 124, 145–147, 150, 167–170, 292, 312–315
Ilyushin, G.D., 117–124
Inclusion compounds, 300

I. Hargittai, B. Hargittai (eds.), *Science of Crystal Structures: Highlights in Crystallography*,
DOI 10.1007/978-3-319-19827-9

Infantes, L., 269–280
Inorganic crystal structures, 39, 311, 324
Inorganic gene, 293, 297
Intrinsic disorder, 174

J
Janner, A., 69–78, 145
Jansen, M., 323–331

K
Karle, I., 23–33
Karle, J., 6, 7, 21, 23–33
Kauffman, G.B., 179–185
Krivovichev, S.V., 291–297
Kuo, K.H., 43–51

L
Layered zeolite precursor, 255–259
Liquid structure, 54, 68
Lo, T., 231–236
Lord, E.A., 81–90, 130

M
Mackay, A.L., 37–41, 43, 44, 46, 47, 51, 53, 69, 87, 113–116, 134, 138, 144, 236, 291, 292, 294, 311, 320, 321
Mackay polyhedra, 44
Madison, A.E., 113–116
Magnetic moments, 173–176
McHenry, M.E., 173–176
Melting point, 239, 241, 243, 244, 247–250
Metal-organic compounds, 300
Microscopy, 38–40, 46, 95, 101, 113–115, 137, 138, 291
Molecular condensed phases, 240
Molecular dimers, 282
Morphogenesis, 38, 321
Motherwell, S., 269–280
Multidimensional space, 310
Multiprotein complexes, 232, 233, 235, 236

N
Nanoclusters, 117–124, 312, 314–316
Nanoparticles, 106, 114, 116, 117, 122, 293, 316
Nanotubes, 81–90, 101
Nanoworld, 114, 117, 293
National Institute of Standards and Technology (NIST), 137–142
Nested fullerenes, 104
Nets, 86–88, 118, 123, 155, 156, 158–163, 232, 256, 287, 294
Neutron scattering, 54, 55, 151, 173
NIST. *See* National Institute of Standards and Technology (NIST)
NLO. *See* Nonlinear optics (NLO)
Nobel Prize, 125, 140, 141, 179, 180
Noncovalent interactions, 214
Nonlinear optics (NLO), 205–211
Non-negativity, 26, 27
NSI (zeolites), 257–259

O
O'Handley, R.C., 173–176
O'Keeffe, M., 155–162
Optically active crystal, 179–185
Organic crystals, 240, 288
Oximes, 269–280

P
Packing, 43, 44, 46–50, 81, 84, 85, 87, 90, 92–98, 113–116, 118, 124, 143, 146–148, 150, 153, 155–162, 165–170, 188, 190, 191, 198, 202, 207, 240, 247–249, 251, 257, 259, 277, 279, 282–284, 287, 288, 292, 293, 300–307, 312, 314, 315, 320, 321, 323
Papilloma, 114–116
Paraffins, 91–96
Pauling, L., 40, 44, 50, 51, 125–131, 135, 146, 282, 321, 323
Paulingite, 291–297, 321
Pellegrini, L., 231–236
Peptide bond, 128, 130
Petroleum wax, 95, 96, 98
Phase problem, 3–17, 19–33, 53, 148
Phason, 144, 151–153
Polycyclic aromatic hydrocarbons, 188
Polyethylene, 91, 96–98
Polyhedra, 39, 40, 44–45, 51, 81, 88–90, 101, 102, 115, 118, 123, 145, 146, 148, 159, 162, 165, 250, 310, 312–315, 324, 330
Polyoma, 114, 116
Pseudorotaxanes, 214–224, 227
Push–pull quinonoids, 205–211
Pyrazoles, 269–280

Q
Quasicrystals (QC), 39, 40, 44, 46, 51, 69, 133–137, 140, 141, 143–153, 165, 168, 170, 173–176
Quasi-equivalence, 113–116
Quasiperiodicity, 143, 148

R
Radhakrishnan, T.P., 205–211
Ravi, M., 205–211
Receptor tyrosyl kinases, 232, 233
Roth, W.J., 255–259

S
Sayre, D., 3–17, 22
Scale rotations, 70, 76
Schon, J.C., 323–331
Scientific discovery, 135, 141
Secondary structure, 270, 271, 275, 277, 279
Self-assembly, 113, 114, 116–118, 120, 121, 214, 216, 219, 220, 222, 224, 226, 292
Shevchenko, V.Ya, 113–124, 291–297, 309–321
Shimoni, L., 187–236
Sibanda, B.L., 231–236
Signal transduction, 234, 235
Slovokhotov, Y.L., 239–253
Solid solutions, 91–95, 98, 248, 325, 329
Solid-state structures, 217, 221–223
Space groups, 9, 30–32, 39, 93, 96, 113, 118, 145, 182, 184, 185, 191, 207, 277, 278, 283, 284, 293, 296, 297, 323, 324, 326, 328
Sphere packing, 44, 47, 49, 81, 84–85, 87, 90, 115, 146, 148, 155–162, 248, 249, 284, 292
Spherical onions, 106, 107, 109, 111
Spontaneous resolution, 179–185
Srinivas, V., 173–176
Stability and symmetry, 103

Steurer, W., 144, 165–170
Stoddart, J.F., 213–227
Structural code, 319
Structure classification, 323, 324
Structure comparison, 324, 326
Structure of proteins, 125–131, 319
Sultania, M., 323–331
Supramolecular chemistry, 214, 227
Supramolecular structures, 214, 299
Symmetry, 8, 9, 21, 30, 32, 44–46, 62, 69–71, 73, 75, 76, 81, 89, 92, 93, 101, 103, 104, 108–110, 113, 114, 118, 123, 130, 134, 135, 138–140, 143–146, 148, 150, 151, 153, 156–159, 162, 165, 166, 170, 183, 210, 216, 225, 232–236, 239–253, 255–259, 264, 266, 267, 273–278, 281, 283–287, 289, 293, 296, 297, 300, 301, 310–312, 323, 324, 326, 330

T

Terrones, G., 101–111
Terrones, H., 87, 101–111
Terrones, M., 101–111
Theory of resonance, 127, 131
Tilings, 19, 81, 87, 89, 114–116, 134, 141, 144, 145, 148, 156, 158, 161, 162, 165–170, 173, 312
Triangulation number, 114

U

Universal optimum, 293

V

Viruses, 113–116, 231, 232, 236, 318, 319, 321

W

Williams, D.J., 6, 38, 213–227

X

X-ray crystallography, 3–17, 19–22, 126–128, 130, 214, 215, 227

Z

Zeolites, 39, 40, 155, 162, 255–259, 293, 295–297

MIX
Papier aus verantwortungsvollen Quellen
Paper from responsible sources
FSC® C105338

If you have any concerns about our products,
you can contact us on
ProductSafety@springernature.com

In case Publisher is established outside the EU,
the EU authorized representative is:
Springer Nature Customer Service Center GmbH
Europaplatz 3, 69115 Heidelberg, Germany

Printed by Libri Plureos GmbH
in Hamburg, Germany